# GENOMICS

# GENOMICS

## The Science and Technology Behind the Human Genome Project

**Charles R. Cantor**
**Cassandra L. Smith**
*Center for Advanced Biotechnology*
*Boston University*
*Boston, Massachusetts*

**A Wiley-Interscience Publication**

**JOHN WILEY & SONS, INC.**

New York • Chichester • Weinheim • Brisbane • Singapore • Toronto

Copyright © 1999 by John Wiley & Sons, Inc. All rights reserved.

Published simultaneously in Canada.

*Library of Congress Cataloging-in-Publication Data*
Cantor, Charles R.
    Genomics : the science and technology behind the human genome
  project / Charles R. Cantor and Cassandra Smith.
        p.   cm.
    Includes index.
    ISBN 0-471-59908-5 (alk. paper)
    1. DNA--Analysis.   2. Nucleotide sequence.   3. Gene mapping.
  I. Smith, Cassandra.   II. Title.
  QP624.C36   1999
  572.8′ 6--dc21         98-40448

Printed in the United States of America.

10 9 8 7 6 5 4 3 2 1

*Dedicated to*
*Charles DeLisi, who started it.*
*Rick Bourke, who made it so much fun*
*to explain it.*

# CONTENTS

## 9  Enhanced Methods for Physical Mapping                                    285

## 10  DNA Sequencing: Current Tactics                                         325

## 11  Strategies for Large-Scale DNA Sequencing                                361

**12   Future DNA Sequencing without Length Fractionation**                394

**13   Finding Genes and Mutations**                433

# PREFACE

This book is an outgrowth of the George Fisher Baker Lecture series presented by one of us (C.R.C.) at Cornell University in the fall of 1992. This author is tremendously grateful to all those at Cornell who made this occasion truly memorable, personally, and most productive, intellectually. Included especially are Jean Fréchet, Barbara Baird, John Clardy, Jerrold Meinwold, Fred McLafferty, Benjamin Widom, David Usher, and Quentin Gibson, among many others. He also wants to express great thanks to Rick Bourke, who, several times provided extraordinary help, without which lectures would have been missed or required rescheduling.

The book also builds in part on the research experience of both authors over the past 15 years. Without the help of numerous talented co-workers and collaborators, we would never have achieved the accomplishments that form some of the work described herein. No book is produced without heroic efforts on the part of support staff who handle endless chores, even in this era of efficient word processing. We are especially grateful to Michael Hirota for hours of patient deciphering of handwritten scribble. A major part of the effort in completing any substantial publication is receiving and dealing with constructive criticism from those who have read early drafts. Joel Graber and Chandran Sabanayagam helped with the illustrations and many other details; Scott Downing, Foual Siddiqi, and many others used preliminary versions as a textbook in courses over the past few years. To them we owe our thanks for taking the time to reveal ambiguities and painful lapses in our logic. Correcting these has, we hope, made the book far more accurate and comprehensible.

Inevitably in a fast-moving field like genome research, parts of this book were out of date even at the time it was submitted for publication. Furthermore, like many books based on a lecture series, this one relies too heavily on the authors' own work to provide a truly fair representation of many parallel or competing efforts in the field. No attempt was made to be comprehensive or representative in citing sources and additional readings. Despite these limitations, by concentration on fundamentals, the authors hope that a vast majority of workers in the field will find this text a useful companion.

CHARLES R. CANTOR<br>
CASSANDRA L. SMITH

*Boston Massachusetts*

# INTRODUCTION

## WHY DNA IS INTERESTING

The topic of this book is the analytical chemistry of DNA. DNA is, arguably, the most interesting of all molecules. Certainly it is the largest, well-defined, naturally occurring molecule. The peculiar nature of DNA—its relatively regular structure consisting of a sequence of just four different kinds of bases, the intense specificity of base-pairing interactions, its high charge density, and the existence of biological machinery to duplicate it and copy it into RNA—has led to a powerful arsenal of methods for DNA analysis. These include procedures for detecting particular DNA sequences, for purifying these sequences, and for immortalizing these sequences.

It probably seems strange for most readers to hear the term analytical chemistry applied to DNA. However, this is what many studies of DNA, including the massive current efforts to map and sequence the human genome, are all about. The basic chemical structure and biological properties of DNA are mostly well understood. However, the complexity of DNA in most samples of biological interest can be staggering. Many experimental efforts, ranging from clinical diagnoses to gene identification, ultimately come down to asking whether a particular DNA sequence is present, where it is located along a larger DNA molecule, and, sometimes, how many copies are present. Such questions are at the heart of traditional analytical chemistry. However, the methods used for DNA are mostly unique to it.

Some might say that the basic chemistry of DNA is rather uninteresting and mostly well worked out. It might seem that DNA bases and the ribose-phosphate backbone are not fertile ground for exotic or innovative chemical modification or manipulation. However, recent focus on these very basic aspects of DNA chemistry has led to the development of a number of novel compounds with interesting chemical characteristics as well as unique potentials for detection and manipulation of nucleic acids.

DNA itself has thus far shown only modest evidence of possessing any intrinsic catalytic activities, although the prospect that more will be discovered in the future is surely plausible. Rather, the fascination of DNA lies in the enormous range of activities made possible by its function as the cell's primary storage of genetic information. In the DNA sequence lies the mechanism for the specific interaction of one DNA (or RNA) molecule with another, or the interaction of DNAs with proteins.

## WHAT THIS BOOK IS ABOUT

In this book we first review the basic chemical and biological properties of DNA. Our intention is to allow all readers to access subsequent chapters, whether they begin this journey as biologists with little knowledge of chemistry, chemists with equally little knowledge of biology, or others (engineers, computer scientists, physicians) with insufficient grounding or recent exposure to either biology or chemistry. We also review the basic biological and chemical properties of chromosomes, the nucleoprotein assemblies that package DNA into manipulatable units within cells. Methods for fractionation of the chromosomes of both simple and complex organisms were one of the key developments that opened up the prospect of large-scale genome analysis.

Next the principles behind the three major techniques for DNA analysis are described. In hybridization, a specific DNA sequence is used to identify its complement, usually among a complex mixture of components. We understand hybridization in almost quantitative detail. The thermodynamics and kinetics of the process allow accurate predictions in most cases of the properties of DNAs, although there are constant surprises. In the polymerase chain reaction (PCR) and related amplification schemes, short DNA segments of known sequence are used to prime the synthesis of adjacent DNA where the sequence need not be known in advance. This allows us, with bits of known sequence and synthetic DNA primers, to make essentially unlimited amounts of any relatively short DNA molecules of interest. In electrophoresis, DNA molecules are separated primarily, if not exclusively, on the basis of size. Electrophoresis is the primary tool used for DNA sequencing and for most DNA purifications and analytical methods. The high charge density, great length, and stiffness of DNA lends it particularly amenable to high-resolution length separations, and some variants of electrophoresis, like pulsed field gel electrophoresis (PFG), seem uniquely suited to DNA analysis. Frequently these three basic techniques are combined such as in the electrophoretic analysis of a PCR reaction or in the use of PCR to speed up the rate of a subsequent hybridization analysis.

We next describe DNA mapping. Maps are low-resolution structures of chromosomes or DNA molecules that fall short of the ultimate detail of the full DNA sequence. Many different types of maps can be made. Cytogenetic maps come from examination of chromosomes, usually as they appear in the light microscope after various treatments. Interspecies cell hybrids formed from intact chromosomes or chromosomes fragmented by radiation are also useful mapping tools. Closer to the molecular level of DNA are restriction maps, which are actually bits of known DNA sequence spaced at known intervals, or ordered libraries, which are sets of overlapping clones that span large DNA regions. Methods for producing all of these types of maps will be described and evaluated.

Totally different flavors of maps emerge from genetic or pseudogenetic analyses. Even with no knowledge about DNA structure or sequence, it is possible to place genes in order along a chromosome by their pattern of inheritance. In the human species direct breeding experiments are impossible for both moral and practical reasons. However, retrospective analysis of the pattern of inheritance in respectable numbers of families is frequently an acceptable alternative. For genetic disease searches, it is particularly useful to study geographically isolated, inbred human populations that have a relatively large preponderance of a disease. Some of these populations, like families studied in Finland, have the added advantage of having religious or government records that detail the genetic history of the population.

Genetically mapped observable traits like inherited diseases or eye color are related, ultimately to DNA sequence, but at the current state of the art they afford a really different view of the DNA than direct sequence analysis. An analogy would be a person whose sense of smell or taste might be so accomplished that these could determine the components of a perfume or a sauce without the need for any conventional chemical analysis. Following the impact of the genes gives us direct functional information without the need for any detailed chemical knowledge of the sequence that is ultimately responsible for it. Occasional properties are inherited through the pattern of modification of DNA bases rather than through the direct base sequence of DNA that we ordinarily consider when talking about genes. As a footnote, one must also keep in mind that it is only the DNA in germ (sperm and egg) cells that is heritable. This means that DNA in somatic (nongermline) cells may undergo nonheritable changes.

Standard techniques for the determination of the base sequence of DNA molecules are well developed and their strengths and weaknesses well understood. Through automated methodology, raw DNA sequence data can be generated by current procedures at a rate of about $10^5$ base pairs per day per scientist. Finished DNA sequence, assembled and reduced in errors by redundancy, accumulates at a tenfold slower rate by standard strategies. Such figures are impressive, and laboratories interested in small DNA targets can frequent accumulate sequence data faster than they can provide samples worth sequencing, or faster than they can analyze the resulting data in a useful fashion. The situation is different when a large-scale sequencing effort is contemplated, encompassing thousands of continuous kilobases (kb) of DNA. Here new strategies are being formulated and tested to make such larger-scale projects efficient.

The task of determining the DNA sequence of even one human genome with $3 \times 10^9$ base pairs is still daunting to any existing technologies. Perhaps these techniques can be optimized to gain one or two orders of magnitude of additional speed, and progress in this direction is described. However, it seems likely that eventually new approaches to DNA sequencing will appear that may largely surpass today's methods, which rely on gel electrophoresis, an intrinsically slow process. Among these methods, mass spectrometry and sequencing by hybridization (SBH, in which data are read essentially by words instead of letters) are described in greatest detail, because they seem highly likely to mature into generally useful and accessible tools.

The ultimate purpose behind making DNA maps and determining DNA sequences is to understand the biological function of DNA. In current terms this means acquiring the ability to find the position of genes of interest with sufficient accuracy to actually possess these genes on DNA fragments suitable for further study and manipulation. Our ability to do this is improving rapidly. One of the basic rationales behind the human genome project is that it will systematically reveal the locations of all genes. This will be far more efficient and cost effective than a continued search for genes one by one. In addition to gene location and identity, DNA can be used to study the occurrence of mutations caused by various environmental agents or by natural, spontaneous, biological processes. This task is still a formidable one. We may need to search for a single altered base among a set of $10^8$ normally inherited ones. The natural variation among individuals confounds this task, because it must be first filtered out before any newly created differences, or mutations, can be identified.

In a typical higher organism, more than 90% of the DNA sequence is not, apparently, translated into protein sequence. Such material is frequently termed *junk*. Sydney Brenner was apparently the first to notice that junk is a more appropriate term than garbage because junk is kept while garbage is discarded, and this noncoding component of the genome has remained with us. A great challenge for the future will be to dissect out any important functional elements disguised within the junk. Much of this so-called junk is repeated DNA sequence, either simple tandem repeats like $(AC)_n$ or interspersed, longer (and less perfect) repeats. Although from a biological standpoint much of the genome appears to be junk, for DNA analysis it is frequently extremely useful because it can be exploited to look at whole classes of DNAs simultaneously. For example, human-specific repeats allow the human DNA component of an interspecies cell hybrid to be detected selectively.

By taking advantage of the properties of particular DNA sequences, it is often possible to perform sequence-specific manipulations such as purification, amplification, or cloning of desired single species or fractions from very complex mixtures. A number of relatively new techniques have been developed that facilitate these processes. Some have implications far beyond the simple chemical analysis of DNA. This feature of DNA analysis, the ability to focus on one DNA species in $10^{10}$, on a desired class of species which represents, say $10^{-5}$ of the whole, or on a large fraction of a genome, say 10%, is what really sets DNA analysis apart from most other systematic analytical protocols.

As large amounts of DNA sequence data accumulate, certain global patterns in these data are emerging. Today more than 1,000 megabases (Mb) of DNA sequence are available in publicly accessible databases. This number will spiral upward to levels we cannot even guess at today. Some of the patterns that have emerged in the existing data are greatly facilitating our ability to find the locations of previously unknown genes, just from their intrinsic sequence properties. Other patterns exist that may have little to do with genes. Some are controversial; others we simply have no insight on at all, yet.

Like it or not, a DNA revolution is upon us. Pilot projects in large-scale DNA sequencing of regions of model organisms have been successfully concluded, and the complete DNA sequence of many bacterial species and one yeast species have been published. Considerable debate reigns over whether to proceed with continued large-scale genomic DNA sequence or just concentrate on genes themselves (isolated through complementary DNA copies [cDNAs] of cellular mRNAs). Regardless of the outcome of these debates, and largely independent of future further advances in DNA sequence analysis, it now seems inevitable that the next half decade will see the successful sequencing and discovery of essentially all human genes. This has enormous implications for the practice of both biological research and medical care. We are in the unique position of knowing, years ahead of time, that ongoing science will have a great, and largely positive, impact on all humanity. However, it behooves us to look ahead, at this stage, contemplate, and be prepared to deal with the benefits to be gained from all this new knowledge and cope with the few currently perceptible downside risks.

# GENOMICS

# 1 DNA Chemistry and Biology

## BASIC PROPERTIES OF DNA

DNA is one of the fundamental molecules of all life as we know it. Yet many of the features of DNA as described in more elementary sources are incorrect or at least misleading. Here a brief but fairly rigorous overview will be presented. Special stress will be given to try to clarify any misconceptions the reader has based on prior introductions to this material.

## COVALENT STRUCTURE

The basic chemical structure of DNA is well-established. It is shown in Figure 1.1. Because the phosphate-sugar backbone of DNA has a polarity, at each point along a polynucleotide chain the direction of that chain is always uniquely defined. It proceeds from the 5′-end via 3′- to 5′-phosphodiester bonds until the 3′-end is reached. The structure of DNA shown in Figure 1.1 is too elaborate to make this representation useful for larger segments of DNA. Instead, we abbreviate this structure by a series of shorthand forms, as shown in Figure 1.2. Because of the polarity of the DNA, it is important to realize that different sequences of bases, in our abbreviations, actually correspond to different chemical structures (not simply isomers). So ApT and TpA are different compounds with, occasionally, rather different properties. The simplest way to abbreviate DNA is to draw a single polynucleotide strand as a line. Except where explicitly stated, this is always drawn so that the left-hand end corresponds to the 5′-end of the molecule.

RNA differs from DNA by having an additional hydroxyl at the 2′-position of the sugar (Fig. 1.1). This has two major implications that distinguish the chemical and physical properties of RNA and DNA. The 2′-OH makes RNA unstable with respect to alkaline hydrolysis. Thus RNA is a molecule intrinsically designed for turnover at the slightly alkaline pH's normally found in cells, while DNA is chemically far more stable. The 2′-OH also restricts the range of energetically favorable conformations of the sugar ring and the phosphodiester backbone. This limits the range of conformations of the RNA chain, compared to DNA, and it ultimately restricts RNA to a much narrower choice of helical structures. Finally the 2′-OH can participate in interactions with phosphates or bases that stabilize folded chain structures. As a result an RNA can usually attain stable tertiary structures (ordered, three-dimensional, relatively compact structures) with far more ease than the same corresponding DNA sequence.

## DOUBLE HELICAL STRUCTURE

The two common base pairs A–T and G–C are well-known, and little evidence for other base interactions within the DNA's double helix exists. The key feature of double-helical DNAs (duplexes), which dominates their properties, is that an axis of symmetry relates the two strands (Box 1.1).

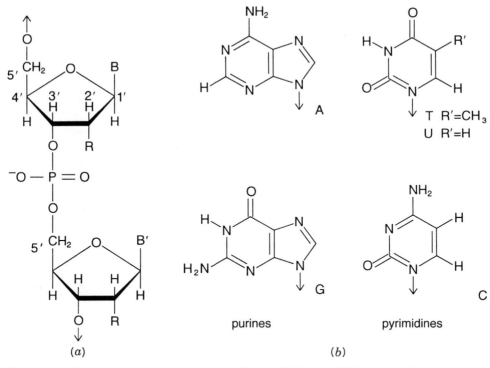

**Figure 1.1**   Structure of the phosphodiester backbone of DNA and RNA, and the four major bases found in DNA and RNA: the purines, adenine (A) and guanine (G), and the pyrimidines, cytosine (C) and thymine (T) or uracil (U). *(a)* In these abbreviated structural formulas, every vertex not occupied by a letter is a carbon atom. R is H in DNA, OH in RNA; B and B′ indicate one of the four bases. *(b)* The vertical arrows show the base atoms that are bonded to the C1′ carbon atoms of the sugars.

In ordinary double-helical DNA this is a pseudo C2 axis, since it applies only to the backbones and not to the bases themselves, as shown in Figure 1.3*a*. However, certain DNA sequences, called *self-complementary*, have a true C2 symmetry axis perpendicular to the helix axis. We frequently abbreviate DNA duplexes as pairs of parallel lines, as shown in Figure 1.3*b*. By convention, the top line almost always runs from 5′-end to 3′-end.

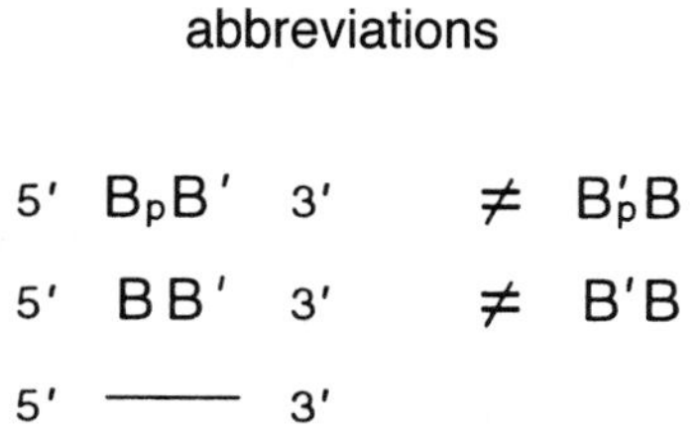

abbreviations

5′  B$_p$B′  3′      ≠  B$_p'$B

5′   B B′   3′      ≠  B′B

5′  ──── 3′

**Figure 1.2**   Three ways in which the structures of DNA and RNA are abbreviated. Note that BpB′ is not the same chemical compound as B′pB.

**BOX 1.1**
**C2 SYMMETRY**

C2 symmetry implies that a structure is composed of two identical parts. An axis of rotation can be found that interchanges these two parts by a 180-degree rotation. This axis is called a *C2 axis.* An example of a common object that has such symmetry is the stitching of a baseball, which is used to bring two figure-8 shaped structures together to make a spherical shell. An example of a well-known biological structure with C2 symmetry is the hemoglobin tetramer which is made up of two identical dimers each consisting of one alpha chain and one beta chain. Another example is the streptavidin tetramer which contains four copies of the same subunit and has three different C2 axes. One of these passes horizontally right through the center of the structure (shown in Fig. 3.26).

Helical symmetry means that a rotation and a translation along the axis of rotation occur simultaneously. If that rotation is 180 degrees, the structure generated is called a *twofold helix* or a *pleated sheet.* Such sheets, called *beta structures,* are commonly seen in protein structure but not in nucleic acids. The rotation that generates helical symmetry does not need to be an integral fraction of 360 degrees; DNA structures have 10 to 12 bases per turn; under usual physiological conditions DNA shows an average of about 10.5 bases per turn. In the DNA double helix, the two strands wrap around a central helical axis at each turn.

Pseudo C2 symmetry means that some aspects of a structure can be interchanged by a rotation of 180 degrees, while other aspects of the structure are altered by this rotation. This process might be imagined as a disk painted with the familiar yin and yang symbols of the Korean flag. Then, except for a color change, a C2 axis perpendicular to the yin and yang exchanges them. The pseudo C2 axes in DNA are perpendicular to the helix axis. They occur in the plane of each base pair (this axis interchanges the position of the two paired bases) and between each set of adjacent base pairs (this axis interchanges a base on one strand with the nearest neighbor of the base to which it is paired to the other strand). Thus for DNA with 10 bases per turn there are 20 C2 axes per turn.

The antiparallel nature of the DNA strands imposed by the pseudosymmetry of their structure means that the bottom strand in this simple representation runs in the opposite direction to our usual convention. Where the DNA sequence permits it, double helices can also be formed by the folding back of a single strand upon itself to make structures called *hairpins* (Fig. 1.3c) or more complex topologies such as structures called *pseudoknots* (Fig. 1.3d).

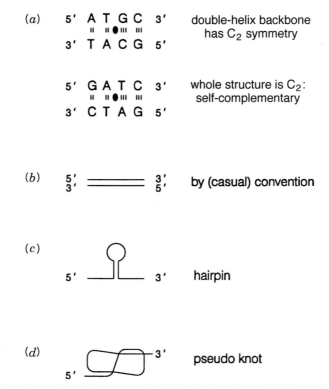

**Figure 1.3**   Symmetry and pseudosymmetry in DNA double helices. *(a)* The vertical lines indicate hydrogen bonds between base pairs. The base pairs and their attached backbone residues can be flipped by a 180° rotation around an axis through the plane of the bases and perpendicular to the helix axis (shown as a filled lens-shaped object). *(b)* Conventional way of representing a double-stranded DNA. *(c)* Example of a DNA hairpin. *(d)* Example of a DNA pseudoknot.

In a typical DNA duplex, the phosphodiester backbones are on the outside of the structure; the base pairs are internal (Fig. 1.4). The structure of the double helix appears to be regular because the A–T base pair fills a three-dimensional space in a manner similar to a G–C base pair. The spaces between the two backbones are called *grooves*. Usually one groove is much broader (the major groove) than the other (the minor groove). The structure appears as a pair of wires held fairly close together and wrapped loosely around a cylinder. Three major classes of DNA helical structures (secondary structures) have been found thus far. DNA B, the structure first analyzed by Watson and Crick, has 10 base pairs per turn. DNA A, which is very similar to the structure almost always present in RNA, has 11 base pairs per turn. Z DNA has 12 base pairs per turn; unlike DNA A and B, it is a left-handed helix. Only a very restricted set of DNA sequences appears able to adopt the Z helical structure. The biological significance of the range of structures accessible to particular DNA sequences is still not well understood. Elementary texts often focus on hydrogen bonding between the bases as a major force behind the stability of the double helix. The pattern of hydrogen bonds in fact is responsible for the specificity of base–base interactions but not their stability. Stability is largely determined by electrostatic and hydrophobic interactions between parallel overlapping base planes which generate an attractive force called *base stacking*.

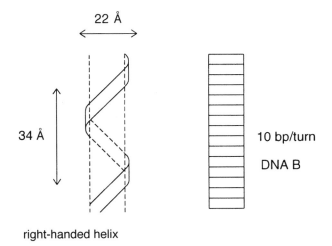

**Figure 1.4**   Schematic view of the three-dimensional structure of the DNA double helix. Ten base pairs per turn fill a central core; the two backbone chains are wrapped around this in a right-handed screw sense. Note that the A–T, T–A, G–C, and C–G base pairs all fit into *exactly* the same space between the backbones.

## METHYLATED BASES

DNA from most higher organisms and from many lower organisms have additional analogues of the normal bases. In bacteria these are principally *N6*-methyl A and 5-methyl C (or 4-methyl C). Higher organisms contain 5-methyl C. The presence of methylated bases has a strong biological effect. Once in place, the methylated bases are apt to maintain their methylated status after DNA replication. This is because the hemi-methylated duplex produced by one round of DNA synthesis is a far better substrate for the enzymes that insert the methyl groups (methylases) than the unmethylated sequence. The result is a form of inheritance (epigenetic) that goes beyond the ordinary DNA sequence: DNA has a memory of where it has been methylated.

The role of these modified bases is understood in some detail. In bacteria these bases mostly arise by a postreplication endogenous methylation reaction. The purpose of methylation is to protect the cellular DNA against endogenous nucleases, directed against the same specific DNA sequences as the endogenous methylases. It constitutes a cellular defense or restriction system that allows a bacterial cell to distinguish between its own DNA, which is methylated, and the DNA from an invading bacterial virus (bacteriophage) or plasmid, which is unmethylated and can be selectively destroyed. Unless the host DNA is replicating too fast for methylation to keep up, it is continually protected by methylation shortly after it is synthesized. However, once lytic bacteriophages are successfully inside a bacterial cell, their DNA will also become methylated and protected from destruction. In bacteria particular methylated sequences function to control initiation of DNA replication, DNA repair, gene expression, and movement of transposable elements.

In bacteria, methylases and their cognate nucleases recognize specific sequences that range in size from 4 to 8 base pairs in length. Each site is independently methylated. In higher organisms the principal, if not the exclusive, site of methylation is the sequence CpG which is converted to $^m$CpG. Although the eukaryotic methylases recognize only a dinucleotide sequence, methylation (or unmethylated) at CpGs appears to be regionally specific, suggesting that nearby $^m$CpG sequences interact. This plays a role in allowing cells with the same exact DNA sequence to maintain stable, different patterns of gene expression (cell differentiation), and it also allows the contributions of the genomes of two parents to be distinguished in offspring, since they often have different patterns of DNA methylation.

The fifth base in the DNA of humans and other vertebrates is 5-methyl C. Its presence has profound consequences for the properties of these DNA. The influence of $^m$C on DNA three-dimensional structure is not yet fully explored. We know, however, that it favors the formation of some unusual DNA structures, like the left-handed Z helix. Its biological importance remains to be elucidated. However, it is on the level of the DNA sequence, the primary structure, where the effect of $^m$C is most profoundly felt. To understand the impact, it is useful to consider why DNA contains the base T (5-methyl U) instead of U, which predominates, overwhelmingly, in RNA. While the base T conveys a bit of extra stability in DNA duplexes because of interactions between the methyl group and nearby bases, the most decisive influence of the methyl group of T is felt in the repair of certain potential mutagenic DNA lesions. By far the most common mutagenic DNA damage event in nature appears to be deamination of C. As shown in Figure 1.5, this yields, in duplex DNA, a U mispaired with a G. Random repair of this lesion would result, 50% of the time, in replacement of the G by an A, a mutation, instead of replacement of the U by a C, restoring the original sequence.

Apparently the intrinsic rate of the C to U mutagenic process is far too great for optimum evolution. Some rate of mutation is always needed; otherwise, a species could not adapt or evolve. Too high a rate can lead to deleterious mutations that interfere with reproduction. Thus the mutation rate must be carefully tuned. Nature accomplishes this for deamination of C by a special repair system that recognizes the G–U mismatch and selectively excises the U and replaces it with a C. This system, centered about an enzyme called *Uracil DNA glycosylase,* biases the repair process in a way that effectively avoids most mutations. However, a problem arises when the base 5-$^m$C is present in the DNA. Then, as shown in Figure 1.5, deamination produces T which is a normally occurring base. Although the G–T mismatch is still repaired with a bias toward restoring the presumptive original $^m$C, the process is not nearly as efficient (nor should it be, since some of the time the G–T mismatch will have come by misincorporation of a G for an A). The result is that $^m$C represents a mutation hotspot within DNA sequences that contain it.

In the DNA of vertebrates, all known $^m$C occur in the sequence $^m$CpG. About 80% of this sequence occurs in the methylated form (on both strands). Strikingly the total occurrence of CpG (methylated or not) is only 20% of that expected from simple binomial statistics based on the frequency of occurrence of the four bases in DNA:

$$\frac{X_{CpG}}{X_C X_G} = 0.2$$

where $X$ indicates the mole fraction. The remainder of the expected CpG has apparently been lost through mutation. The occurrence of the product of the mutation, TpG, is elevated, as expected. This is a remarkable example of how a small bias, over the evolutionary time scale, can lead to dramatic alteration in properties. Presumably the rate of mutation of $^m$CpG's continues to slow as the target size decreases. There must also be considerable functional constraints on the remaining $^m$CpG's that prevent their further loss.

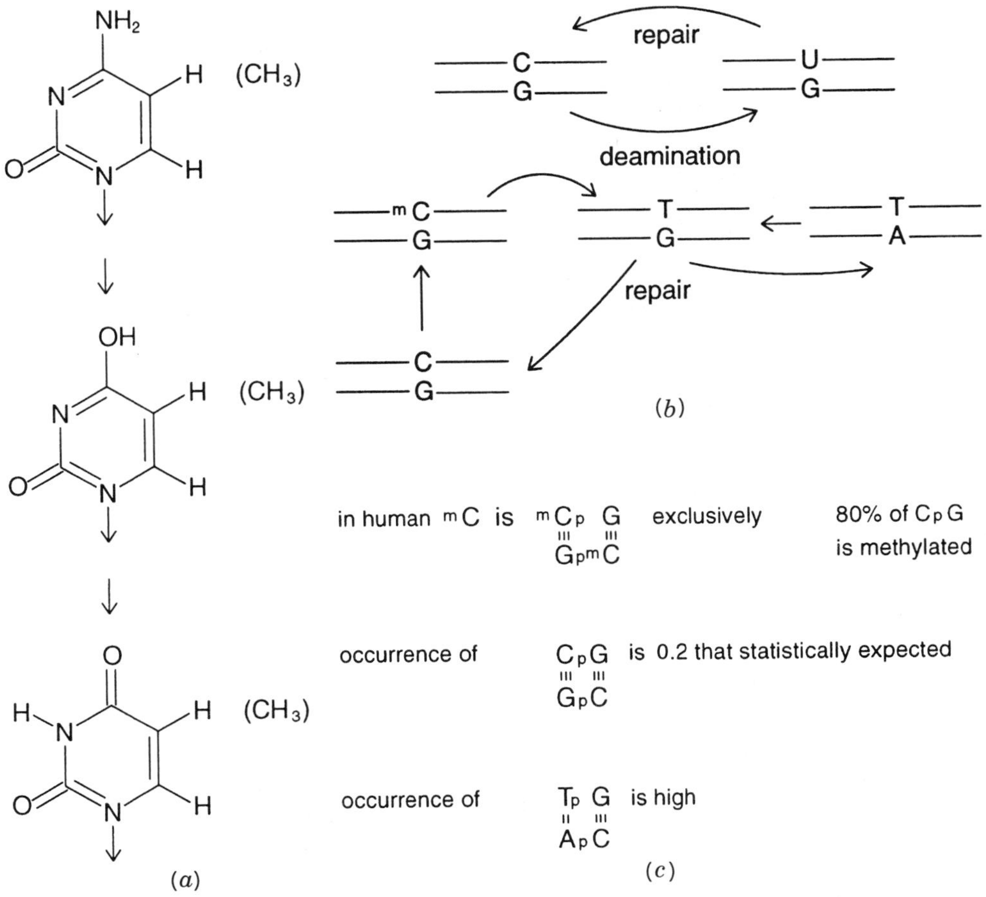

**Figure 1.5**  Mutagenesis and repair processes that alter DNA sequences. *(a)* Deamination of C and ${}^{m}$C produce U and T, respectively. *(b)* Repair of uracil-containing DNA can occur without errors, while repair of mismatched T–G pairs incurs some risk of mutagenesis. *(c)* Consequences of extensive ${}^{m}$CpG mutagenesis in mammalian DNA.

The vertebrate immune system appears to have learned about the striking statistical abnormality of vertebrates. Injections of DNA from other sources with a high G + C content, presumably with a normal ratio to CpG, act as an adjuvant; that is, these injections stimulate a generally heightened immune response.

## PLASTICITY IN DNA STRUCTURE

Elementary discussions of DNA dwell on the beauty and regularity of the Watson-Crick double helix. However, the helical structure of DNA is really much more complex than this. The Watson-Crick structure has 10 base pairs per turn. DNA in solution under physiological conditions shows an average structure of about 10.5 base pairs per turn, roughly halfway between the canonical Watson-Crick form and the A-type helix with a larger diameter and tilted base pairs which are characteristic of RNA. In practice, these are just average forms. DNA is revealed to be fairly irregular by methods that do not have to average over long expanses of structure. DNA is a very plastic molecule with a backbone easily distorted and with optimal geometry very much influenced by its local sequences. For example, certain DNA sequences, like properly spaced sets of ApA's promote extensive curvature of the backbone. Thus, while base pairs predominate, the angle between the base pairs, the extent of their stacking (which holds DNA together) above and below neighbors, their planarity, and their disposition relative to helix axis can vary substantially. Almost all known DNA structures can be viewed in detail by accessing the Nucleic Acid Database, NDB <http://ndbserver.rutgers.edu/>.

We do not really know enough about the properties of proteins that recognize DNA. One extreme view is that these proteins look at the bases directly and, if necessary, distort the helix into a form that fits well with the structure of protein residues in contact with the DNA. The other extreme view has a key role played by the DNA structure with proteins able to recognize structural variants, without explicit consideration of the sequence that generated them. These views have very different implications for proteins that might recognize classes of DNA sequences rather than just distinct single sequences. We are not yet able to decide among these alternative views or to adopt some sort of compromise position. The structures of the few protein-nucleic acid complexes known can be viewed in the NDB.

## DNA SYNTHESIS

Our ability to manufacture specific DNA sequences in almost any desired amounts is well developed. Nucleic acid chemists have long learned and practiced the powerful approach of combining chemical and enzymatic syntheses to accomplish their aims. Automated instruments exist that perform stepwise chemical synthesis of short DNA strands (oligonucleotides) principally by the phosphoramidite method. Synthesis proceeds from the 3'-end of the desired sequence using an immobilized nucleotide as the starting material (Fig. 1.6a). To this are added, successively, the desired nucleotides in a blocked, activated form. After each condenses with the end of the growing chain, it is deblocked to allow the next step to proceed. It is a routine procedure to synthesize several compounds 20 nucleotides long in a day. Recently instruments have been developed that allow almost a hundred compounds to be made simultaneously. Typical instruments produce about a thousand times the amount of material needed for most biological experiments. The cost is about $0.50 to $1.00 per nucleotide in relatively efficient settings. This arises primarily from the costs of the chemicals needed for the synthesis. Scaling down the process will reduce the cost accordingly, and efforts to do this are a subject of intense interest. For certain strategies of large-scale DNA analysis, large numbers of different oligonucleotides are required. The resulting cost will be a significant factor in evaluating the merits of the overall scheme. The currently used synthetic schemes make it very easy to incorporate unusual or modified nucleotides at desired places in the sequence, if appropriate derivatives are available. They also make it very easy to add, at the ends of the DNA strand, other functionalities like chemically reactive alkyl amino or thiol groups or useful biological ligands like biotin, digoxigenin, or fluorescein. Such derivatives have important uses in many analytical application, as we will demonstrate later.

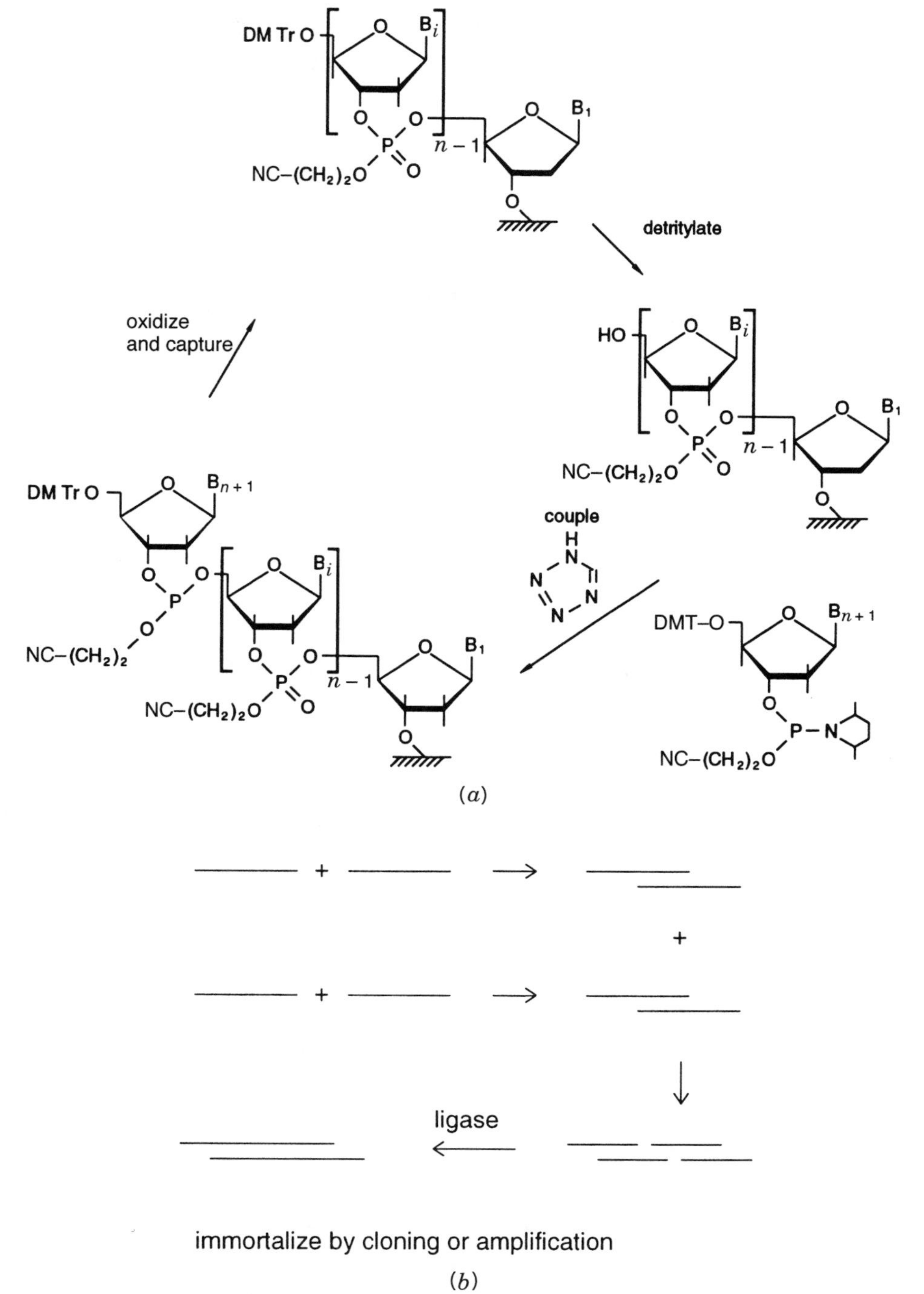

**Figure 1.6**  DNA synthesis by combined chemical and enzymatic procedures. *(a)* Phosphoramidite chemistry for automated solid state synthesis of DNA chains. *(b)* Assembly of separately synthesized chains by physical duplex formation and enzymatic joining using DNA ligase.

**BOX 1.2**
**SIMPLE ENZYMATIC MANIPULATION OF DNAs**

The structure of a DNA strand is an alternating polymer of phosphate and sugar-based units called nucleosides. Thus the ends of the chain can occur at phosphates (p) or at sugar hydroxyls (OH).

Polynucleotide kinase can specifically add a phosphate to the 5'-end of a DNA chain.

$$5' \text{ HO–ApTpCpG–OH } 3' \xrightarrow[\text{kinase}]{\text{ATP}} 5' \text{ pApTpCpG–OH } 3'$$

Phosphatases remove phosphates from one or both ends.

$$5' \text{ pApTpCpGp } 3' \xrightarrow[\substack{\text{alkaline} \\ \text{phosphatase}}]{} 5' \text{ HO–ApTpCpG–OH } 3'$$

DNA ligases will join together two DNA strands that lie adjacent along a complementary template. These enzymes require that one of the strands have a 5'-phosphate:

$$\begin{array}{l} \quad\quad\quad\quad \text{HO p}\\ 5' \text{ GpCpCpT GpTpCpCpA } 3' \\ 3' \text{ CpGpGpApCpApGpGpA } 5' \end{array} \rightarrow \begin{array}{l} 5' \text{ GpCpCpTpGpTpCpCpA } 3' \\ 3' \text{ CpGpGpApCpApGpGpA } 5' \end{array}$$

DNA ligase can also fuse two double-stranded DNAs at their ends provided that 5'-phosphates are present:

$$\begin{array}{l} 5'\text{———}3' \\ 3'\text{———}p5' \end{array} + \begin{array}{l} 5'p\text{———}3' \\ 3'\text{———}5' \end{array} \rightarrow \begin{array}{l} 5'\text{———}p\text{———}3' \\ 3'\text{———}p\text{———}5' \end{array}$$

This reaction is called blunt-end ligation. It is not particularly sensitive to the DNA sequences of the two reactants.

Restriction endonucleases cleave both strands of DNA at or near the site of a specific sequence. They usually cleave at all sites with this particular sequence. The products can have blunt-ends, 3'-overhangs, or 5'-overhangs, as shown by the examples below:

$$\begin{array}{l} 5'\text{—pApCpGpTp—}3' \\ 3'\text{—pTpGpCpAp—}5' \end{array} \rightarrow \begin{array}{l} 5'\text{—pApC–OH} \\ 3'\text{—pTpGp} \end{array} + \begin{array}{l} \text{pGpTp—}3' \\ \text{HO–CpAp—}5' \end{array}$$

$$\begin{array}{l} 5'\text{—pApCpGpTp—}3' \\ 3'\text{—pTpGpCpAp—}5' \end{array} \rightarrow \begin{array}{l} 5'\text{—pApCpGpT–OH} \\ 3'\text{—p} \end{array} + \begin{array}{l} \text{p—}3' \\ \text{HO–TpGpCpAp—}5' \end{array}$$

$$\begin{array}{l} 5'\text{—pApCpGpTp—}3' \\ 3'\text{—pTpGpCpAp—}5' \end{array} \rightarrow \begin{array}{l} 5'\text{—OH} \\ 3'\text{—pTpGpCpAp} \end{array} + \begin{array}{l} \text{pApCpGpT—}3' \\ \text{HO—}5' \end{array}$$

Restriction enzymes always leave 5'-phosphates on the cut strands. The resulting fragments are potential substrates for DNA ligases. Most restriction enzymes cleave at sites with C2 symmetry like the examples shown above.

*(continued)*

**BOX 1.2** *(Continued)*

Exonucleases remove nucleotides from the ends of DNA. Most cleave single nucleotides one at a time off one end of the DNA. For example, Exonuclease III carries out steps like

$$5'\ ApCpGp \longrightarrow pTpApA\ \ 3'$$
$$3'\ TpGpCp \longrightarrow pApTpT\ \ 5'$$

$$5'\ Ap\ \ CpGp \longrightarrow pTpA\text{--}OH\ 3'$$
$$3'\ HO\text{--}GpCp \longrightarrow pApTpT\ \ 5'$$

$$5'\ ApCpGp \longrightarrow pT\text{--}OH\ \ \ 3'$$
$$3'\ HO\text{--}\ Cp \longrightarrow pApTpT\ \ 5'$$

and continues until there are no double-stranded regions left.

Terminal transferases add nucleotides at the 3′-end of a DNA strand. They require that this end not have a 3′-phosphate. No template is required; hence the sequence of the incorporated residues is random. However, if only a single type of pppdN is used, then a set of homopolymeric products is produced

$$GpApTpCpA + pppdT \xrightarrow[\text{tranferase}]{} GpApTpCpA\ (pT)n$$

where *n* is variable. Some DNA polymerases (Box 1.6) also have a terminal transferase activity (see Box 1.5).

Separately synthesized DNA strands can be combined to yield synthetic double helices. The specificity and strength of base pairing will ensure the formation of the correct duplexes under almost all circumstances if the two strands can form 8 or more base pairs. As shown in Figure 1.6*b,* synthetic DNA duplexes can be strung together, and nicks in the polynucleotide backbone can be sealed with the enzyme DNA ligase. This enzyme requires a 5′-phosphate group and a free 3′-OH in order to form a phosphodiester bond (Box 1.2). There are two commonly used forms of the enzyme: the species isolated from *Escherichia coli* uses NAD as an energy source, while the bacteriophage T7 enzyme requires ATP. Thermostable ligases isolated from thermophilic organisms are also available. Their utility will be demonstrated later.

Once a duplex DNA is available, it can be immortalized by cloning it into an appropriate vector (see Box 1.3). Duplexes (or single strands) can also be amplified in vitro by methods such as the polymerase chain reaction (Chapter 4). The result is that for any desired species, a chemical synthesis needs only to be done once. Thereafter biological or enzymatic methods usually suffice to keep up the supply of product. It remains to be seen in the future, as larger stocks of specific synthetic DNA sequences accumulate worldwide, whether it is worthwhile to set up a distribution system to supply the potentially millions of different compounds to interested users, or whether it will be simpler and cheaper to manufacture a needed compound directly on site.

By taking advantage of the specificity of base pairing, it is possible to synthesize and assemble more complex DNA structures than those that predominate in nature. Naturally occurring DNAs are either circles with no ends or linear duplexes with two protected ends.

**BOX 1.3**
**CLONING DNAs IN BACTERIAL PLASMIDS**
**AND BACTERIOPHAGE**

Cloning is the process of making many identical cells (or organisms) from a single precursor. If that precursor contains a target DNA molecule of interest, the cloning process will amplify that single molecule into a whole population. Most single-cell organisms are clonal. Their progeny are identical replicas. Cloning can also be done by manipulating single, immortal, or immortalizable cells of higher organisms including plants and animals. Here we concentrate on bacterial cloning, which can be carried out in a large number of different species but often involves the favorite laboratory organism *Escherichia coli.*

Cloning DNA requires three components: the target DNA of interest, a vector DNA that will accommodate the target as an insertion and that contains a replication origin for it to be propagated indefinitely, and a host cell that will replicate the vector and its insert. The simplest early cloning systems used two types of vectors, either plasmids, which are small DNAs that can replicate independently of the host cell DNA, or bacteriophages, which also have independently replicating DNA systems. Plasmids are usually double-stranded circular DNAs. Bacteriophages can be circular or linear, single stranded, or double stranded. Here we illustrate the steps in cloning double strands.

The vector is usually designed so that it has a unique cutting site for a particular restriction enzyme. Digestion linearizes the vector and produces two ends, usually with the same overhanging sequence. The target is cut with the same enzyme, or alternatively, with a different enzyme that nevertheless leaves the same overhangs. A complex target will yield many fragments, only a few of which may be of interest.

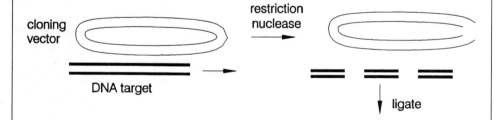

When the vector and target are mixed, they can anneal and form suitable substrates of DNA ligase. Depending on the relative concentrations used, and whether the 5'-phosphate ends produced by the restriction nuclease are left intact or are removed with a phosphatase, a variety of different products combining targets and vectors will be formed. A number of tricks exist to bias the ligation in favor of particular 1:1 target-vector adducts. For example, in the scheme above a single target and vector can come together in two different polarities. More complex schemes allow the polarity to be preselected.

*(continued)*

**BOX 1.3** (*Continued*)

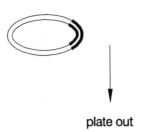

plate out

Next the vector-target adduct must be reintroduced into the bacterial cells. Depending on the vector, different procedures are used. If the vector is a plasmid, the bacterial cells are made permeable by chemical, enzymatic, electrical, or mechanical procedures. When mixed with DNA, these bacteria take up some of the vector before the gaps in their surfaces can be resealed. The process is called *transformation* or *transfection.* Very similar procedures can be used to introduce DNA into other kinds of cells. The efficiencies vary widely depending on the type of vector, type of cell, and the procedures used for permeabilization.

If the vector is a bacteriophage, it is usually preferable to repackage the vector-target adduct in vitro and then allow the assembled cells to infect the host cells naturally. This increases the efficiency of delivering DNA into the host cells, both because bacteriophage insertion is often quite effective and because there is no need to permeabilize the host cells, a process that frequently kills many of them.

Traditionally microbiological screening systems are used to detect host cells propagating cloned DNA. If a dilute suspension of bacteria is allowed to coat the surface of a culture dish, colonies will be observed after several days of growth, each of which represents cloned progeny arising from a single cell. The cloning process can be used to introduce markers that aid in the selection or screening of bacteria carrying plasmids of potential interest. For example, the plasmid may carry a gene that confers resistance to an antibiotic so that only host cells containing the plasmid will be able to grow on a culture medium in which that antibiotic is placed. The site at which the vector is cut to introduce the target can disrupt a gene that yields a colored metabolic product when the precursor of that product is present in the medium. Then vectors that have been designated as having no target insert will still yield colored colonies, while those containing a target insert will appear white.

When a bacteriophage vector is used, the same sets of selection and screening systems can be used. If the bacteriophage shows efficient lysis, that is, if infected cells disrupt

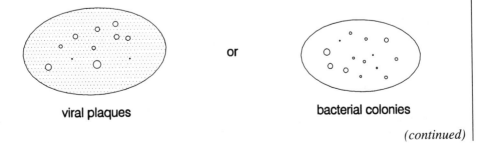

viral plaques          or          bacterial colonies

*(continued)*

**BOX 1.3** *(Continued)*

and release bacteriophage, then a convenient method of cloning is to grow a continuous layer of susceptible bacteria and then cover the culture with a thin layer of dilute bacteriophage suspension. Individual bacteriophage will infect cells, leading to continual cycles of bacteriophage growth and cell lysis until visible holes in the bacterial lawn, called bacteriophage plaques, are observed. Individual colonies or plaques are selected for further study by picking, literally by sampling with a sterile toothpick or other sharp object, and then reinnoculated, respectively, into a permissive growth medium or strain.

A key variable in the choice of vector is the range of target sizes that can be accommodated. Plasmids generally allow a broad range of target sizes, while bacteriophages, because of constraints in the packaging process, generally accept only a narrow range of target sizes. The earliest cloning vectors were efficient for relatively small targets, under 10 kb. Progressively interest has turned to ever larger targets, specialized vector/host systems needed to propagate such targets are described later (Boxes 2.3 and 8.2).

A second key variable in the choice of cloning vector is the copy number. Some plasmids and bacteriophages can grow in tens of thousands of copies per *E. coli* cell. This greatly facilitates recovery and purification of cloned DNA inserts. However, if DNA fragments toxic to the host are the desired targets, a high copy number vector system will make it more difficult for them to be propagated.

The major advantages of cloning, as compared to the in vitro replication systems discussed in Chapter 4, are the ability to handle very large numbers of samples in parallel and the absence of a need for any prior knowledge about the desired target sequences. The major disadvantages of cloning are that it is a relatively time-consuming process difficult to automate, and some DNA sequences are not cloneable either because they are toxic to particular hosts, or they are unstable in those hosts and fragments rearrange or delete.

However, sequences can be designed that associate to give specific structures with three, four, or more ends. These are called DNA junctions (Fig. 1.7). Similar four-stranded structures are actually seen in cells as intermediate steps in genetic recombination. However, these so-called Holliday structures are unstable, since the participating strands are pairs of identical sequences. The location of the junction can move around by a process called *branch migration.* Synthetic junctions can be designed that are stable because their DNA sequences do not allow branch migration. At low salt concentrations these structures become relatively flat, as indicated in Figure 1.7; at higher salt concentrations they form an X-shaped structure with one pair of co-axially stacked helices crossing over a second pair of co-axially stacked helices.

Still other types of DNA structures can be formed by taking advantage of the fact that in the double helix, specific hydrogen bond donor and acceptor sites on the bases remain exposed in the major groove of the double helix. Under appropriate conditions this allows the association of a third DNA strand in a sequence-specific manner to form a DNA triplex. It produces a DNA structure with four ends. The great utility of triplexes, which we will exploit in considerable detail later (Chapter 14), is that triplexes can be formed

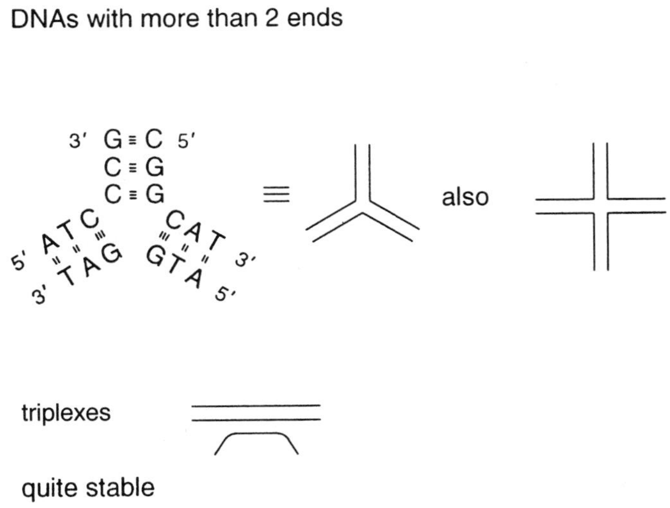

DNAs with more than 2 ends

**Figure 1.7**   DNA structures with more than two ends.

and broken under conditions that do not particularly alter the stability of duplexes but facilitate the sequence-specific manipulation of DNA.

## DNA AS A FLEXIBLE SET OF CHEMICAL REAGENTS

Our ability to synthesize and replicate almost any DNA is encouraging novel applications
of DNAs. Two examples will be sketched here: the use of DNAs as aptamers and the use
of DNAs in nanoengineering. Aptamers are molecules selected to have high affinity binding to a preselected target. The plasticity of single-stranded nucleic acids polymers combined with the affinity of complementary bases to base pair provides an enormous potential for variation in three-dimensional (tertiary) structure. These properties have been
taken advantage of to identify nucleic acids that bind tightly to specific ligands. In this
approach random DNA or RNA libraries are made by synthesizing 60 to 100 base variable compositions. DNAs and RNAs are ideal aptamers because powerful methods exist
to work with very complex mixtures of species and to purify from these mixtures just the
molecules with high affinity for a target from which to characterize the common features
of these classes of molecules. These cycles of design and optimization form the heart of
any engineering process. (See Chapter 14 for a detailed discussion.)

The goal, in nanoengineering is to make machines, motors, transducers, and tools at
the molecular level. It is really the ultimate chemical synthesis, since entire arrays of molecules (or atoms) must be custom designed for specific mechanical or electromagnetic
properties. The potential advantages of using DNAs for such purposes is due to several
factors: the great power to synthesize DNAs of any desired length and sequence, the ability to make complex two- and three-dimensional arrays using Holliday junctions, and the
formation of structures of accurate length by taking advantage of the great stiffness of
duplex DNA. (See Box 1.4.) Proteins can be anchored along the DNA at many points
to create arrays with more complex properties. The disadvantages of using DNA and

proteins for nanoengineering is that these structures are mechanically easier to deform than typical atomic or molecular solids; they usually require an aqueous environment, and they have relatively limited electrical or mechanical properties. There are two potential ways to circumvent the potential disadvantages of DNA in nanoengineering. One is to use DNA (and proteins) to direct the assembly of other types of molecules with the desired properties needed to make engines or transducers. A second is to use the DNA as a resist: to cast a solid surface or volume around it, remove the DNA, and then use the resulting cavity as a framework for the placement of other molecules. In both applications DNA is really conceived of as the ultimate molecular scaffold, with adjustable lengths and shapes. Time will tell if this fantasy can be ever realized in practice. A simple test of the use of DNA to control the spacing of two proteins is described in Box 1.4.

Our powerful ability to manipulate DNA is really just hinted at by some of the above discussion. We know most of the properties of short DNA molecules; our ability to make quantitative predictions about the properties of unknown sequences is not bad. What makes the problem of managing cellular DNAs difficult is their enormous size. The DNAs of small viruses have thousands of base pairs. Bacterial chromosomal DNA molecules are typically 1 to 10 million base pairs (0.3–3.0 mm) long. Human chromosomal DNAs range in size from about 0.5 to $2.5 \times 10^8$ base pairs. The largest DNA molecules in nature may approach a billion base pairs, which corresponds to molecular weights of almost a trillion Daltons. These are indeed large molecules; just describing their sequence in any detail, if we knew it, would be a formidable task. What is impressive is that all DNAs, from the largest to the smallest, can be analyzed by a small number of very similar physical and genetic techniques (Chapters 3–5). However, it is the recent development of some physical methods that have moved biological experimentation beyond the study of the relatively few organisms with well-developed genetic systems. The results of these experiments indicate that the biology of a large number of organisms seems to fall along very similar lines.

**BOX 1.4**
**POTENTIAL USE OF DNA IN NANOENGINEERING**

As a test case for nanoengineering with DNA, we have explored the use of DNA as a spacer of known length between two sites with biological functionality. The immunological problem which motivated this work was a desire to understand the complex set of molecular interactions that occurs when an antigen-presenting cell is recognized by a T lymphocyte. Antigen is presented as peptide bound to a cell surface molecule known as a major histocompatibility molecule, in complex with associated accessory membrane proteins. The detection is done by the clonotypic T cell receptor: a complex membrane protein that also can associate with other accessory proteins. Our current picture of this interaction is summarized in Figure 1.8. To refine this picture, we need to determine the structure of the complex, to decide what elements in this structure pre-exist on the cell surface in associated form in each of the participating cells before they become engaged, and to monitor the fate of the components during and after engagement. Because the complex of proteins involved in antigen presentation is so

*(continued)*

**BOX 1.4** *(Continued)*

large, it is difficult to study cellular interaction by short-range methods such as energy transfer or crosslinking. Thus we must seek to make longer molecular rulers that can span distances of up to 100 Å or more.

The idea is to make pairs of antibodies or antibody combining sites separated by DNA spacers. Monoclonal antibodies exist for many chemical structures (epitopes) on the molecules present on the surfaces of T cells or antigen-presenting cells. The two approaches used to link such antibodies are shown schematically in Figure 1.9. The key point in both approaches is that separate, complementary DNA single strands are conjugated to particular antibodies or fragments. Then pairs of these conjugates are mixed, and the double-helix formation directs the specific production of heterospecific antibody conjugates. The length of the DNA double helix is 3.4 Å per base pair. Thus a 32-base DNA duplex will be about 100 Å long. It is expected to behave as an extremely rigid structure with respect to lateral bending, but it can undergo torsional twisting. Thus, while the distance between the tethered antibodies should be relatively fixed, the angle between them may be variable. This should help both sites reach their targets simultaneously on the same cell surface. The actual utility of such compounds remains to be established, but it is interesting that it is possible to make such constructs relatively easily.

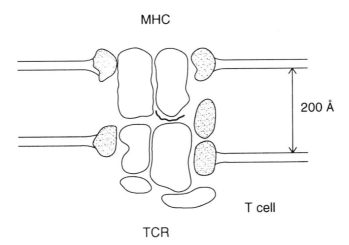

antigen presenting cell

MHC

200 Å

T cell

TCR

questions about assemblies:
    structure
    existence prior to engagement
    co-internalization

**Figure 1.8** Schematic illustration of the recognition of an antigen-presenting cell by a T lymphocyte, a key step in the immune response. The antigen-presenting cell has on its surface molecules of the major histocompatibility complex (MHC) which have bound an antigen (black chain). The T cell has specific receptors (TCR's) that recognize a particular MHC molecule and its bound peptide. Both cells have accessory proteins (dotted) that assist the recognition.

*(continued)*

**BOX 1.4** *(Continued)*

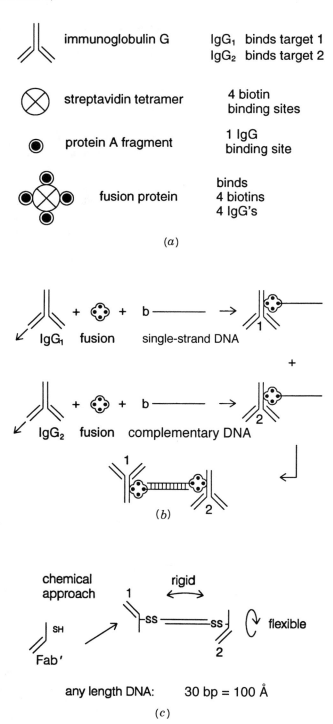

**Figure 1.9** Use of DNA molecules as spacers to construct long molecular rulers. *(a)* Symbols used. *(b)* Noncovalent coupling. *(c)* Covalent coupling.

## BASIC DNA BIOLOGY

DNA is the essential store of genetic information in the cell. That information must be duplicated to be passed to daughter cells by the process known as *replication.* It must be read out as an RNA copy by the process known as *transcription* so that this copy (after editing in many cases) can be used to direct the synthesis of proteins.

DNA replication is limited by the fact that all of the known enzymes that can faithfully copy the information encoded in DNA have very restricted properties. Such enzymes are called *DNA polymerases.* They all require a 3′-ended nucleic acid (or a protein substituting for this 3′-end) as a primer, and a template strand, which is copied. They cannot initiate synthesis de novo but can only elongate a strand defined by a primer (Figure 1.10). They can, for the most part, copy a DNA strand all of the way to its end. Accurate replication cannot depend on base pairing specificity alone because the thermodynamic difference between the stability of a single base pair and a mismatch is not very great (Chapter 3). Editing is used to correct any misincorporated bases. Some of this editing is carried out by the DNA polymerase itself. A newly incorporated mispaired base can be excised by the 3′-exonuclease activity of most DNA polymerases. Some enzymes also have a 5′-exonuclease activity that degrades any DNA strands in front of a wave of new synthesis. This process is called *nick translation.* Other enzymes that lack this activity will displace one strand of a duplex in order to synthesize a new strand. This process is called *strand displacement.* These enzyme activities are illustrated in Box 1.5.

Special procedures are used to synthesize the primers needed to start DNA replication. For example, RNA primers can be used and then degraded to result in a complete DNA strand. Both strands of DNA must be replicated. The unidirectional mode of DNA polymerases makes this process complicated, as shown in Figure 1.11. In a single-replication fork the synthesis of one strand (the leading strand), once initiated, can proceed in an uninterrupted fashion. The other strand (the lagging strand) must be made in a retrograde fashion from periodically spaced primers, and the resulting fragments are then stitched together. Most replication processes employ not a single fork but a bidirectional pair of replication forks, as shown in Figure 1.11.

Any DNA replication process poses an interesting topological problem, whether it is carried out by a cell or in the laboratory. Because the two DNA strands are wound about a common axis, the helix axis, they are actually also twisted around each other. They cannot be unwound without rotation of the structure, once for each helix turn. Thus DNA must spin as it is being replicated. In the bacterium *E. coli,* for example, the rate of replication is fast enough to make a complete copy of the chromosomal DNA in about 40 minutes. Since the chromosome contains almost 5 million base pairs of DNA, the required replication rate is $1.2 \times 10^5$ base pairs per minute. Since *E. coli* uses a pair of bidirectional replication forks, each must move at $6 \times 10^4$ bases per minute. Thus each side of the replication fork must unwind $6 \times 10^3$ helical turns per minute: it must rotate at $6 \times 10^3$ rpm.

### DNA polymerase only goes one way

$$5' \xrightarrow{\hspace{2cm}} 3'$$
$$3' \underline{\hspace{2.5cm}} 5'$$

**Figure 1.10**  Primer extension by DNA polymerase in a template-directed manner. The newly synthesized strand is shown as a dashed arrow.

**BOX 1.5**
**PROPERTIES OF DNA POLYMERASES**

DNA polymerases play the central role in how cells replicate accurate copies of their DNA molecules. These enzymes also serve to illustrate many of the basic properties of other enzymes that make an RNA copy of DNA, a DNA copy of an RNA, or replicate RNAs. In the laboratory DNA polymerases are an extraordinarily useful tool for many of the most common ways in which DNAs are manipulated or analyzed experimentally. All DNA polymerases can extend a primer along a template in a sequence specific manner.

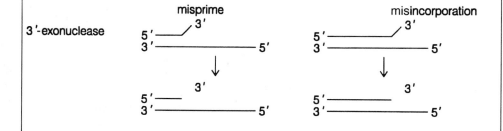

Most polymerases have one or more additional activities. A 3'-exonuclease activity will prevent the use of mispaired primers. This activity, essentially looks backward as the polymerase proceeds, and if an incorrect base is inserted, the polymerase pauses to remove it. In the absence of pppdN's, the 3'-exonuclease activity will progressively shorten the primer by removing nucleotides one at a time, even though they are correctly base-paired to the template.

Many DNA polymerases have a 5'-exonuclease activity. This degrades any DNA ahead of the site of chain extension. The result is the progressive migration of a nick in the strand complementary to the template. Thus this activity is often called nick translation.

*(continued)*

**BOX 1.5** (*Continued*)

One function of nick translation is to degrade primers used at earlier stages of DNA synthesis. Frequently these primers are RNA molecules. Thus the nick translation, 5′-exonuclease activity ensures that no RNA segments remain in the finished DNA. This activity also is used as part of the process by which DNA damaged by radiation or chemicals is repaired.

The final DNA polymerase activity commonly encountered is strand displacement. This is observed in some mechanisms of DNA replication.

strand displacement

In most DNA replication, however, a separate enzyme, DNA helicase, is used to melt the double helix (separate the base-paired strands) prior to chain extention. Some DNA polymerases also have a terminal transferase activity. For example, *Taq* polymerase usually adds a single nontemplated A onto the 3′-ends of the strands that it has synthesized.

A large library of enzymes exists for manipulating nucleic acids. Although several enzymes may modify nucleic acids in the same or a similar manner, differences in catalytic activity or in the protein structure may lead to success or failure in a particular application. Hence the choice of a specific enzyme for a novel application may require an intimate knowledge of the differences between the enzymes catalyzing the same reaction. Given the sometimes unpredictable behavior of enzymes, empirical testing of several similar enzymes may be required.

More details are given here about one particularly well-studied enzyme. DNA polymerase I, isolated from *Escherichia coli* by Arthur Kornberg in 1963, established much of the nomenclature used with enzymes that act on nucleic acids. This enzyme was one of the first enzymes involved in macromolecular synthesis to be isolated, and it also displays multiple catalytic activities. DNA polymerase I replicates DNA, but it is mostly a DNA repair enzyme rather than the major DNA replication enzyme in vivo. This enzyme requires a single-stranded DNA template to provide instructions on which DNA sequence to make, a short oligonucleotide primer with a free 3′-OH terminus to specify where synthesis should begin, activated precursors (nucleoside triphosphates), and a divalent cation like $MgCl_2$. The primer oligonucleotide is extended at its 3′-end by the addition, in a 5′- to 3′-direction, of mononucleotides that are complementary to the opposite base on the template DNA. The activity of DNA polymerase I is distributive rather than processive, since the enzyme falls off the template after incorporating a few bases.

*(continued)*

**BOX 1.5** (*Continued*)

Besides the polymerase activity, DNA polymerase I has two activities that degrade DNA. These activities are exonucleases because they degrade DNA from either the 5'- or the 3'-end. Both activities require double-stranded DNA. The 3'-exonuclease proofreading activity is the reverse reaction of the 5'- to 3'-polymerase activity. This activity enhances the specificity of extension reaction by removing a mononucleotide from the 3'-primer end when it is mismatched with the template base. This means that the specificity of the DNA polymerase I extension reaction is enhanced from $\sim 10^{-8}$ to $\sim 10^{-9}$. Both the extension and 3'-exonuclease activity reside in the same large, proteolytic degradation fragment of DNA polymerase, also called the Klenow fragment.

The 5'- to 3'-exonuclease activity is quite different. This activity resides in the smaller proteolytic degradation fragment of DNA polymerase I. It removes oligonucleotides containing 3–4 bases, and its activity does not depend on the occurrence of mismatches. A strand displacement reaction depends on a concerted effort of the extension and 5'-exonuclease activity of DNA polymerase I. Here the extension reaction begins at a single-stranded nick; the 5'-exonuclease activity degrades the single-stranded DNA annealed to the template ahead of the 3'-end being extended, thus providing a single-stranded template for the extension reaction. The DNA polymerase I extension reaction will also act on nicked DNA in the absence of the 5'-exonuclease activity. Here the DNA polymerase I just strand displaces the annealed single-stranded DNA as it extends from the nick.

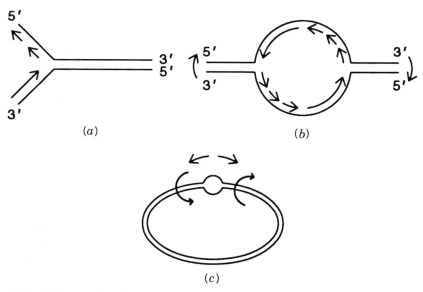

**Figure 1.11**   Structure of replication forks. *(a)* A single fork showing the continuously synthesized leading strand and discontinuously synthesized lagging strand. *(b)* A double fork, common in almost all genomic DNA replication. *(c)* Topological aspects of DNA replication. Thin arrows show translational motion of the forks; bold arrows show the required DNA rotation around the forks.

Unfortunately, the rotations generated by the two forks do not cancel each other out. They add. If this rotation were actually allowed to occur across massive lengths of DNA, the cell would probably be stirred to death. Instead, optional topoisomerases are used to restrict the rotation to regions close to the replication fork. Topoisomerases can cut and reseal double-stranded DNA very rapidly without allowing the cut ends to diffuse apart. Thus the rotations can let the torque generated by helix unwinding to be dissipated. (The actual mechanism of topoisomerases is more complex and indirect, but the outcome is the same as we have stated.)

The information stored in the sequence of DNA bases comprises inherited characteristics called genes. Early in the development of molecular biology, observed facts could be accounted for by the principle that one gene (i.e., one stretch of DNA sequence) codes for one protein, a linear polymer containing a sequence composed of up to 20 different amino acids, as shown in Figure 1.12a. A three-base sequence of DNA directs the incorporation of one amino acid; hence the genetic code is a triplet code. The gene would define both the start and stop points of actual transcription and the ultimate start and stop points of the translation into protein. Nearby would be additional DNA sequences that regulate the nature of the gene expression: when, where, and how much mRNA should be made. A typical gene in a prokaryote (a bacterium or any other cell without a nucleus) is one to two thousand base pairs. The resulting mRNA has some upstream and downstream untranslated regions; it encodes (one or more) proteins with several hundred amino acids.

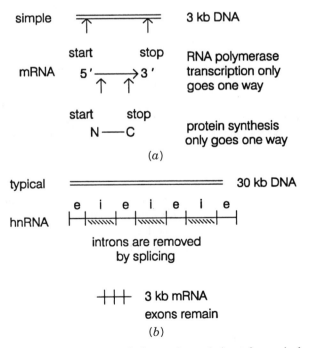

**Figure 1.12** What genes are. (a) Transcription and translation of a typical prokaryotic gene. N and C indicate the amino and carboxyl ends of the peptide backbone of a protein. (b) Transcription and translation of a typical eukaryotic gene. Introns (i) are removed by splicing leaving only exons (e).

multiple reading frames

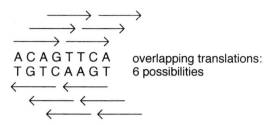

ACAGTTCA        overlapping translations:
TGTCAAGT        6 possibilities

**Figure 1.13**  Six possible reading frames (arrows) for a stretch of DNA sequence.

The genes of eukaryotes (and even some genes in prokaryotes) are much larger and more complex. A typical mammalian gene might be 30,000 base pairs in length. Its regulatory regions can be far upstream, downstream, or buried in the middle of the gene. Some genes are known that are almost 100 times larger. A key difference between prokaryotes and eukaryotes is that most of the DNA sequence in eukaryotic genes is not translated (Fig. 1.12*b*). A very long RNA transcript (called *hnRNA,* for heterogeneous nuclear RNA) is made; then most of it is removed by a process called *RNA splicing.* In a typical case several or many sections of the RNA are removed, and the remaining bits are resealed. The DNA segments coding for the RNA that actually remain in the mature translated message are called *exons* (because they are expressed). The parts excised are called *introns.* The function of introns, beyond their role in supporting the splicing reactions, is not clear. The resulting eukaryotic mRNA, which codes for a single protein, is typically 3 kb in size, not much bigger than its far more simply made prokaryotic counterpart.

Now that we know the DNA structure and RNA transcription products of many genes, the notion of one gene one protein has to be broadened considerably. Some genes show a pattern of multiple starts. Different proteins can be made from the same gene if these starts affect coding regions. Quite common are genes with multiple alternate splicing patterns. In the simplest case this will result in the elimination of an exon or the substitution of one exon for another. However, much more complicated variations can be generated in this way. Finally DNA sequences can be read in multiple reading frames, as shown by Figure 1.13. If the sequence allows it, as many as six different (but not independent) proteins can be coded for by a single DNA sequence depending on which strand is read and in what frame. Note that since transcription is unidirectional, in the same direction as replication, one DNA strand is transcribed from left to right as the structures are drawn, and the other from right to left.

Recent research has found evidence for genes that lie completely within other genes. For example, the gene responsible for some forms of the disease neurofibromatosis is an extremely large one, as shown in Figure 1.14. It has many long introns, and they are tran-

bottom strand
read

big intron

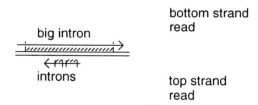

introns

top strand
read

**Figure 1.14**  In eukaryotes some genes can lie within other genes. A small gene with two introns coded for by one of the DNA strands lies within a single intron of a much larger gene coded for by the other strand. Introns are shown as hatched.

scribed off of the opposite strand used for the transcription of the type one neurofibromatosis gene. The small gene is expressed, but its function is unknown.

## GENOME SIZES

The purpose of the human genome project is to map and sequence the human genome and find all of the genes it contains. In parallel, the genomes of a number of model organisms will also be studied. The rationale for this is clear-cut. The human is a very poor genetic organism. Our lifespan is so long that very few generations can be monitored. It is unethical (and impractical) to control breeding of humans. As a result one must examine inheritance patterns retrospectively in families. Typical human families are quite small. We are a very heterogeneous outbred species, with just the opposite genetic characteristics of highly inbred, homogeneous laboratory strains of animals used for genetic studies. For all these reasons experimental genetics is largely restricted to model organisms. The gold standard test for the function of a previously unknown gene is to knock it out and see the resulting effect, in other words, determine the phenotype of a deletion. For organisms with two copies of their genome, like humans, this requires knocking out both gene copies. Such a double knockout is extremely difficult without resorting to controlled breeding. Thus model organisms are a necessary part of the genome project.

Considerable thought has gone into the choice of model organisms. In general, these represent a compromise between genome size and genetic utility. *E. coli* is the best-studied bacterium; its complete DNA sequence became available early in 1997. *Saccharomyces cerevisiae* is the best studied yeast, and for that matter the best studied single-cell eukaryotic organism. Its genetics is exceptionally well developed, and the complete 12,068 kb DNA sequence was reported in 1996, the result of a worldwide coordinated effort for DNA sequencing. *Caenorhabditis elegans,* a nematode worm has very well-developed genetics; its developmental biology is exquisitely refined. Every cell in the mature organism is identified as are the cell lineages that lead up to the mature adult. The last invertebrate canonized by the genome project is the fruit fly *Drosophila melanogaster.* This organism has played a key role in the development of the field of genetics, and it is also an extraordinarily convenient system for studies of development. The fruit fly has an unusually small genome for such a complex organism; thus the utility of genomic sequence data is especially apparent in this case.

For vertebrates, if a single model organism must be selected, the mouse is the obvious choice. The size of the genome of *Mus musculus* is similar to that of humans. However, its generation time is much shorter, and the genetics of the mouse is far easier to manipulate. A number of inbred strains exist with relatively homozygous but different genomes; yet these will crossbreed in some cases. From such interspecific crosses very powerful genetic mapping tools emerge, as we will describe in Chapter 6. Mice are small, hence relatively inexpensive to breed and maintain. Their genetics and developmental biology are relatively advanced. Because of efforts to contain the projected costs of the genome project, no other "official" model organisms exist. However, many other organisms are of intense interest for genome studies; some of these are already under active scrutiny. These include maize, rice, *Arabidopsis thaliana,* rats, pigs, cows, as well as a number of simpler organisms.

In thinking of which additional organisms to subject to genome analysis, careful attention must be given to what is called the G-value paradox. Within even relatively similar classes of organisms, the genome size can vary considerably. Furthermore, as the data in Table 1.1 reveal, there is not a monotomic relationship between genome size and our

TABLE 1.1   Genome Sizes (base pairs)

| | |
|---|---|
| Bacteriophage lambda | $5.0 \times 10^4$ |
| *Escherichia coli* | $4.6 \times 10^6$ |
| Yeasts | $12.0 \times 10^6$ |
| *Giardia lamblia* | $14.0 \times 10^6$ |
| *Drosophila melanogaster* | $1.0 \times 10^8$ |
| Some hemichordates | $1.4 \times 10^8$ |
| Human | $3.0 \times 10^9$ |
| Some amphibians | $8.0 \times 10^{11}$ |

Note: These are haploid genome sizes. Many cells will have more than one copy of the haploid genome.

view of how evolutionarily advanced a particular organism is. Thus, for example, some amphibians have genomes several hundred times larger than the human. Occasional organisms like some hemichordates or the puffer fish have relatively small genomes despite their relatively recent evolution. The same sort of situation exists in plants.

In planning the future of genome studies, as attention broadens to additional organisms, one must decide whether it will be more interesting to examine closely related organisms or to cast as broad a phylogenetic net as funding permits. Several organisms seem to be of particular interest at the present time. The fission yeast *Schizosaccharomyces pombe* has a genome the same size as the budding yeast *S. cerevisiae*. However, these two organisms are as far diverged from each other, evolutionarily, as each is from a human being. The genetics of *S. pombe* is almost as facile as that of *S. cerevisiae*. Any features strongly conserved in both organisms are likely to be present throughout life as we know it. Both yeasts are very densely packed with genes. The temptation to compare them with full genomic sequencing may be irresistible. Just how far genome studies will be extended to other organisms, to large numbers of different individuals, or even to repeated samplings of a given individual will depend on how efficient these studies eventually become. The potential future need for genome analysis is almost unlimited, as described in Box 1.6.

---

**BOX 1.6**
**GENOME PROJECT ENHANCEMENTS**

| DNA Sequencing Rate: bp Per Person Per Day | Accessible Targets |
|---|---|
| $10^6$ | One human, five selected model organisms |
| | Organisms of commercial value |
| $10^7$ | Selected diagnostic DNA sequencing |
| $10^8$ | Human diversity (see Chapter 15) |
| | $5 \times 10^9$ individuals $\times$ 6 to $12 \times 10^6$ differences = 3 to $6 \times 10^{16}$ |
| | Full diagnostic DNA sequencing |
| $10^9$ | Environment exposure assessment |

## NUMBERS OF GENES

It has been estimated that half of the genes in the human genome are central nervous system specific. For such genes, one must wonder how adequate a model the mouse will be for the human. Even if there are similar genes in both species, it is not easy to see how the counterparts of particular human phenotypes will be found in the mouse. Do mice get headaches, do they get depressed, do they have fantasies, do they dream in color? How can we tell? For such reasons it is desirable, as the technology advances to permit this, to bring into focus the genomes of experimental animals more amenable to neurophysiological and psychological studies. Primates like the chimp are similar enough to the human that it should be easy to study them by starting with human material as DNA probes. Yet the differences between humans and chimps are likely to be of particular interest in defining the truly unique features of our species. Other vertebrates, like the rat, cat, and dog, while more distant from the human, may also be very attractive genome targets because their physiologies are very convenient to study, and in some cases they display very well-developed personality traits. Other organisms, such as the parasitic protozoan, *Giardia lamblia* or the blowfish, fugu, are eukaryotes of particular interest because of their comparatively small genome sizes.

The true goal of the genome project is to discover all of the genes in an organism and make them available in a form convenient for future scientific study. It is not so easy, with present tools and information, to estimate the number of genes in any organism. The first complete bacterial genome to be sequenced is that of *H. influenzae* Rd. It has 1,830,137 base pairs and 1743 predicted protein coding regions plus six sets of three rRNA genes and numerous genes for other cellular RNAs like tRNA. *H. influenzae* is not as well studied as *E. coli,* and we do not yet know how many of these coding regions are actually expressed. For the bacterium *E. coli,* we believe that almost all genes are expressed and translated to at least a detectable extent. In two-dimensional electrophoretic fractionations of *E. coli* proteins, about 2500 species can be seen. An average *E. coli* gene is about 1 to 2 kb in size; thus the 4.6 Mb genome is fully packed with genes. Yeasts are similarly packed. Further details about gene density are given in Chapter 15.

In vertebrates the gene density is much more difficult to estimate. An average gene is probably about 30 kb. In any given cell type, 2d electrophoresis reveals several thousand protein products. However, these products are very different in different cell types. There is no way to do an exhaustive search. Various estimates of the total number of human genes range from $5 \times 10^4$ to $2 \times 10^5$. The true answer will probably not be known until long after we have the complete human DNA sequence, because of the problems of multiple splicing patterns and genes within genes discussed earlier. However, by having cloned and sequenced the entire human genome, any section of DNA suspected of harboring one or more genes will be easy to scrutinize further.

## SOURCES AND ADDITIONAL READINGS

Alivisatos, A. P., Jonsson, K. P., Peng, X., Wilson, T. E., Loweth, C. J., Bruchez, M. P., and Schultz, P. G. 1996. Organization of "nanocrystal molecules" using DNA. *Nature* 382: 609–611.

Berman, H. M. 1997. Crystal studies of B-DNA: The answers and the questions. *Biopolymers* 44: 23–44.

Berman, H. M., Olson, W. K., Beveridge, D. L., Westbrook J., Gelbin, A., Demeny, T., Hsieh, S.-H., Srinivasan, A. R., and Schneider, B. 1992. The Nucleic Acid Database: A comprehensive relational database of three-dimensional structures of nucleic acids. *Biophysical Journal* 63: 751–759.

Cantor, C. R., and Schimmel, P. R. 1980. *Biophysical Chemistry.* San Francisco: W. H. Freeman, ch. 3 (Protein structure) and ch. 4 (Nucleic acid structure).

Garboczi, D. N., Ghosh, P., Utz, U., Fan, Q. R., Biddison, W. E, and Wiley, D. C. 1996. Structure of the complex between human T-cell receptor, viral peptide and HLA-A2. *Nature* 384: 134–141.

Hartmann, B., and Lavery, R. 1996. DNA structural forms. *Quarterly Review of Biophysics* 29: 309–368.

Klinman, D. A., Yi, A., Beaucage, S., Conover, J., and Krieg, A. M. 1996. CpG motifs expressed by bacterial DNA rapidly induce lymphocytes to secrete IL-6, IL-12, and IFN-g. *Proceeding of the National Academy of Sciences USA* 93: 2879–2883.

Lodish, H., Darnell, J., and Baltimore, D. 1995. *Molecular Cell Biology,* 3rd. ed. New York: Scientific American Books.

Mao, C., Sun, W., and Seeman, N. C. 1997. Assembly of Borromean rings from DNA. *Nature* 386: 137–138.

Mirkin, C. A., Letsinger, R. L., Mucic, R. C., and Storhoff, J. J. 1996. A DNA-based method for rationally assembling nanoparticles into macroscopic materials. *Nature* 382: 607–609.

Niemeyer, C. M., Sano, T., Smith, C. L., and Cantor, C. R. 1994. Oligonucleotide-directed self-assembly of proteins: Semisynthetic DNA-streptavidin hybrid molecules as connectors for the generation of macroscopic arrays and the construction of supramolecular bioconjugates. *Nucleic Acids Research* 22: 5530–5539.

Saenger, W. 1984. *Principles of Nucleic Acid Structure.* New York: Springer-Verlag.

Timsit, H. Y., and Moras, D. 1996. Cruciform structures and functions. *Quarterly Review of Biophysics* 29: 279–307.

# 2 A Genome Overview at the Level of Chromosomes

## BASIC PROPERTIES OF CHROMOSOMES

Chromosomes were first seen by light microscopy, and their name reflects the deep color they take on with a number of commonly used histological stains. In a cell a chromosome consists of a single piece of DNA packaged with various accessory proteins. Chromosomes are the fundamental elements of inheritance, since it is they that are passed from cell to daughter cell, from parent to progeny. Indeed a cell that did not need to reproduce would not have to keep its DNA organized into specific large molecules. Some single-cell organisms, like the ciliate Tetrahymena, actually fragment a working copy of their DNA into gene-sized pieces for expression while maintaining an unbroken master copy for reproductive purposes.

## BACTERIAL CHROMOSOMES

Bacteria generally have a single chromosome. This is usually a circular DNA duplex. As shown in Figure 2.1, the chromosome has at least three functional elements. The replication origin (ori) is the location of the start of DNA synthesis. The termination (ter) region provides a mechanism for stopping DNA synthesis of the two divergent replication forks. Also present are *par* sequences which ensure that chromosomes are partitioned relatively uniformly between daughter cells. A description of a bacterial chromosome is complicated by the fact that bacteria are continually replicating and transcribing their DNA.

In rapidly growing organisms, a round of replication is initiated before the previous round is completed. Hence the number of copies of genomic DNA depends on how rapidly the organism is growing and where in the genome one looks. Genes near the origin are often present at several times the copy number of genes near the terminus. In general, this seems to have little effect on the bacterium. Rather, bacteria appear to take advantage of this fact. Genes whose products are required early in the replication cycle, or in large amounts (like the ribosomal RNAs and proteins), are located in the early replicated regions of the chromosomes. Although the bacterial chromosome is relatively tolerant of deletions (involving nonessential genes) and small insertions, many large rearrangements involving inversion or insertions are lethal. This prohibition appears to be related to conflicts that rise between convergent DNA replication forks and the transcription machinery of highly expressed genes.

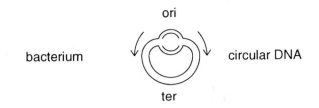

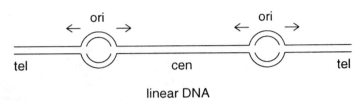

**Figure 2.1**  Basic functional elements in chromosomes: ori (replication origin), tel (telomere), cen (centromere), ter (termination region). Little is known about eukaryotic termination regions.

Bacteria will frequently harbor additional DNA molecules smaller than their major chromosome. Some of these may be subject to stringent copy number control and to orderly partitioning, like the major chromosome. These low copy number plasmids use the same DNA replication machinery as the chromosome, and they are present in a copy number that is equal to the chromosomal copy number. An example is the naturally occurring F+ DNA of *E. coli* which represents about 2% of the DNA present in the cell.

In some bacteria essential genes are located on two genetic elements. For instance, in *Rhodobacteria sphaeroides* the genes encoding the ribosomal RNAs (rRNAs) are located on a 0.9 Mb chromosome, whereas the remainder of the genome is located on a 3 Mb chromosome. In Pseudomonas species, many genes that code for catabolic (degradative) enzymes are located on large extrachromosomal plasmids. Plasmids containing antibiotic resistance genes have been isolated from many species. In part, it is the rapid transfer of plasmids, sometimes even between different genera, that accounts for the rapid development of large populations of antibiotic resistant bacteria.

The control of replication of other, usually smaller plasmids, is more relaxed. These plasmids can have a very high intracellular copy number, and they are usually the focus of recombinant DNA cloning experiments. These plasmids use a DNA replication mechanism that is distinct from that used by the chromosomes. In fact selective inhibition of chromosomal replication machinery focuses the cell replication machinery on producing more plasmid such that it is possible to increase the copy number of these plasmids to about 1000 copies per cell. Selection of growth conditions, which depend on genes carried by the plasmid, can also be used to increase or decrease its copy number. Some plasmids do not contain a par functioning region and are not partitioned in an orderly fashion to daughter cells. This means that their inheritance is subject to statistical fluctuations that may lead to significant instabilities especially with low copy plasmids. Since the genome is defined as all of the DNA in a cell, plasmids and other extrachromosomal DNA elements must be counted as part of it.

Bacterial chromosomes contain bound protein molecules that are essential for normal growth. Some of these promote an organization or packaging of DNA similar to that seen in higher organisms. However, the proteins that appear to be responsible for packaging are present in such small amounts that can only interact with about 20% of the genomic DNA. Furthermore the packaging does not seem to be as orderly or as stable as the packaging of DNA in eukaryotic chromosomes. Bacterial chromosomal DNA is organized into topological constrained domains (Box 2.1) that average 75 kb in size. The way in which this occurs, and its functional consequences, are not yet understood in a rigorous way.

---

**BOX 2.1**
**TOPOLOGICAL PROPERTIES OF DNA**

Because the two strands of DNA twist around a common axis, for unbent DNA, each turn of the helix is equivalent to twisting one strand 360° around the other. Thus, when the DNA is circular, if one strand is imagined to be planar, the other is wrapped around it once for each helix turn. The two circular strands are thus linked topologically. They cannot be pulled apart except by cutting one of the strands. A single nick anywhere in either strand removes this topological constraint and allows strand separation. The topological linkage of the strands in circular DNA leads to a number of fascinating phenomena, and the interested reader is encouraged to look elsewhere for detailed descriptions of how these are studied experimentally and how they are analyzed mathematically. (Cantor and Schimmel, 1980; Cozzarelli and Wang, 1990).

For the purpose of this book, the major thing the reader must bear in mind is that physical interactions between DNA strands can be blocked by topological constraints. If a linear strand of DNA contacts a surface at two points, the region between these points is topologically equivalent to a circle that runs through the molecule and then through the surface. Hybridization of a complementary strand to this immobilized strand will require twisting the former around the latter. This may be difficult if the latter is close to the surface, and it will be impossible if the former is circular itself.

Cells deal with the topological constraints of DNA double helices by having a series of enzymes that relax these constraints. Type I topoisomerases make a transient single-stranded nick in DNA, which allows one strand to rotate about the other at that point. Type II topoisomerases make a transient double-strand nick and pass an intact segment of the duplex through this nick. Though it is less obvious, this has the effect of allowing the strands to rotate 720° around each other. When DNA is replicated or transcribed, the double helix must be unwound ahead of and rewound behind the moving polymerases. Topoisomerases are recruited to enable these motions to occur.

In condensed chromatin, loops of 300 Å fiber are attached to a scaffold. Although the DNA of mammalian chromosomes is linear, these frequent attachment points make each constrained loop topologically into a circle. Thus topoisomerases are needed for the DNA in chromatin to function. Type II topoisomerases are a major component of the proteins that make up the chromosome scaffold.

## CHROMOSOMES OF EUKARYOTIC ORGANISMS

All higher organisms usually have linear chromosomal DNA molecules, although circles can be produced under special circumstances. These molecules have a minimum of three functional features, as shown in Figure 2.1. Telomeres are specialized structures at the ends of the chromosome. These serve at least two functions. They provide a mechanism by which the ends of the linear chromosomes can be replicated. They stabilize the ends. Normal double-strand DNA ends are very unstable in eukaryotic cells. Such ends could be the result of DNA damage, such as that caused by X rays, and could be lethal events. Hence very efficient repair systems exist that rapidly ligate ends not containing telomeres together. If several DNAs are broken simultaneously in a single cell, the correct fragment pairs are unlikely to be reassembled, and one or more translocations will result. If the ends are not repaired fast enough by ligation, they may invade duplex DNA in order to be repaired by recombination. The result, even for a single, original DNA break, is a re-arranged genome.

Centromeres are DNA regions necessary for precise segregation of chromosomes to daughter cells during cell division. They are the binding site for proteins that make up the kinetochore, which in turn serves as the attachment site for microtubules, the cellular organelles that pull the chromosomes apart during cell division.

Another feature of eukaryotic chromosomes is replication origins. We know much less about the detailed structural properties of eukaryotic origins than prokaryotic ones. What is clear from inspecting the pattern of DNA synthesis along chromosomes is that most chromosomes have many active replication origins. These do not necessarily all initiate at the same time, but a typical replicating chromosome will have many active origins. The presence of multiple replication origins allows for complete replication of entire human genome in only 8 hours. It is not known how these replication processes terminate.

## CENTROMERES

In the yeast, *S. cerevisiae,* the centromere has been defined by genetic and molecular experiments to reside in a small DNA region, about 100 bp in length. This region contains several A–T rich sequences that may be of key importance for function. The centromere of *S. cerevisiae* is very different in size and characteristics than the centromeres of more advanced organisms, or even the yeast *S. pombe.* This is perhaps not too surprising in view of the key role that the centromere plays in cell division. Unlike these other species, *S. cerevisiae* does not undergo symmetrical cell division. Instead, it buds and exports one copy of each of its chromosomes to the daughter cell. In species that produce two nominally identical daughter cells, centromeres appear to be composed mostly of DNA with tandemly repeating sequences. The most striking feature of these repeats is that the number of copies can vary widely. A small repeating sequence, $(GGAAT)_n$, has recently been found to be conserved across a wide range of species. This conservation suggests that the sequence may be a key functional element in centromeres. While not yet proved, physical studies on this DNA sequence reveal it to have rather unusual helical properties that at least make it an attractive candidate for a functional element. As shown in Figure 2.2, the G-rich strand of the repeat has a very stable helical structure of its own, although the detailed nature of this structure is not yet understood.

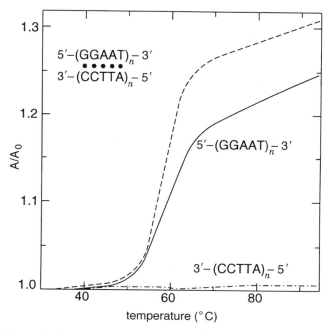

**Figure 2.2**  Evidence for the presence of some kind of unusual helical structure in a simple cen-tromeric repeating sequence. The relative amount of absorbance of 260 nm UV light is measured as a function of temperature. Shown are results for the two separated strands of a duplex, and the du-plex itself. What is unusual is that one of the separated strands shows a change in absorbance that is almost as sharp and as large as the intact duplex. This indicates that it, alone, can form some sort of helical structure. (From Moyzis et al., 1988.)

The other repeats in the centromeres are much longer tandemly repeated structures. In the human these were originally termed *satellite DNA* because they were originally seen as shoulders or side bands when genomic DNA fragments were fractionated by base com-position by equilibrium ultracentrifugation on CsCl gradients (see Chapter 5). For exam-ple, the alpha satellite of the african green monkey is a 171 base pair tandem repeat. Considerable effort has gone into determining the lengths and sequences of some of these satellites, and their organization on the chromosome. The results are complex. The re-peats are often not perfect; they can be composed of blocks of different lengths or differ-ent orientation. The human alpha satellite, which is very similar in composition (65% identity) to the african green monkey satellite, does not form a characteristic separate band during density ultracentrifugation. Thus the term satellite has evolved to now include tandemly repeated DNA sequences which may be of the same composition of the majority of genome. An example of a satellite sequence is shown in Figure 2.3.

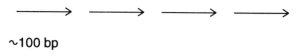

~100 bp

**Figure 2.3**  Example of a tandemly repeating DNA sequence in centromeric DNA.

The implications of the specific sequences of the repeats, and their organization, for centromere function are still quite cloudy.

The size of centromeres, judged by the total length of their simple sequence blocks, varies enormously among species, and even among the different chromosomes contained in one cell, without much indication that this is biologically important. Centromeres in *S. pombe* are only $10^4$ to $10^5$ bases in size, while those in the human can be several Mb. Centromeres on different human chromosomes appear to be able to vary in size widely, and within the population there is great heterogeneity in the apparent size of some centromeres. It is as though nature, having found a good thing, doesn't care how much of it there is as long as it is more than a certain minimum.

## TELOMERES

Telomeres in almost all organisms with linear chromosomes are strikingly similar. They consist of two components as shown in Figure 2.4. At the very end of the chromosome is a long stretch of tandemly repeating sequence. In most organisms the telomere is dominated by the hexanucleotide repeat $(TTAGGG)_n$. The repeating pattern is not perfect; several other sequences can be interspersed. In a few species the basic repeat is different, but it always has the characteristic that one strand is T and G-rich. The G + T rich strand is longer than the complementary strand, and thus the ends of the chromosome have a protruding $3'$-end strand of some considerable length. This folds back on itself to make a stable helical structure. The best evidence suggests that this structure is four stranded, in analogy to the four-strand helix made by aggregates of G itself. Pairs of telomeres might have to associate in order to make this structure. Alternatively, it could be made by looping the $3'$-end of one chromosome back on itself three times. The details not withstanding, these structures are apparently effective in protecting the ends of the chromosomes from attack by most common nucleases.

Next to the simple sequence telomeric repeats, most chromosomes have a series of more complex repeats. These frequently occur in blocks of a few thousand base pairs. Within the blocks there may be some tandemly repeated sequence more complex than the hexanucleotide telomeric repeat. The blocks themselves are of a number of different types, and these are distributed in different ways on different chromosomes. Some unique sequences, including genes, may occur between the repeating sequences. It is not clear if any chromosome really has a unique telomere, or if this matters. Some researchers feel that the sub-telomeric repeats may play a role in positioning the ends of the chromosomes at desired places within the nucleus. Whether and how this information might be coded by the pattern of blocks on a particular chromosome remains to be determined. At least some subtelomeric sequences vary widely from species to species.

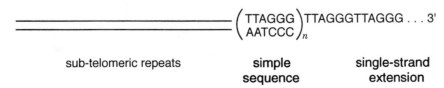

**Figure 2.4**    Structure of a typical telomere.

## DYNAMIC BEHAVIOR OF TELOMERES

The actual length of the simple telomeric repeating DNA sequence is highly variable both within species and between species. This appears to be a consequence, at least in part, of the way in which telomeres are synthesized and broken down. Telomeres are not static structures. If the growth of cells is monitored through successive generations, telomeres are observed to gradually shrink and then sometimes lengthen considerably. Nondividing cells and some cancer cells appear to have abnormal telomere lengths. At least two mechanisms are known that can lead to telomere degradation. These are shown in Figure 2.5a. In one mechanism the single-strand extension is subject to some nuclease cleavage. This shortens that strand. The alternate mechanism is based on the fact that the 5'-ended strand must serve as a starting position for DNA replication. This presumably occurs by the generation of an RNA primer that is then extended inward by DNA replication. Because of the tandem nature of the repeated sequence, the primer can easily be displaced inward before synthesis continues. This will shorten the 5'-strand. Both of these mechanisms seem likely to occur in practice.

A totally different mechanism exists to synthesize telomeres and to lengthen existing telomeres (Fig. 2.5b). The enzyme telomerase is present in all cells with telomeres. It is a ribonucleoprotein. The RNA component is used as a template to direct the synthesis of the 3'-overhang of the telomere. Thus the telomere is lengthened by integral numbers of repeat units. It is not known how the complex and subtle variations seen in this simple sequence arise in practice. Perhaps there is a family of telomerases with different templates. More likely, some of the sequences are modified after synthesis, or the telomerase may just be sloppy.

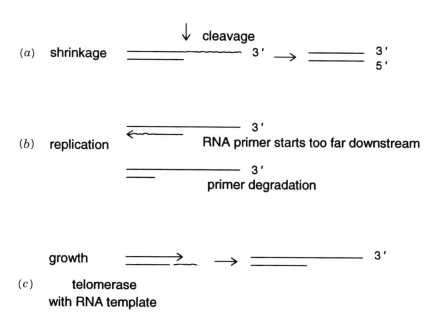

**Figure 2.5** Dynamics of telomeric simple repeating sequences. Mechanisms for telomere shrinkage: (a) Nuclease cleavage; (b) downstream priming by telomerase; (c) mechanism of telomere growth.

The total length of telomeric DNA in the human ranges from 10 to 30 kb. Because of its heterogeneity, fragments of DNA cut from any particular telomere by restriction enzymes have a broad size range and appear in gel electrophoretic size fractionations as broad, fuzzy bands. This is sufficiently characteristic of telomere behavior to constitute reasonable evidence that the band in question is telomeric. In mice, the length of telomeres is much longer, typically 100 kb. We don't know what this means. In general, the message from both telomeres and centromeres is that eukaryotic cells apparently feel no pressure to minimize the size of their genome or the sizes of these important functional elements. Unlike viruses, which must fit into small packages for cell escape and reinfection, and bacterial cells which are under selection pressure in rich media to replicate as fast as they can, eukaryotic cells are not mean and lean.

## CHROMATIN AND THE HIGHER-ORDER STRUCTURE OF CHROMOSOMES

A eukaryotic chromosome is only about half DNA by weight. The remainder is protein, mostly histones. There are five related, very basic histone proteins that bind tightly to DNA. The rest is a complex mixture called, loosely, *nonhistone chromosomal proteins.* This mixture consists of proteins needed to mediate successive higher-order packaging of the DNA, proteins needed for chromosome segregation, and proteins involved in gene expression and regulation. An enormous effort has gone into characterizing these proteins and the structures they form. Nevertheless, for the most part their role in higher-order structure or function is unknown.

At the lowest level of chromatin folding, 8 histones (two each of four types) assemble into a globular core structure that binds about 140 bp of DNA forming it into a coiled structure called the nucleosome (Fig. 2.6). Not only are nucleosomes very similar in all organisms from yeast to humans but in addition the four core histones are among the most evolutionarily conserved proteins known. Nucleosomes pack together to form a filament that is 100 Å in diameter and is known by this name. The details of the filament are different in different species because the lengths of the spacer DNA between the nucleosomes varies. In turn the 100 Å filament is coiled upon itself to make a thicker structure, called the 300 Å fiber. This appears to be solenoidal in shape. Stretches of solenoid containing on average 50 to 100 kb of DNA are attached to a protein core. In condensed metaphase chromosomes, this core appears as a central scaffold of the chromosome; it can be seen when the chromosome is largely stripped of other proteins and examined by electron microscopy (Fig. 2.7). A major component of this scaffold is the enzyme topoisomerase II (Box 2.1), which probably serves a role analogous to DNA gyrase in *E. coli* of acting as a swivel to circumvent any topological problems caused by the interwound nature of DNA strands.

At other stages in the cell cycle, the scaffold proteins may actually attach to the nuclear envelope. The chromosome is then suspended from this envelope into the interior of the nucleus. During mitosis, if this picture is correct, the chromosomes are then essentially turned inside out, separated into daughters, and reinverted. The topological domains created by DNA attachment to the scaffold at 50 to 100 kb intervals are similar in size to those seen in bacteria, where they appear to be formed by much simpler structures.

hierarchy of structures

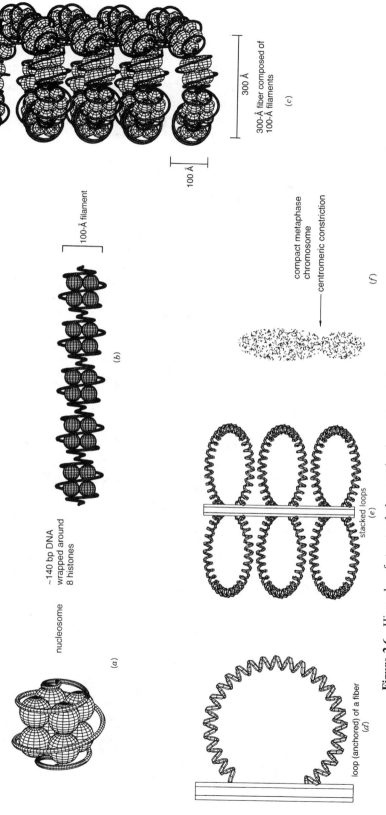

nucleosome

~140 bp DNA
wrapped around
8 histones

*(a)*

*(b)*

100-Å filament

300-Å fiber composed of
100-Å filaments

300 Å

100 Å

*(c)*

loop (anchored) of a fiber

*(d)*

stacked loops

*(e)*

compact metaphase
chromosome

centromeric constriction

*(f)*

**Figure 2.6** Hierarchy of structural elements in chromatin and chromosomes: *(a)* The nucleosome; *(b)* a 100 Å fila-ment; *(c)* a 300 Å solenoid. *(d)* chromosome loop anchored to a protein scaffold; *(e)* successive loops stacked along the scaffold; *(f)* the appearance of a condensed metaphase chromosome. The path of the scaffold is unknown, but it is not straight.

The most compact form of chromosomes occurs in metaphase. This is reasonable because it is desirable to pass chromosomes efficiently to daughter cells. Surely this is facilitated by having more compact structures than the unfolded objects seen in Figure 2.7. Metaphase chromosomes appear to consist of stacks of packed 300 Å chromatin fiber loops. Their structures are quite well controlled. For example, they have a helical polarity that is opposite in the two sister chromatids (the chromosome pairs about to separate in cell division). When the location of specific genes on metaphase chromosomes is examined (as described later), they appear to have a very fixed position within the morphologically visible structure of the chromosome. A characteristic feature of all metaphase chromosomes is that the centromere region appears to be constricted. The reason for this is not known. Also for unknown reasons some eukaryotic organisms, such as the yeast *S. cerevisiae* do not have characteristic condensed metaphase chromosomes.

The higher-order structure of chromatin and chromosomes poses an extraordinary challenge for structural biologists because they are so complex and because these structures are so large. The effort needed to reveal the details of these structures may not be worthwhile. It remains to be shown whether the details of much of the structure actually matter for any particular biological function. One extreme possibility is that this is all cheap packaging; that most of it is swept away whenever the underlying DNA has to be uncovered for function in gene expression or in recombination. We do not yet know if this is the case, but our ability to manipulate DNAs and chromosomes has grown to the point where it should soon be possible to test such notions explicitly. It would be far more elegant and satisfying if we uncover sophisticated mechanisms that allow DNA packaging and unpackaging to be used to modulate DNA function.

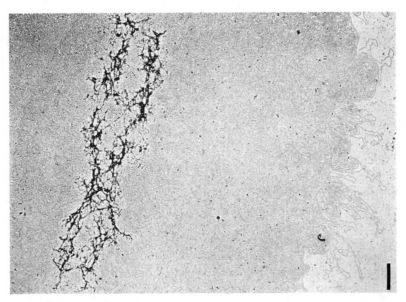

**Figure 2.7**   Electron micrograph of a single mammalian chromosome, denatured to remove most of the protein and allow the DNA to expand. The X-shaped structure is the protein scaffold that defines the shape of the condensed chromosome. The DNA appears as a barely resolvable mass of fiber covering almost the entire field.

## CHROMOSOMES IN THE CELL CYCLE

Bacteria are organisms without nuclei. They are continuously synthesizing DNA and dividing if food is plentiful. In contrast, nucleated cells are often quiescent. Sometimes they are even frozen forever in a nondividing state; examples are cells in the brain or heart muscle. Most eukaryotic cells proceed through a similar cycle of division and DNA synthesis, illustrated in Figure 2.8. Cell division is called *mitosis.* It occurs at the stage labeled M in the figure. After cell division, there is a stage, G1, during which no DNA synthesis occurs. Initiation of DNA synthesis, triggered by some stimulus, transforms the cell to the S phase. Not all of the DNA is necessarily synthesized in synchrony. Once synthesis is completed, another resting phase ensues, G2. Finally, in response to a mitogenic stimulus, the cell enters metaphase, and mitosis occurs in the M stage. In different cell types the timing of the cycle, and the factors that induce its progression, can vary widely.

Only in the M stage are the chromosomes compact and readily separable or visualizable under the light microscope. In other cell cycle stages, most portions of chromosomes are highly extended. The extended regions are called *euchromatin.* Their extension appears to be a prerequisite for active gene expression. This is reasonable considering the enormous steric and topological barriers that would have to be overcome to express a DNA sequence embedded in the highly condensed chromatin structure hierarchy. There are regions, called *heterochromatin,* unusually rich in simple repeated sequences that do not decondense after metaphase but instead remain condensed throughout the cell cycle. Heterochromatin is characteristic of centromeres but can occur to different extents in other regions of the genome. It is particularly prevalent on the human Y chromosome which contains the male sex determining factor but relatively few other genes. Heterochromatic regions are frequently heterogeneous in size within a species. They are generally sites where little or no gene expression occurs.

In most cases the level of expression of a gene does not depend much on its position in the genome, so long as the cis-acting DNA regions needed for regulation are kept reasonably near the gene. In fact typical eukaryotic genes are bracketed by sequences, such as enhancers or nuclear scaffold sites that eliminate transcriptional cross talk between adjacent genes. However, there are some striking exceptions to this rule, called *positional variation.* Using genetic or molecular methods, genes can be moved from euchromatic regions to heterochromatic regions. This usually results in their being silenced. Silencing also sometimes occurs when genes are placed near to telomeres (Fig. 2.9). The mechanism of this latter silencing is not understood.

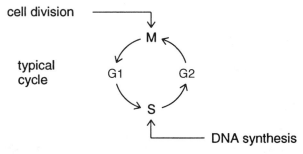

**Figure 2.8**   The cell division cycle in almost all eukaryotic cells. M means a cell undergoing mitosis; S means a cell in the act of DNA synthesis.

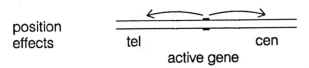

position
effects

**Figure 2.9**   Differences in chromatin affect gene expression. Genes transposed to heterochromatic or telomeric regions are sometimes silenced.

## GENOME ORGANIZATION

There are very visible patterns of gene organization. Many genes occur in families, such as globins, immunoglobins, histones, or zinc finger proteins. These families presumably arose mostly by gene duplications of a common precursor. Subsequent evolutionary divergence led to differences among family members, but usually sufficient traces remain of their common origin through conserved sequences to allow family members to be identified. An alternative mechanism for generating similar sets of genes is convergent evolution. While examples of this are known, it does not appear to be a common mechanism.

The location of gene families within a genome offers a fascinating view of some of the processes that reshape genomes during evolution. Some families are widely dispersed throughout the genome such as zinc finger proteins, although these may have preferred locations. Other families are tightly clustered. An example is the globin genes shown in Figure 2.10. These lie in two clusters: one on human chromosome 11 and one on human chromosome 16. Each cluster has several active genes and several pseudogenes, which may have been active at one time but now are studded with mutations that make them unable to express functional protein. Some families like the immunoglobulins are much more complex than the globin family.

When metaphase chromosomes are stained in various different ways and examined in the light microscope, a distinct pattern of banding is seen. An example is shown in Figure 2.11*a* for human chromosomes. The same bands are usually seen with different stains, implying that this pattern is a reflection of some general intrinsic property of the chromosomes rather than just an idiosyncratic response to a particular dye or a particular staining protocol. Not all genomes show such distinct staining patterns as the human, but most higher organisms do.

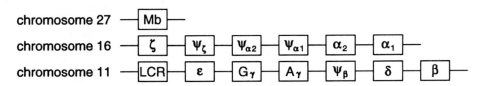

**Figure 2.10**   Genomic organization of the human globin gene family. Hemoglobins expressed in the adult are alpha and beta; hemoglobins expressed in the embryo are gamma and delta; hemoglobins expressed in the early embryo are zeta and eta. Myoglobin is expressed throughout development. Gene symbols preceded by psi are pseudogenes, no longer capable of expression.

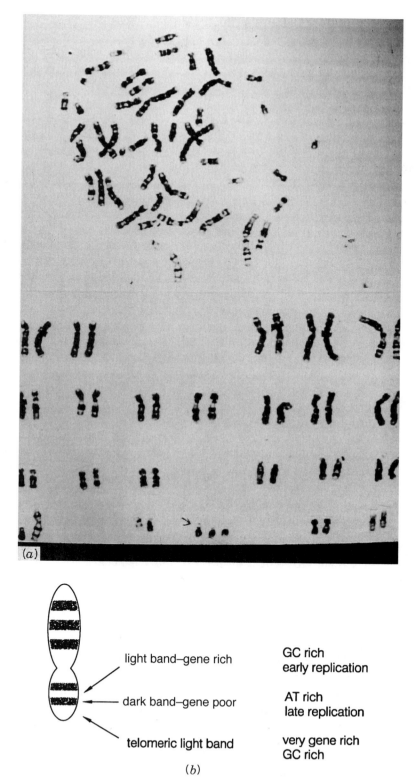

light band—gene rich       GC rich
early replication

dark band—gene poor       AT rich
late replication

telomeric light band       very gene rich
GC rich

(b)

**Figure 2.11** Chromosome banding. (a) A typical preparation of banded chromosomes. Cells are arrested in metaphase and stained with Geimsa. Individual chromosomes are identified by the pattern of dark and light bands, and rearranged manually for visual convenience. The particular individual in this case is a male because there is one X and one Y chromosome, and he has Down's syndrome because there are three copies of chromosome 21. (b) Typical properties of bands.

The molecular origin of the stained bands is not known with certainty. Dark bands seen with a particular stain, Geimsa, appear to be slightly richer in A + T, while light bands are slightly richer in G + C. It is not clear how these base composition differences can yield such dramatic staining differences directly. At one time the light and dark bands were thought to have different densities of DNA packing. As progress is made in mapping extensive regions of the human genome, we can compare the DNA content in different regions. Thus far, although there is still some controversy, not much strong evidence for significant differences in the DNA packing density of light and dark bands can be found. The most tenable hypothesis that remains is that the bands reflect different DNA accessibility to reagents, perhaps as a result of different populations of bound nonhistone proteins.

While the physical origin of chromosome bands is obscure, the biological differences that have been observed between bands are dramatic. There are two general phenomena (Fig. 2.11b). Light Geimsa bands are rich in genes, and they replicate early in the S phase of the cell cycle. Dark Geimsa bands are relatively poor in genes, and they are late replicating. Finer distinctions can be made. Certain light bands, located adjacent to telomeres, are extremely rich in genes and have an unusually high G + C content. An example is the Huntington's disease region at the tip of the short arm of human chromosome 4.

The appearance of chromosome bands is not fixed. It depends very much on the method that was used to prepare the chromosome. Different procedures focused on analyzing chromosomes from earlier and earlier stages in cell division yield more elongated chromosomes that reveal increasing numbers of bands. In general, it is customary to work with chromosomes that show a total of only about 350 bands spanning the entire human genome because the more extended forms are more difficult to prepare reproducibly. Some examples are shown in Figure 2.12. One particular annoyance in studying chromosome banding patterns is that it complicates the naming of bands. Unfortunately, the nomenclature in common use is based on history. Early workers saw few bands and named them outward from the centromere as p1, p2, etc., for the short (p = petit) arm and q1, q2, etc., for the long arm (q comes after p in the alphabet). When a particular band could be resolved into multiplets, its components were named q21, q22, etc. If in later work, with more expanded chromosomes, additional sub-bands could be seen, these were renamed as q21.1, q21.2, etc. More expansion led to more names as in q21.11, q21.12. This nomenclature is not very systematic; it is certainly not a unique naming system, and it risks obfuscating the true physical origins of the bands. However, we appear to be stuck with it. Like the Japanese system for assigning street addresses in the order in which houses were constructed, it is wonderful for those with a proper historical perspective, but treacherous for the newcomer.

Humans have 22 pairs of autosomes and two sex chromosomes (XX or XY). Their DNAs range in size from chromosome 1, the largest with 250 Mb to chromosome 21, the smallest, with 50 Mb. One of each pair of autosomes and one sex chromosome is inherited from each parent. The overall haploid DNA content of a human cell is $3 \times 10^9$ bp. At 660 Da per base pair, this leads to a haploid genome molecular weight of about $2 \times 10^{12}$. The chromosomes are distinguishable by their size and unique pattern of stained bands. A schematic representation of each, in relatively compact form, is given in Figure 2.13. There are a few interesting generalizations from this genome overview. All human telomeres, except Yq, 19q, and 3p, are Geimsa light bands. The ratio of light to dark banding on different chromosomes can vary quite a bit from 19 which is mostly light, and appears to have a very large number of genes, to chromosomes 3 and 13 which are mostly dark, and are presumably relatively sparse in genes.

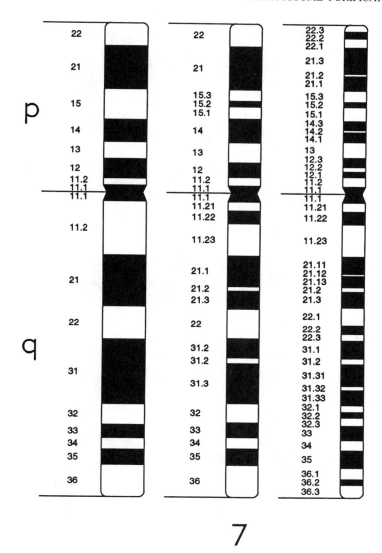

**Figure 2.12**  Example of how different chromosome preparations change the number of bands visible and the appearance of these bands. Three different levels of band resolution for human chromosome 7 are shown schematically. Also illustrated is the way these bands are numbered.

## CHROMOSOME PURIFICATION

The past decade has seen tremendous strides in our ability to purify specific human chromosomes. Early attempts, using density gradient sedimentation, never achieved the sort of resolution necessary to become a routine analytical or preparative technique. The key advance was the creation of fluorescence activated flow sorters with sufficient intensity to allow accurate fluorescence determinations on single metaphase chromosomes. The fluorescence activated flow sorter originally was developed for intact cells, hence the name FACS (fluorescence activated cell sorter). However, it was soon found to be applicable for chromosomes, especially if more powerful lasers were used, and these were focused more tightly.

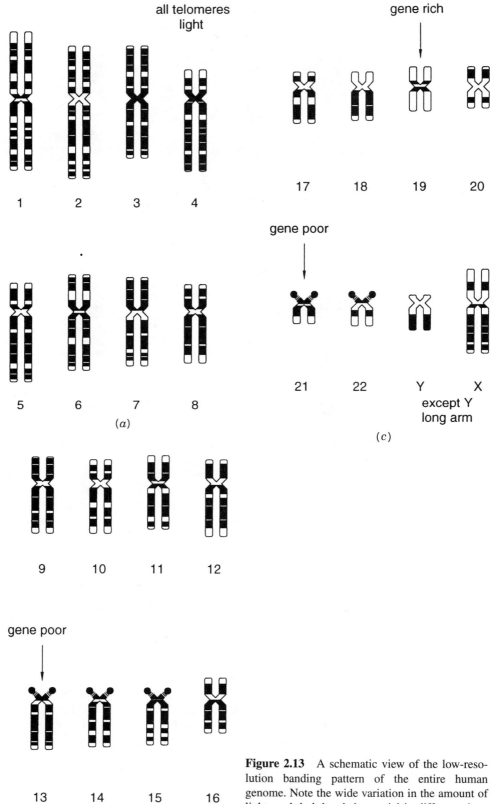

**Figure 2.13** A schematic view of the low-resolution banding pattern of the entire human genome. Note the wide variation in the amount of light- and dark-banded material in different chromosomes.

FACS instruments can be used to determine a profile of chromosome sizes or other characteristics, by pulse height analysis of the emission from large numbers of chromosomes, or they can be used for actual fractionation, one chromosome at a time, as shown schematically in Figure 2.14.

In FACS, fluorescently stained metaphase chromosomes are passed in a collimated flowing liquid stream, one at a time past a powerful focused laser beam. After passing the laser, the stream is broken into uniform droplets by ultrasonic modulation. Each emission pattern is captured and integrated, and the resulting pulse height is stored as an event. If the resulting signal falls between certain preset limits, a potential is applied to the liquid stream just before the chromosome-containing droplet breaks off. This places a net charge on that droplet, and its path can then be altered selectively by an electric field. The result is the physical displacement of the droplet, and its chromosome, to a collection vessel. The circuitry must be fast enough to analyze the emission pattern of the chromosomes and relay this information before the droplet containing the desired target is released. In practice, more than one colored dye is used, and the resulting emission signal is detected at several different wavelengths and angles and analyzed by several-parameter logic. This produces an improved ability to resolve the different human chromosomes. The ideal pattern expected from single parameter analysis is shown in Figure 2.15. Each peak should show the same area, since (neglecting sex chromosomes) each is present in unit stoichiometry.

Real results are more complex as shown by the example in Figure 2.16. Some chromosomes are very difficult to resolve, and appear clustered together in an intense band. The most difficult to distinguish are human chromosomes 9 to 12. In general, larger chromosomes are more fragile and more easily broken than smaller chromosomes. Thus they appear in substoichiometric amounts, and debris from their breakage can contaminate fractions designed to contain only a particular small chromosome. The other limitation of chromosome purification by FACS is that it is a single molecule method. Typical sorting rates are a few thousand chromosomes per second. Even if the yield of a particular chromosome were perfect, this would imply the capture of only a few hundred per second. In practice, observed yields are often much worse than this. Several high-speed sorters have been constructed that increase the throughput by a factor of 3 to 5.

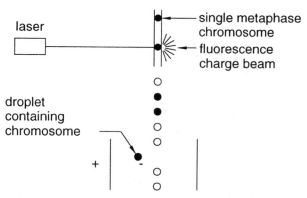

**Figure 2.14** Schematic illustration of the purification of metaphase chromosomes (shown as black dots) by fluorescence activated flow-sorting (FACS).

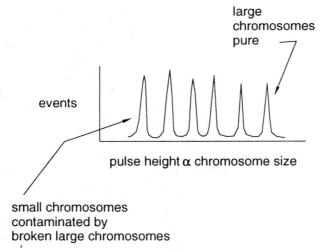

**Figure 2.15** Ideal one-dimensional histogram expected for flow-sorted human chromosomes.

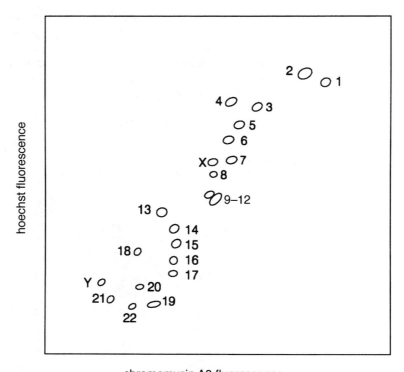

chromomycin A3 fluorescence

**Figure 2.16** An example of actual flow analysis of human chromosomes. Two different fluorescence parameters have been used to try to resolve the chromosomes better. Despite this, four chromosomes, 9 through 12, are unresolved and appear as a much more intense peak than their neighbors. (Provided by the Lawrence Livermore National Laboratory, Human Genome Center.)

However, these are not yet generally available instruments. Thus FACS affords a way of obtaining fairly pure single chromosome material, but usually not in as large quantities as one would really like to have. Alternatives to FACS purification of chromosomes are still needed. One possibility is discussed in Box 2.2.

The problem of contamination of small chromosomes with large, and the problem of resolution of certain chromosomes, can be circumvented by the use of rodent-human hybrid cells. These will be described in more detail later. In ideal cases they consist of a single human chromosome in a mouse or hamster background. However, even if more than one human chromosome is present, they are usually an improved source of starting material. FACS is performed on hybrids just as on pure human cells. Windows (bounds on particular fluorescence signals) are used to select the desired human chromosome. Although this will be contaminated by broken chromosome fragments, these latter will be of rodent origin. The degree of contamination can be easily assessed by looking for rodent-specific DNA sequences.

FACS-sorted chromosomes can be used directly by spotting them onto filters, preparing DNA in situ, and using the resulting filters as hybridization targets for particular DNA sequences of interest (Chapter 3). In this way the pattern of hybridization of a particular DNA sequence allows its chromosome assignment. This procedure is particularly useful when the probe of interest shows undesirable cross-hybridization with other human or with rodent chromosomes. However, for most applications, it is necessary to amplify the flow-sorted human chromosome material. This is done either by variants of the polymerase chain reaction, as described in Chapter 4, or by cloning the DNA from the sorted chromosome into various vectors. Plasmids, bacteriophages like lambda (Box 1.2), P1, or cosmids (Box 2.3), and bacterial or yeast artificial chromosomes (BACs or YACs, Box 8.1) have all been used for this purpose. Collections of such clones are called single chromosome libraries. While early libraries were often heavily contaminated and showed relatively uneven representation of the DNA along the chromosome, more recently-made libraries appear to be much purer and more representative.

Single-chromosome libraries represent one of the most important resources for current genome studies. They are readily available in the United States from the American Type Culture Collection (ATCC). The first chromosome-specific libraries consisting of small clones were constructed in plasmid vectors. A second set of chromosome-specific libraries consists of larger 40 kb cosmid clones. One way in which such libraries are characterized is by their coverage, the probability that a given region is included on at least one clone. If the average insert size cloned in the library is $N$ base pairs, the number of clones is $n$, and the size of the chromosome is C base pairs, the redundancy of the library is just $Nn/C$. Assuming that the library is a random selection of DNA fragments of the chromosome, one can compute from the coverage the probability that any sequence is represented in the library. Consider a region on the chromosome. The probability that it will be contained on the first clone examined is $N/C$. The probability that it will not be contained on this clone is $1 - N/C$. After n clones have been picked at random the probability that none of them will contain the region selected is $(1 - N/C)n$. Thus we can write that the fraction, f, of the chromosome covered by the library is

$$f = 1 - \left(1 - \frac{N}{C}\right)^{n}$$

## BOX 2.2
## PROSPECTS FOR ELECTROPHORETIC PURIFICATION
## OF CHROMOSOMES

In principle, it should be possible to use agarose gel electrophoresis to purify chromo-somes. DNA molecules up to about 50,000 bp in size are well resolved by ordinary agarose electrophoresis; while larger DNAs, up to about 10 Mb in size, can be frac-tionated effectively by pulsed field gel (PFG) electrophoresis. Secondary pulsed elec-trophoresis (SPFG), where short intense pulses are superimposed on the normally slowly varying pulses in PFG, expands the fractionation range of DNA even further (see Chapter 5). An ideal method of chromosome fractionation would be fairly gen-eral; it would allow one to capture the chromosome of interest and discard the remain-der of the genome. One approach to such a scheme exploits the fact that genetic vari-ants can be found in which the desired chromosome is a circular DNA molecule. Bacterial chromosomes are naturally circular. Eukaryotic chromosomes can become circles by recombination between the simple telomeric repeating sequences or sub-telomeric repeats, as shown in Figure 2.17. Many cases of individuals with circular human chromosomes are picked up by cytogenetic analysis. In most cases the circle produces no direct deleterious phenotype because all that is lost is telomeric sequence.

It has been known for a long time that DNA circles larger than 20 kb have a very difficult time migrating in agarose gels under conventional electrophoretic conditions. The explanation is presumably entrapment of the DNA on agarose fibers, as shown in Figure 2.18. At the typical field strengths used for electrophoresis, a linear molecule, once entrapped, can slip free again by moving along its axis, but a circle is perma-nently trapped because of its topology. Changing field directions helps larger circles to move, which is consistent with this picture. Thus field strengths can be found where all of the linear chromosomes in a sample will migrate fairly rapidly through the gel, while circles stay at the origin. For example, in PFG the 4.6 Mb circular *E. coli* chro-mosomal DNA does not move, but once the chromosome is linearized by a single X-ray break, it moves readily. A mutant circular chromosome II of *S. pombe,* which is 4.8 Mb in size, does not move, while the normal linear chromosome moves readily at low electrical field strengths.

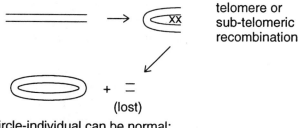

**Figure 2.17**   Generation of circular chromosomal DNA molecules by homologous recombina-tion at telomeric or sub-telomeric repeated DNA sequences.

*(continued)*

**BOX 2.2** *(Continued)*

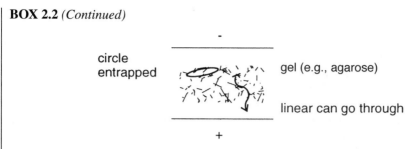

circle
entrapped

gel (e.g., agarose)

linear can go through

**Figure 2.18**    Entrapment of a circular DNA molecule on agarose fibers.

The linear DNA molecules that make up intact human chromosomes are so large that they do not appear able to enter agarose at all under any PFG conditions so far tried. It will be shown in Chapter 5 that these molecules do apparently enter the gel under SPFG conditions. No size fractionation is seen, but the molecules migrate well. We reasoned that under these conditions a circular human chromosomal DNA would be unable to enter agarose; if this were the case we would have a simple bulk procedure for human chromosome purification. Thus far we have experimented to no avail with a cell line containing a chromosome 21 circle. This material seems to co-migrate in the gel with ordinary linear chromosome 21 DNA. The most likely explanation is that under the conditions we used, the molecule has been fragmented—by physical forces during the electrophoresis itself, by nuclease contamination, or, much less likely, the molecule (unlike the morphological appearance of the chromosome) was never a circle to begin with. We will need to explore a wider range of conditions, and look at other circular human chromosomes.

We can arrange this to solve for n and thus determine the number of clones needed to achieve a fractional coverage of f.

$$n = \frac{log(1-f)}{log(1-N/C)}$$

Typical useful libraries will have a redundancy of two- to tenfold.

In practice, however, most libraries are shown to be over-represented in some genome regions, under-represented in others, and totally missing certain DNA segments. Cloning biases can arise from many reasons. Some bacterial strains carry restriction nucleases that specifically degrade some of the methylated DNA sequences found in typical mammalian cells. Some mammalian sequences if expressed produce proteins toxic to the host cell. Others may produce toxic RNAs. Strong promoters, inadvertently contained in a high-copy number clone, can sequester the host cell's RNA polymerase, resulting in little or no growth. DNA sequences with various types of repeated sequences can recombine in the host cell, and in many cases this will lead to loss of the DNA stretch between the repeats. The inevitable result is that almost all libraries are fairly biased.

## BOX 2.3
## PREPARATION OF SINGLE CHROMOSOME LIBRARIES IN COSMIDS AND P1

Because flow-sorting produces only small amounts of purified single chromosomes, procedures for cloning this material must be particularly efficient. The ideal clones will also have relatively large insert capacities so that the complexity of the library, namely the number of clones needed to represent one chromosome equivalent of insert DNA, can be kept within reasonable bounds. The earliest single-chromosome libraries were made in bacteriophage or plasmid vectors (see Box 1.3), but these were rapidly supplanted by cosmid vectors. Libraries of each human chromosome in cosmids have been made and distributed by a collaboration between Lawrence Livermore National Laboratory and Los Alamos National Laboratory. These libraries are available today at a nominal cost from the American Type Culture Collection. Gridded filter arrays of clones from most of these libraries have also been made by various genome centers. Interrogation of these filters by hybridization with a DNA probe or interrogation of DNA pools with PCR primers will identify clones that contain specific DNA sequences. The use of the same arrays by multiple investigators at different sites facilitates coordination of a broad spectrum of genome research.

Cosmids are chosen as vectors for single chromosome libraries because they have relatively large inserts. A cosmid clone consists of two ends of bacteriophage lambda DNA (totaling about 10 kb in length) with all of the middle of the natural vector removed. The ends contain all of the sequence information needed to package DNA into viruses. Hence a cloned insert can replace the central 40 kb of lambda. Recombinant molecules are packaged in vitro using extracts from cells engineered to contain the proteins needed for this reaction. Lambda DNA packaged into a bacteriophage head is a linear molecule, but the ends have 12 base complementary 5'-extensions. Once the virus infects an *E. coli* cell, the two ends circularize and are ligated together as the first step in the viral life cycle. The 5'-extensions are called *COS sites,* for cohesive ends, and the name of this site has been carried over to the vectors that contain them (COSmids). Cosmids propagate in *E. coli* cells as low-copy plasmids.

Bacteriophage P1 offers another convenient large insert cloning system. P1 packages its DNA by a headful mechanism that accommodates about 90 kb. Hence, if the target DNA is much larger than 90 kb, it will be cut into adjacent fragments as it is packaged. A typical P1 cloning vector is shown below. The vector is equipped with bacteriophage SP6 and T7 promoters to allow strand-specific transcription of the insert. The resulting clones are called P1 artificial chromosomes (PACs).

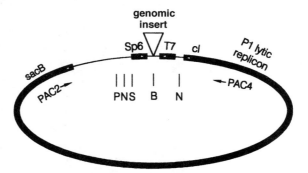

PAC cloning vector. In the circle are shown cutting sites for some restriction nucleases and also locations of known sequences, PAC2, PAC4 suitable for PCR amplifications of the insert (see Chapter 4).

## CHROMOSOME NUMBER

There are two determinants of the total number of chromosomes in a cell. The first is the number of different chromosomes. The second is the number of copies of a particular chromosome. In general, in a species the number of different autosomes (excluding sex chromosomes) is preserved, and the number of copies of each chromosome is kept in a constant ratio, although the absolute number can vary in different cells. We use the term *ploidy* to refer to the number of copies of the genome. Yeast cells are most typically haploid with one genome copy, but after mating they can be diploid, with two copies. Human gametes are haploid; somatic cells vary between diploid and tetraploid depending on what stage in the cell cycle they are in.

When actively growing eukaryotic cells are stained for total DNA content and analyzed by FACS, a complex result is seen, as shown in Figure 2.19. The three peaks represent diploid G1 cells, tetraploid G2 and M cells, and intermediate ploidy for cells in S phase. Plant cells frequently have much higher ploidy. In specialized tissues much higher ploidy is occasionally seen in animal cells as, for example, in the very highly polyploid chromosomes of Drosophila salivary glands. When such events occur, it is a great boon for the cytogeneticist because it makes the chromosomes much easier to manipulate and to visualize in detail in the light microscope.

Aneuploidy is an imbalance in the relative numbers of different chromosomes. It is often deleterious and can lead to an altered appearance, or phenotype. Ordinarily gene dosage is carefully controlled by the constant ratios of different segments of DNA. An extra chromosome will disturb this balance because its gene products will be elevated. In the human the most commonly seen aneuploidy is Down's syndrome: trisomy chromosome 21. The result is substantial physical and mental abnormalities, although the individuals survive. Trisomy 13 is also seen, but this trisomy leads to even more serious deformations and the individuals do not survive long beyond birth. Other trisomies are not seen in live births because fetuses carrying these defects are so seriously damaged that they do not survive to term.

In the human (and presumably in most other diploid species) monosomies, loss of an entire chromosome, are almost always fatal. This is due to the presence of recessive lethal alleles. For example, imagine a chromosome that carries a deletion for a gene that codes for an essential enzyme. As long as the corresponding gene on the homologous chromosome is intact, there may be little phenotypic effect of the haploid state of that gene.

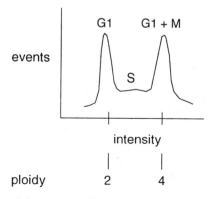

**Figure 2.19**   FACS analysis of the amount of DNA in a population of rapidly growing and dividing cells. Peaks corresponding to particular cell cycle stages are indicated.

However, monosomy will reveal the presence of any recessive lethal allele on the remaining chromosome. The result is not compatible with normal development.

Cell ploidy is maintained by mitosis, for somatic cells (see Box 2.4), and it is altered by meiosis (see Box 6.1) during gametogenesis. Errors during these processes or errors made during DNA damage and repair can lead to cells that themselves are aneuploid or that produce aneuploid offspring. Cells with defects in their division cycle frequently accumulate abnormal chromosome complements. This is particularly dramatic in many late-stage tumors which frequently have large numbers of different chromosome abnormalities.

---

**BOX 2.4**
**MITOSIS**

During cell division, events must be carefully controlled to ensure that each daughter cell receives one copy of each of the two homologous chromosomes, namely receives one copy of the paternal genome and one copy of the maternal genome. An obvious potential source of confusion is the similarity of the two parental genomes. The way in which this confusion is effectively avoided is shown by the schematic illustration of some of the steps in mitosis in Figure 2.20. Our example considers a cell with only two chromosome types. At the G1 phase the diploid cell contains one copy of each parental chromosome. These are shown condensed in the figure, for clarity, but remember that they are not condensed except at metaphase. After DNA synthesis the cell is tetraploid; there are now two copies of each parental genome. However, the two copies are paired; they remain fused at the centromere. We call these structures *sister chromatids*. Hence there is no chance for the two parental genomes to mingle. During mitosis the paired sister chromatids all migrate to the metaphase plate of the cell. Microtubules form between each centromere and the two centrioles that will segregate into the two daughter cells. As the microtubules shrink, each pair of sister chromatids is dragged apart so that one copy goes to each daughter cell.

Errors can occur; one type is called nondisjunction. The sister chromatids fail to separate so that one daughter gets both sister chromatids; the other gets none. Usually such an event will be fatal because of recessive lethal alleles present on the sole copy of that chromosome present in the daughter. One additional complication bears mention. While sister chromatids are paired, a process called sister chromatid exchange can occur. In this form of mitotic recombination, DNA strands from one sister invade the other; the eventual result is a set of DNA breaks and reunions that exchanges material between the two chromatids. Thus each final product is actually a mosaic of the two sisters. Since these should be identical anyway (except for any errors made in DNA synthesis) this process has no phenotypic consequences. We know of its existence most compellingly through elegant fluorescence staining experiments conceived by Samuel Latt at Harvard Medical School. He used base analogues to distinguish the pre-existing and newly synthesized sister chromatids, and a fluorescent stain that showed different color intensities with the two base analogues. Thus each chromatid exchange point could be seen as a switch in the staining color, as shown schematically in Figure 2.21.

*(continued)*

**BOX 2.4** *(Continued)*

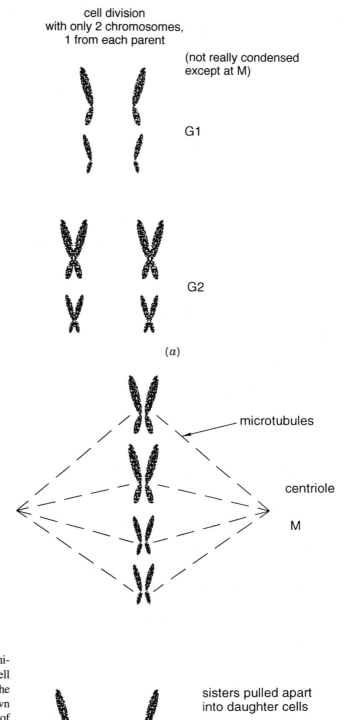

cell division
with only 2 chromosomes,
1 from each parent

(not really condensed
except at M)

G1

G2

*(a)*

microtubules

centriole

M

**Figure 2.20** Steps in mitosis, the process of cell division. For simplicity the chromosomes are shown condensed at all stages of the cell cycle. In actuality, they are condensed only during metaphase.

sisters pulled apart
into daughter cells

*(b)*

*(continued)*

**BOX 2.4** *(Continued)*

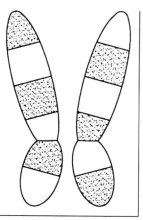

**Figure 2.21**  Visualization of sister chromatid exchange, by fluorescence quenching. Newly synthesized DNA was labeled with 5-bromoU which quenches the fluorescence of the acridine dye used to stain the chromosomes at metaphase.

Partial aneuploidy can arise in a number of different ways. The results are often severe. One common mechanism is illustrated in Figure 2.22. Reciprocal chromosome translocations are fairly common, and they are discovered by genetic screening because the offspring of such individuals frequently have genetic abnormalities. The example, shown in the figure, is a reciprocal translocation between chromosome 5 and chromosome 20. Such translocations can occur by meiotic or mitotic recombination. Unless the break points interrupt vital genes, the translocation results in a normal phenotype because all of the genome is present in normal stoichiometry. This illustrates once again that the arrangement of genes on the chromosomes is not usually critical.

Now consider the result of a mating between the individual with a reciprocal translocation and a normal individual. Fifty percent of the children will have a normal dosage of all of their chromosomes. Half of these will have a totally normal genotype because they will have received both of the normal homologs originally present in the parent with the reciprocal translocation. Half will have received both abnormal chromosomes from that parent; hence their genome will still be balanced. The remaining 50% of the offspring will show partial aneuploidy. Half of these will be partially trisomic for chromosome 20, partially monosomic for chromosome 5. The other half will be partially monosomic for chromosome 20, partially trisomic for chromosome 5.

## UNUSUAL CHARACTERISTICS OF SEX CHROMOSOMES AND MITOCHONDRIA

In mammals a female carries two copies of the X chromosome; males have one X and one Y. However, this simple difference in karyotype (the set of chromosomes) has profound effects that go beyond just the establishment of sex. The first thing to consider is why we need sex at all. In species with just one sex, each organism can reproduce clonally. The offspring of that organism may be identical. If the organism inhabits a wide ecological range, different selection processes will produce a geographical pattern of genetic differences, but there is no rapid way to combine these in response to a shifting environment. Sex, on the other hand, demands continual outbreeding, so it leads to much more efficient mixing of the gene pool of a species.

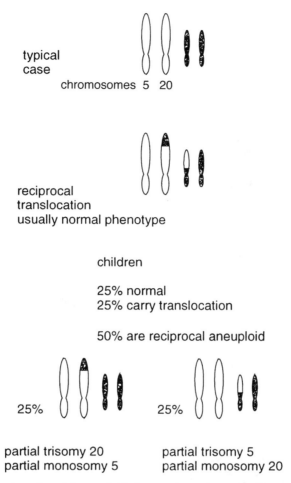

typical
case

chromosomes 5  20

reciprocal
translocation
usually normal phenotype

children

25% normal
25% carry translocation

50% are reciprocal aneuploid

25%

25%

partial trisomy 20          partial trisomy 5
partial monosomy 5        partial monosomy 20

**Figure 2.22**   Generation of partial aneuploidy by a reciprocal translocation, followed by segregation of the rearranged chromosomes to gametes and, after fertilization, generation of individuals that are usually phenotypically abnormal.

We are so used to the notion of two sexes, that it often goes unquestioned why two and only two? Current speculation is that this is the result of the smallest component of mammalian genomes, the mitochondrial DNA. Mitochondria have a circular chromosome, much like the typical bacterial genome from which they presumably derived. This genome codes for key cellular metabolic functions. In the human it is 16,569 kb in size, and the complete DNA sequence is known. A map summarizing this sequence is shown in Figure 2.23. Mitochondria have many copies of this DNA. What is striking is that all of an individual's mitochondria are maternally inherited. The sperm does contain a few mitochondria, and these can enter the ovum upon fertilization, but they are somehow destroyed.

Bacterial DNAs carry restriction nucleases that can destroy foreign (or incompatible) DNA (see Box 1.2). Perhaps, from their bacterial origin, mitochondria also have such properties.

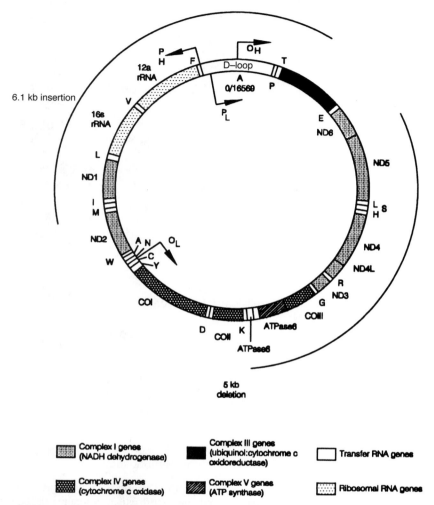

**Figure 2.23**  Map of human mitochondrial DNA. The tRNAs are indicated by their cognate amino acid letter code. The genes encoded by the G-rich heavy (H) strand are on the outside of the circle, while those for the C-rich light (L) strand are on the inside. The H- and L-strand origins ($O_H$ and $O_L$) and promoters ($P_H$ and $P_L$) are shown. The common 5-kb deletion associated with aging is shown outside the circle. (Adapted from Wallace, 1995.)

If they do, this would explain why one sex must contribute all of the mitochondria. It can be used as an argument that there should only be two sexes. In fact, however, cases are known where organisms have more than two sexes. The slime mold, *Physarum polycephalum*, has 13 sexes. However, these turn out to be hierarchical. When two sexes mate, the higher one on the hierarchy donates its mitochondria to the offspring. This ensures that only one parental set of mitochondria survive. So one important thing about sex is who you get your mitochondria from.

In the human and other mammals, the Y chromosome is largely devoid of genes. The long arm is a dark G band (Fig. 2.13), and the short arm is small. However, an exception is the gene-rich tip of the short arm, which is called the *pseudoautosomal region*.

---

**BOX 2.5**
**MORE ON MITOCHONDRIAL DNA**

The pure maternal inheritance of mitochondria makes it very easy to trace lineages in human populations, since all of the complexities of diploid genetics are avoided. The only analogous situation is the Y chromosome which must be paternally inherited. One region of the mitochondrial DNA, near the replication origin, codes for no known genes. This region shows a relatively fast rate of evolution. By monitoring the changes in the DNA of this region, Allen Wilson and his coworkers have attempted to trace the mitochondrion back through human prehistory to explore the origin of human ethnic groups and their geographic migrations. While considerable controversy still exists about some of the conclusions, most scientists feel that they can trace all existing human groups to a single female progenitor who lived in Africa some 20,000 years ago.

The mitochondrion has recently been implicated in studies by Norman Arnheim, Douglas Wallace, and their coworkers as a major potential site of accumulated damage that results in human aging. In certain inherited diseases a large deletion occurs in mitochondrial DNA. This deletion drops out more than 5 kb of the genome between two repeated sequence elements (Fig. 2.23). It presumably arises by recombination. Small amounts of similar deletions have been detected in aging human tissue, particularly in cells like muscle, heart, and brain that undergo little or no cell division. While the full significance of these results remains to be evaluated, on the surface these deletions are striking phenomena, which provide potential diagnostic tools for what may be a major mechanism of aging and a way to begin to think rationally about how to combat it. The mitochondrion is the site of a large amount of oxidative reactions; these are known to be able to damage DNA and stimulate repair and recombination. Hence it is not surprising that this organelle should be a major target for DNA aging.

---

This region is homologous to the tip of the short arm of the X chromosome. A more detailed discussion of this region will be presented in Chapter 6. There are also a few other places on X and Y where homologous genes exist. Beyond this, most of the X contains genes that have no equivalent on the Y. This causes a problem of gene dosage. The sex chromosomes are unbalanced because a female will have two copies of all these X-linked genes while the male will have only one. The gene dosage problem is solved by the process known as X-inactivation.

Mature somatic cells of female origin have a densely staining condensed object called a Barr body. This object is absent in corresponding cells of male origin. Eligibility of female athletes competing in the Olympics used to be dependent on the presence of a Barr body in their cells. Mary Lyon first demonstrated that the Barr body is a highly condensed X chromosome. Since we know that condensed chromatin is inactive in expression, this suggests that in the female one of the two X chromosomes is inactivated. This process, X-inactivation, occurs by methylation of C. It covers the entire X chromosome except for the pseudoautosomal region and other genes that are homologous on X and Y. The exact mechanism is still not understood in detail, but it seems to be a process that is nucleated at some X-specific sequences and then diffuses (except where barriers limit its spread). Cells with translocations between the X chromosome and autosomes are known; in these cases the inactivation can spread onto part of the adjacent autosome fragment.

If X-inactivation occurred at the single-cell stage of an embryo; one of the two parental Xs would be lost, and males and females would have similar sex-linked genetic properties. However, X-inactivation occurs later in embryogenesis. When it occurs, the two parental X chromosomes have an equal probability of inactivation. The resulting female embryo then becomes a mosaic with half the cells containing an active paternal X chromosome, half an active maternal X chromosome. When these two chromosomes carry distinguishable markers, this mosaicism is revealed in patterns of somatic differences in clones of cells that derive from specific embryonic progenitors. One spectacular example is the tortoise shell cat (Fig. 2.24). This X chromosome of this animal can carry two different color coat alleles. The male is always one color or the other because it has one allele or the other. The female can have both alleles and will inactivate each in a subset of embryonic ectodermal cells. As

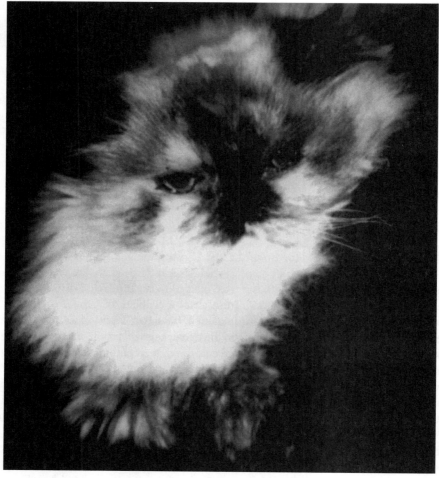

**Figure 2.24**  A tortoise shell Himalayan female cat. A gene responsible for overall development of skin pigment is temperature sensitive. As a result, pigmentation occurs only in regions where the animal is normally cold such as the tips of the ears, nose, and paws. These colored regions are mottled because the cat is a mosaic of different color alleles because of random X-inactivation early in development. (Photograph courtesy of Chandran Sabanayagam.)

these cells multiply and differentiate, a two-colored animal emerges; the pattern of color distribution reveals the clonal origin of the cells that generated it. The tortoise shell phenotype is only seen in the female. Thus there are truly fundamental differences between the sexes; females are mosaics in all their tissues for most of the genes on the X chromosome; males are monosomic.

Abnormalities in sex chromosome balance reveal the interplay between factors directed by genes on the X and Y chromosomes. An individual with only one X chromosome is called XO. This individual is female, but she is short and has other marked abnormalities. It is not clear how much of this is due to deleterious recessive alleles on the X and how much is due to gene dosage. The abnormality is called Turner's syndrome. An XXY individual is male, but he has some feminine characteristics such as breasts. This syndrome is called Kleinfelter's. This individual will have a Barr body because X-inactivation occurs in cells with more than one X chromosome. An individual with XYY is a male, mostly normal, but tall.

## SYNTENY

In general, from some of the items discussed in the last few sections, we can conclude that the relative number of chromosomes (or even parts of chromosomes) matters a great deal, but the actual number of different chromosomes is not terribly important. A classic case is two closely related deer that are phenotypically extremely similar. The Reeves Muntjak has only 3 autosomes plus sex chromosomes. These autosomes are enormous. The Indian Muntjak has 22 autosomes plus sex chromosomes, a number far more typical of other mammals. The two deer must have very similar genomes, but these are distributed differently. The chromosome fusion events that resulted in the Reeves Muntjak must be fairly recent in evolution. Their major consequence is that the two species cannot interbreed because the progeny would have numerous chromosome imbalances.

The Muntjak example suggests that chromosome organization can vary, in a dramatic way, superficially, without much disruption of gene organization and gene content. This notion is fully borne out when the detailed arrangement of genes is observed in different species. For example, in most closely related species, like human and chimp, the detailed appearance of chromosome bands is nearly identical. Humans and chimps in fact show only two regions of significant chromosome morphological differences in the entire genome. These regions are clearly of some interest for future study, since they could reveal hints of genes that differ significantly in the two species.

When more distantly related organisms are compared, their chromosomes appear, superficially, to be very different. For example, the mouse has 19 autosomes; the human has 22. All of the mouse chromosomes are acrocentric: Their centromeric constrictions occur at one end of the chromosome. In essence each mouse chromosome has only one arm. In contrast, most of the human chromosomes have an internally located centromere; in many cases it is near the center and the two arms are comparable in size. Only five human chromosomes are acrocentric, and each of these is a special case: Chromosomes 13, 14, 15, 21, and 22 all have tiny short arms containing large numbers of tandemly repeated rDNA genes (see Fig. 2.13). However, when the detailed arrangement of genes along chromosomes is compared in the mouse and human, a much more conserved structural pattern is seen.

Synteny is the term used to describe similar arrangements of genes on maps, whether these are genetic maps or chromosome morphology. Most regions of the mouse and human genomes display a high degree of synteny. This is summarized in Figure 2.25,

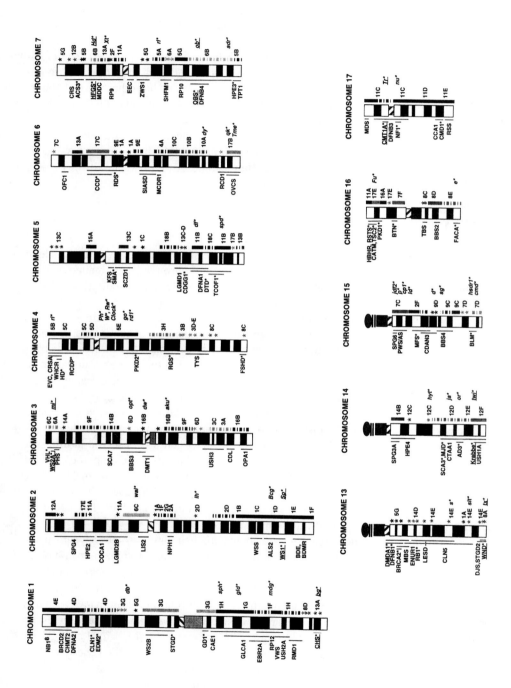

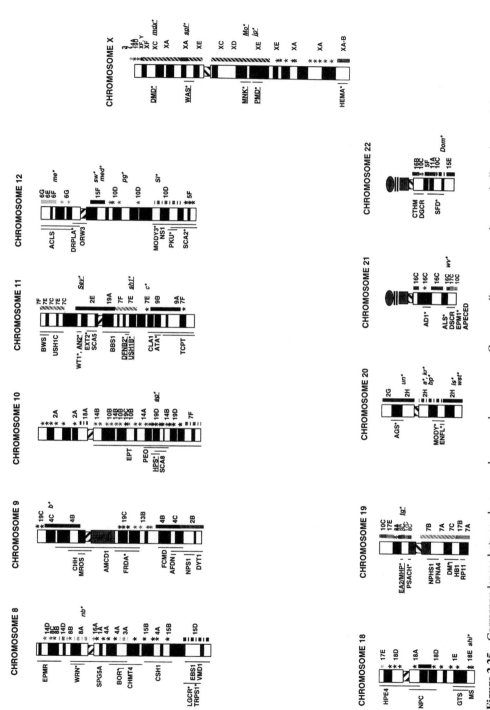

**Figure 2.25** Correspondences between human and mouse chromosomes. Corresponding mouse regions are indicated on each human chromosome. (Kindly provided by Lisa Stubbs; Carver and Stubbs, 1997.)

which reveals that a typical mouse chromosome has genes that map to two to four human chromosomes, and vice versa. If we look at genes that are neighbors in the mouse, these are almost always neighbors in the human. Figure 2.26 illustrates this for mouse chromosome 7: The corresponding human 19q regions contain blocks of the same genes in the same order. This is extremely useful because it means that genetic or physical mapping data in one species can usually be applied to other species. It speeds, enormously, the search for corresponding genes in different species. The superficial differences between the mouse and human genome have presumably been caused largely by gross chromo-

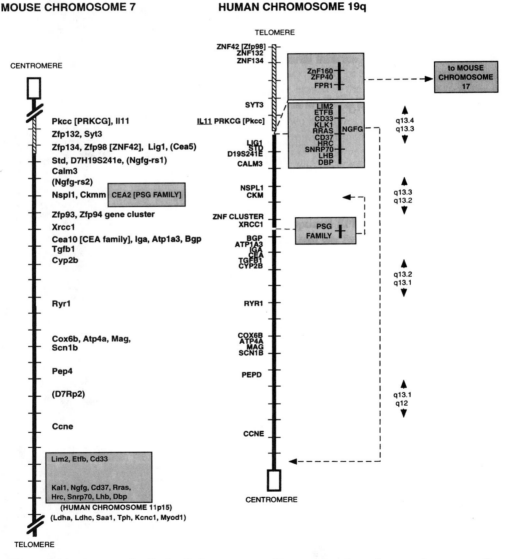

**Figure 2.26**   An example of a detailed synteny map. Shown are sections of mouse chromosome 7 corresponding to regions of human chromosome 19q which contain similarly ordered blocks of markers. (Kindly provided by Lisa Stubbs; Stubbs et al., 1996.)

some rearrangements in both species. It has been estimated that only a few hundred such rearrangements will ultimately account for all of the differences we see in the contemporary versions of these chromosomes.

## SOURCES AND ADDITIONAL READINGS

Buckle, V. J., Edwards, J. H., Evans, E. P., Jonasson, J. A., Lyon, M. F., Peters, J., and Searle, A. G. 1984. Chromosome maps of man and mouse II. *Clinical Genetics* 1: 1–11.

Cantor, C. R., and Schimmel, P. R. 1980. *Biophysical Chemistry.* San Francisco: W. H. Freeman, chapter 24.

Carver, E. A., and Stubbs, L. 1997. Zooming in on the human-mouse comparative trap: Genome conservation re-examined on a high resolution scale. *Genome Research* 7: 1123–1137.

Cozzarelli, N. R., and Wang, J. C., eds. 1990. *DNA Topology and Its Biological Effects.* Cold Spring Harbor, NY: Cold Spring Harbor Laboratory Press.

Harrington, J. J., Van Bokkeln, G., Mays, R. W., Gustashaw, K., and Willard, H. F. 1997. Formation of de novo centromeres and construction of first-generation human artificial microchromosomes. *Nature Genetics* 15: 345–355.

Holmquist, G. P. 1992. Review article: Chromosome bands, their chromatin flavors, and their functional features. *American Journal of Human Genetics* 51: 17–37.

König, P., and Rhodes, D. 1997. Recognition of telomeric DNA. *Trends in Biochemical Sciences* 22: 43–47.

Levy, M. Z., Allsopp, R. C., Futcher, A. B., Greider, C. W., and Harley, C. B. 1992. Telomere end-replication problem and cell aging. *Journal of Molecular Biology* 225: 951–960.

Manuelidis, L. 1990. A view of interphase chromosomes. *Science* 250: 1533–1540.

Moyzis, R. K., Buckingham, J. M, Cram, L. S., Dani, M., Deaven, L. L., Jones, M. D., Meyne, J., Ratliff, R. L., and Wu, J. R. 1988. A highly conserved repetitive DNA sequence, (TTAGGG)$n$, present at the telomeres of human chromosomes. *Proceedings of the National Academy of Sciences USA* 85: 6622–6626.

Niklas, R. B. 1997. How cells get the right chromosomes. *Science* 275: 632–637.

Therman, E. 1986. *Human Chromosomes: Structure, Behavior, Effects,* 2nd ed. New York: Springer-Verlag.

Saccone, S., De Sario, A., Wiegant, J., Raap, A. K., Valle, G. D., and Bernardi, G. 1993. Correlations between isochores and chromosomal bands in the human genome. *Proceedings of the National Academy of Sciences USA* 90: 11929–11933.

Stubbs, L., Carver, E. A., Shannon, M. E., Kim, J., Geisler, J., Generoso, E. E., Stanford, B. G., Dunn, W. C., Mohrenweiser, H., Zimmermann, W., et al. 1996. Detailed comparative map of human chromosome 19q and related regions of the mouse genome. *Genomics* 35: 499–508.

Wallace, D. C. 1995. Mitochondrial DNA variation in human evolution, degenerative disease, and aging. *American Journal of Human Genetics* 57: 201–223.

# 3 Analysis of DNA Sequences by Hybridization

## BASIC REQUIREMENTS FOR SELECTIVITY AND SENSITIVITY

The haploid human genome is $3 \times 10^9$ base pairs, and a typical human cell, as described in the last chapter, is somewhere between diploid and tetraploid in DNA content. Thus each cell has about $10^{10}$ base pairs of DNA. A single base pair is 660 Da. Hence the weight of DNA in a single cell can be calculated as $10^{10} \times 660 / (6 \times 10^{23}) = 10^{-11}$ g or 10 pg. Ideally we would like to be able to do analyses on single cells. This means that if only a small portion of the genome is the target for analysis, far less than 10 pg of material will need to be detected. By current methodology we are in fact able to determine the presence or absence of almost any 20-bp DNA sequence within a single cell, such as the sequence ATTGGCATAGGAGCC-CATGG. This analysis takes place at the level of single molecules. Two requirements must be met to perform such an exquisitely demanding analysis. There must be sufficient experimental sensitivity to detect the presence of the sequence. This sensitivity is provided by either chemical or biological amplification procedures or by a combination of these procedures. There must also be sufficient experimental selectivity to discriminate between the desired, true target sequence and all other similar sequences, which may differ from the target by as little as one base. That specificity lies with the intrinsic selectivity of DNA base pairing, itself.

The target of a 20-bp DNA sequence is not picked casually. Twenty bp is just about the smallest DNA length that has a high probability, a priori, of being found in a single copy in the human genome. This can be deduced as follows from simple binomial statistics (Box 3.1).

For simplicity, pretend that the human genome contains equal amounts of the four bases, A, T, C, and G, and that the occurrences of the bases are random. (These constraints will be relaxed elsewhere in the book when some of the unusual statistical properties of natural DNAs need to be considered explicitly. Then the expected frequency of occurrence of any particular stretch of DNA sequence, such as $n$ bases beginning as ATCCG . . ., is $4^{-n}$. The average number of occurrences of this particular sequence in the haploid human genome is $3 \times 10^9 \times 4^{-n}$. For a sequence of 16 bases, $n = 16$, the average occurrence is $3 \times 10^9 \times 4^{-16}$ which is about 1. Thus such a length will tend to be seen as often as not by chance; it is not long enough to be a unique identifier. There is a reasonable chance that the sequence 16 bases long will occur several times in different places in the genome. Choosing $n = 20$ gives an average occurrence of about 0.3%. Such sequences will almost always be unique genome landmarks. One corollary of this simple exercise is that it is a very futile exercise to look at random for the occurrence of particular 20-mers in the sequence of a higher organism unless there is good a priori reason for suspecting the presence of these sequences. This means that sequences of length 20 or more can be used as unique identifiers (see Box 3.2).

**BOX 3.1**
**BINOMIAL STATISTICS**

Binomial statistics describe the probable outcome of events like coin flipping, events that depend on a single random variable. While a normal coin has a 50% chance of heads or tails with each flip, we will consider here the more general case of a weighted coin with two possible outcomes with probabilities $p$ (heads) and $q$ (tails). Since there are no other possible outcomes $p + q = 1$. If $N$ successive flips are executed, and the outcome is a particular string, such as *hhhhttthhh*, the chance of this particular outcome is $p^n q^{N-n}$, where $n$ is the number of times heads was observed. Note that all strings with the same numbers of heads and tails will have the same a priori probability, since in binomial statistics each event does not affect the probability of subsequent events. Later in this book we will deal with cases where this extremely simple model does not hold. If we care only about the chance of an outcome with $n$ heads and $N - n$ tails, without regard to sequence, the number of such events is $N!/(n!)(N - n)!$, and so the fraction of times this outcome will be seen is $(p^n q^{N-n})N!/(n!)(N - n)!$

A simple binomial model can also be used to estimate the frequency of occurrence of particular DNA base sequences. Here there are four possible outcomes (not quite as complex as dice throwing where six possible outcomes occur). For a particular string with $n_A$ A's, $n_C$ C's, $n_G$ G's and $n_T$ T's, and a base composition of $X_A$, $X_C$, $X_G$, and $X_T$ the chance occurrence of that string is $X_A^{n_A} X_C^{n_C} X_G^{n_G} X_T^{n_T}$. The number of possible strings with a particular base composition is $N!/(n_A!n_C!n_G!n_T!)$, and by combining this with the previous term, the probability of a string with a particular base composition can easily be computed. Incidentally, the number of possible strings of length $N$ is $4^N$, while the number of different base compositions of this length is $(N + 3)!/(N!3!)$.

The same statistical models can be used to make estimates that two people will share the same DNA sequences. Such estimates are very useful in DNA-based identity testing. Here we consider just the simple case of two allele polymorphisms. In a particular place in the genome, suppose that a fraction of all individuals have one base, $f_1$, while the remainder have another, $f_2$. The chance that two individuals share the same allele is $f_1^2 + f_2^2 = g^2$. If a set of $M$ two-allele polymorphisms $(i, j, k, \ldots)$ is considered simultaneously, the chance that two individuals are identical for all of them is $g_i^2 g_j^2 g_k^2 \ldots$ By choosing $M$ sufficiently large, we can clearly make the overall chance too low to occur, unless the individuals in question are one and the same. However, two caveats apply to this reasoning. First, related individuals will show a much higher degree of similarity than predicted by this model. Monozygotic twins, in principle, should share an identical set of alleles at the germ-line level. Second, the proper allele frequencies to use will depend on the racial, ethnic, and other genetic characteristics of the individuals in question. Thus it may not always be easy to select appropriate values. These difficulties notwithstanding, DNA testing offers a very powerful approach to identification of individuals, paternity testing, and a variety of forensic applications.

**BOX 3.2**
**DNA SEQUENCES AS UNIQUE SAMPLE IDENTIFIERS**

The following table shows the number of different sequences of length $n$ and compares these values to the sizes of various genomes. Since genome size is virtually the same as the number of possible short substrings, it is easy to determine the lengths of short sequences that will occur on average only once per genome. Sequences a few bases longer than these lengths will, for all practical purposes, occur either once or not at all, and hence they can serve as unique identifiers.

| LENGTH | NUMBER OF SEQUENCES | GENOME, GENOME SIZE (BP) |
|---|---|---|
| 8 | $6.55 \times 10^4$ | Bacteriophage lambda, $5 \times 10^4$ |
| 9 | $2.60 \times 10^5$ | |
| 10 | $1.05 \times 10^6$ | |
| 11 | $4.20 \times 10^6$ | *E. coli, $4 \times 10^6$* |
| 12 | $1.68 \times 10^7$ | *S. cerevisiae, 1.3 $\times 10^7$* |
| 13 | $6.71 \times 10^7$ | |
| 14 | $2.68 \times 10^8$ | All mammalian mRNAs, $2 \times 10^8$ |
| 15 | $1.07 \times 10^9$ | |
| 16 | $4.29 \times 10^9$ | Human haploid genome, $3 \times 10^9$ |
| 17 | $1.72 \times 10^{10}$ | |
| 18 | $6.87 \times 10^{10}$ | |
| 19 | $2.75 \times 10^{11}$ | |
| 20 | $1.10 \times 10^{12}$ | |

## DETECTION OF SPECIFIC DNA SEQUENCES

DNA molecules themselves are the perfect set of reagents to identify particular DNA sequences. This is because of the strong, sequence-specific base pairing between complementary DNA strands. Here one strand of DNA will be considered to be a target, and the other, a probe. (If both are not initially available in a single-stranded form, there are many ways to circumvent this complication.) The analysis for a particular DNA sequence consists in asking whether a probe can find its target in the sample of interest. If the probe does so, a double-stranded DNA complex will be formed. This process is called *hybridization,* and all we have to do is to discriminate between this complex and the initial single-stranded starting materials (Fig. 3.1*a*).

The earliest hybridization experiments were carried out in homogeneous solutions. Hybridization was allowed to proceed for a fixed time period, and then a physical separation was performed to capture double-stranded material and discard single strands. Hydroxyapatite chromatography was used to do this discrimination because conditions could be found in which double-stranded DNA bound to a column of hydroxylapatite, while single strands were eluted (Fig. 3.1*b*). The amount of double-stranded DNA could be quantitated by using a radioisotopic label on the probe or the target, or by measuring the bulk amount of DNA captured or eluted. This method is still used today in select cases, but it is very tedious because only a few samples can be conveniently analyzed simultaneously.

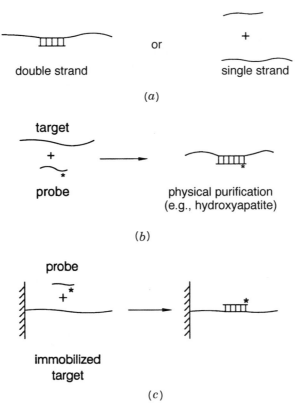

**Figure 3.1** Detecting the formation of specific double-stranded DNA sequences. *(a)* The problem is to tell whether the sequence of interest is present in single-stranded (s.s.) or double-stranded (d.s.) form. *(b)* Physical purification by methods such as hydroxyapatite chromatography. *(c)* Physical purification by the use of one strand attached to an immobilized phase. A label is used to detect hybridization of the single-stranded probe.

Modern hybridization protocols immobilize one of the two DNAs on a solid support (Fig. 3.1*c*). The immobilized phase can be either the probe or the target. The complementary sample is labeled with a radioisotope, a fluorescent dye, or some other specific moiety that later allows a signal to be generated in situ. The amount of color, or radioactivity, on the immobilized phase is measured after hybridization, for a fixed period, and subsequent washing of the solid support to remove adsorbed, but nonhybridized, material. As will be shown later, an advantage of this method is that many samples can be processed in parallel. However, there are also some disadvantages that will become apparent as we proceed.

## EQUILIBRIA BETWEEN DNA DOUBLE AND SINGLE STRANDS

The fraction of single- and double-stranded DNA in solution can be monitored by various spectroscopic properties that effectively average over different DNA sequences. Such measurements allow us to view the overall reaction of DNA single strands.

Ultraviolet absorbance, circular dichroism, or the fluorescence of dyes that bind selectively to duplex DNA can all be used for this purpose. If the amount of double-stranded (duplex) DNA in a sample is monitored as a function of temperature, the results typically obtained are shown in Figure 3.2. The DNA is transformed from double strands at low temperature, rather abruptly at some critical temperature, to single strands. The process, for long DNA, is usually so cooperative that it can be likened to the melting of a solid, and the transition is called *DNA melting*. The midpoint of the transition for a particular DNA sample is called the melting temperature, $T_m$. For DNAs that are very rich in the bases G + C, this can be 30 or 40°C higher than for extremely (A + T)-rich samples. It is such spectroscopic observations, on large numbers of small DNA duplexes that have allowed us to achieve a quantitative understanding of most aspects of DNA melting.

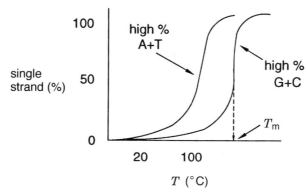

**Figure 3.2**  Typical melting behavior for DNA as a function of average base composition. Shown is the fraction of single-stranded molecules as a function of temperature. The midpoint of the transition is the melting temperature, $T_m$.

The goal of this section is to define conditions that allow sequence-specific analyses of DNA using DNA hybridization. Specificity means the ratio of perfectly base-paired duplexes to duplexes with imperfections or mismatches. Thus high specificity means that the conditions maximize the amount of double-stranded perfectly base-paired complex and minimize the amount of other species. Key variables are the concentrations of the DNA probes and targets that are used, the temperature, and the salt concentration (ionic strength).

The melting temperature of long DNA is concentration *in*dependent. This arises from the way in which $T_m$ is defined. Large DNA melts in patches as shown in Figure 3.3a. At $T_m$, the temperature at which half the DNA is melted, (A + T)-rich zones are melted, while (G + C)-rich zones are still in duplexes. No net strand separation will have taken place because no duplexes will have been completely melted. Thus there can be no concentration dependence to $T_m$.

In contrast, the melting of short DNA duplexes, DNAs of 20 base pairs or less, is effectively all or none (Fig. 3.3b). In this case the concentration of intermediate species, partly single-stranded and partly duplex, is sufficiently small that it can be ignored. The reaction of two short complementary DNA strands, *A* and *B*, may be written as

$$A + B = AB$$

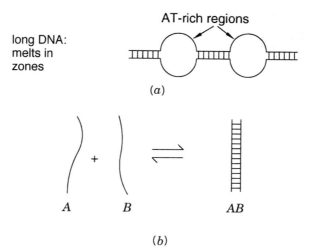

(b)

**Figure 3.3**    Melting behavior of DNA. *(a)* Structure of a typical high molecular weight DNA at its melting temperature. *(b)* Status of a short DNA sample at its melting temperature.

The equilibrium (association) constant for this reaction is defined as

$$K_a = \frac{[AB]}{[A][B]}$$

Most experiments will start with an equal concentration of the two strands (or a duplex melted to give the two strands). The initial total concentration of strands (whatever their form) is $C_T$. If, for simplicity, all of the strands are initially single stranded, their concentrations is

$$[A]_o = [B]_o = \frac{C_T}{2}$$

At $T_m$ half the strands must be in duplex. Hence the concentrations of the different species at $T_m$ will be

$$[AB] = [A] = [B] = \frac{C_T}{4}$$

The equilibrium constant at $T_m$ is

$$K_a = \frac{[AB]}{[A][B]} = \frac{C_T/4}{(C_T/4)^2} = \frac{4}{C_T}$$

Do not be misled by this expression into thinking that the equilibrium constant is concentration dependent. It is the $T_m$ that is concentration dependent. The equilibrium constant is temperature dependent. The above expression indicates the value seen for the equilibrium constant, $4/C_T$, at the temperature, $T_m$. This particular $T_m$ occurs when the equilibrium is observed at the total strand concentration, $C_T$.

A special case must be considered in which hybridization occurs between two short single strands of the same identical sequence, $C$. Such strands are self-complementary. An example is GGGCCC which can base pair with itself. In this case the reaction can be written

$$2C = C_2$$

The equilibrium constant, $K_a$, becomes

$$K_a = \frac{[C_2]}{[C]^2}$$

At the melting temperature half of the strands must be duplex. Hence

$$[C_2] = [2C] = \frac{C_T}{4}$$

where $C_T$ as before is the total concentration of strands. Thus we can evaluate the equilibrium expression at $T_m$ as

$$K_a \frac{C_T/4}{(C_T/2)^2} = \frac{1}{C_T}$$

As before, what this really means is that $T_m$ is concentration dependent. In both cases simple mass action considerations ensure that $T_m$ will increase as the concentration is raised.

The final case we need to consider is when one strand is in vast excess over the other instead of both being at equal concentrations. This is frequently the case when a trace amount of probe is used to interrogate a concentrated sample or, alternatively, when a large amount of probe is used to interrogate a very minute sample. The formation of duplex can be written as before as $A + B = AB$, but now the initial starting conditions are

$$[B]_o \gg [A]_o$$

Effectively the total strand concentration, $C_T$, is thus simply the initial concentration of the excess strand: $B_o$. At $T_m$,

$$[AB] = [A]$$

Thus the equilibrium expression at $T_m$ becomes

$$K_a = \frac{[AB]}{[A][B]} = \frac{1}{C_T}$$

The melting temperature is still concentration dependent.

The importance of our ability to drive duplex formation cannot be underestimated. Figure 3.4 illustrates the practical utility of this ability. It shows the concentration dependence of the melting temperature of two different duplexes. We can characterize each re-

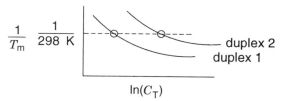

**Figure 3.4** The dependence of the melting temperature, $T_m$, of two short duplexes on the total concentration of DNA strands, $C_T$.

action by its melting temperature and can attempt to extract thermodynamic parameters like the enthalpy change, $\Delta H$, and the free energy change, $\Delta G$, for each reaction. However, with the $T_m$'s different for the two reactions, if we do this at $T_m$, the thermodynamic parameters derived will refer to reactions at two different temperatures. There will be no way, in general, to compare these parameters, since they are expected to be intrinsically temperature dependent. The concentration dependence of melting saves us from this dilemma. By varying the concentration, we can produce conditions where the two duplexes have the same $T_m$. Now thermodynamic parameters derived from each are comparable. We can, if we wish, choose any temperature for this comparison. In practice, 298 K has been chosen for this purpose.

## THERMODYNAMICS OF THE MELTING OF SHORT DUPLEXES

The model we will use to analyze the melting of short DNA double helices is shown in Figure 3.5. The two strands come together, in a nucleation step, to form a single pair. Double strands can form by stacking of adjacent base pairs above or below the initial nucleus until a full duplex has zippered up. It does not matter, in our treatment, where the initial nucleus forms. We also need not consider any intermediate steps beyond the nucleus and the fully duplex state. In that state, for a duplex of $n$ base pairs, there will be $n - 1$ stacking interactions (Fig. 3.6). Each interaction reflects the energetics of stacking two adjacent base pairs on top of each other. There are ten distinct such interactions, as ApG/CpT, ApA/TpT, and so on (where the slash indicates two complementary antiparallel strands). Because their energetics are very different, we must consider the DNA sequence explicitly in calculating the thermodynamics of DNA melting.

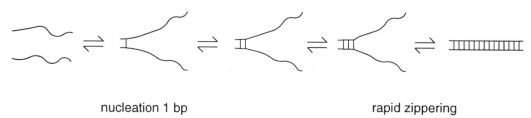

nucleation 1 bp                    rapid zippering

**Figure 3.5** A model for the mechanism of the formation of duplex DNA from separated complementary single strands.

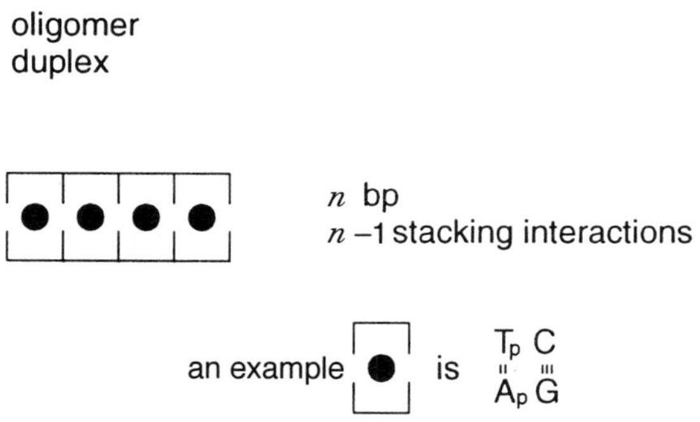

oligomer
duplex

$n$ bp
$n - 1$ stacking interactions

an example   ▢●   is   $\overset{\text{T}_\text{p}}{\underset{\text{A}_\text{p}}{\underset{\text{\tiny III}}{\overset{\text{\tiny II}}{|}}}}\overset{\text{C}}{\underset{\text{G}}{}}$

**Figure 3.6**   Stacking interactions in double-stranded DNA.

For each of the ten possible stacking interactions, we can define a standard $\Delta G_\text{s}^0$ for the free energy of stacking, a $\Delta H_\text{s}^0$ for the enthalpy of stacking, and a $\Delta S_\text{s}^0$ for the entropy of stacking. These quantities will be related at a particular temperature by the expression

$$\Delta G_\text{s}^0 = \Delta H_\text{s}^0 - T\Delta S_\text{s}^0$$

For any particular duplex DNA sequence, we can compute the thermodynamic parameters for duplex formation by simply combining the parameters for the competent stacking interactions plus a nucleation term. Thus

$$\Delta G^0 = \Delta G_\text{nuc}^0 + \Sigma\Delta G_\text{s}^0 + \Delta g_\text{sym}$$
$$\Delta H^0 = \Delta H_\text{nuc}^0 + \Sigma\Delta H_\text{s}^0$$

and similarly for the entropy, where the sums are taken over all of the stacking interactions in the duplex, where $\Delta g_\text{sym} = 0.4$ kcal/mol if the two strands are identical; otherwise, $\Delta g_\text{sym} = 0$. The equilibrium constant for duplex formation is given by

$$K = ks_1 s_2 s_3 \ldots s_{n-1}$$

where $k$ is the equilibrium constant of nucleation, related to the $\Delta G_\text{nuc}^0$ by

$$\Delta G_\text{nuc}^0 = - RT \ln k$$

and each $s_i$ is the microscopic equilibrium constant for a particular stacking reaction, related to the $\Delta G_i^0$ for that reaction by

$$\Delta G_i^0 = - RT \ln s_i$$

The key factor involved in predicting the stability of DNA (and RNA) duplexes is that all of these thermodynamic parameters have been measured experimentally. One takes advantage of the enormous power available to synthesize particular DNA sequences, combines complementary pairs of such sequences, and measures their extent of duplex as a function of temperature and concentration. For example, we can study the properties of

$A_8/T_8$ and compare these with $A_9/T_9$. Since the only difference between these complexes is an extra ApA/TpT stacking interaction, the differences will yield the thermodynamic parameters for that interaction. Other sequences are more complex to handle, but this has all been accomplished.

It is helpful to choose a single standard temperature and set of environmental conditions for the tabulation of thermodynamic data: 298 K has been selected, and $\Delta G_s^0$'s at 298 K in 1 M NaCl are listed in Table 3.1. Enthalpy values for the stacking interactions can be obtained in two ways: either by direct calorimetric measurements or by examining the temperature dependence of the stacking interactions. $\Delta H_s^0$'s are also listed in Table 3.1. So are $\Delta S_s^0$'s, which can be calculated from the relationship $\Delta G_s^0 = \Delta H_s^0 - T\Delta S_s^0$. From these data, thermodynamic values at other temperatures can be estimated as shown in Box 3.3. The effects of salt are well understood and are described in detail elsewhere (Cantor and Schimmel, 1980).

The results shown in Table 3.1 make it clear that the effects of the DNA sequence on duplex stability are very large. The average $\Delta H_s^0$ is $-8$ kcal/mol; the range is $-5.6$ to $-11.9$ kcal/mol. The average $\Delta G_s^0$ is $-1.6$ kcal/mol with a range of $-0.9$ to $-3.6$ kcal/mol. Thus the DNA sequence must be considered explicitly in estimating duplex stabilities. The two additional parameters needed to do this concern the energetics of nucleation. These are relatively sequence independent, and we can use average values of $\Delta G_{nuc}^0 = +5$ kcal/mol (except if no G–C pairs are present, then $+6$ kcal/mol should be used) and $\Delta H_{nuc}^0 = 0$.

For estimating the stability of perfectly paired duplexes, these nucleation parameters and the stacking energies in Table 3.1 are used. Table 3.2 shows typical results when calculated and experimentally measured $\Delta G^0$'s are compared. The agreement in almost all cases is excellent, and the few discrepancies seen are probably within the range of typical experimental errors. The approach described above has been generalized to predict the thermodynamic properties of triple helices, and presumably it will also serve for four-stranded DNA structures.

**TABLE 3.1   Nearest-neighbor Stacking Interactions in Double-stranded DNA**

| | Nearest-neighbor Thermodynamics | | |
|---|---|---|---|
| Interaction | $-\Delta H^\circ$ (kcal/mol) | $-\Delta S^\circ$ (cal/Kmol) | $-\Delta G^\circ$ (kcal/mol) |
| AA/TT | 9.1 | 24.0 | 1.9 |
| AT/TA | 8.6 | 23.9 | 1.5 |
| TA/AT | 6.0 | 16.9 | 0.9 |
| CA/GT | 5.8 | 12.9 | 1.9 |
| GT/CA | 6.5 | 17.3 | 1.3 |
| CT/GA | 7.8 | 20.8 | 1.6 |
| GA/CT | 5.6 | 13.5 | 1.6 |
| CG/GC | 11.9 | 27.8 | 3.6 |
| GC/CG | 11.1 | 26.7 | 3.1 |
| GG/CC | 11.0 | 26.6 | 3.1 |

*Source:* Adapted from Breslauer et al. (1986).

**BOX 3.3**
**STACKING PARAMETERS AT OTHER TEMPERATURES**

In practice, it turns out to be a sufficiently accurate approximation to assume that the various $\Delta H_s^0$'s are independent of temperature. This is a considerable simplification. It allows direct integration of the van't Hoff relationship to compute thermodynamic parameters at any desired temperature from the parameters measured at 298 K. Starting from the usual relationship for the dependence of the equilibrium constant on temperature,

$$\frac{d(\ln K)}{d(1/T)} = \frac{\Delta H^0}{R}$$

we can integrate this directly:

$$\int d(\ln K) = \frac{\Delta H^0}{R} \int d\left(\frac{1}{T}\right)$$

With limits of integration from $T_0$, which is 298 K, to any other melting temperature, $T_m$, the result is

$$\ln K(T_m) - \ln K(T_0) = \frac{\Delta H^0}{R}\left(\frac{1}{T_m} - \frac{1}{T_0}\right)$$

However, the first term of the left-hand side of the equation, for a pair of complementary oligonucleotides, is just $4/C_T$. The second term on the left-hand side is equal to $\Delta G(T_0)/RT_0$. Inserting these values and rearranging gives a final, useful expression for computing the $T_m$ of a duplex at any temperature from data measured at 298 K. All of the necessary data are summarized in Table 3.1.

$$\frac{T_m}{T_0} = \frac{\Delta H^0}{\Delta H^0 - \Delta G^0(T_0) + RT_0 \ln(C_T/4)}$$

## THERMODYNAMICS OF IMPERFECTLY PAIRED DUPLEXES

In contrast to the small number of discrete interactions that must be considered in calculating the energetics of perfectly paired DNA duplexes, there is a plethora of ways that duplexes can pair imperfectly. We have available model compound data on most of these so that estimates of the energetics can be made. However, the large number of possibilities precludes a complete analysis of imperfections in the context of all possible sequences, at least for the present.

The simplest imperfection is a dangling end, as shown in Figure 3.7a. If both ends are dangling, their contributions can be treated separately. From available data it appears that a dangling end contributes $-8$ kcal/mol on average to the overall $\Delta H$ of duplex formation, and $-1$ kcal/mol to the overall $\Delta G$ of duplex formation. The large enthalpy arises

**TABLE 3.2  Predicted and Observed Stabilities
(free energy of formation) of Various
Oligonucleotide Duplexes**

| | Comparison of Calculated and Observed $\Delta G$ (kcal/mol) | | |
|---|---|---|---|
| | Oligomeric Duplex | $-\Delta G$pred | $-\Delta G$obs |
| 1 | GCGCGC | 11.1 | 11.1 |
| | CGCGCG | | |
| 2 | CGTCGACG | 11.2 | 11.9 |
| | GCAGCTGC | | |
| 3 | GAAGCTTC | 7.9 | 8.7 |
| | CTTCGAAG | | |
| 4 | GGAATTCC | 9.3 | 9.4 |
| | CCTTAAGG | | |
| 5 | GGTATACC | 6.7 | 7.4 |
| | CCATATGG | | |
| 6 | GCGAATTCGC | 16.5 | 15.5 |
| | CGCTTAAGCG | | |
| 7 | CAAAAAG | 6.1 | 6.1 |
| | GTTTTTC | | |
| 8 | CAAACAAAG | 9.3 | 10.1 |
| | GTTTGTTTC | | |
| 9 | CAAAAAAAG | 9.9 | 9.6 |
| | GTTTTTTTC | | |
| 10 | CAAATAAAG | 8.5 | 8.5 |
| | GTTTATTTC | | |
| 11 | CAAAGAAAC | 9.3 | 9.5 |
| | GTTTCTTTG | | |
| 12 | CGCGTACGCGTACGCG | 32.9 | 34.1 |
| | GCGCATGCGCATGCGC | | |

Note: Calculations use the equations given in the text and the measured thermodynamic values given in Table 3.1.

*Source:* Adapted from Breslauer et al. (1986).

because the first base of the dangling end can still stack on the last base pair of the duplex. Note that there are two distinct types of dangling ends: a 3′-overhang and a 5′-overhang. At the current level of available information, we treat these as equivalent. Simple dangling ends will arise whenever a target and a probe are different in size.

The next imperfection, which leads to considerable destabilization, is an internal mismatch. As shown in Figure 3.7b, this leads to the loss of two internal $\Delta H_s^0$'s and two internal $\Delta G_s^0$'s. Apparently this is empirically compensated by some residual stacking either between the bases that are mispaired and the duplex borders or between the bases themselves. Whatever the detailed mechanism, the result is to gain back about $-8$ kcal/mol in $\Delta H$. There is no effect on the $\Delta G$. A larger internal mismatch is called an *internal loop.* Considerable data on the thermodynamics of such loops exist for RNA, and much less for DNA. In general, such structures will be far less stable than the perfect duplex because of the loss of additional free energies and enthalpies of stacking.

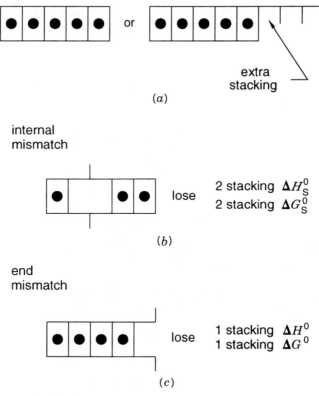

**Figure 3.7** Stacking interactions in various types of imperfectly matched duplexes. *(a)* Dangling ends. *(b)* Internal mismatch. *(c)* Terminal mismatch.

A related imperfection is a bulge loop in which two strands of unequal size come together so that one is perfectly paired while the other has an internal loop of unpaired residues. There can also be internal loops in which the single-stranded regions are of different length on the two strands. Again the data describing the thermodynamics of such structures are available mostly just for RNAs (Cantor and Schimmel, 1981). There are many complications. See, for example, Schroeder et al. (1996).

A key imperfection that needs to be considered to understand the specificity of hybridization is a terminal mismatch. As shown in Figure 3.7c, this results in the loss of one $\Delta H_s^0$ and one $\Delta G_s^0$. Such an external mismatch should be less destabilizing than an internal mismatch because one less stacking interaction is disrupted. It is not clear, at the present time, how much any stacking of the mismatch to the end of the duplex partially compensates for the lost duplex stacking. It may be adequate to model an end-mismatch like a dangling end. Alternatively, one might consider it like an internal mismatch with only one lost set of duplex stacking interactions. Both of these cases require no correction to the predicted $\Delta G$ of duplex formation, once the lost stacking interactions have been accounted for. Therefore at present it is simplest to concentrate on $\Delta G$ estimates and ignore $\Delta H$ estimates. More studies are needed in this area to clear up these uncertainties. In Chapter 12 we will discuss attempts to use hybridization of short oligonucleotides to infer

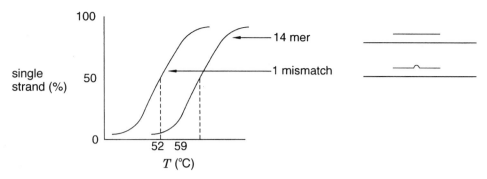

**Figure 3.8** Equilibrium melting behavior of a perfectly matched oligonucleotide duplex and a duplex with a single internal mismatch.

sequence information about a larger DNA target. As interest in this approach becomes more developed, the necessary model compound data will doubtlessly be accumulated. Note that there are 12 distinct terminal mismatches, and their energetics could depend not only on the sequence of the neighboring duplex stack but also on whether one or both strands of the duplex are further elongated. A large number of model compounds will need to be investigated.

Despite the limitations and complications just described, our current ability to predict the effect of a mismatch on hybridization is quite good. A typical example is shown in Figure 3.8. Note that the melting transitions of oligonucleotide duplexes with 8 or more base pairs are quite sharp. This reflects the large $\Delta H_s^0$'s and the all-or-none nature of the reaction. The key practical concern is finding a temperature where there is good discrimination between the amount of perfect duplex formed and the amount of mismatched duplex. The results shown in Figure 3.8 indicate that a single internal mismatch in a 14-mer leads to a 7°C decrease in $T_m$. Because of the sharpness of the melting transitions, this can easily be translated into a discrimination of about a factor of ten in amount of duplex present.

## KINETICS OF THE MELTING OF SHORT DUPLEXES

The major issue we want to address here is whether, by kinetic studies of duplex dissociation, one can achieve a comparable or greater discrimination between the stability of different duplexes than by equilibrium melting experiments. In practice, equilibrium melting experiments are technically difficult because one must discriminate between free and bound probe or target. If the analysis is done spectroscopically, it can be done in homogeneous solution, but this usually requires large amounts of sample. If it is done with a radioisotopic tag (or a fluorescent label that shows no marked change on duplex formation), it is usually done by physical isolation of the duplex. This assumes that the rate of duplex dissociation is slow compared with the rate of physical separation, a constraint that usually can be met.

In typical hybridization formats a complex is formed between probe and target, and then the immobilized complex is washed to remove excess unbound probe (Fig. 3.9).

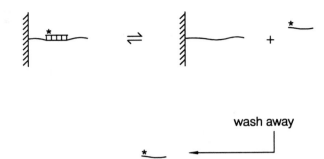

**Figure 3.9**   Schematic illustration of the sort of experiment used to measure the kinetics of duplex melting.

The complex is stored under conditions where little further dissociation occurs, and afterward the amount of probe bound to the target is measured. A variation on this theme is to allow the complex to dissociate for a fixed time period and then to measure the amount of probe remaining bound. It turns out that this procedure gives results very similar to equilibrium duplex formation. The reason is that the temperature dependence for the reaction between two long single strands to form a duplex is very slight. The activation energy for this reaction has been estimated to be only about 4 kcal/mol. Figure 3.10 shows a hypothetical reaction profile for interconversion between duplex and single strands. If we assume that the forward and reverse reaction pathways are the same, then the activation energy for the dissociation of the double strand will be just $\Delta H + 4$ kcal/mol. Since $\Delta H$ is much larger than 4 kcal/mol for most duplexes, the temperature dependence of the reaction kinetics will mirror the equilibrium melting of the duplex.

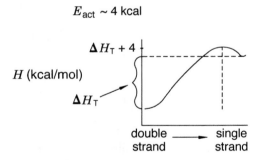

**Figure 3.10**   Thermodynamic reaction profile for oligonucleotide melting. Shown is the enthalpy, *H*, as a function of strand separation.

It is simple to estimate the effect of temperature on the dissociation rate. At $T_m$ we expect an equation of the form

$$k_m = A \exp\left(-\frac{[\Delta H + 4]}{RT_m}\right)$$

to apply, where $A$ is a constant, and the exponential term reflects the number of duplexes that have enough energy to exceed the activation energy. At any other temperature, $T_d$,

$$k_d = A \exp\left( - \frac{[\Delta H + 4]}{RT_d} \right)$$

The rate constants $k_m$ and $k_d$ are for duplex melting at temperatures $T_m$ and $T_d$, respectively. We define the time of half release at $T_d$ to be $t_{wash}$, and the time of half release at $T_m$ to be $t_{1/2}$. Then simple algebraic manipulation of the above equations yields

$$\ln\left( \frac{t_{wash}}{t_{1/2}} \right) = \frac{(\Delta H + 4)(1/T_d - 1/T_m)}{R}$$

This expression allows us to calculate kinetic results for particular duplexes once measurements are made at any particular temperature.

## KINETICS OF MELTING OF LONG DNA

The kinetic behavior of the melting of long DNA is very different from that of short duplexes in several respects. Melting near $T_m$ cannot normally be considered an all-or-none reaction. Melting experiments at temperatures far above $T_m$ reveal an interesting complication. The base pairs break rapidly, and if the sample is immediately returned to temperatures below $T_m$, they reform again. However, if a sufficient time at high temperature is allowed to elapse, now the renaturation rate can be many orders of magnitude slower. What is happening is shown schematically in Figure 3.11. Once the DNA base pairs are broken, the two single strands are still twisted around each other. The twist represented by each turn of the double helix must still be unwound. As the untwisting begins, the process greatly slows because the coiled single strands must rotate around each other. Experiments show that the untwisting time scales as the square of the length of the DNA. For molecules 10 kb or longer, these unwinding times can be very long. This is probably one of the reasons why people have trouble performing conventional polymerase chain reaction (PCR) amplifications on high molecular weight DNA samples. The usual PCR protocols allow only 30 to 60 seconds for DNA melting (as discussed in the next chapter.) This is far too short for strand untwisting of large DNAs.

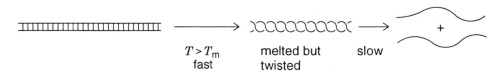

$T > T_m$      melted but      slow
fast        twisted

**Figure 3.11**    Kinetic steps in the melting of very high molecular weight DNA.

## KINETICS OF DOUBLE-STRAND FORMATION

The kinetics of renaturation of short duplexes are simple and straightforward. The reaction is essentially all or none; nucleation of the duplex is the rate-limiting step, and it comes about by intermolecular collision of the separated strands. Duplex renaturation kinetic measurements can be carried out at any temperature sufficiently below the melting temperature so that the reverse process, melting, does not interfere. The situation is much more complicated when longer DNAs are considered. This is illustrated in Figure 3.12a, where the absorbance increase upon melting is used to measure the relative amounts of single- and double-stranded structures. If a melted sample, once the strands have physically unwound, is slowly cooled, full recovery of the original duplex can be achieved. However, depending on the sample, the rate of duplex formation can be astoundingly slow, as we will see. If the melted sample is cooled too rapidly, the renaturation is not complete. In fact, as shown in Figure 3.12b, true renaturation has not really occurred at all. Instead, the separated single strands start to fold on themselves to make hairpin duplexes, analogous to the secondary structures seen in RNA. If the temperature falls too rapidly, these structures become stable; they are kinetically trapped and, effectively, can never dissociate to allow formation of the more stable perfect double-stranded duplex.

To achieve effective duplex renaturation, one needs to hold the sample at a temperature where true duplex formation is favored at the expense of hairpin structures so that these rapidly melt, and the unfolded strands are free to continue to search for their correct partners. Alternatively, hairpin formation is kept sufficiently low so that there remains a sufficiently unhindered DNA sequence to allow nucleation of complexes with the correct partners.

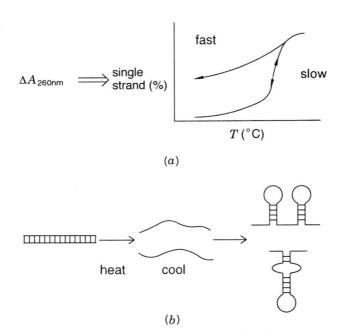

**Figure 3.12**   Melting and renaturation behavior of DNA as a function of the rate of cooling. *(a)* Fraction of initial base pairing as a function of temperature. *(b)* Molecular structures in DNA after rapid heating and cooling.

    In practice, renaturations are carried out most successfully when the renaturation temperature is $(T_m - 25)°C$. In free solution the simple model, shown in Figure 3.5, is sufficient to account for renaturation kinetics under such conditions. It is identical to the picture used for short duplexes. The rate-limiting step is nucleation, when two strands collide in an orientation that allows the first base pair to be formed. After this, rapid duplex growth occurs in both directions by a rapid zippering of the two strands together. This model predicts that the kinetics of renaturation should be a pure second-order reaction. Indeed, they are. If we call the two strands $A$ and $B$ we can describe the reaction as

$$A + B \rightarrow AB$$

The kinetics will be governed by a second-order rate constant $k_2$:

$$\frac{d[AB]}{dt} = k_2[A][B]$$

$$\frac{-d[A]}{dt} = \frac{-d[B]}{dt} = k_2[A][B]$$

    In a typical renaturation experiment, the two strands have the same initial concentration: $[A]_o = [B]_o = C_s$, where $C_s$ is $\frac{1}{2}$ the initial total concentration of DNA strands,

$$C_s = \frac{[A]_o + [B]_o}{2}$$

At any subsequent time in the reaction, simple mass conservation indicates that the concentrations of single strands remaining will be $[A] = [B] = C$, where $C$ is the total instantaneous concentration of each strand at time $t$. This allows us to write the rate equation for the renaturation reaction as in terms of the disappearance of single strands as

$$\frac{-dC}{dt} = k_2 C^2$$

We can integrate this directly from the initial concentrations to those at any time $t$,

$$-\int \frac{dC}{C^2} = \int_0^t k_2 dt$$

which becomes

$$\frac{1}{C} - \frac{1}{C_s} = k_2 t$$

and can be rearranged to

$$\frac{C_s}{C} = 1 + k_2 t C_s$$

It is convenient to rearrange this further to

$$\frac{C}{C_s} = f_s = \frac{1}{(k_2 t C_s + 1)}$$

where $f_s$ is the fraction of original single strands remaining.

When $f_s = 0.5$, we call the time of the reaction the half time, $t_{1/2}$. Thus

$$k_2 t_{1/2} C_s = 1 \quad \text{or} \quad t_{1/2} = \frac{1}{k_2 C_s}$$

The key result, for a second-order reaction, is that we can adjust the halftime to any value we choose by selecting an appropriate concentration.

The concentration that enters the equations just derived is the concentration of unique, complementary DNA strands. For almost all complex DNA samples, this is not a concentration we can measure directly. Usually the only concentration information readily available about a DNA sample is the total concentration of nucleotides, $C_0$. This is obtainable by direct UV absorbance measurements or by less direct but more sensitive fluorescence measurements on bound dyes. To convert $C_0$ to $C_s$, we have to know something about the sequence of the DNA. Suppose that our DNA was the intact genome of an organism, with $N$ base pairs, as shown schematically in Figure 3.13. The concentration of each of the two strands at a fixed nucleotide concentration, $C_0$, is just

$$C_s = \frac{C_0}{2N}.$$

Now suppose that the genome is divided into chromosomes, or the chromosomes into unique fragments, as shown in Figure 3.13. This does not change the concentration of any of the strands of these fragments. (An important exception occurs if the genome is randomly broken instead of uniquely broken; see Cantor and Schimmel, 1980.) Thus we can write the product of $k_2$, the concentration and half-time, as

$$k_2 t_{1/2} C_s = \frac{k_2 t_{1/2} C_0}{2N} = 1$$

and this can be rearranged to

$$C_0 t_{1/2} = \frac{2N}{k_2}$$

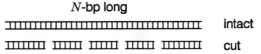

**Figure 3.13**   Schematic illustration of intact genomic DNA, or genomic DNA digested into a set of nonoverlapping fragments.

This key result means that the rate of genomic DNA reassembly depends linearly on the genome size. Since genome sizes vary across many orders of magnitude, the resulting renaturation rates will also. Here we have profound implications for the design and execution of experiments that use genomic DNA. Molecular biologists have tended to lump the two variables $C_0$ and $t_{1/2}$ together, and they talk about the parameter $C_0t_{1/2}$ as "a half cot." We will use this term, but don't try to sleep on it.

Some predicted kinetic results for the renaturation of genomic DNA that illustrate the renaturation behavior of the different samples are shown in Figure 3.14. These results are calculated for typical renaturation conditions with a nucleotide concentration of $1.5 \times 10^{-4}$ M. Note that under these conditions a 3-kb plasmid will renature so quickly that it is almost impossible to work with the separate strands under conditions that allow renaturation. In contrast, the $3 \times 10^9$ bp human genome is predicted to require 58 days to renature under the very same conditions. In practice, this means never; much higher concentrations must be used to achieve the renaturation of total human DNA.

The actual renaturation results seen for human DNA are very different from the simple prediction we just made. The reason is a significant fraction of human DNA is highly repeated sequences. This changes the renaturation rate of these sequences. The concentration of strands of a sequence repeated $m$ times in the genome is $mC_0/2N$. Thus this sequence will renature much faster:

$$C_0t_{1/2} = \frac{2N}{mk_2}$$

The same equation also handles the case of heterogeneity in which a particular sequence occurs in only a fraction of the genomic DNA under study. In this case $m < 1$, and the renaturation proceeds slower than expected. Typical renaturation results for human (or any mammalian) DNA are shown in Figure 3.15. There are three clearly resolved kinetic

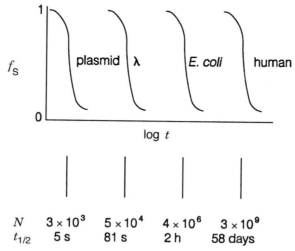

**Figure 3.14**  Renaturation kinetics expected for different DNA samples at a total initial nucleotide concentration of $1.5 \times 10^{-4}$ M at typical salt and temperature conditions. The fraction of single strand remaining, $f_s$, is plotted as a function of the log of the time, $t$. Actual experiments conform to expectations except for the human DNA sample. See Figure 3.15.

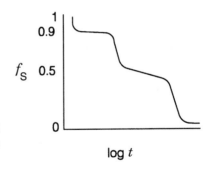

**Figure 3.15** Renaturation kinetics actually seen for total human genomic DNA. The parameters plotted are the same as in Figure 3.14.

phases to the reaction. About 10% of the DNA renatures almost immediately. It turns out this is very simple sequence DNA, some of which is self-complementary, and can renature by folding back on itself to make a hairpin. Other single-stranded components of this class form novel DNA ordered structures like some triplet-repeating sequences (see Chapter 13). About 40% of the DNA renatures $10^5$ to $10^6$ times as fast as expected for single-copy human DNA. This consists mostly of highly repeated sequences like the human *Alu* repeat. The effective $C_0t_{1/2}$ for this DNA is about 1. The remaining half of the genome does show the expected renaturation times for single-copy DNA.

The repeated sequences that renature fast, but intermolecularly, are mostly interspersed with single-copy DNA. Thus the length of the DNA fragments used in a renaturation experiment will have a profound effect on the resulting products that one might purify at various times during the reaction. Repeats that come together fast will usually be flanked by single-copy sequences that do not correspond. If one purifies double stranded material, it will be a mixture of duplex and single strands, as shown in Figure 3.16. Upon further renaturation the dangling single strands will find their partners, but only at the cost of generating a complex crosslinked network of DNA (Fig. 3.16*b*). Eventually this will act as such a barrier to diffusion that some of the strands will never find their mates.

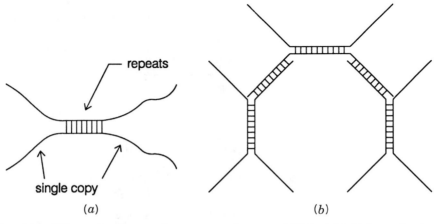

(a)                                        (b)

**Figure 3.16** Effects of repeats on the structure of renatured DNA. *(a)* Initial complex between two DNAs that share partial sequence complementarity at a repeat. *(b)* Subsequent complexes formed at $C_0t$'s sufficiently large to allow duplex formation between single copy sequences.

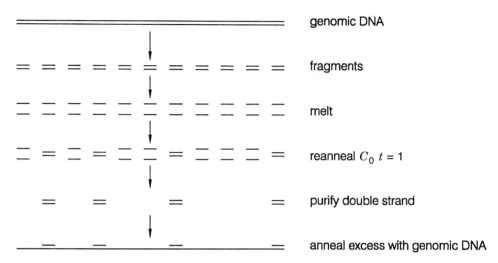

**Figure 3.17**    Blocking repeats by prehybridization with excess purified repeated DNA.

There are two solutions to this problem. One is to cut the genomic DNA down to sizes of a few hundred base pairs so that most single-copy sequences are essentially removed from the repeats. The other alternative, if larger DNA is needed for a particular application, is to purify repeated DNA as small fragments and pre-anneal it at high concentrations to single-stranded genomic DNA. This will block the repeat sites, as shown in Figure 3.17, and then the flanking single strands will be free to find their complements without network formation.

One could ask why not cut genomic DNA down to tiny fragments for all reannealing or hybridization experiments. Indeed many current schemes for analyzing DNA with hybridization do use short oligonucleotide probes. These have one disadvantage. Their intrinsic $k_2$ is small. Recall that $k_2$ really represents the nucleation reaction for duplex formation. This is the rate of formation of the first correct base pair that can zip into the duplex structure. The number of potential nucleation sites increases as $L$, the length of the strands. There is a second way in which length affects $k_2$; this is the excluded volume of the coiled DNA single strands which retards the interpenetration needed for duplex formation. This depends on $L^{-1/2}$. When these two effects are combined, the result is that $k_2$ depends on $L^{1/2}$. This discourages the use of small fragments for hybridization unless they are available at very high concentrations. Ultimately, if too small fragments are used, there will also be an effect on the stability of the resulting duplexes.

## COMPLEXITY

In thinking about DNA hybridization, it is useful to introduce the notion of DNA complexity. This is the amount of different DNA sequence present in a sample. For a DNA sample with repeated sequence and heterogeneities, the complexity is

$$\frac{\Sigma_i N_i}{m_i}$$

where the sum is taken over every type of DNA species present; $m_i$ is the number of times a species of size $N_i$ is repeated. The complexity will determine the hybridization kinetics of the slowest renaturing species in the sample.

## HYBRIDIZATION ON FILTERS

In contemporary DNA work, hybridization of probes in solution to targets immobilized on filters is much more common than duplex formation in homogeneous solution. Two basic formats are most frequently used; these are summarized in Figure 3.18. In a dot blot, a sample of DNA from a cell, a viral plaque, or a sorted chromosome is immobilized (nominally) on the surface of a filter. It is anchored there by heating or UV crosslinking. The exact mechanism of this immobilization is unknown, but it is important, especially for procedures in which the same filter is going to be subjected to many serial hybridizations. Generally, the DNA sample is denatured by alkali or heating to temperatures near 100°C just before immobilization so that what is stuck to the filter is largely single-stranded material. A number of different types of filters are in common use; the most frequent are nitrocellulose and nylon.

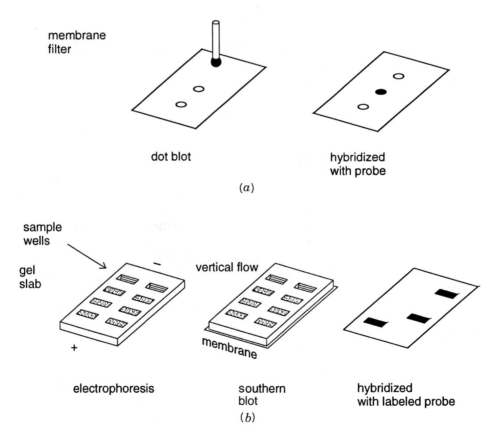

**Figure 3.18**  Hybridization on filters. *(a)* A dot blot, in which DNA is spotted directly on a filter. *(b)* A Southern blot in which a filter replica is made of DNA after a gel-electrophoretic separation.

In the Southern blot format, a mixture of DNA species is size-separated by electrophoresis on polyacrylamide or agarose gels (see Chapter 5). A replica of the gel-separated material is made by transfer of the DNA through the plane of the separation gel to a parallel filter, as shown in Figure 3.18. In this way the targets of the subsequent hybridization reaction can be characterized by their electrophoretic properties. Several methods exist for transfer of the gel sample to the filter. These include bulk liquid flow through the capillary network of a stacked set of paper sheets, vacuum transfer of the fluid phase of the gel, pressure transfer of the fluid phase of the gel, or electrophoretic transfer of the DNA by using electrodes perpendicular to the plane of the original gel separation. All of these methods work.

Several procedures exist where the transfer step can be avoided. One is prehybridization of labeled probes to samples before electrophoresis. Another is drying the separation gel and using it directly as a target for hybridization with radiolabeled probes. These procedures have never achieved the popularity of the Southern blot, mainly because they are less amenable to successful serial probings.

There are two basic advantages of filter hybridization compared with homogeneous phase hybridization. The target samples are highly confined. Therefore many different samples can be reacted with the same probe simultaneously. The second advantage is that the original target molecules are confined in separate microscopic domains on the filter surface. Thus they cannot react with each other. This largely avoids the formation of entangled arrays, and it simplifies some of the problems caused by interspersed repeated DNA sequences. Hybridization at surfaces does, however, produce some complications. The immobilized target DNA is presumably attached to the surface at several places. This introduces a number of obstacles to hybridization. The first of these is the potential for nonspecific attraction or repulsion of the probe DNA to the surface. More serious are the topological boundaries imposed by the points of surface-DNA attachment. These greatly reduce the effective length of the target DNA. As measured by the relative kinetics of solution and immobilized targets, the latter appear to have an effective length of only ten bases. The topological barriers will also affect the ability of DNA's with nonlinear topologies to hybridize (Fig. 3-19).

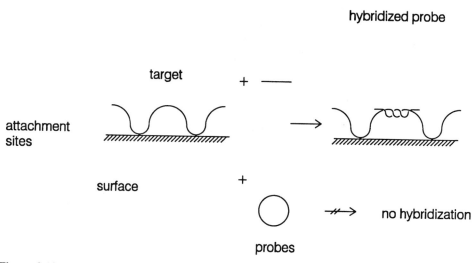

**Figure 3.19** Topological obstacles to filter hybridization as a result of frequent DNA attachment points.

Formation of double strands requires twisting of one strand around the other (see Box 2.1). Thus a circular probe (e.g., circular single-stranded DNA from a bacteriophage M13 clone) will be incapable of forming extensive double helix with an immobilized DNA target. An ingenious use of this topological constraint to improve the detection of specific hybridization signals is the padlock probe approach described in Box 3.4.

---

**BOX 3.4**
**PADLOCK PROBES**

Ulf Landegren and his collaborators recently demonstrated that DNA topology can be used to improve the stringency of detection of correctly hybridized DNA probes. A schematic view of the procedure that they have developed is shown in Figure 3.20. Consider the hybridization of a linear single-stranded probe to a single-stranded circular target. The probe is specially designed so that it's 3'- and 5'-terminal sequences are complementary to a continuous block of DNA sequence in the target. Probes like this can always be made synthetically (or they can be made by circularization with an arbitrary flanking sequence by ligation and then cleavage within the probe sequence in a

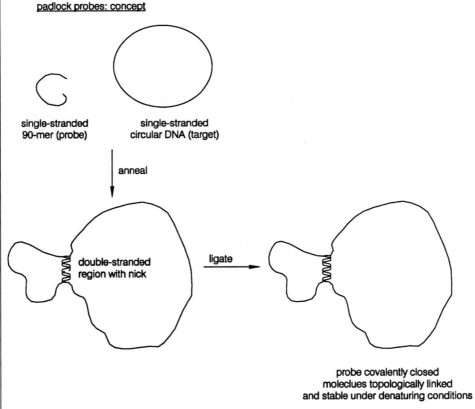

**Figure 3.20**    Example of the use of padlock probes. Here ligation is used to create a topologically linked probe-target complex in solution. The same approach can be used to link probes onto DNA targets immobilized on filters. (Drawing provided by Fouad Siddiqi.)

*(continued)*

**BOX 3.4** *(Continued)*

way similar to the production of jumping probes described in Chapter 8). When the probe is hybridized to the target, a substrate for DNA ligase is created. After ligation the probe and target form a continuous stretch of double-stranded DNA. If this duplex is greater than 10 bp in length, the probe and target strands are topologically linked. Thus, as described in Box 2.1, although strong denaturants can melt the double-stranded region, the two resulting single strands cannot physically separate.

When padlock probes are used in a filter hybridization format, a single-stranded circular target is not required. Because the target is fixed on the filter in multiple places, as shown in Figure 3.19, each interior segment is topologically equivalent to circular DNA. Thus padlock probes that are ligated after hybridization to such segments will become topologically linked with them. This permits very stringent washing conditions to be used to remove any probe DNA that is hybridized but not ligated, or probe DNA that is nonspecifically adsorbed onto the surface of the filter.

One variant on blot hybridization is the reverse dot blot. Here the probe is immobilized on the surface while the labeled target DNA is introduced in homogeneous solution. This really does not change the advantages and disadvantages of the blot approach very much. Another variation on hybridization to blotted DNA samples is in-gel hybridization. Here DNA samples are melted and reannealed in hydrated gels, sometimes in between two distinct electrophoresis fractionations. Such hybridization schemes have been developed to specifically detect DNA fragments that have rearranged in the genome, by Michio Oishi, and to isolate classes of DNA repeats, by Igor Roninsen.

The kinetics of hybridization of DNA on filters or in dried gels (Fig. 3.21) can be described by a variation of the same basic approach used for solution hybridization kinetics. The basic mechanism of the reaction is still bimolecular. At long distances the diffusion of probe DNA in solution toward the target will be more efficient than the diffusion of the immobilized target toward the probe, but these bulk motions do not really matter that much to the rate. What matters is diffusion once the two DNAs are in each others' neighborhood, and here we can think of them as still showing equivalent behavior. Usually, however, the relative concentrations of the probe and sample are very different. Most often the initial probe concentration, $[P]_0$, is much greater that the initial target concentration $[T]_0$, and we say that the probe drives the hybridization. The amount of free probe remains essentially constant during the reaction. Starting from the same basic second-order rate equation, we can write

$$\frac{-d[T]}{dt} = k_2[P][T] = k_2[P]_0[T]$$

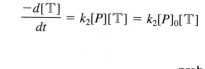

**Figure 3.21** Kinetics of hybridization of a probe to a target DNA immobilized on a filter at initial concentrations of $[P]_0$ and $[T]_0$, respectively.

Because $[P]_0$ is constant, the resulting kinetics appear as pseudo–first order rather than second order. The rate equation can be integrated as

$$\int \frac{-d[T]}{[T]} = k_2[P]_0 \int_0^t dt$$

from the initial concentration of target, $[T]_0$, at time zero, to the concentration $[T]$, at any later time $t$, to give

$$-\ln\left[\frac{[T]}{[T]_0}\right] = k_2[P]_0 t \quad \text{or}$$

$$\frac{[T]}{[T]_0} = \exp\left(-k_2[P]_0 t\right)$$

The halftime of the reaction is defined as the time at which $[T]/[T]_0 = \frac{1}{2}$. This can be written as

$$t_{1/2} = \frac{\ln 2}{k_2[P]_0} = \frac{\ln 2}{k_2 m C_0 / 2N}$$

where $N$ is the genome size of the probe, $C_0$ is the probe concentration in nucleotides, and $m$ reflects any repeated sequences or heterogeneity. Note that if the probe is single stranded, the factor of two should be removed from the denominator. This allows us to write

$$C_0 t_{1/2} = 2 \ln 2 \left(\frac{N}{mk_2}\right)$$

The equation is almost the same as the definition of $C_0 t_{1/2}$ for ordinary homogeneous solution hybridization.

Occasionally the probe contains a trace radiolabel. The target is in great excess, and it drives the hybridization reaction. The above equations are exactly the same except that the role of target and probe are reversed, and $mC_0/2N$ refers to *target* DNA.


## SENSITIVE DETECTION

Hybridization frequently involves the detection of very small amounts of target DNA. Thus it is essential that the probe be equipped with a very sensitive detection system. The most conventional approach is to use a radiolabeled probe. [32]P is the most common choice because its short half-life and high energy create a very intense radiodecay signal. The key to the use of [32]P is to incorporate the label at very high specific activity at many places in the probe. High specific activity means that many of the phosphorous atoms in the probe are actually [32]P and not [31]P. The most common way to do this is random priming of DNA polymerase by short oligonucleotides to incorporate [alpha [32]P]-labeled deoxynucleoside triphosphates ([[32]P]dpppN's) at many places in the chain. There is a trade-off between priming at too many places, which gives probes too short to hybridize well, and priming too infrequently, which runs the risk of not sampling the probe uniformly. In principle, the highest specific activity probes would be made by using all four $[^{32}P]$dpppN's. In practice, however, only one or two $[^{32}P]$dpppN's are usually employed. Attempts to make even hotter probes are usually plagued by technical difficulties.

The short half-life of $^{32}P$ is responsible for its intense signal, but it also leads the major disadvantage of this detection system: The probes, once made, are rapidly decaying. A second disadvantage of very high energy radioisotopes like $^{32}P$ is that the decay produces a track in a photographic film or other spatially resolved detectors that can be significantly longer than the spatial resolution of the target sample. This blurs the pattern seen in techniques like autoradiographic detection of high-resolution DNA sequencing gels or in situ hybridization analysis (Chapter 7) of gene location on chromosomes. To circumvent this problem, lower-energy radioisotopes like $^{35}S$-labeled thio-derivatives of dpppN's can be incorporated into DNA. However, to avoid the disadvantages of radioisotopes altogether, many alternative detection systems have been explored. A few will be discussed here. The principal concern with each is the ratio of the target signal to the background signal. The advantage of radioactive detection is that there is no intrinsic background from the sample (just a counting background due to cosmic rays and the like). This advantage disappears when chemical or spectroscopic detection methods are used.

After $^{32}P$, fluorescence is currently the most popular method for DNA detection, with chemiluminescent methods (Box 3.5) rapidly developing as strong alternatives. The ultimate sensitivity of fluorescent detection allows one to work with single fluorescent molecules. To do this, the dye label is pumped continually with an intense exciting beam of light, until it is finally bleached in some photochemical reaction. Clearly the more times the dye can absorb before it is destroyed, and the higher a fraction of dye decays that emit a photon, the greater is the sensitivity. Both of these factors indicate that the ideal dye will have a fluorescence quantum yield (probability of de-excitation by light emission) as close to 1.0 as possible and a photodestruction quantum yield as low as possible.

---

**BOX 3.5**
**CHEMILUMINESCENT DETECTION**

In chemiluminescence, a chemical reaction is used to generate a product that is in an electronically excited state. The product subsequently relaxes to the ground state by emitting a photon. The photon is detected, and the accumulation of such events can be used to make an image of the target. The advantage of this approach is that since there is no exciting light, there is no background of stray light from the exciting beam, nor is there any background of induced fluorescence from the filter or other materials used in the hybridization and blotting. An example of some of the chemical steps used in current chemiluminescent detection of DNA is shown in Figure 3.22. These methods have sensitivity comparable to those with the highest specific activity radioisotopes used, and they are more rapid. However, they often do not allow convenient serial reprobings of the filter; this is a distinct disadvantage, and the possibility for multicolor detection, well developed with fluorescence, is not yet common with chemiluminescence.

*(continued)*

**BOX 3.5** *(Continued)*

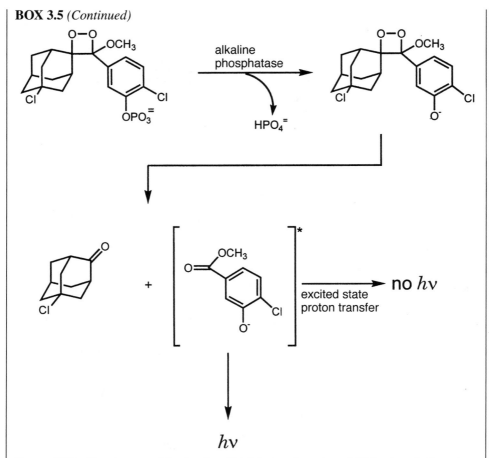

**Figure 3.22**  Chemistry used in the chemiluminescent detection of DNA. An alkaline phosphatase-conjugated probe is used for hybridization. Then it is presented with the substrate CDP-star$^{tm}$. Upon dephosphorylation of the substrate by alkaline phosphatase, a metastable phenolate anion intermediate is formed that decomposes and emits light at a maximum wavelength of 466 nm.

While such single molecule detection is achievable, it is intrinsically noisy, as is any single molecule method. For this reason it is helpful to supplement fluorescent detection schemes by intermediate chemical amplification steps. Thus a single target molecule produces or is coupled to a large number of intermediates, and each of these in turn becomes the target for a fluorescence detection system. Two general schemes for chemical amplification are shown in Figure 3.23. Catalytic amplification uses an enzyme coupled to the probe. After hybridization the enzyme is allowed to catalyze a reaction that releases a fluorescent or chemiluminescent product, or an intensely colored product. This must be done under circumstances where the products do not diffuse too far away from the site of their creation; otherwise, the spatial resolution of the blot will be degraded. In stoichiometric amplification, potential diffusion problems are avoided by building up a physical complex of molecules attached to the original probe. One way to accomplish this is by successive rounds of complementary polyvalent regents, sites, or ligands. This eventually generates a large network that can be coupled to many detector molecules.

Catalytic

target-probe-enzyme

invisible substrate

visible product

Stoichiometric

target-probe

D
D
D
D
D

multivalent
site

**Figure 3.23**  Two types of chemical amplification systems. Catalytic amplification; stoichiometric amplification using two complementary multivalent reagents. D is a molecule that allows detection, such as a fluorescent dye.

There are a number of different ligands and polyvalent reagents that have been widely used for nucleic acid analysis. Several of the reagents are illustrated in Figure 3.24. Potentially the most powerful, in several respects, is the combination of biotin and streptavidin or avidin. Biotin can be attached to the 5'- or 3'-ends of nucleic acids, as discussed earlier, or it can be incorporated internally by various biotinylated dpppN analogues. Streptavidin and avidin bind biotin with a binding constant $K_a$ approaching $10^{15}$ under physiological conditions. The three-dimensional structure of streptavidin is known from two X-ray crystallographic studies (Figure 3.25). Since avidin and streptavidin are tetravalent, there is a natural amplification route. After attaching single streptavidins to biotinylated probes, the three remaining biotin-binding sites on each protein molecule can be coupled to a polyvalent biotinylated molecule such as a protein with several biotins. Then more streptavidin can be added, in excess; the network is grown by repeats of such cycles (Fig. 3.26a). In practice, for most nucleic acid applications, streptavidin, the protein product from the bacterium *Streptomyces avidinii,* is preferred over egg white avidin because it has an isoelectric point near pH 7, which reduces nonspecific binding to DNA, and because it has no carbohydrate, which reduces nonspecific binding to a variety of components and materials in typical assay configurations.

Most other stoichiometric amplification systems use monoclonal antibodies with high affinity for haptens that can easily be coupled to DNA. Digoxigenin and fluorescein are the most frequent choices. After the antibody has been coupled to the hapten-labeled DNA, amplification can be achieved by sandwich techniques in which a haptenated second antibody, specific for the first antibody, is allowed to bind, and then additional cycles of monoclonal antibody and haptenated second antibody are used to build up a network. An example in the case of fluorescein is shown in Figure 3.26b. Other haptens that have proved useful for such amplification systems are coumarin, rhodamine, and dinitrophenyl. In all of these cases a critical concern is the stability of the network. While anti-

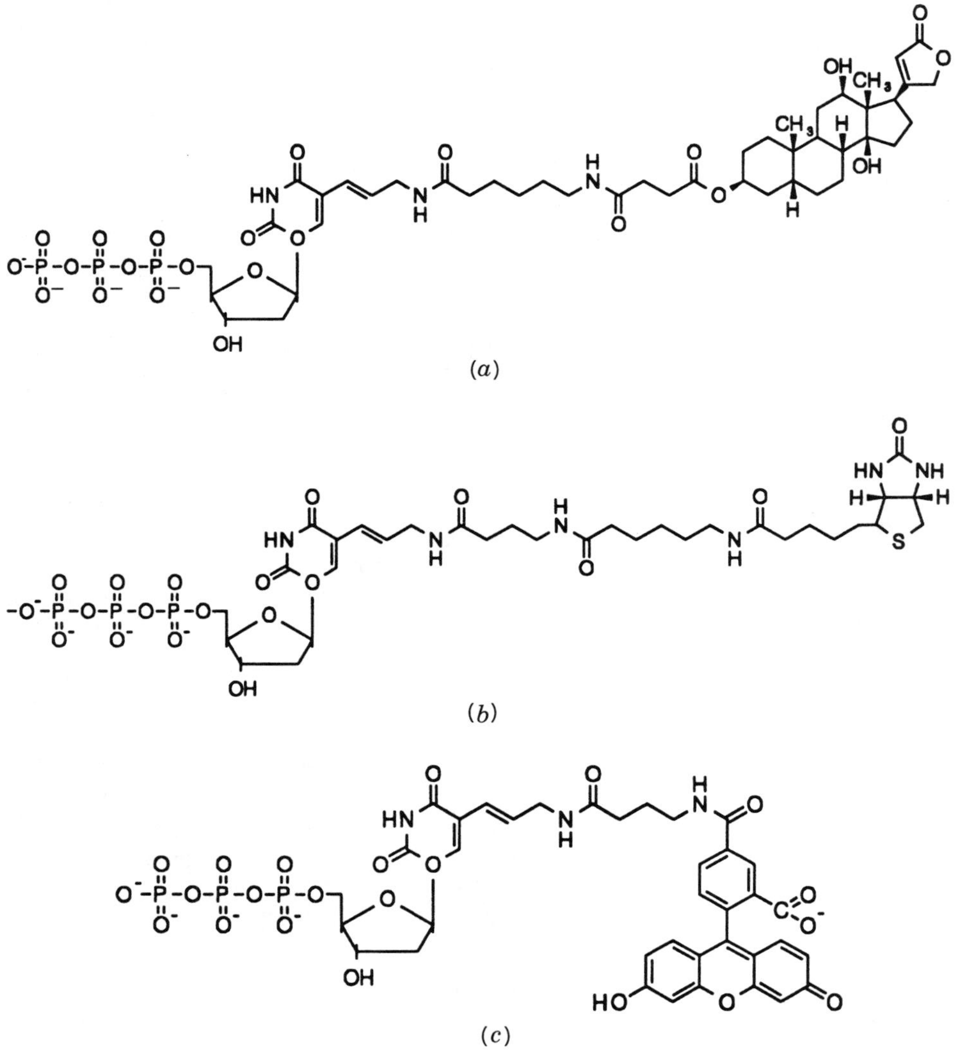

**Figure 3.24** Three ligands used in stoichiometric amplification systems. *(a)* Digoxigenin. *(b)* Biotin. *(c)* Fluorescein. All are shown as derivatives of dpppU, but other derivatives are available.

bodies and their complexes with haptens are reasonably stable, streptavidin-biotin complexes are much more stable and can survive extremes of temperature and pH in ways comparable to DNA. Parenthetically, a disadvantage of the streptavidin system is that the protein and its complexes are so stable that it is very difficult to reverse them to generate free DNA again, if this is needed. Even greater degrees of signal amplification can be achieved by using dendrimers as described in Box 3.6.

All of these amplification systems work well, but they do not have the same power of sample multiplication that can be achieved when the amplification is carried out directly at the DNA level by enzymatic reactions. Such methods are the subject of the next chapter. Ultimately sample amplification systems can be combined with color-generating amplification systems to produce exquisitely sensitive ways of detecting multiple DNA samples, sometimes in multiple colors.

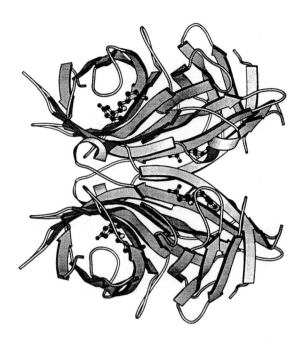

**Figure 3.25** The three-dimensional structure of streptavidin. Four bound biotin molecules are shown in boldface. (Illustration created by Sandor Vajda using protein coordinates provided by Wayne Hendrickson.)

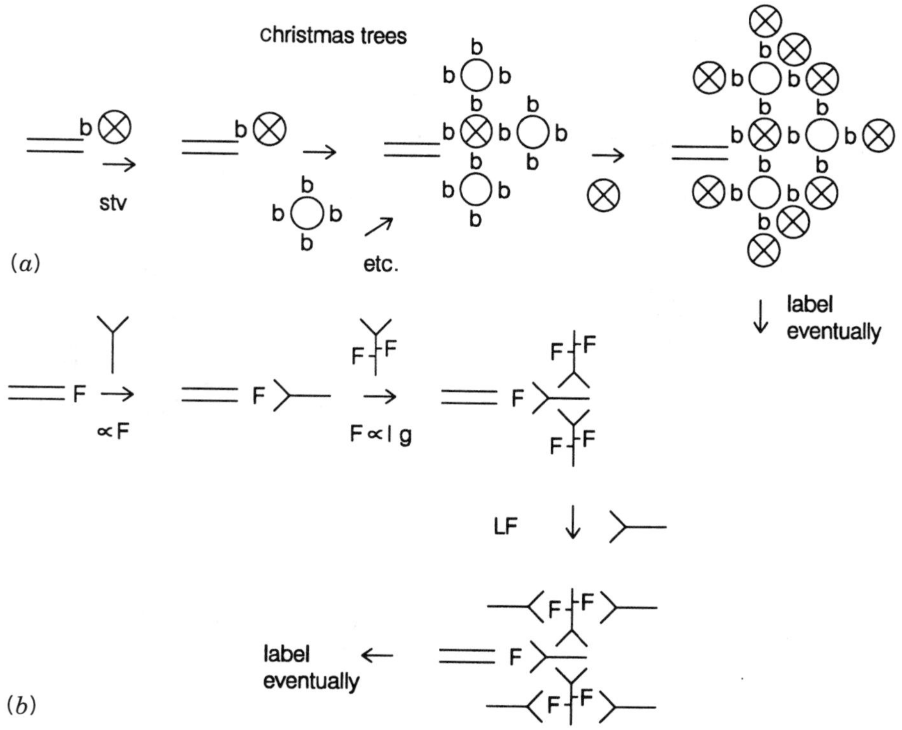

**Figure 3.26** Detailed structural intermediates formed by two methods for stoichiometric amplification. (a) Streptavidin and some other protein containing multiple attached biotin (b) residues. (b) A monoclonal antibody directed against fluorescein (F) and a fluorescinated polyclonal antibody specific for the monoclonal antibody.

**BOX 3.6**
**DENDRIMERIC DNA PROBES**

Dendrimers are a chemical amplification system that allows large structures to be constructed by systematic elaboration of smaller ones. A traditional dendrimer is formed by successive covalent additions of branched reactive species to a starting framework. Each layer added grows the overall mass of the structure considerably. The process is a polyvalent analogue of the stoichiometric amplification schemes described in Figure 3.26.

Recently schemes have been designed and implemented to construct dendrimeric arrays of DNA molecules. Here branched structures are used to create polyvalency, and base-pairing specificity is used to direct the addition of each successive layer. The types of structures used and the complexity of the products that can be formed are illustrated schematically in Figure 3.27. These structures are designed so that each layer presents equal amounts of two types of single-stranded arms for further complexation. Ultimately one type of arm is used to identify a specific target by base pairing, while the other type of arm is used to bind molecules needed for detection. Dendrimers could be built on a target layer by layer, or they can be preformed with specificity selected for each particular target of interest. The latter approach appears to offer a major increase in sensitivity in a range of biological applications including Southern blots, and in situ hybridization.

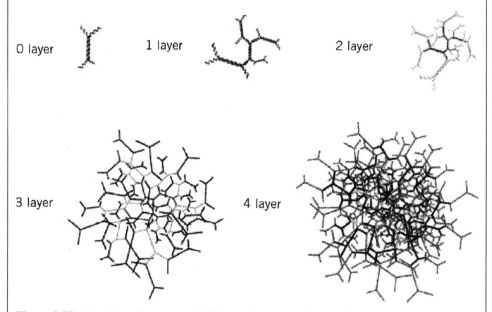

**Figure 3.27**    Dendrimer layer growth. Figure also appears in color insert. (Illustration provided by Thor Nilsson.)

## SOURCES AND ADDITIONAL READINGS

Breslauer, K. J., Franz, R., Blöcker, H., and Marky, L. A. 1986. Predicting DNA duplex stability from the base sequence. *Proceedings of the National Academy of Sciences USA* 83: 3746–3750.

Cantor and Schimmel. 1980. *Biophysical Chemistry* III. San Francisco: W. H. Freeman, pp. 1226–1238.

Nilsson, M., Malmgren, H., Samiotaki, M., Kwiatkowski, M., Chowdhary, B. P., and Landegren, U. 1994. Padlock probes: Circularizing oligonucleotides for localized DNA detection. *Science* 265: 2085–2088.

Yokota, H., and Oishi, M. 1990. Differential cloning of genomic DNA: Cloning of DNA with an altered primary structure by in-gel competitive reassociation. *Proceedings of the National Academy of Sciences USA* 87: 6398–6402.

Roberts, R. W., and Crothers, D. M. 1996. Prediction of the Stability of DNA triplexes. *Proceedings of the National Academy of Sciences USA* 93: 4320–4325.

Roninson, I. B. 1983. Detection and mapping of homologous, repeated and amplified DNA sequences by DNA renaturation in agarose gels. *Nucleic Acids Research* 11: 5413–5431.

SantaLucia Jr., J., Allawi, H. T., and Seneviratne, P. A. 1996. Improved nearest-neighbor parameters for predicting DNA duplex stability. *Biochemistry* 35: 3555–3562.

Schroeder, S., Kim, J., and Turner, D. H. 1996. G–A and U–U mismatches can stabilize RNA internal loops of three nucleotides. *Biochemistry* 35: 16015–16109.

Sugimoto, N., Nakano, S., Yoneyama, M., and Honda, K. 1996. Improved thermodynamic parameters and helix initiation factor to predict stability of DNA duplexes. *Nucleic Acids Research* 24: 4501–4505.

Sugimoto, N., Nakano, S., Katoh, M., Matsumura, A., Nakamuta, H., Ohmichi, T., Yoneyama, M., and Sasaki, M. 1995. Thermodynamic parameters to predict stability of RNA/DNA hybrid duplexes. *Biochemistry* 34: 11211–6.

Wetmur, J. G. 1991. DNA probes: Applications of the principles of nucleic acid hybridization. *Critical Reviews in Biochemistry and Molecular Biology* 26: 227–259.

# 4 Polymerase Chain Reaction and Other Methods for In Vitro DNA Amplification

## WHY AMPLIFY DNA?

The importance of DNA signal amplification for sensitive detection of DNA through hybridization was discussed in the previous chapter. Beyond mere sensitivity, there are two basic reasons why direct amplification of DNA is a vital part of DNA analysis. First, DNA amplification provides a route to an essentially limitless supply of material. When amplification is used this way, as a bulk preparative procedure, the major requirement is that the sample be uniformly amplified so that it is not altered, distorted, or mutated by the process of amplification. Only if these constraints can be met, can we think of amplification as a true immortalization of the DNA.

The second rationale behind DNA amplification is that selective amplification of a region of a genome, chromosome, or sample provides a relatively easy way to purify that segment from the bulk. Indeed, if the amplification is sufficient in magnitude, the DNA product becomes such an overwhelming component of the amplification mixture that the starting material is reduced to a trivial contaminant for most applications.

Amplification can be carried out in vivo by growing living cells (Box 1.2) or in vitro by using enzymes. There are several overwhelming advantages of in vitro amplification. Any possible toxic effects of a DNA target on the host cell are eliminated. There is no need to purify the amplified material away from the host genome or the vector used for cloning. Base analogues can be used that would frequently be unacceptable to a living cell system. Samples can be manipulated by automated methods that are far easier to implement in vitro than in vivo. The major limitation of existing in vitro amplification methods until recently is that they were restricted to relatively short stretches of DNA, typically less than 5 kb. New long polymerase chain reaction (PCR) procedures have extended the range of in vitro amplification up to about 20 kb. For longer targets than this, in vivo cloning methods must still be used.

## BASIC PRINCIPLES OF THE POLYMERASE CHAIN REACTION (PCR)

What makes PCR a tool of immense power and flexibility is the requirement of DNA polymerases for pre-existing DNA primers. Thus DNA polymerases cannot start DNA chains de novo; a primer can be used to determine where, along a DNA template, the synthesis of the complement of that stand begins. It is this primer requirement that allows the selective amplification of any DNA region by using appropriate, specific DNA primers.

Once started, a DNA polymerase like *E. coli* DNA polymerase I (pol I) will proceed in the 3′- to 5′-direction until it has copied all the way to the 5′-end of the template. The simplest amplification scheme for in vivo DNA amplification is successive cycles of priming, chain extension, and product denaturation, shown schematically in Figure 4.1. If these steps are carried out efficiently, the result is a linear increase in the amount of product strand with increasing cycles of amplification. This scheme, called *linear amplification,* is very useful in preparing DNA samples for DNA sequencing. Here chain terminating analogues of dpppN's are added in trace amounts to produce a distribution of products with different chain lengths (see Chapter 10). When linear amplification is used in this context, it is called *cycle sequencing.*

A step up in complexity from linear amplification is the use of two antiparallel primers, as shown in Figure 4.2*a*. These primers define the region to be amplified, since after the first few cycles of DNA synthesis, the relative amount of longer DNA sequences that contain the region spanned by the primer becomes insignificant. The target DNA is denatured, and two antiparallel primers are added. These must be in sufficient excess over target that renaturation of the original duplex is improbable, and essentially all products are primers annealed to single-stranded templates. The first cycle of DNA synthesis copies both of the original template strands. Hence it doubles the number of targets present in the starting reaction mixture. Each successive round of DNA denaturation, primer binding, and chain extension will, in principle, produce a further doubling of the number of target molecules. Hence the amount of amplified product grows exponentially, as $2^n$, where $n$ is the number of amplification cycles. This is the basic design of a typical PCR procedure. Note that only the DNA sequence flanked by the two primers is amplified (Fig. 4.2).

Early PCR protocols employed ordinary DNA polymerases such as the Klenow fragment of *E. coli* DNA polymerase I (a truncated version of the natural protein with its 5′-exonuclease activity removed). The difficulty with this approach is that these enzymes

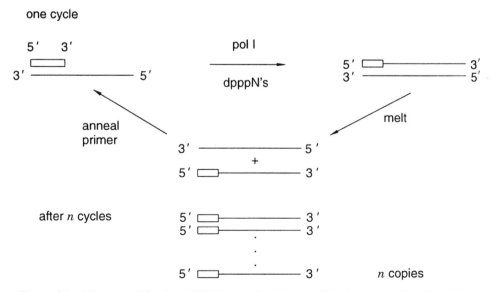

**Figure 4.1**   Linear amplification of DNA by cycles of repeated in vitro synthesis and melting.

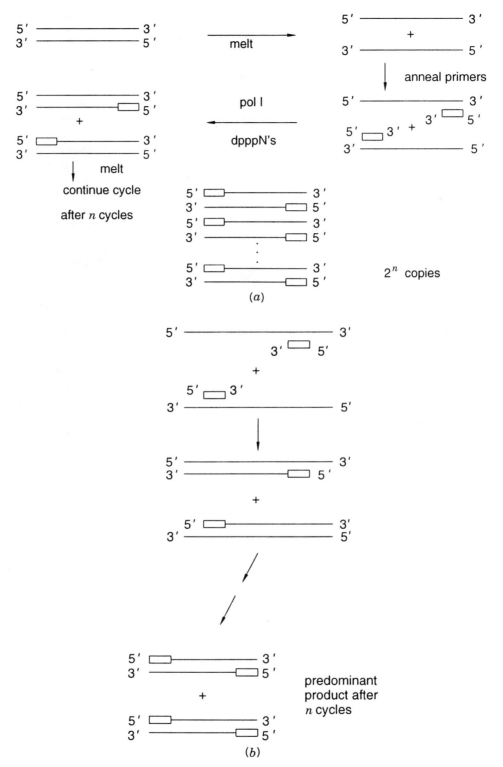

**Figure 4.2** Exponential amplification of DNA in vitro by the use of two antiparallel primers. *(a)* Polymerase chain reaction (PCR) with successive cycles of DNA synthesis and melting. *(B)* Only the DNA flanked by the primers is amplified.

are easily denatured by the high temperatures needed to separate the two strands of duplex DNA. Unlike DNA, however, proteins like DNA polymerases, once denatured, are generally reluctant to renature. Thus, with each cycle of amplification, it was necessary to add fresh polymerase for efficient performance. This problem was relieved when DNA polymerases from thermophilic organisms became available. The most widely used of these enzymes in PCR is from *Thermis acquaticus,* the *Taq* polymerase. This enzyme has an optimal temperature for polymerization in the range of 70 to 75°C. It extends DNA chains at a rate of about 2 kb per minute. Most important, it is fairly resistant to the continual cycles of heating and cooling required for PCR. The half-life for thermal denaturation of *Taq* polymerase is 1.6 h at 95°C, but it is less than 6 minutes at 100°C. When very high denaturation temperatures are needed, as in the PCR amplification of very G + C rich DNA, it is sometimes useful to employ even more thermal-stable polymerases such as enzymes isolated from organisms that grow in the superheated water (temperature above 100°C) near geothermal vents. Such enzymes are called *vent polymerases.* Examples are the enzyme from *Thermococcus litoralis* with a half-life of 1.8 h at 100°C and the enzyme from *Pyrococcus furiosis* with a half-life of 8 h at 100°C. However, these enzymes are not as processive as *Taq* polymerase. This means that they tend to fall off their template more readily, which makes it more difficult to amplify long templates.

   With *Taq* polymerase typical PCR cycle parameters are

| | |
|---|---|
| DNA denaturation | 92–96°C for 30 to 60 seconds |
| Primer annealing | 55–60°C for 30 seconds |
| Chain extension | 72°C for 60 seconds |

The number of cycles used depends on the particular situation and sample concentrations, but typical values when PCR is efficient are 25 to 30 cycles. Thus the whole process takes about an hour. Typical sample and reagent concentrations used are

| | |
|---|---|
| Target | $< 10^{-15}$ mol, or down to as little as 1 mol |
| Primer | $2 \times 10^{-11}$ mol |
| dpppN's | $2 \times 10^{-8}$ mol of each |

These concentrations imply that the reaction will eventually saturate once the primer and dpppN's are depleted.

   There are a few peculiarities of the *Taq* polymerase that must be taken into consideration in designing PCR procedures or experimental protocols based on PCR. *Taq* polymerase has no 3′-proofreading exonuclease activity. Thus it can, and does, misincorporate bases. We will say more about this later. *Taq* polymerase does have a 5′-exonuclease activity. Thus it will nick translate, which is sometimes undesirable. However, mutants exist that remove this activity, and they can be used if necessary. The most serious problem generally encountered is that *Taq* polymerase can add an extra nontemplate-coded A to the 3′-end of DNA chains, as shown in Figure 4.3*a*. It can also, perhaps using a related activity, make primer dimers, which may or may not contain additional uncoded residues as shown in Figure 4.3*b*. This is a serious problem because once such dimeric primers are created, they are efficient substrates for further amplification. These primer dimers are a major source of artifacts in PCR. However, they are usually short and can be removed by gel filtration or other sizing methods to prevent their interference with subsequent uses of the PCR reaction product.

**Figure 4.3** Artifacts introduced by the use of *Taq* DNA polymerase. *(a)* Nontemplated terminal adenylation; *(b)* primer dimer formation.

(a)                                  (b)

The efficiency of typical PCR reactions can be impressive. Let us consider a typical case where one chooses to amplify a short, specific region of human DNA. This sort of procedure would be useful, for example, in examining a particular region of the genome for differences among individuals, without having to clone the region from each person to be tested. To define this region, one needs to know the DNA sequence that flanks it, but one does not need know anything about the DNA sequence between primers. For unique, efficient priming, convergent primers about 20 bp long are used on each side of the region. A typical amplification would start with $1\mu g$ of total human genomic DNA. Using Avogadro's number, $6 \times 10^{23}$, the genome size of $3 \times 10^9$ bp, and the 660 Da molecular weight of a single base pair, we can calculate that this sample contains

$$\frac{10^{-6} \times 6 \times 10^{23}}{3 \times 10^9 \times 6.6 \times 10^2} = 3 \times 10^5 \text{ copies}$$

of the genome or $3 \times 10^5$ molecules of any single-copy genomic DNA fragments containing our specific target sequence. If PCR were 100% efficient, and 25 cycles of amplification were carried out, we would expect to multiply the number of target molecules by $2^{25} = 6.4 \times 10^7$. Thus, after the amplification, the number of targets should be

$$6.4 \times 10^7 \times 3 \times 10^5 = 2 \times 10^{13} \text{ molecules}$$

If the product of the DNA amplification is 200 bp, it will weigh

$$200 \text{ bp} \times \frac{(6.6 \times 10^2 \text{ Da/bp}) \times (2 \times 10^{13})}{6 \times 10^{23} \text{ g/Da}} = 4\mu g$$

If such efficient amplification could be achieved, the results would be truly impressive. This means that one would be sampling just a tiny fraction of the genome, and in a scant hour of amplification, the yield of this fraction would be such as to amount to 80% of the total DNA in the sample. Thus one would be able to purify any DNA sample.

In practice, the actual PCR efficiencies typically achieved are not perfect, but they are remarkably good. We can define the efficiency, $E$, for $n$ cycles of amplification by the ratio of product, $P$, to starting material, $S$, as

$$\frac{P}{S} = (1 + E)^n$$

Actual efficiencies turn out to be in the range of 0.6 to 0.9; this is impressive. Such high efficiencies immediately raise the notion of using PCR amplification as a quantitative tool to measure not just the presence of a particular DNA sequence in a complex sample but to determine its amount. In practice, this can be done, but it is not always reliable; it usually requires coamplification with a standard sample, or competition with known amounts of a related sample.

# NOISE IN PCR: CONTAMINATION

As in any high-gain amplification system, any fluctuations in conditions are rapidly magnified, especially when they occur early in the reaction. There are many sources of noise in PCR. One extremely common source is variations in the temperature at different positions in typical thermal cycling blocks. A typical apparatus for PCR uses arrays of tube holders (or microtitre plate holders) that can be temperature controlled by heating elements and a cooling bath, by thermoelectric heating and cooling, by forced convection, or by switching among several pre-equilibrated water baths. A number of different fundamental designs for such thermal cyclers are now available, and the serious practitioner would be well advised to look carefully at the actual temperature characteristics of the particular apparatus used. For extremely finicky samples, such as those that need very G + C rich primers, it may be necessary to use the same sample well each time to provide reproducible PCR. It is for reasons like this that PCR, while it has revolutionized the handling of DNA in research laboratories, has not yet found broad acceptance in clinical diagnostic laboratories despite its potential power.

A major source of PCR noise appear to lie with characteristics of the samples themselves. The worst problem is undoubtedly sample contamination. If the same sample is used repetitively in PCR assays, the most likely source of contamination is the PCR reaction product from a previous assay. With PCR we are dealing with a system that amplifies a DNA sequence by $10^7$ fold. Thus, even if there is one part per million in carry over contamination; it will completely dominate the next round of PCR. A second major source of contamination is DNA from organisms in dust or from DNA shed by the experimenter, in the form of dander, hair follicles, sweat, or saliva. This problem is obviously of greatest significance when the samples to be amplified contain human sequences.

The basic cure for most contamination is to carry out PCR under typical biological containment procedures and use good sterile technique including plugged pipetmen tips, to prevent contamination by aerosols, and laminar flow hoods, to minimize the ability of the investigator to inadvertently contaminate the sample. However, these approaches are not always sufficient to deal with the contamination caused by previous PCR experiments on similar samples. This situation is extremely frustrating because it is not uncommon for neophytes to have success with procedures that then progressively deteriorate as they gain more experience, since the overall level and dispersal of contaminant in the laboratory keeps rising. There are general solutions to this problem, but they have not yet seen widespread adoption. One helpful procedure is to UV irradiate all of the components of a PCR reaction before adding the target. This will kill double-stranded DNA contaminants in the polymerase and other reagents. A more general solution is to exploit the properties of the enzyme, Uracil DNA glycosylase (DUG), which we described in Chapter 1. This enzyme degrades DNA at each incorporated dU. Normally these incorporations are rare mutagenic events, and the lesion introduced into the DNA is rapidly repaired.

DUG is used in PCR by carrying out the initial PCR amplification using dpppU instead of dpppT. The amplification product can be characterized in the normal way; the properties of DNA with T fully substituted by dU are not that abnormal except for a lower melting temperature and inability to be recognized by many restriction nucleases. Next, when a subsequent PCR reaction needs to be done, the sample, including the target, is treated with DUG prior to thermal cycling. This will destroy any carryover from the previous PCR because there will be so many dU's removed that the resulting DNA will be incapable of replication. If the second PCR is also performed with dpppU, its subsequent carryover can also be prevented by a DUG treatment, and this procedure can be repeated *ad libertum*.

## PCR NOISE: MISPRIMING

Typical PCR conditions with two convergent primers offer a number of possible unintended primed amplifications. These are illustrated in Figure 4.4. If the primers are not chosen wisely, one of the two primers may be able to act alone to amplify DNA as shown in Figure 4.4*b*. Alternatively, the two convergent primers may have more than one site in the target that allows amplification. There is no way to plan for these events, unless the entire sequence of the sample is known. However, the chances of such accidental, unintended, but perfect priming can be minimized by using long enough primers so that the probability of such a coincidental match in DNA sequence is very small.

A more serious and more common occurrence is mispriming by inexact pairing of the primer with the template. Note that if the 3′-end of the primer is mispaired with the template, this is unlikely to lead to amplification, and thus there will be little harm. However, if the 5′-end of the primer is mispaired, the impact is much more serious, as shown in Figure 4.4*c* and *d*. If primer annealing is carried out under insufficiently stringent conditions, once elongation is allowed to start, a 5′ mispaired primer may still be able to lead to DNA synthesis. In the next round of PCR the incorrect elongated product from this synthesis will serve as a template if it contains a sequence complementary to any of the primers in the solution. However, when this synthesis extends past the original mispaired primer, the sequence that is made is now the precise complement of that primer. From this round on, no stringency conditions will discriminate between the desired product and the misprinted artifact. Thus more than one product will amplify efficiently in subsequent steps. The key is to prevent the mispriming in the first place. If the misprimed sequence is nearly identical to the primed sequence, the most obvious way to solve this problem is to change primers.

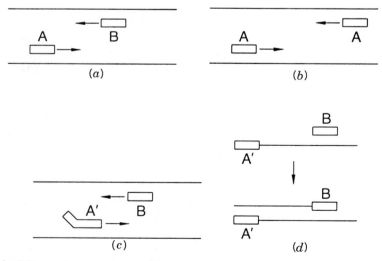

**Figure 4.4** Effects of mispriming on PCR reaction products. *(a)* Desired product. *(b)* Product formed by an inverted repeat of one primer. *(c)* Product formed by nonstringent annealing. *(d)* After the second round of DNA synthesis, the product in *(c)* is now a perfect match to the primer for subsequent rounds of amplification.

Hopefully, another sequence near by on the desired template will not have its near mate somewhere else in the sample. However, there are more general cures for some mispriming as shown below.

In typical PCR reactions the denatured target and all of the other components are mixed at room temperature, and then the cycles of amplification are allowed to start. This has the risk that polymerase extensions of the primers may start before the reaction is heated to the optimal temperature for the elongation step. If this occurs, it enhances the risk of mispriming, since room temperature is far from a stringent enough annealing temperature for most primers. This problem can easily be avoided by what is called *hot start PCR.* Here the temperature is kept above the annealing temperature until all the components, including the denatured target, have been added. This avoids most of the mispriming during the first step; and it is always the first step in PCR that is most critical for the subsequent production of undesired species.

A very powerful approach to favoring the amplification of desired products and eliminating the amplification of undesired products is nested PCR. This can be done whenever a sufficient length of known sequence is available at each end of the desired target. The process is shown schematically in Figure 4.5. Two sets of primers are used; one is internal to the other. Amplification is allowed to proceed for half the desired rounds using the external primers. Then the primers are switched to the internal primers, and amplification is allowed to continue for the remaining rounds. The only products that will be present at high concentration at the end of the reaction are those that can be amplified by both sets of primers. Any sequence that can inadvertently be amplified by one set is most unlikely to be a target for the second set, since, in general, there is no relationship or overlap of the sequences used as the two sets of primers. With nested priming it is possible to carry out many more than 30 rounds of amplification with relatively little background noise. Hence this procedure is to be especially recommended when very small samples are used and large numbers of amplifications are needed to produce desired amounts of product.

Instead of full nesting, sometimes it is desirable or necessary to use a dual set of primers on one side of the target but only a single set on the other. This is called *hemi-nesting,* and it is still much safer and generally yields much cleaner products than no nesting at all. Particularly elegant versions of hemi-nesting have been demonstrated where the two nested primers can both be introduced at the start of the PCR at a temperature at which only the external primer functions well. Halfway through the amplification cycles, the annealing temperature is shifted so that now the internal primer becomes by far the favored one.

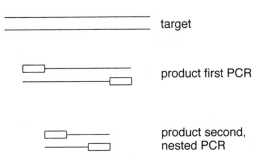

**Figure 4.5**    The use of nested primers to increase the specificity of PCR amplification.

## MISINCORPORATION

We mentioned earlier that *Taq* polymerase has no 3' editing exonuclease. Because of this it has a relatively high rate of misincorporation when compared with many other DNA polymerases. Thus the products of PCR reactions accumulate errors. The misincorporation rate of *Taq* polymerase has been estimated as $1.7 \times 10^{-4}$ to $5 \times 10^{-6}$ per nucleotide per cycle. The error rate depends quite a bit on the reaction conditions, especially on the concentration of dpppN's used, and on the sequence of the target. The impact of these mispairing rates on the product are straightforward to calculate. At any site in the target, the fraction of correct bases after n cycles of amplification will be

$$X_{corr} = (1 - X_{mis})^n \sim 1 - nX_{mis}$$

where $X_{mis}$ is the fraction of misincorporation rate at that site for a single cycle. For 30 cycles the fraction of misincorporated bases at any site will be $30X_{mis} = 5.1 \times 10^{-3}$ to $1.5 \times 10^{-4}$, using the numbers described above. Thus at any site the correct sequence is still overwhelmingly predominant.

However, if one asks, instead, for the amplification of a DNA of length, $L$, how many incorrect bases will each DNA product molecule have after $n$ steps, the number is $LnX_{mis}$. With $L = 1000$, this means that products will have from 0.15 to 5 incorrect bases. How serious a problem is this? It depends on the use to which the DNA will be put. As a hybridization probe, these errors are likely to be invisible. If one sequences the DNA directly, the errors will still be invisible (except for the rare case where a misincorporation occurred in the first round or two of the amplification and then was perpetuated). This is because the errors are widely distributed at different sites on different molecules, and sequencing sees only the average occupant of each site. However, if the PCR products are cloned, the impact of misincorporation is much more serious. Now, since each clone is the immortalization of a single DNA molecule, it will contain whatever particular errors that molecule had. In general, it is a hazardous idea to clone PCR products and then sequence them. The sequences will almost always have errors. Similarly PCR starting from single DNA molecules is fine for most analyses, but one cannot recommend it for sequencing because, once again, the products are likely to show a significant level of misincorporation errors.

## LONG PCR

A number of factors may limit the ability of conventional PCR to amplify long DNA targets. These include depurination of DNAs at the high temperatures used for denaturation, inhibition of the DNA polymerase by stable intramolecular secondary structure in nominally single-stranded templates, insufficient time for strand untwisting during conventionally used denaturation protocols, as described in Chapter 3, and short templates. The first problem can be reduced by using increased pH's to suppress purine protonation, a precursor to depurination. The second problem can be helped somewhat by adding denaturants like dimethyl sulfoxide (DMSO). The third problem can be alleviated by using longer denaturation times. The last problem can be solved by preparing DNA in agarose (Chapter 5). However, the most serious obstacle to the successful PCR amplification of long DNA targets rests in the properties of the most commonly used DNA polymerase, the *Taq* polymerase.

*Taq* polymerase has a significant rate of misincorporation as described above. However, it lacks a 3′ proofreading exonuclease activity. Once a base is misincorporated at the 3′-end, the chances that the extension will terminate at this point become markedly enhanced. This premature chain termination ultimately leads to totally ineffective PCR amplification above DNA sizes of 5 to 10 kb. To circumvent the problem of premature chain termination, Wayne Barnes and coworkers have added trace amounts of a second thermally stable DNA polymerase like *Pfu,* Vent, or Deep Vent that possesses a 3′-exonuclease activity. This repairs any terminal mismatches left by *Taq* polymerase, and then the latter can continue chain elongation. With such a two-enzyme procedure, successful PCR amplification of DNA targets in the 20 kb to 40 kb range are now becoming common.

## INCORPORATING EXTRA FUNCTIONALITIES

PCR offers a simple and convenient way to enhance or embroider the properties of DNA molecules. As shown in Figure 4.6, primers can be pre-labeled with radioisotopes, biotin, or fluorescent dyes. Thus the ends of the amplified targets can be derivatized as an intrinsic part of the PCR reaction. This is extremely convenient for many applications. Since two primers are chosen, two different labels or tags can be used. One frequent and every effective strategy is to put a capture tag on one primer—like a biotin—and a detection tag on the other—like a fluorophore. After the amplification, the product is captured and analyzed. Only double-stranded material that is the result of amplification that incorporated both of the primers should be visible.

PCR can also be used to modify the DNA sequence at the ends of the target. For example, as shown in Figure 4.6*b*, the primer can overhang the ends of the desired target. As successive amplification cycles are carried out, the DNA duplexes that accumulate will contain the target sequence flanked by the additional segments of primer sequence. This has a number of useful applications. It allows any restriction sites needed to be built into the primer. Then, as shown in the figure, after the PCR reaction the product can be cleaved at these sites for subsequent ligation or cloning steps.

Another use for overhanging primers arises in circumstances where the original amount of known sequence is too short or too imperfect to allow efficient amplification. This problem arises, for example, when a primer is made to an imperfectly repeating sequence. The usual desire in such experiments is to amplify many different copies of the repeat (e.g., to visualize human DNAs among a background of rodent DNA in a hybrid cell as illustrated in Chapter 14), but few of the repeats match the primer well enough to really give good amplification. By having an overhanging primer, after the first few rounds of amplification, the complements to the primer sequence now contain the extra overhang (Fig. 4.6*c*). The resulting template-primer complexes are much more stable and amplify much more effectively.

## SINGLE-SIDED PCR

A major limitation in conventional PCR is that known DNA sequence is needed on both sides of the desired target. It is frequently the case, as shown in Figure 4.7*a*, that a known sequence is available only at one place within the desired target, or at one end of it.

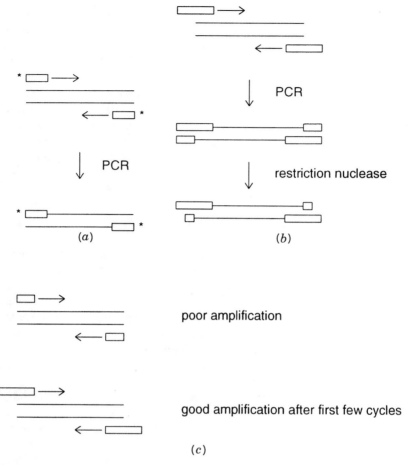

**Figure 4.6** Introducing extra functionalities by appropriately designed primers. *(a)* Incorporation of a 5′-terminal label (asterisk). *(b)* Incorporation of flanking restriction sites, useful for subsequent cloning. *(c)* Compensating for less than optimal length initial sequence information.

This problem typically arises when one is trying to walk by from a known region of the genome into flanking unknown regions. Then one starts with a bit of known sequence at the extreme edge of the charted region, and now the goal is to make a stab into the unknown. There is still not a consensus on the best way to do this, but there have been a few successes, and many failures.

Consider the simple scheme shown in Figure 4.7*b*. The known sequence is somewhere within a restriction fragment of a length suitable for PCR. One can ligate arbitrary bits of known DNA sequence, called *splints* or *adapters,* onto the ends of the DNA fragment of interest. Now, in principle, the use of the one known primer and one of the two ligated splint primers will allow selective amplification of one side of the target. This amplification process will work as expected. However, the complication arises not from the target but from all the other molecules in the sample. They too are substrates for amplification using two copies of the splint primer (Fig. 4.7*c*).

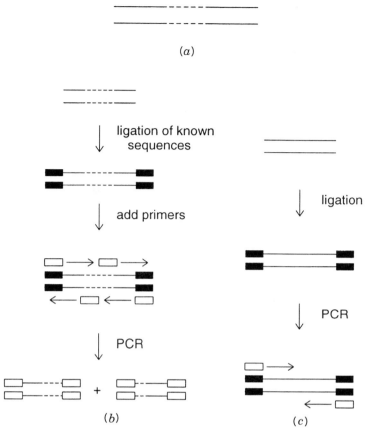

**Figure 4.7**  Single-sided PCR. *(a)* General situation that requires this method; dashed line represents the only known sequence in the target. *(b)* A potentially simple scheme. *(c)* Unwanted amplification products that defeat the scheme in *(b)*.

Thus the desired product will be overwhelmed with undesirable side products. One must either separate these away (e.g., by using a capture tag on the known sequence primer) or find a way of preventing the undesired products from amplifying in the first place (like suppression PCR, discussed later in this chapter). Sometimes it is useful to do both, as we will illustrate below.

In capture PCR, one can start with the very same set of primers shown for the unsuccessful example in Figure 4.7. However, the very first cycle of the amplification is performed using only the sequence-specific primer with a capture tag, like biotin. Then, before any additional rounds of amplification are executed, the product of the first cycle is physically purified using streptavidin-coated magnetic beads or some other strepatvidin-coated surface (Fig. 4.8). Now the splint primer is added, along with nonbiotinylated specific primer, and the PCR amplification is allowed to proceed normally. When the procedure is successful, very pure desired product is achieved, since all of the potential side product precursors are removed before they are able to be amplified.

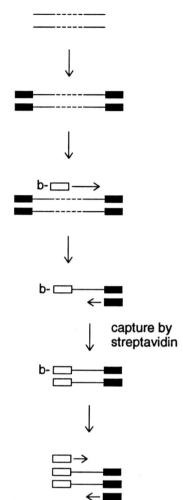

**Figure 4.8**   Capture PCR, where the known sequence is used to purify the target DNA from a complex sample before further rounds of amplification.

An alternative, general approach to single-sided PCR is to use a splint and primer combination designed so that the primer will work only after the splint has been replicated once, and the splint can only be replicated by synthesis initiated at the known bit of sequence. Two versions of this are shown in Figure 4.9. In one case a splint is used that is dephosphorylated, so it cannot ligate to itself, and the double-stranded region is short and A + T rich (Fig. 4.9a). After this is ligated to the target, the sample is heated to melt off the short splint strand. Next one cycle of DNA synthesis is allowed, using only the primer in the known target sequence. This copies the entire splint and produces a template for the second splint-specific primer. Both primers are now added, and ordinary PCR is allowed to proceed. However, only those molecules replicated during the first PCR cycle will be substrates for subsequent cycles. In the second version, called *bubble PCR,* the splint contains a noncomplementary segment of DNA which forms an interior loop (Fig. 4.9b). As before, the first round of PCR uses only the sample-specific primer. This copies the mispaired template strand faithfully so that when complementary primer is added to it, normal PCR can ensue. These procedures are reported to work well. It is worth noting that they can easily be enhanced by attaching a capture tag to the target-specific primer in the

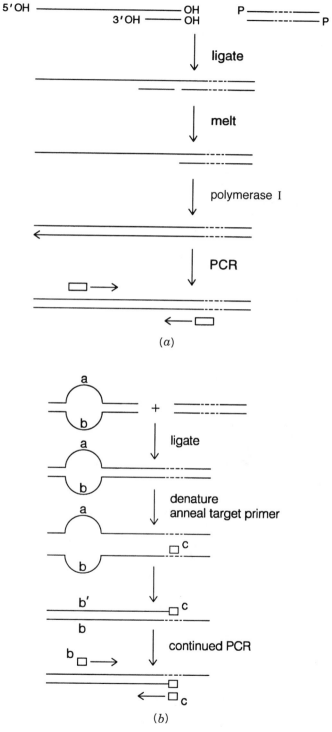

**Figure 4.9** Two single-sided PCR schemes that use a linker that must be replicated before the complementary primer will function; dashed lines indicate known target sequence. *(a)* use of an appropriately designed dephosphorylated linker. *(b)* Bubble PCR.

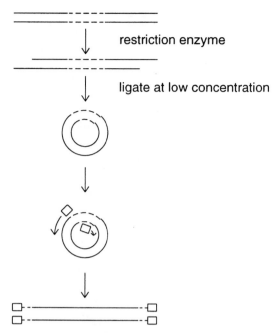

**Figure 4.10**   Inverse PCR, using primers designed to be extended outward on both sides of the known sequence (dashed lines).

first step. This will allow purification, as in capture PCR, before the reaction is allowed to continue.

A third variation on the single-sided PCR theme is inverse PCR. This is extremely elegant in principle. Like many such novel tricks, it was developed independently and simultaneously by several different research groups. Inverse PCR is shown schematically in Figure 4.10. Here the target is cut with a restriction enzyme to generate ends that are easily ligatable. Then it is diluted to very low concentrations and exposed to DNA ligase. Under these conditions the only products expected are DNA circles. To perform PCR, two primers are chosen within the known sequence, but they are oriented to face outward. Successful PCR with these primers should produce a linear product in which two, originally separate, segments of the unknown sequence are now fused at a restriction site and lie in between the two bits of known sequence. In practice, this procedure has not seen widespread success. One difficulty is that it is not easy to obtain good yields of small double-stranded circles by ligation. A second problem is the topological difficulties inherent in replication of circular DNA. The polymerase complex must wind through the center of the DNA circle once for each turn of helix. In principle, this latter problem could be removed by cleaving the known sequence with a restriction enzyme to linearize the target prior to amplification.

## REDUCING COMPLEXITY WITH PCR

PCR allows any desired fraction of the genome to be selectively amplified if one has the primers that define that fraction. The complexity of a DNA sample was defined in

Chapter 3. It is the total amount of different DNA sequence. For single-copy DNA the complexity determines the rate of hybridization. Thus it can be very useful to selectively reduce the complexity of a sample, since this speeds up subsequent attempts to analyze that sample by hybridization. The problem is that in general, one rarely has enough information about the DNA sequences in a sample to choose a large but specific subset of it for PCR amplification.

A powerful approach has been developed to use PCR to selectively reduce sample complexity without any prior sequence knowledge at all. This approach promises to greatly facilitate genetic and physical mapping of new, uncharted genomes. It is based on the use of short, random (arbitrary) primers. Consider the use of a *single* oligonucleotide primer of length $n$. As shown in Figure 4.11, this primer can produce DNA amplification only if its complementary sequence exists as an inverted repeat, spaced within a distance range amenable to efficient PCR. The inverted repeat requires that we specify a DNA sequence of $2n$ bases. For a statistically random genome of $N$ base pairs, the probability of this occurring at any particular place is $N4^{-2n}$, which is quite small for almost any $n$ large enough to serve as an effective primer. However, any placement close enough for PCR will yield amplification products of the two primer sites. If $L$ is the maximum practical PCR length, the probability that some observable PCR product will be seen is $LN4^{-2n}$. It is instructive to evaluate this expression for a mammalian genome with $N = 3 \times 10^9$ bp. For $L = 2000$ the results are

| OLIGONUCLEOTIDE LENGTH | NUMBER OF PCR PRODUCTS | TOTAL AMOUNT OF AMPLIFIED DNA |
|:---:|:---:|:---:|
| 8 | 1500 | $1.5 \times 10^6$ bp |
| 9 | 100 | $1.0 \times 10^5$ bp |
| 10 | 6 | $6.0 \times 10^3$ bp |

These results make it clear that by using single arbitrary short primers, we can sample useful discrete subsets of a genome. Each different choice of primer will presumably give a largely nonoverlapping subset. The complexity of the reaction products can be controlled by the primer length to give simple or complex sets of DNA probes. This method has been used, quite successfully, to search for new informative polymorphic genetic markers in plants. It has been called *RAPD mapping,* which is short for randomly amplified polymorphic DNA. The idea is to amplify as large a number of bands as can be clearly analyzed by a single electrophoretic lane and then to compare the patterns seen in a diverse set of individuals.

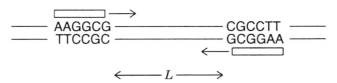

**Figure 4.11** Sampling a DNA target (reducing complexity) with a single short oligonucleotide primer (RAPD method).

## ADDITIONAL VARIANTS OF THE BASIC PCR REACTION

Here we will illustrate a number of variations on the basic PCR theme that increase the utility of the technique. The first of these is a convenient approach to the simultaneous analysis of a number of different genomic targets. This approach is called *multiplex PCR,* and it is carried out in a reverse dot blot format. The procedure is illustrated schematically in Figure 4.12. Each target to be analyzed is flanked by two specific primers. One is ordinary; the other is tagged with a unique 20-bp overhanging DNA sequence. PCR can be carried out separately (or in one pot if conditions permit), and then the resulting products pooled. The tagged PCR products are next hybridized to a filter consisting of a set of spots, each of which contains the immobilized complementary sequence of one of the tags. The unique 20-bp duplex formed by each primer sequence will ensure that the corresponding PCR products become localized on the filter at a predetermined site. Thus the overall results of the multiplex PCR analysis will be viewed as positive or negative signals at specific locations on the filters. This approach, where amplified products are directed to a known site on a membrane for analysis, dramatically simplifies the interpretation of the results and makes it far easier to automate the whole process.

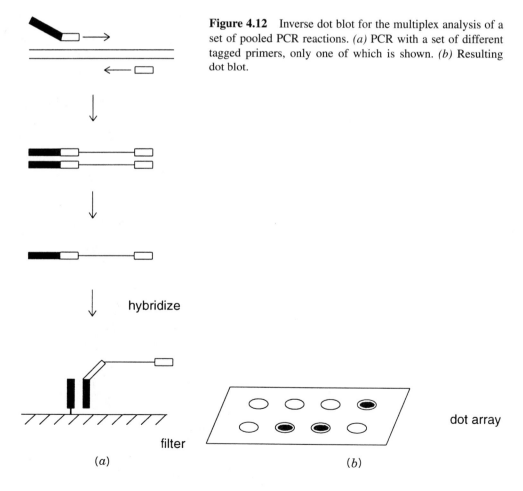

**Figure 4.12**   Inverse dot blot for the multiplex analysis of a set of pooled PCR reactions. *(a)* PCR with a set of different tagged primers, only one of which is shown. *(b)* Resulting dot blot.

hybridize

filter

dot array

*(a)*

*(b)*

For the approach shown in Figure 4.12, it would be far more useful to produce single-stranded product from the PCR reaction, since only one of the two strands of the amplified target can bind to the filter; the other strand will actually act as a competitor. Similarly, for DNA sequencing, it is highly desirable to produce a single-stranded DNA product. Presence of the complementary strand presents an unwanted complication in the sequencing reactions, and it can also act as a potential competitor. To produce single strands in PCR, a very simple approach called *asymmetric PCR* can be used. Here ordinary PCR is carried out for a few less than the usual number of cycles, say 20. Then one primer is depleted or eliminated, and the other is allowed to continue through an additional 10 cycles of linear PCR. The result is a product that is almost entirely single stranded. Clearly one can have whichever strand one wants by the appropriate choice of primer.

For diagnostic testing with PCR, one needs to distinguish different alleles at the DNA sequence level. A general approach for doing this is illustrated in Figure 4.13. It is called *allele-specific PCR*. In the case shown in Figure 4.13, we have a two-allele polymorphism. There is a single base difference possible, and we wish to know if a given individual has two copies of one allele, or the other, or one copy of each. The goal is to distinguish among these three alternatives in a single, definitive test. To do this, two primers are constructed that have 3'-ends specific for each of the alleles. A third general primer is used somewhere downstream where the two sequences are identical. The key element of allele-specific PCR is that because *Taq* polymerase does not have a 3'-exonuclease, it cannot use or degrade a primer with a mispaired 3'-terminus. Thus the allele-specific primers will only amplify the allele to which they correspond precisely. Analysis of the results is simplified by using primers that are tagged either by having different lengths or by having different colored fluorescent dyes. With length tagging, electrophoretic analysis will show different size bands for the two different alleles, and the heterozygote will be revealed as a doublet. With color tagging, the homozygotes will show, for example, red or green fluorescence, while the heterozygote will emit both red and green fluorescence, which our eye detects as yellow.

Color can also be used to help monitor the quantitative progress of a PCR reaction. Colored primers would not be useful unless their color were altered during the chain extension process. A similar problem holds for the potential use of colored dpppN's. The best approach to date is a slightly complicated strategy, which takes advantage of the 5'-exonuclease activity of thermostable DNA polymerases (Heid et al. 1996). A short oligonucleotide probe containing two different fluorescent labels is allowed to hybridize downstream from one of the PCR primers. The two dyes are close enough that the emission spectrum of the pair is altered by fluorescence resonance energy transfer. As the primer is extended, the polymerase reaches this probe and degrades it.

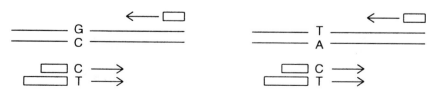

**Figure 4.13**   Allele-specific PCR used in genetic analysis. Different primer lengths allow the results of amplification of the two different alleles to be distinguished.

A color change is produced because the nucleotide products of the digestion diffuse too far away from each other for energy transfer to be efficient. Thus each chain extension results in the same incremental change in fluorescence. This procedure is called the TaqMan™ assay. A related spectroscopic trick, called molecular beacons, has recently been described in which a hybridization probe is designed as an oligonucleotide hairpin with different fluorescent dyes at its 3'- and 5'-ends (Kramer, 1996). In the hairpin these are close enough for efficient energy transfer. When the probe hybridizes to a longer target to form a duplex, its ends are now separated far apart in space, and the energy transfer is eliminated.

Hairpins can also be used to produce selective PCR amplification. In suppression PCR, long GC-rich adapters are ligated onto the ends of a mixture of target fragments (Diatchenko et al., 1996). When the ligation products are melted, the ends of the resulting single strands can form such stable hairpins that these ends become unaccessible for shorter primers complementary to the adapter sequences. However, if molecules in the target mixture contain a known internal target sequence, this can be used to initiate PCR. Chain extension from a primer complementary to the internal sequence will produce a product with only a single adapter. This will now allow conventional PCR amplification with one internal primer and one adapter primer.

A final PCR variant is called *DNA shuffling* (Stemmer, 1994). Here the goal is to enhance the properties of a target gene product by in vitro recombination. Suppose that a series of mutant genes exist with different properties; the goal is to combine them in an optimal way. The genes are randomly cleaved into fragments, pooled, and the resulting mixture is subjected to PCR amplification using primers flanking the gene. Random assembly of overlapping fragments will lead to products that can be chain extended until full length reassembled genes are produced. These then support exponential PCR amplification. The resulting populations of mutants are cloned and characterized by some kind of screen or selection in order to concentrate those with the desirable properties. This new method appears to be extremely promising. A very interesting alternative method to shuffle DNA segments uses catalytic RNAs (Mikheeva and Jarrell, 1996).

## TOTAL GENOME AMPLIFICATION METHODS

A frequently encountered problem in biological research is insufficient amounts of sample. If the sample is a cultured cell or microorganism, the simplest solution is to grow more material. However, many samples of interest cannot be cultured. For example, many differentiated cells cannot be induced to divide without destroying or altering their phenotype. Sperm cells are incapable of division. Most microorganisms cannot be cultured by any known technique—we know of their existence only because we can see their cells or detect aspects of their DNA. Fossil samples and various clinical biopsies are other examples of rare materials with insufficient DNA for convenient analysis. Finally sorted chromosomes (Chapter 2) present the challenge of a very useful resource for which there is always more demand than supply.

In each of these cases mentioned above, one could use a particular set of primers to amplify any given known DNA region of interest. However, once this were done, the rest of the sample would be lost for further analysis. Instead, what would be useful is an amplification method that first samples all of the DNA in the rare material. This can then be stockpiled for future experiments of a variety of types including more specific PCR, when needed. The issue is how to do this in such a way that the stockpile represents a complete, or at least a relatively complete and even sampling of the original sample. The danger of course is that the sample will consist of a set of regions with very different amplification efficiencies with any particular set of primers. After PCR the stockpile will now be a highly distorted version of the original, and future experiments will all be plagued by this distorted view.

One approach to PCR sampling of an entire genome is the method of primer extension preamplification (PEP). This was designed to be used on a single cell or single sperm. The detailed rationale for PEP will become apparent when genetics by single sperm PCR is discussed in Chapter 7. PEP is illustrated schematically in Figure 4.14a. A mixture of all possible $4^n$ primers of length $n$ is generated by automated oligonucleotide synthesis, using at each step all four nucleotides rather than just a single one. This extremely complex mixture is then used as a primer. Although the concentration of any one primer is vanishingly small, there are always enough primers present that any particular DNA segment has a reasonable chance of amplification. Norman Arnheim and his coworkers have reported reasonable success at using this approach with $n = 15$ (Arnheim and Ehrlich, 1993). They use 50 cycles of amplification and estimate that at least 78% of the genome will be amplified to 30 or more copies by this method.

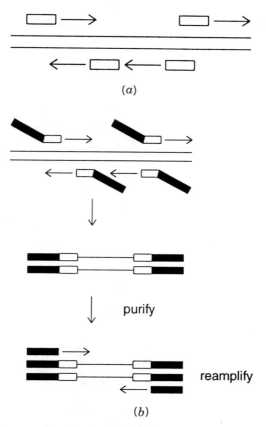

Figure 4.14  Methods for total genome PCR. *(a)* Primer extension preamplification (PEP). *(b)* Tagged random primer PCR (T-PCR).

When we attempted to use the PEP method, we ran into the difficulty that a large number of template-independent sequences were amplified in the reaction mixture. This is shown in the results of Figure 4.15, which illustrates the pattern of hybridization seen when PEP-amplified *S. pombe* chromosome 1 is hybridized to an arrayed cosmid library providing a fivefold coverage of the *S. pombe* genome. Almost all of the clones are detected with comparable intensities, even though only about 40% of them should contain material from chromosome 1. We reasoned that the complex set of long primers might allow for very significant levels of primer dimers (Fig. 4.3*b*) to be produced, and since the primers represented all possible DNA sequence, their dimers would also represent a broad population of sequences. Thus, when used in hybridization, this mixture should detect almost everything, which, indeed, it seems to do.

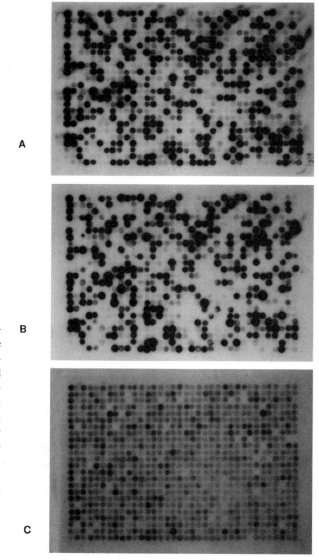

A

B

**Figure 4.15** Examples of amplification of *S. pombe* chromosome I DNA. *(a)* A schematic of an experiment in which directly labeled chromosome I or amplified chromosome I is used as a hybridization probe against an arrayed library of the entire *S. pombe* genome cloned into cosmids. *(b)* Actual hybridization results for labeled chromosome I *(top)*, chromosome I labeled after T-PCR amplification *(center)*, chromosome I labeled after PEP amplification *(bottom)*. (From Grothues et al., 1993.)

C

Three variations were introduced to circumvent the primer dimer problem in PEP. Together these constitute an approach we call *T-PCR,* for tagged random primer PCR (Grothues et al., 1993). Our primers consist of all $4^9$ nonanucleotides. Their shorter length and smaller complexity should be an advantage compared with the $4^{15}$ compounds used in ordinary PEP. Each primer was equipped at its 5'-end with a constant 17 base sequence; this is the tag. Thus the actual primers used were

$$GTTTTCCCAGTCACGACN_9$$

where N is a mixture of A, T, G, and C. After a few rounds of PCR, the resulting mixture was fractionated by gel filtration, and material small enough to be primer dimers was discarded. Then the remaining mixture was used as a target for amplification with only the tag sequence as a primer (Fig. 4.14*b*). As shown in Figure 4.15, this yielded reaction products that produced a pattern of hybridization with the *S. pombe* cosmid array almost identical to that seen with directly labeled chromosome 1. Thus we feel that T-PCR offers very good prospects for uniformly sampling a complex DNA sample. In our hands this approach has been successful thus far with as little as $10^{-12}$ g DNA, which corresponds to less than a single human cell.

Quite a few variations on this approach have been developed by others. One example is degenerate oligonucleotide-primed PCR (DOP–PCR) described by Telenius et al. (1992). Here primers are constructed like

$$AAGTCGCGGCCGCN_6ATG$$

with a six base totally degenerate sequence flanked by a long 5' unique sequence and a specific 3 to 6 base unique sequence. The 3'-sequence serves to select a subset of potential PCR start points. The degenerate sequence acts to stabilize the primer-template complex. The constant 5'-sequence can be used for efficient amplification in subsequent steps just as the tag sequence is used in T-PCR. It is not yet certain how to optimize whole genome PCR methods for particular applications. Issues that must be considered include the overall efficiency of the amplification, the uniformity of the product distribution, and the fraction of the original target that is present in the final amplified product. A recently published DOP–PCR protocol (Cheung and Nelson, 1996) looks particularly promising.

## APPLICATION OF PCR TO DETECT MOLECULES OTHER THAN DNA

A natural extension of PCR is its use to detect RNA. Two general approaches for doing this are summarized in Figure 4.16. In one, which is specific for polyadenylated mRNA, an oligo-dT primer is used, with reverse transcriptase, to make a DNA copy of the RNA. Then conventional PCR can be used to amplify that DNA. In the other approach, which is more general, random $(dN)_n$'s are used to prime reverse transcriptase to make initial DNA copies, and then ordinary PCR ensues.

Less obvious is the use of PCR to detect antigens or other non-nucleic acid molecules. We originally demonstrated the feasibility of this approach, which should have a broad

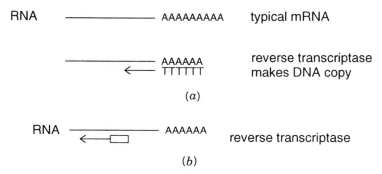

**Figure 4.16** Two methods for PCR amplification of RNA. *(a)* Use of reverse transcriptase with an oligo-dT primer. *(b)* Use of reverse transcriptase with short random oligonucleotide primers.

range of applications, and it should be generalizable to almost any class of molecule. The basic principle of what we call *immuno-PCR* (i-PCR) is shown in Figure 4.17. DNA is used as the label to indirectly tag an antibody. Then the DNA is detected by ordinary PCR.

In the test case, shown in Figure 4.17, the antibody is allowed to detect an immobilized antigen and to bind to it in the conventional way (Fig. 4.18). Then the sample is exhaustively washed to remove free antibody. Next a molecule is added that serves to couple DNA to the bound antibody. That molecule is a chimeric protein fusion between the protein streptavidin and two domains of staphylococcal protein A. The chimera was made by conventional genetic engineering methods and expressed as a gene fusion in *E. coli*. After purification the chimeric protein is fully active. Its properties are summarized in Table 4.1. The chimera is a tetramer. It is capable of binding four immunoglobulin G's and four biotins.

After the chimera is bound to the immobilized antibody, any unbound excess material is removed, and now biotinylated DNA is added (Fig. 4.18). In our hands, end-biotinylated linearized pUC19 was used. This was prepared by filling in the ends of a restriction enzyme-digested plasmid with biotinylated dpppU, but it could just as easily have been made by PCR with biotinylated primers. The biotinylated DNA binds to the immobilized chimera.

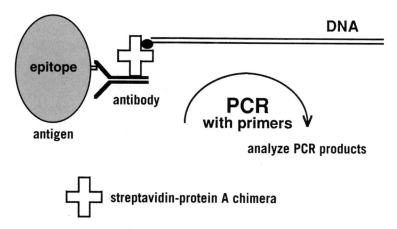

**Figure 4.17** Basic scheme for immuno-PCR: Detection of antigens with DNA-labeled antibodies.

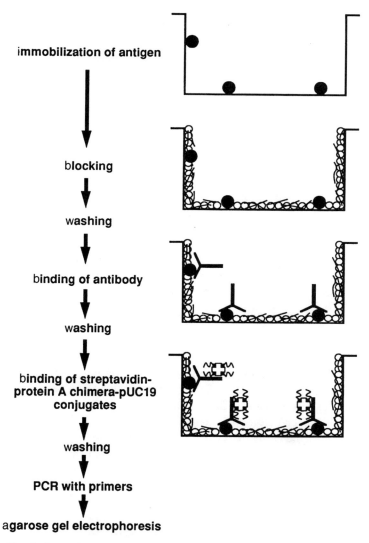

**immobilization of antigen**

**blocking**

**washing**

**binding of antibody**

**washing**

**binding of streptavidin-
protein A chimera-pUC19
conjugates**

**washing**

**PCR with primers**

**agarose gel electrophoresis**

**Figure 4.18**   Detailed experimental flow chart for implementing immuno-PCR to detect an immobilized antigen.

**TABLE 4.1   Streptavidin-Protein A Chimera**

| | |
|---|---|
| Expression vector: | pTSAPA-2 |
| Amino acid residues: | 289 per subunit |
| Subunits: | 4 (subunit tetramer) |
| Molecular mass: | 31.4 kDa per subunit |
| | 126 kDa per molecule |
| Biotin binding: | 4 per molecule |
| | 1 per subunit |
| IgG binding: | 4 per molecule |
| | 1 per subunit |
| | (human IgG) |

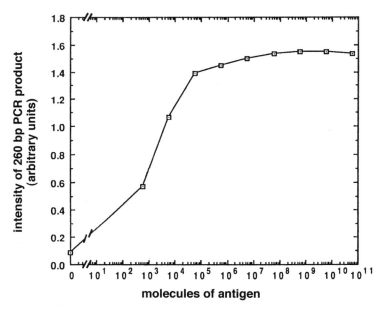

**Figure 4.19** Result of detection of tenfold serial dilutions of an antigen, bovine serum albumin, by immuno-PCR. (Taken from Sano et al., 1992.)

Excess unbound DNA is carefully removed by extensive washing. Now the entire sample is subjected to PCR using primers specific for the particular DNA tag. Typical results are shown for a set of serial dilutions of antigen in Figure 4.19. The results indicate that i-PCR promises to be an antigen-detection system of unparalleled sensitivity. Less than 600 molecules of antigen could be easily detected without any effort to optimize the system. This is $10^5$ times more sensitive than conventional immunoassays. A major advantage of i-PCR is that the DNA molecule used is purely arbitrary. It can be changed at will to prevent the buildup of laboratory contaminants. It need not correspond (indeed it should not correspond) to any sequences found in the samples. Thus there should be no interference from sample DNA. Finally a number of different DNA labels could be detected simultaneously, which would open the way for multiplex PCR detection of several antigens simultaneously. Such an application has recently been reported (Hendrickson et al., 1995).

## DNA AMPLIFICATION WITHOUT THERMAL CYCLING AND OTHER ALTERNATIVES TO PCR

From a practical viewpoint it is difficult to fault PCR. If there is any step that is tedious, it is the need for stringent control at several different temperatures. It would be nice to eliminate this requirement. From a commercial standpoint, existing PCR patents create quite a powerful band of protection around this technology and make the notion of potential competing technologies quite attractive as lucrative business ventures. Taken together, these considerations have fueled a number of attempts to create alternate DNA amplification procedures. Several of these have been shown to be practical. Some appear to be very attractive alternates to PCR for certain applications. None yet have shown the generality or versatility of PCR. The degree of amplification achievable by these methods is quite impressive, but it is still considerably less than that seen with conventional PCR (Table 4.2).

**TABLE 4.2   Comparison of Various In Vitro Nucleic Acid Amplification Procedures**

| Method | Amplified Species | Temperature Used (°C) | Target-specific Probes Needed | Amplification Extent |
|---|---|---|---|---|
| PCR | Target | 50–98 cycle | 2 or more | $10^{12}$ |
| QβR | Probe | 37 isothermal | 1 | $10^{9}$ |
| LCR | Probe | 50–98 cycle | 4 | $10^{5}$ |
| 3SR | Target | 42 isothermal | 2 | $10^{10}$ |
| SDA | Target | 37 isothermal | 4 | $10^{7}$ |

*Source:* Adapted from Abramson and Myers (1993).

Isothermal self-sustained sequence replication (3SR) is illustrated in Figure 4.20. In this technique an RNA target is the preferred starting material. DNA targets can always be copied by extending a primer containing a promoter site for an enzyme like T7 RNA polymerase (and then that enzyme is used to generate an RNA copy of the original DNA). The complementary DNA strand of the RNA is synthesized by Avian myeloblastosis virus (AMV) reverse transcriptase (RT) using a primer that simultaneously introduces a promoter of T7 RNA polymerase. AMV RT contains an intrinsic RNase H activity. This activity specifically degrades the RNA strand of an RNA-DNA duplex. Thus, as AMV RT synthesizes the DNA complement, it degrades the RNA template. The result is a single-stranded DNA complement of the original RNA. Now a second primer, specific for the target sequence, is used to prime the RT to synthesize a double-stranded DNA (Fig. 4.20a). When this is completed, the resulting duplex now contains an intact promoter for T7 RNA polymerase so that enzyme can, rapidly, synthesize many RNA copies. These RNAs are the complement of the original RNA target (Fig. 4.20b).

Now, in a cyclical process, the RT makes DNA complements of the RNAs, degrading them in the process by its RNaseH activity. RT then turns the single-stranded DNAs into duplexes. These duplexes in turn serve as templates for T7 RNA polymerase to make many more copies of single-stranded RNA. The key point is that all these reactions can be carried on simultaneously at a constant temperature. A substantial level of amplification is observed, and in principle, many of the same tricks and variations of PCR can be implemented through the 3SR approach. Primer nesting does appear to be more difficult, and it is not clear how well this technique will work in multiplexing.

A method that is similar in spirit but rather different in detail is strand displacement amplification (SDA). This is illustrated in Figure 4.21. It is based on the peculiarities of the restriction endonuclease *Hinc* II which recognizes the hexanucleotide sequence and cleaves it, as shown below:

$$GTTGAC \qquad GTT + GAC$$
$$CAACTG \rightarrow CAA \quad CTG$$

The key feature of this enzyme exploited in SDA is the effect of alpha thio-substituted phosphates on the enzyme. These can be introduced into DNA by the use of alpha-S-dpppA. When this is incorporated into the top strand of the recognition sequence, there is no effect. However, in the bottom strand the thio derivatives inhibit cleavage (Fig. 4.21a). How this peculiarity is used for isothermal amplification is illustrated in Figure 4.21b.

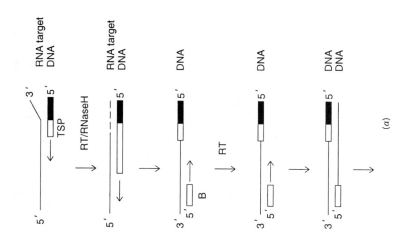

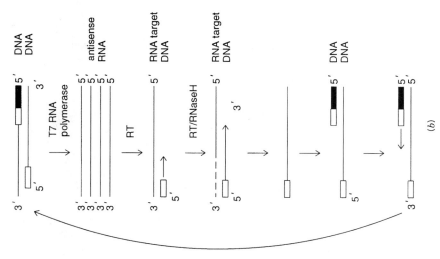

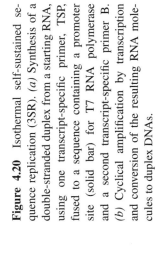

**Figure 4.20** Isothermal self-sustained sequence replication (3SR). (*a*) Synthesis of a double-stranded duplex from a starting RNA, using one transcript-specific primer, TSP, fused to a sequence containing a promoter site (solid bar) for T7 RNA polymerase and a second transcript-specific primer B. (*b*) Cyclical amplification by transcription and conversion of the resulting RNA molecules to duplex DNAs.

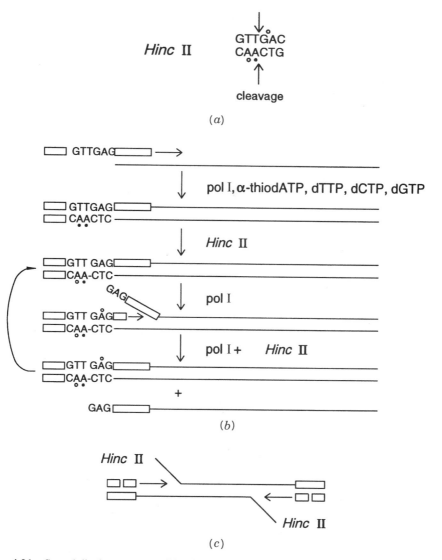

**Figure 4.21** Strand displacement amplification (SDA). A *Hinc* II site with alpha thio DNA derivatives that block *(solid circles)* and do not block *(open circles)* cleavage. *(b)* Linear amplification by strand displacement from one *Hinc* II cleavage site. *(c)* Exponential amplification from two *Hinc* II cleavage sites.

A DNA polymerase I mutant with no 5'-exonuclease activity is used. This leads to strand displacement. Consider first the effect of this enzyme on the target-primer complex shown in the figure. The primer has a potential *Hinc* II site overhanging the template. The polymerase extends both the template and the primer, incorporating alpha-S-A. The top strand of the resulting duplex can be cleaved by *Hinc* II; the bottom strand is resistant. This creates a target for the polymerase that can strand displace most of the top strand and continually make copies of it, resulting in linear amplification.

If primers are established with overhanging *Hinc* II sites on both sides of a target duplex, each strand can be made by linear amplification (Fig. 4.21*c*). Now, however, each newly synthesized strand can anneal with the original primer to form new complexes capable of further *Hinc* II cleavage and DNA polymerase-catalyzed amplification. Thus the overall system will show exponential amplification in the presence of excess primers.

In order to use SDA, one must have the desired target sequence flanked by *Hinc* II sites (or sites for any other restriction enzyme that might display similar properties). These sites can be introduced by using primers flanking the target sequence and tagged with additional 5'-sequences containing the desired restriction enzyme cleavage sites. These flanking primers are used in a single cycle of conventional PCR; then SDA initiates spontaneously, and the amplification can be continued isothermally.

Overall, SDA is a very clever procedure that combines a number of tricks in DNA enzymology. It would appear to have some genuinely useful applications. However, SDA as originally described seems unlikely to become a generally used method because the resulting products have alpha-S-A, which is not always desirable, and the primers needed are rather complex and idiosyncratic. Recently variants on this scheme were developed that have fewer restrictions.

Other modes of DNA polymerase-based amplification are still in their infancy, including rolling circle amplification (Fire and Xu, 1995) and protein-primed DNA amplification (Blanco et al., 1994). Alternate schemes for DNA amplification have been developed that avoid the use of DNA polymerase altogether. Foremost among these is the ligase chain reaction (LCR). This is illustrated in Figure 4.22. The target DNA is first denatured. Then two oligonucleotides are annealed to one strand of the target. Unlike PCR, these two sequences must be adjacent in the genome, and they correspond to the same DNA strand.

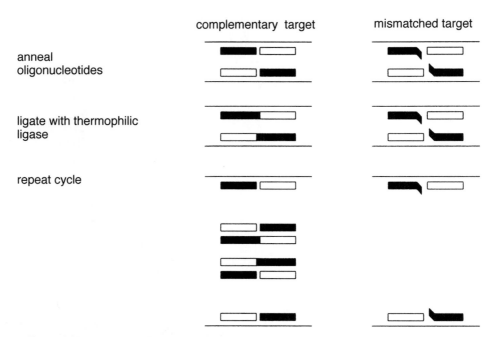

**Figure 4.22** Example of allele-specific amplification using the ligase chain reaction (LCR).

If they match the target sequence exactly, DNA ligase will efficiently seal them together provided that a 5'-phosphate is present to form the phosphodiester bond between them. A complementary set of oligonucleotides can be used to form a ligation product directed by the other DNA strand. The overall result is to double the number of DNA strands. Both duplexes are melted, more oligonucleotides anneal, and the process can be continued indefinitely. With continual thermal cycling, the result is exponential amplification of the target. This is most easily detected in automated systems by using a capture tag on one of the oligonucleotides to be ligated and a color-producing tag on the other. This procedure is obviously limited to small DNA target sequences, but it could form a powerful alternative to allele-specific PCR. Hybrid amplification procedures that combine LCR and polymerase extension reactions also appear to be very promising.

The final amplification scheme we will discuss is carried out by the enzyme $Q\beta$ replicase. This occurs strictly at the RNA level. Appropriate RNA targets can be made by subcloning DNA samples into vectors that embed the desired targets within $Q\beta$ sequences and place them all downstream from a T7 RNA polymerase promoter so that an RNA copy can be made to start the $Q\beta$ replication process ($Q\beta$R). A much more general approach is to construct two separate RNA probes that can anneal to adjacent sequences on a target RNA. In the presence of T4 DNA ligase, the two probes will become covalently joined. Neither prone alone is a substrate for $Q\beta$ replicase. However, the ligation product is a substrate and is efficiently amplified (Fig. 4.23).

$Q\beta$ has an unusual mode of replication. No primer is needed. No double-stranded intermediate is formed. The enzyme recognizes specific secondary structure features and sequence elements on the template, and then makes a complementary copy of it. That copy dissociates from the template as it is made, and it folds into its own stable secondary structure which is a complement of that of the template. This structure also can serve as a template for replication. Thus the overall process continually produces both strands as targets, much in the manner of a dance in which the two partners move frenetically but never stay in contact for an extended period.

The usual mode of $Q\beta$ replication is very efficient. It is not uncommon to make $10^7$ to $10^8$ copies of the original target. One can start from the single molecule level. However, the system is not that easy to manipulate; $Q\beta$ replicase itself is a complex four-subunit enzyme not that commonly available. The procedures needed to prepare the DNA target for $Q\beta$ replication are somewhat elaborate, and there are considerable restrictions on what sorts of RNA insertions can be tolerated by the polymerase. For all these reasons the $Q\beta$ amplification system is most unlikely to replace PCR as a general tool for DNA analysis. It may, however, find unique niches for analyses where the idiosyncrasies of the system do not interfere, and where very high levels of amplification at constant temperature are needed.

## FUTURE OF PCR

In this chapter we have illustrated myriad variations and potential applications of PCR. In viewing these, it is important to keep in mind that PCR is a young technique. It is by no means clear that today's versions are the optimal ones or the most easily adaptable ones for the large-scale automation eventually needed for high-throughput genome analysis. Much additional thought needs to be given on how best to format PCR for widespread use and how to eliminate many of the current glitches and irreproducibility inherent in such a high-gain amplification system.

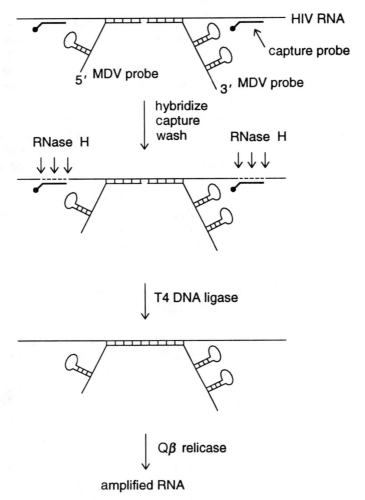

**Figure 4.23** Example of the use of Qβ to detect a target, HIV RNA, by ligation. Here two specific RNA probes are designed to be complementary to adjacent target sequences. Each probe alone contains insufficient secondary structure features to support Qβ replicase amplification. Two additional end-biotinylated DNA capture probes are used to allow a streptavidin-based solid phase purification of the probe-target complexes and then release of these complexes by digestion with RNAse H which specifically cleaves RNA-DNA duplexes. The released complexes are treated with T4 DNA ligase, which will work on a pure double-stranded RNA substrate. Ligation produces an RNA that is now recognized and amplified by Qβ replicase. (Adapted from Kramer and Tyagi, 1996.

## SOURCES AND ADDITIONAL READINGS

Abramson R. D., and Myers, T. W. 1993. Nucleic acid amplification technologies. *Current Biology* 4:41–47.

Arnheim, N., and Erlich, H. 1992. Polymerase chain reaction strategy. *Annual Review of Biochemistry* 61:131–156.

Barany, F. 1991. Genetic disease detection and DNA amplification using cloned thermostable ligase. *Proceedings of the National Academy of Sciences USA* 88:189–193.

Barnes, W. 1994. PCR Amplification of up to 35-kb DNA with high fidelity and high yield from bacteriophage templates. *Proceedings of the National Academy of Sciences USA* 91:2216–2220.

Blanco, L., Lazaro, J. M., De Vega, M., Bonnin, A., and Salas, M. 1994. Terminal protein-primed DNA amplification. *Proceedings of the National Academy of Sciences USA* 91: 12198–12202.

Bloch, W. 1991. A biochemical perspective of the polymerase chain reaction. *Biochemistry* 30: 2735–2747.

Caetano-Annolés, G. 1996. Scanning of nucleic acids by in vitro amplification: New developments and applications. *Nature Biotechnology* 14: 1668–1674.

Cheng, S., Fockler, C., Barnes, W. M., and Higuchi, R. 1994. Effective amplification of long targets from cloned inserts and human genomic DNA. *Proceedings of the National Academy of Sciences USA* 91: 5695–5699.

Cheung, V. G., and Nelson, S. F. 1996. Whole genome amplification using a degenerate oligonucleotide primer allows hundreds of genotypes to be performed on less than one nanogram of genomic DNA. *Proceedings of the National Academy of Sciences USA* 93: 14676–14679.

Chou, W., Russell, M., Birch, D. E., Raymond, J., and Block, W. 1992. Prevention of pre-PCR mispriming and primer dimerization improves low-copy-number amplifications. *Nucleic Acids Research* 20: 1717–1723.

Cobb, B. D., and Clarkson, J. M. 1994. A Simple procedure for optimising the polymerase chain reaction (PCR) using modified Taguchi methods. *Nucleic Acids Research* 22: 3801–3805.

Diatchenko, L., Lau, Y. F., Campbell, A. P., Chenchik, A., Moqadam, F., Huang, B., Lukyanov, S., Lukyanov, K., Gurskaya, N., and Sverdlov, E. D. 1996. Suppression subtractive hybridization: A method for generating differentially regulated or tissue-specific cDNA probes and libraries. *Proceedings of the National Academy of Sciences USA* 93: 6025–6030.

Erlich, H. A., Gelfand, D., and Sninsky, J. J. 1991. Recent advances in the polymerase chain reaction. *Science* 252: 1643–1651.

Fahy, E., Kwoh, D. Y., and Gingeras, T. R. 1991. Self-sustained sequence replication (3SR): An isothermal transcription-based amplification system alternative to PCR. *PCR Methods and Applications* 1: 25–33.

Fire, A., and Xu, S., 1995. Rolling replication of short DNA circles. *Proceedings of the National Academy of Sciences USA* 92: 4641–4645.

Grothues, D., Cantor, C. R., and Smith, C. L. 1993. PCR amplification of megabase DNA with tagged random primers (T-PCR). *Nucleic Acids Research* 21:1321–1322.

Heid, C. A., Stevens, J., Livak, K. J., and Williams, P. M. 1996. Real time quantitative PCR. *Genome Research* 6: 986–994.

Hendrickson, E. R., Hatfield-Truby, T. M., Joerger, R. D., Majarian, W. R., and Ebersole, R. C. 1995. High sensitivity multianalyte immunoassay using covalent DNA-labeled antibodies and polymerase chain reaction. *Nucleic Acids Research* 23: 522–529.

Lagerström, M., Parik, J., Malmgren, H., Stewart, J., Pettersson, U., and Landegren, U. 1991. Capture PCR: Efficient amplification of DNA fragments adjacent to a known sequence in human and YAC DNA. *PCR Methods and Applications* 1: 111–119.

Landegren, U. 1992. DNA probes and automation. *Current Opinion in Biotechnology* 3: 12–17.

Lizardi, P. M., Guerra, C. E., Lomeli, H., Tussie-Luna, I., and Kramer, F. R. 1988. Exponential amplification of recombinant-RNA hybridization probes. *Biotechnology* 6: 1197–1202.

Mikheeva, S., and Jarrell, K. A. 1996. Use of engineered ribozymes to catalyze chimeric gene assembly. *Proceedings of the National Academy of Sciences USA* 93: 7486–7490.

Nickerson, D. A., Kaiser, R., Lappin, S., Stewart, J., Hood, L., and Landegren, U. 1990. Automated DNA diagnostics using an ELISA-based oligonucleotide ligation assay. *Proceedings of the National Academy of Sciences USA* 87: 8923–8927.

Sano, T., and Cantor, C. R. 1991. A streptavidin-protein A chimera that allows one-step production of a variety of specific antibody conjugates. *Bio/Technology* 9: 1378–1381.

Sano, T., Smith, C. L., and Cantor, C. R. 1992. Immuno-PCR: Very sensitive antigen detection by means of specific antibody-DNA conjugates. *Science* 258: 120–122.

Siebert, P. D., Chenchik, A., Kellogg, D. E., Lukyanov, K. A., and Lukyanov, S. A. 1995. An improved PCR method for walking in uncloned genomic DNA. *Nucleic Acids Research* 23: 1087–1088.

Stemmer, W. P. C. 1994. Rapid evolution of a protein *in vitro* by DNA shuffling. *Nature* 370: 389–391.

Telenius, H., Carter, N. P., Bebb, C. E., Nordenskjöld, M., Ponder, B. A. J., and Tunnacliffe, A. 1992. Degenerate oligonucleotide-primed PCR: General amplification of target DNA by a single degenerate primer. *Genomics* 13: 718–725.

Telenius, H., Pelmear, A., Tunnacliffe, A., Carter, N. P., Behmel, A., Ferguson-Smith, M. A., Nordenskjöld, M., Pfragner, R., and Ponder, B. A. J. 1992. Cytogenetic analysis by chromosome painting using DOP-PCR amplified flow-sorted chromosomes. *Genes, Chromosomes and Cancer* 4: 257–263.

Tyagi, S., and Kramer, F. R. 1996. Molecular beacons: Probes that fluoresce upon hybridization. *Nature Biotechnology* 14: 303–308.

Tyagi, S., Landedren, U., Tazi, M., Lizardi, P. M., and Kramer, F. R. 1996. Extremely sensitive, background-free gene detection using binary probes and $Q\beta$ replicase. *Proceedings of the National Academy of Sciences USA* 93: 5395–5400.

Walker, G. T., Little, M., Nadeau, J. G., and Shank, D. D. 1992. Isothermal in vitro amplification of DNA by a restriction enzyme/DNA polymerase system. *Proceedings of the National Academy of Sciences USA* 89: 392–396.

White, T. J. 1996. The future of PCR technology: diversification of technologies and applications. *Trends in Biotechnology* 14: 478–483.

Wittwer, C. T., Herrmann, M. G., Moss, A. A., and Rasmussen, R. P. 1997. Continuous fluorescence monitoring of rapid cycle DNA amplification. *BioTechniques* 22: 130–138.

# 5 Principles of DNA Electrophoresis

## PHYSICAL FRACTIONATION OF DNA

The methods we have described thus far all deal with DNA sequences. PCR allows, in principle, the selective isolation of any short DNA sequence. Hybridization allows, in principle, the sequence-specific capture of almost any DNA strand. However, to analyze the results of these powerful sequence-directed methods, we usually resort to length-dependent fractionations. There are two reasons for this. Separation of DNAs by length were developed before we had much ability to manipulate DNA sequences. Hence the methods were familiar and validated, and it was natural to incorporate them into most protocols for using DNAs. Second, our ability to fractionate DNA by length is actually remarkably good. The fact that DNAs are stiff and very highly charged greatly facilitates these fractionations. In this chapter we briefly review DNA fractionations other than electrophoresis. The bulk of the chapter will be spent on trying to integrate the maze of experimental and modeling observations that constitute our present-day knowledge of the principles that underlie DNA electrophoresis. Practical aspects of DNA separations are treated in Boxes 5.1 and 5.2.

## SEPARATION OF DNA IN THE ULTRACENTRIFUGE

Velocity or equilibrium ultracentrifugation of DNA represents the only serious alternative to DNA electrophoresis. For certain applications it is a powerful tool. However, for most applications, the resolution of ultracentrifugation just isn't high enough to compete with electrophoresis. DNA can be separated by size in the ultracentrifuge by zonal sedimentation. Commonly density gradients of small molecules like sucrose are employed to prevent convection caused by gravitational instabilities. Sucrose gradient sedimentation is tedious because ultracentrifuges typically allow only half a dozen samples to be analyzed simultaneously.

One unique and very useful application of the ultracentrifuge to DNA fractionation is the separation of DNA fragments of different base composition by equilibrium density gradient centrifugation. There are slight density differences between A–T and G–C base pairs. These densities match different CsCl concentrations. Thus, when DNA fragments are subjected to centrifugation in an equilibrium gradient of CsCl, different species will show different buoyant densities. That is, they will concentrate in different regions of the CsCl gradient where their density matches that of the bulk fluid. A schematic example is shown in Figure 5.1. Because of these density differences, it is possible to obtain fractions of DNA physically isolated on the basis of their average base composition. The small intrinsic effects of base composition on density can be enhanced by the use of density shift ligands that bind differentially to A + T or G + C rich DNA. This allows much higher-resolution density fractionation of DNA.

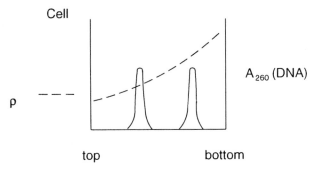

**Figure 5.1** Schematic illustration of separation *(from top to bottom of a tube)* of DNAs of different base composition by density gradient equilibrium ultracentrifugation; $\rho$ is density, Absorbance $(A_{260})$ is proportional to DNA concentration.

The utility of such preparative methods will be illustrated in Chapter 15, when the biological properties of isochores, specific density fractions of DNA, are discussed.

The other aspect of DNA structure that has been conveniently probed by ultracentrifugation is topological properties. Velocity sedimentation can be used to discriminate among linear DNAs, relaxed circles, and supercoiled circles. Equilibrium density gradient separations are also very powerful for this discrimination because, under the right circumstances, supercoils will bind very different amounts of intercalating dyes, like ethidium bromide, than bound by relaxed circles or linear duplexes (see Cantor and Schimmel, 1980, ch. 24). This leads to large density shifts. Thus such fractionations have played a major role in early studies of DNA topology, and they have served as major tools for the separation of plasmid DNAs key in many recombinant DNA experiments. Once again, however, as the need for dealing with very large numbers of samples in parallel grows, these centrifugation methods have to be replaced by others. The capacities of existing centrifuges simply cannot cope with large numbers of DNA samples, and loading and unloading centrifuges is a procedure that is very difficult to automate efficiently. In addition even the best protocols for DNA size or topology separation by ultracentrifugation have far lower resolution than typical electrophoretic methods.

## ELECTROPHORETIC SIZE SEPARATIONS OF DNA

In contemporary methods of DNA analysis, electrophoresis dominates size fractionations for all species from oligonucleotides with a few bases up to intact chromosomal DNAs with sizes in excess of $10^7$ bp. Rarely does a single method prove to be so powerful across such a broad size range. The size resolution for optimized separations is a single base up to more than 1000 bases for single-stranded DNAs, a few percent in size for double strands up to about 10 kb, and progressively lower resolution as DNA sizes increase until, for species in excess of 5 Mb, only about 10% size resolution is available.

Size is the overwhelming factor that determines the separation power of double-stranded DNA electrophoresis. In general, double-stranded DNA fractionations are carried out in dilute aqueous buffer near neutral pH. Base composition and base sequence have almost no effect on the gross separation patterns. There is a significant effect of base

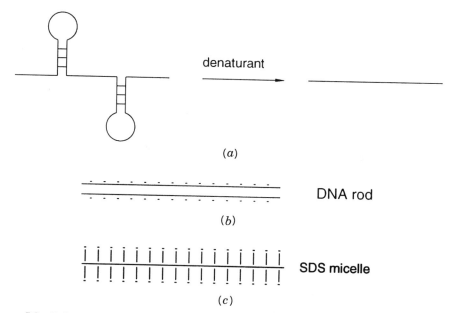

**Figure 5.2** Behavior of highly charged molecules in electrophoresis. *(a)* Structure of single-stranded DNA without denaturants, or in the presence of strong denaturants where charge and friction are both proportional to length. *(b)* Structure of short double-stranded DNA, where charge and friction are both proportional to length. *(c)* Proteins, denatured by SDS, form micelles where charge and friction are both proportional to length.

sequence in the special case of sequences that lead to pronounced DNA bending. The most common example are AA's or TT's spaced at intervals comparable to the helical repeat distance. When this occurs, the slight bends at each AA are in phase, and the overall structure can be markedly curved.

Electrophoretic fractionations of single-stranded DNA are mostly carried out in the presence of strong denaturants such as high temperature or high concentrations of urea or formamide. Under these conditions most of the potential secondary structure of the DNA strands is eliminated, and size dominates the electrophoretic separations (Fig. 5.2a). Sometimes, however, single-stranded DNA electrophoresis is carried out under conditions that allow substantial secondary structure formation. In this case it is the extent of folded structure that tends to dominate the separation pattern. We will discuss these cases in more detail in Chapter 13 when we describe the single-stranded–conformational polymorphism (SSCP) method of looking for mutations at the DNA level.

## ELECTROPHORESIS WITHOUT GELS

Essentially all DNA electrophoresis is carried out within gel matrices. Crosslinked polyacrylamide is used for short single-stranded DNAs, agarose is used for very large double-stranded DNAs, and certain other specialized gel matrices have been optimized for particular intermediate separations. To understand the key role played by the gel in these separations, it is instructive to consider DNA electrophoresis without gels. The key parameter measured in electrophoresis is the mobility, $\mu$. This is the velocity per unit field. In one dimension,

$$\mu = \frac{v}{E}$$

where the velocity, $v$, is measured in cm/s and the electric field, **E,** is measured in volts/cm. Double-stranded DNA behaves in electrophoresis as a free-draining coil. This means that each segment of the chain is able to interact with the solvent in a manner essentially independent from any of the others. Under these conditions the frictional coefficient of the molecule, $f$, felt by the coil as it moves through the fluid is proportional to the length, $L$, of the coil:

$$f = \alpha_1 L$$

DNA has a constant charge per unit length. There is one negative charge for each phosphate; a significant fraction of this charges is effectively screened by bound counterions, but the net result is still a charge $\mathbf{Z}$ proportional to $L$ (Fig. 5.2$b$):

$$\mathbf{Z} = \alpha_2 L$$

The net steady state velocity in electrophoresis is the result of equal, opposite electrostatic forces accelerating the molecule, $\mathbf{ZE,}$ and frictional forces, $f\mathbf{v}$, retarding the motion. This lets us set $\mathbf{ZE} = f\mathbf{v}$, where $\mathbf{Z}$ is the net charge on the molecule. It can be rearranged to

$$\mathbf{v} = \frac{\mathbf{ZE}}{f} = \frac{\alpha_2 L\mathbf{E}}{\alpha_1 L} = \frac{\alpha_2 \mathbf{E}}{\alpha_1}$$

Thus from simple considerations we are led to the conclusion that the mobility of DNA in electrophoresis should be independent of size. Indeed this prediction was verified by the work of Norman Davidson and his collaborators more than 20 years ago. Electrophoresis of DNA in free solution fails to achieve any size fractionation at all. Why, then is electrophoresis such a powerful tool in fractionating DNA. The answers will all have to lie with the ways in which DNA molecules interact with gels under the influence of electrical fields.

In passing, it is worth noting that the problem of size-independent electrophoretic mobility is not limited to DNA molecules. A common form of protein electrophoresis is the fractionation of proteins denatured by the detergent sodium dodecyl sulphate (SDS). For proteins without disulphide crosslinks, SDS denaturation produces a highly extended protein chain saturated by bound SDS molecules, as shown in Figure 5.2$c$. This produces a tubular micellar structure that superficially resembles a DNA double helix in size and shape and charge characteristics. SDS-protein micelles have an approximately constant charge per unit length, and they behave as free-draining structures. As a result, like DNA molecules, proteins in SDS show a size-independent electrophoretic mobility in the absence of a gel. The great power of SDS electrophoresis to fractionate proteins also rests in the nature of the interaction of the SDS micelles with the gel. Here, however, the similarities end. The SDS micelles seem to be mostly sieved by the gels; the interaction of DNA with the gels turns out to be much more complex than this.

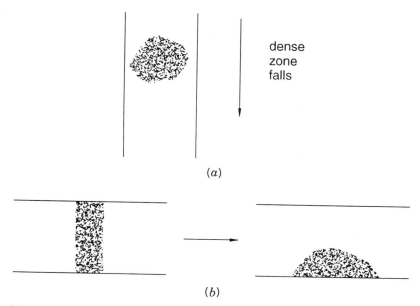

**Figure 5.3**   Convective instabilities in electrophoresis. *(a)* Vertical electrophoresis. *(b)* Horizontal electrophoresis.

Electrophoresis in free solution is not a common technique, for reasons that actually have little to do with the above considerations. Almost all electrophoresis is done with gels or other support matrices, even when the molecules involved do not behave as free draining. Why use a gel at all under these circumstances? The reason is to prevent convective instabilities. Placement of an electrical field across a conductive solution leads to significant current flow and significant heating. This heating produces nonuniformities in temperature, and convective solvent motion will result from these, much as convection patterns are established if water is heated on the surface of a stove. The presence of bands of dissolved solute molecules leads to regions with local bulk density differences, as illustrated in Figure 5.3. These dense zones are gravitationally unstable. If the electrophoresis is carried out vertically, the dense zone can simply fall through the solution like a droplet of mercury in water. If the electrophoresis is horizontal, any local fluctuations will lead to collapse of the zone as shown in the figure. Gels are one way to minimize the effects of these gravitational and convective instabilities. Another is to use density gradients generated by high concentrations of uncharged molecules, such as sucrose or glycerol. In practice, however, density gradient electrophoresis is a cumbersome and rarely used technique.

## MOTIONS OF DNA MOLECULES IN GELS

It is extremely useful to contrast the effect of a gel on two types of macromolecular separation processes: gel filtration or molecule sieve chromatography, and gel electrophoresis. Both of these techniques are illustrated schematically in Figure 5.4. In both cases the gel matrix potentially acts as a sieve. In gel filtration, particles of gel matrix are suspended in a bulk fluid phase. Molecules too large to enter the gel matrix see the gel particles as solid objects.

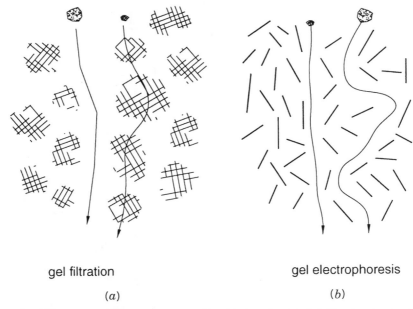

gel filtration

gel electrophoresis

(a)

(b)

**Figure 5.4**   Motion *(shown by arrows)* of molecules in gels. *(a)* Gel filtration (molecular sieve chromatography) where large molecules elute more rapidly than small molecules. *(b)* Gel electrophoresis where small molecules migrate more rapidly than large molecules.

As bulk liquid flow occurs, these large molecules move through the fluid interstices between the gel particles. They can find relatively straight paths, and thus their net velocity is rapid. In contrast, smaller molecules can enter the gel matrix. They experience a far larger effective volume as they move through the gel, since they can meander through both the gel matrix and the spaces between the particles. As a result their net motion is considerably retarded. Thus large species elute more rapidly than small species in gel filtration chromatography.

In typical gel electrophoresis the entire sample is a continuous gel matrix. There is no free volume between gel particles. All molecules must move directly through the gel matrix; thus the pore sizes must be large enough to accommodate this. Generally the gel matrices used have a very wide range of pore sizes. As a result relatively small molecules can move through almost all the pores. They experience the gel matrix as an open network, and they can take relatively straight paths under the influence of the electrical field. In contrast, larger molecules can pass through only a restricted subset of the gel pores. To find this subset, they have to take longer paths. Thus, even though their local velocity is the same as that of the small molecules, their net effective motion in the direction of the field is much slower. This means that small molecules will move much faster than large ones in gel electrophoresis.

## COMPLEX EFFECTS OF GEL STRUCTURE AND BEHAVIOR

The picture described above is the classical view of molecule motions in gels. It successfully explains the behavior of proteins and small DNAs under conditions where sieving is, in fact, the dominant solute-gel interaction. The details of the gel structure and its interaction with the solute are not considered at all. However, for DNA electrophoresis we now know that the details of the gel structure, its behavior, and its interactions with the solute matter considerably. Unfortunately, our current knowledge about the structure of gels, and the ways macromolecules interact with them, is very slight. For example, experiments have shown that the gels used for typical DNA separations can respond directly to electrical fields. The result is to change the orientation of gel fibers in order to minimize their interaction energy with the applied field. In general, macroscopic measures of gel fiber orientation show that these direct gel-field effects occur at field strengths higher than those typically employed is most electrophoresis.

When DNA is present, applied electrical fields lead to changes in the gel not seen in the absence of DNA. These changes presumably reflect distortion and orientation of the gel caused by motion of the DNA and direct gel-DNA interactions. It is not known if DNA has any indirect effect on the interaction between the gel and the electrical field. What is even less clear is whether the effects of DNA electrophoresis through the gel matrix lead to any irreversible changes in the gel structure. Certainly the overall forces of interaction between a highly charged macromolecule and an obstructive stationary phase could be considerable. We know that gels can usually not be reused more than a few times for electrophoresis without a serious degradation in their performance. This is attributed to some breakdown in necessary gel properties. How much of this is due to electrochemical attack on the gel matrix, and how much to DNA-mediated damage, remains unknown.

The buffer system used to cast the gel, and used in the actual gel electrophoresis itself, matters a great deal. For example, agarose gels cast or run in Tris (trihydroxymethyl-aminomethane) acetate and Tris borate behave very differently. This is believed to reflect the direct interaction of borate ion with cis hydroxyls on the agarose fibers. Finally the specific monomers and crosslinkers used in gels, like polyacrylamide, can have a profound effect on gel running speed, resolution, and the ability of the gel to discriminate unusual DNA structures. For simplicity, we will ignore all of these complications in the discussion that follows.

In general, polyacrylamide-like matrices are used for small, single-stranded DNAs in the size range of 1 to 2 kb. They are also used for small double-stranded DNAs. Agaroses, with much larger pores, are used for larger double-stranded DNAs, ranging from 1 kb to more than 10 Mb in size. These gel systems were not specifically designed to handle nucleic acids, and there is no particular reason to think that they are anywhere optimal for these materials. However, some alternative materials are clearly undesirable for DNA separations because they cannot be made sufficiently free of nucleases or because they demonstrate nonspecific adsorption of DNA molecules.

The ultimate solution for DNA separations may be to use micro-fabricated separation matrices. Here one uses techniques like microlithography to manufacture a custom-designed surface or volume containing specific obstacles to modulate DNA movement and prevent gross convection. Robert Austin at Princeton University has recently demonstrated the electrophoresis of DNA on a microlithographic array. He constructed a regular array of posts spaced at $1\text{-}\mu$ intervals, sticking up from a planer surface. In the presence of an electrical field, DNAs in solution move through these posts (Fig. 5.5). The much larger DNA molecules tend to get hooked on the posts; they eventually are pulled off and net motion through the array continues. This carefully controlled situation appears to mimic a number of key features of DNA electrophoresis in natural gels, as will soon be demonstrated. Whether such a regular array can actually outperform the separation characteristics of natural gels remains to be seen.

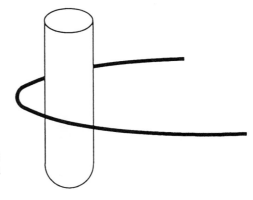

**Figure 5.5**   Example of DNA behavior seen during electrophoresis of DNA molecules on a microlithographic array of posts.

## BIASED REPTATION MODEL OF DNA BEHAVIOR IN GELS

A typical DNA molecule in a polyacrylamide or agarose gel will extend across tens to thousands of discrete pores or channels. The notion of treating this as a simple sieving problem clearly makes no sense. A number of groups including Lumpkin and Zimm, Slater and Noolandi, and Lerman and Frisch, have adapted the models of condensed polymer phases originally developed by Pierre DeGennes to explain DNA gel electrophoresis. DeGennes coined the term *reptation* to explain how a polymer diffuses through a condensed polymer solution or a gel. The gel defines a tube in which a particular DNA molecule slithers (Fig. 5.6a). Diffusion of DNA in the tube is restricted to sliding forward or backward. A net motion in either direction means that the head or tail of the DNA has entered a new channel of the pore network, and the remainder of the molecule has followed.

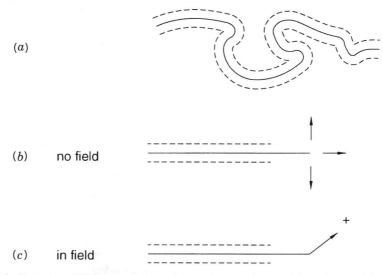

(a)

(b)   no field

(c)   in field

**Figure 5.6**   Behavior of DNA in gels according to the reptation model. *(a)* The path in the gel defines a tube. *(b)* With no field, motion of the head (or tail) of the DNA occurs by random diffusion. *(c)* With an electrical field, the direction chosen by the head is biased.

This redefines the tube. The model has the DNA behaving much in the same way as a snake in a burrow; hence the term reptation from the word reptile.

In polymer diffusion, when the head of the chain emerges from its tube, it makes a random choice from among the pores available to it (Fig. 5.6b). The elegant simplification offered by the reptation model is that to explain DNA motion, one need only consider what is happening at the head and the tail. The remainder of the complex network of pores, and the detailed configuration of the DNA (except for the overall length of the tube) can be ignored. Indeed this model works very well to explain polymer diffusion in condensed phases. To adapt the reptation model to electrophoresis, it was proposed that the influence of the electrical field biases the choice of pores made by the head of the DNA molecule (Fig. 5.6c). In this biased reptation model, one only needs to consider the detailed effect of the field on the head; the remainder of the chain follows, and its interaction with the gel and the field can be modeled in a simple way that does not require knowledge of the detailed configuration of the DNA or the gel pores.

The biased reptation model makes the following predictions: For relatively small DNAs (small for a particular combination of field strength and gel matrix), diffusion tends to balance out most of the effects of the applied field. The DNA molecule retains a chain configuration that is highly coiled, and the ends have a wide choice of pores. For much larger DNAs, the electrical field dominates over random diffusion. The molecule becomes highly elongated and highly oriented. The ends have a relatively small choice of pores, and there is more frictional interaction between the DNA and the pores. The predicted result of this on electrophoretic behavior is shown in Figure 5.7. Most experiments conform very well to this prediction. At a given field strength, up to some critical DNA length, the mobility drops progressively with DNA size. However, once a point is reached where the DNAs are fully oriented, they all move at the same speed. For typical agarose gel electrophoresis conditions, this plateau occurs at around 10 kb. Fully oriented molecules in gels display a constant charge and friction per unit length; thus their electrophoretic mobility is size independent.

Early attempts to circumvent the problems of DNA orientation and the resultant loss in electrophoretic resolution above a size threshold were not very successful. Lowering the field strength increased the separation range but at the great cost of much longer experimental running times. Lowering the gel concentration increased pore size and allowed larger molecules to be handled, but the resulting gels were extremely difficult to use because of their softness and fragility.

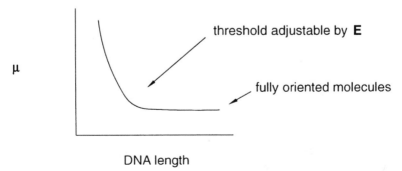

**Figure 5.7**  Dependence of electrophoretic mobility on DNA length predicted by the biased reptation model and observed under ordinary gel electrophoretic conditions.

## PULSED FIELD GEL ELECTROPHORESIS (PFG)

A totally new method for electrophoresis, PFG, was originally conceived as a way to circumvent the limitations on ordinary gel electrophoresis caused by DNA orientation. However, PFG has had a much broader impact because, as attempts to unravel the mechanism of this technique have progressed, they have revealed that the fundamental picture of DNA electrophoresis offered by the biased reptation model is totally inadequate, even for ordinary electrophoresis. The original rationale for PFG is shown in Figure 5.8. Imagine DNA molecules moving in a gel, fully elongated and oriented in response to the field. Suppose that the field direction is suddenly switched. After a while the DNA molecules will find themselves moving in the new field direction and fully oriented once again. However, to achieve this, they have to reorient, and this presumably requires passing through an intermediate state where the molecules are less elongated. Without considering the detailed mechanism of this reorientation (which is still unknown), it seems intuitively reasonable to guess that larger DNA molecules will take longer to reorient than smaller ones. Therefore, if periodically alternating field directions were used, larger DNAs should display slower net motion than smaller ones. This is because they will spend a larger percentage of each field cycle in reorientation, without much productive net motion (Fig. 5.9). Some examples of the power of PFG to separate large DNA molecules are illustrated in Box 5.1.

In retrospect, it is not surprising that PFG enhanced the ability to separate larger DNAs. It is still surprising that the effect is so dramatic. The first PFG experiments demonstrated the ability to resolve, easily, DNA molecules up to 1 Mb in size. Subsequent refinements have pushed these limits a factor of ten or more higher. Thus PFG has expanded the domain of useful electrophoretic separations of DNA by a factor of 100 to 1000. We still are not completely sure why.

There are countless variations on the type of PFG apparatus used, and the details of the electrical field directions employed. With rare exceptions, square wave field pulses have

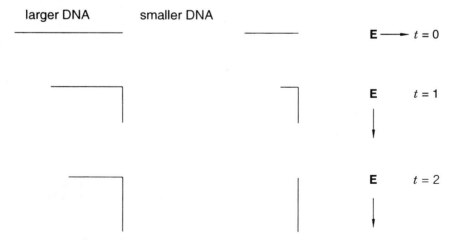

**Figure 5.8**  Original rationale for the development of pulsed field gel electrophoresis. The notion was that longer molecules would reorient more slowly in the gel in response to a change in electrical field direction.

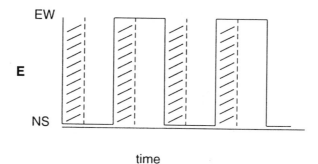

**Figure 5.9** Effect of reorientation time on PFG mobility, predicted from very simple consideration of the fraction of each pulse time needed for reorientation *(striped sections)*.

been used. These can be characterized by two parameters: the pulse time, which is the period during which the field direction is constant, and the field strength, during that pulse. In most general modes of PFG, the field strength is always constant. Commonly two alternate field directions are used; the net motion of the DNA is along the resultant of these two field directions as shown in Figure 5.10a. The usual angle between the two applied fields is 120° Changes in field direction can be accomplished by switching between pairs (or multiples) of electrodes, by physically rotating the electrodes relative to the sample, or by physically rotating the sample relative to a fixed pair of electrodes. The applied fields can be homogeneous, in which case the DNA molecules will move in straight paths, or inhomogeneous. In the latter case, where field gradients exist, DNA molecules do not move in straight paths. This makes it more difficult to compute the mobility of a sample or to compare, quantitatively, results on samples in different lanes, that is, samples with different starting positions in the gel. However, field inhomogeneities, properly applied, can produce band sharpening because one can create a situation where the leading edge of a zone of DNA is always moving slightly slower than the trailing edge.

---

**BOX 5.1**
**EXAMPLES OF PFG FRACTIONATIONS OF LARGE DNA MOLECULES**

The cohesive ends of bacteriophage lambda DNA were discussed in Box 2.3. These allow intramolecular circularization, but at high concentration, linear concatenation-ization is thermodynamically preferred. (For a quantitative discussion, see Cantor and Schimmel, 1980.) Variants of lambda with different sizes and other viral DNAs with similar cohesive ends are also known. These samples provide sets of molecules with known lengths spaced at regular intervals. Such concatemers are the primary size standards used in most PFG work. An example of what a PFG separation of such molecules looks like is shown in Panel A.

*(continued)*

**BOX 5.1** *(Continued)*

### Panel A. Separations of Concatemeric Assemblies of Bacteriophage DNAs

An overview of the DNA in a whole microbial genome can be gained by digestion of that genome with a restriction nuclease that has relatively rare recognition sites. This produces a set of discrete fragments that can be displayed by PFG size fractionation. An estimate of the total genome size can be made reliably by adding the sizes of the fragments. Variations in different strains show up readily as evidenced by the example in Panel B.

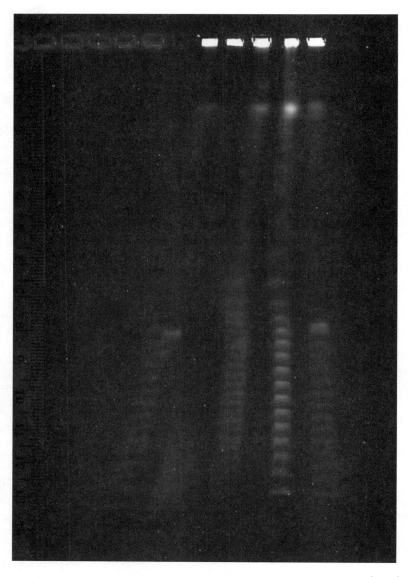

*(continued)*

**BOX 5.1** *(Continued)*

## Panel B. Separations of Total Restriction Nuclease Digests of Different Strains of *Escherichia coli*

Microorganisms, like yeasts, that contain linear chromosomal DNAs can be analyzed by PFG to yield a molecular karyotype. Any major rearrangements of these chromosomes are usually revealed by shifts in the size or number of chromosomes in the patterns of chromosomes that hybridize to a particular DNA probe. Since each chromosome is a genetic linkage group (Chapter 6), an initial PFG analysis (e.g., the examples shown in Panel C for yeasts) provides an instant overview of the genetics of an organism and greatly facilitates subsequent gene mapping by pure physical procedures.

*(continued)*

**BOX 5.1** *(Continued)*

**Panel C. Separations of Yeast Chromosomes**

In part (a), lane 2 is lambda DNA concatemer; lane 3 is some of the smaller *S. cerevisial* chromosomal DNAs. In part (b), lane 2 is *S. pombe* chromosomal DNAs; lane 3 is the largest *S. cerevisial* chromosomal DNAs.

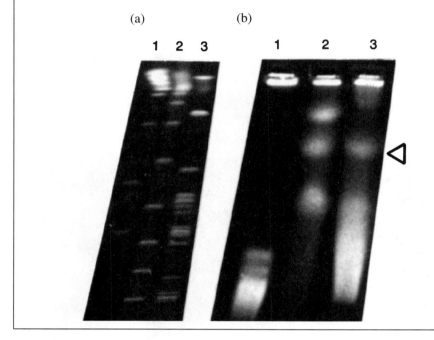

A number of variants of PFG exist where multiple field directions are used. These are generally not employed for routine PFG analyses. One convenient type of PFG apparatus uses multiple point electrodes, each individually adjusted by a computer-controlled power supply. The Poisson equation, $(\nabla)^2\phi = \rho$, can be used to compute the electrostatic potential, $\phi$, in the gel, by recognizing that the free charge, $\rho$, is zero everywhere in the gel, and

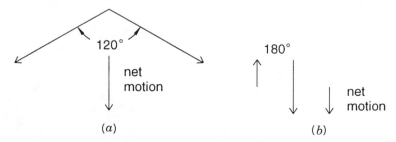

**Figure 5.10**   Two common experimental arrangements for PFG. *(a)* Field directions that alternate 120*(a)*. *(b)* FIGE where field directions alternate by 180*(a)*.

using as boundary conditions the voltages set at the electrodes. Then the electrical field, **E**, at each position in the gel can be calculated as $E = \text{grad } \phi$. This permits a single apparatus can be used for multiple field shapes and directions without the need to physically move or rewire numerous electrodes (Fig. 5.11). A popular variant of this approach uses voltages set at individual electrodes to ensure very homogeneous field shapes and thus very straight lanes. This version is called the contour-clamped homogeneous electrical fields (CHEF).

One version of PFG is basically different. It uses 180° angles between the applied fields (Fig. 5.10*b*). In this technique, called field inversion gel electrophoresis (FIGE), either the length of the forward and backward pulses must be different or their field strengths must be different; otherwise, there will be no net DNA motion at all. FIGE is quite popular because the apparatus needed for it is very simple and because FIGE can achieve very high resolution separations under some conditions. However, the effect of DNA size on FIGE mobility is complex, as we will demonstrate, and it is also rather sensitive to overloading by too high sample concentrations.

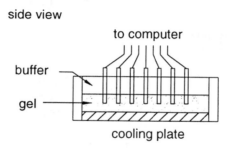

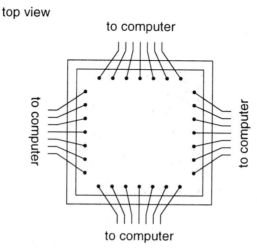

**Figure 5.11**   Schematic of a contemporary PFG apparatus in which computer-controlled individual electrodes can be used to generate a variety of field shapes in a single apparatus.

## MACROSCOPIC BEHAVIOR OF DNA IN PFG

The results of a large number of systematic studies on DNA mobilities in PFG are summarized in Figures 5.12 through 5.14. The major variables explored have been DNA size, the field strength, the angle between the fields, and the pulse time. Other parameters are also known to be important, such as the gel concentration (and the details of the type of agarose used), the buffer type, the ionic strength, and the temperature. These will not be considered further. The effect of angle is relatively slight in the range around 120°. However, regular oscillation of homogeneous electrical fields between two sets of parallel electrodes 90° apart does not produce PFG fractionations. More complex sets of 90° pulses, with varying durations or field strengths, or multiple directions, have been shown to be effective under some circumstances.

The PFG behavior of DNAs up to about a Mb in size with 120° alternating fields is shown in Figure 5.12 as a function of pulse time. A very simple picture suffices to explain these results, but it begs the question of the details of how DNA moves in a gel. At very long pulse times, the process of reorientation should require an insignificant fraction of each pulse period. As a result the DNA moves by essentially ordinary electrophoresis, in a zigzag pattern centered along the average of the two field directions. Because ordinary electrophoresis is dominant, there is no net effect of DNA size on mobility; the DNAs are essentially fully oriented almost all the time. At very short pulse times, the field changes direction much more rapidly than the DNA molecules can reorient. They experience a constant net field which is just the vector sum of the two distinct applied fields. They move in response to this net average field by ordinary electrophoresis. Since they become oriented and elongated, their net mobility is size independent and in fact is the same as their mobility at very long pulse times.

At intermediate pulse times, DNA reorientation processes occupy a significant fraction of each pulse period. This leads to a marked decrease in overall electrophoretic mobility. The pulse time at which mobility is a minimum increases roughly linearly with DNA size. Larger DNAs have a progressively broader response to pulse time. The result is that one can find pulse times that afford very good resolution of particular DNA size classes.

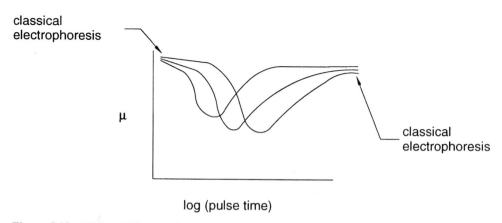

**Figure 5.12**   PFG mobility as a function of pulse time for DNAs of different size, observed using 120° alternate field directions.

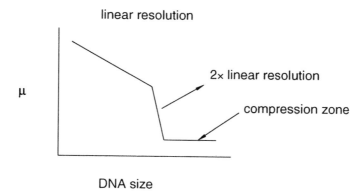

**Figure 5.13**   Dependence of PFG mobility on DNA size, with 120° alternation in field direction.

Resolution is most easily gauged by plotting mobility as a function of DNA size for a fixed pulse time. This is illustrated in Figure 5.13. Relatively short DNAs have a mobility that decreases linearly with DNA size. Above a sharp size threshold, the slope of this linear decrease doubles. In this size zone the resolution of PFG is particularly high. At even larger DNA sizes, the mobility of DNA becomes size independent. This size range is called the *compression zone*. The DNAs are still migrating in the gel, but there is no size resolution. The compression zone is useful where, for example, one wishes to purify all DNAs above a certain critical size. From the simple picture of PFG described above, the compression zone should consist of DNAs with sizes too big to reorient during the pulse time. However, the simple picture fails to explain the zone with especially high resolution. Indeed no model of PFG has yet explained this.

The dependence of FIGE mobility on DNA size is shown in Figure 5.14. It is dramatically different from the behavior seen in ordinary PFG. With FIGE, the mobility of DNA is not a monotonic function of its size. Very large and very small molecules move at comparable speeds. Molecules with intermediate sizes move slower, and the retardation of the slowest species is quite marked. A simple explanation of FIGE behavior, based on the notions we have explored thus far would say that small DNAs orient rapidly with each ong forward pulse and move efficiently. Large DNAs are never disoriented by the short

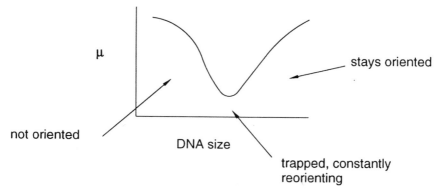

**Figure 5.14**   DNA mobility as a function of size in typical FIGE using only a single pulse time.

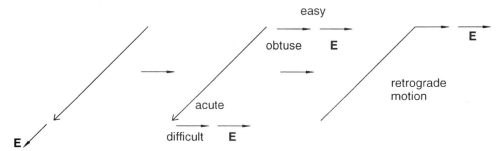

**Figure 5.15**    The ratchet model for PFG, proposed by Edwin Southern of Oxford University.

backward pulses; thus they remain oriented and move efficiently during the forward pulses. Intermediate size DNAs never achieve a configuration that allows efficient motion for an appreciable fraction of the forward pulse cycle.

The complex dependence of FIGE mobility on DNA size makes it difficult to use FIGE, at any particular set of conditions, to fingerprint a population of different DNA sizes. To circumvent this problem, one can progressively vary either the FIGE pulse times or field strengths during the course of an experiment. The use of such programs produces a superposition of a spectrum of FIGE experiments in which an approximate monotonic decrease in DNA mobility with increasing size is restored.

Early in the development of PFG, a very simple model was put forth by Southern that effectively explained why 90° angles might be ineffective, and why mobility should decrease linearly with DNA size. This ratchet model of DNA motion in gels is shown in Figure 5.15. In the ratchet model the head of the moving DNA is oriented along the applied field direction. When the field direction is switched, the molecule attempts to reorient by a reptation motion. If this were led by the head, it would require that the chain bend through an acute angle. It would seem easier, instead, for the tail to lead the reorientation, since that would require bending the chain through a much less sharp obtuse angle. This leads to retrograde motion until the entire chain has changed orientation. The ratchet model leads directly to a linear dependence of mobility on DNA size because the mobility decreases with the length of the retrograde motion. It also justifies the poor performance of angles 90° and smaller, since these eliminate the need for retrograde motion. However, the ratchet model does not easily account for FIGE, nor for the zone of enhanced resolution in PFG. Finally the ratchet model predicts that molecules greater than a specific size will not move at all, contrary to the observation that the molecules in the compression zone move, albeit slowly.

## INADEQUACY OF REPTATION MODELS FOR PFG

Three specific observations of macroscopic DNA behavior are very difficult to reconcile with any type of biased reptation model such as the ratchet or simple reorientation pictures described above. The first of these are measurements of the field strength dependence of PFG mobility, especially for relatively short DNAs. These showed $\mu$ proportional to $E^2$, where $E$ is the field strength in volts/cm. This behavior clearly reflects a complex mechanism, since $\mu$ must be an odd function of $E$ to ensure net motion in a particular field direction, while $E^2$ is an even function that implies no net migration direction.

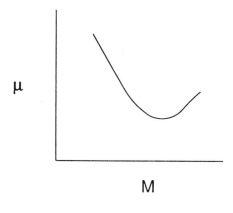

**Figure 5.16** Behavior of DNA in ordinary electrophoresis at high field strengths in dense gels: Mobility, $\mu$, as a function of molecular weight, M.

The second perplexing observation is DNA trapping. For DNAs with sizes above a particular threshold, at high enough fields no PFG motion could be observed. The threshold size decreased with increasing field strength. This DNA trapping implies more complex DNA-gel interactions than contained in simple reptation models. A final, devastating observation was made in ordinary DNA electrophoresis in high concentration gels at very high field strengths. Here it was observed that, under some conditions, the mobility of DNA was no longer a monotonically decreasing function of DNA size. Instead, at a particular DNA size, a minimum in mobility is observed, just as in FIGE, even though in these experiments a constant, uniform applied electrical field was employed (Figure 5.16).

Three different approaches have been used to examine in detail the nature of DNA motions in gels under the influence of electrical fields. All of these produced unexpected results, totally inconsistent with simple reptation pictures. This was true, even in ordinary electrophoresis with constant fields, or pulsed fields in a single direction. Several different groups reported, almost simultaneously, the detection, by UV fluorescence microscopy, of single DNA molecules undergoing gel electrophoresis. In order to do these experiments, the DNAs were prestained with an intercalating dye like acridine orange or ethidium bromide. These dyes bind at every other base pair and increase the overall length of the DNA by about 50%. Studies of macroscopic DNA electrophoresis, or PFG, with and without bound dyes, indicate that there is little perturbation of the behavior of the DNA by the dyes, except for the predictable consequences of this increase in length.

When observed in the microscope at constant field strength, the motion of DNA molecules was very irregular. In the absence of a field, the molecules were coiled and moderately compact (Figure 5.17). This is expected from random walk models of polymer chain statistics. On application of an electrical field, the molecules elongate, orient, and move parallel to the field. However, the head soon collides with some obstacle in the gel. It stops moving, but the rest of the molecule does not. As a result the elongated chain collapses; the tail may even overtake the head. A very condensed configuration is formed around the obstacle. Eventually one end of the DNA works free and starts to move, pulling some of the chain with it. If the DNA is still attached to the obstacle, both ends may pull free, and the result is a very elongated, tethered structure. Finally the motion of one end dominates; the DNA slips free of the obstacle and starts to run as an elongated aligned structure until it impacts on the next obstacle. Thus the DNA spends most of its time entrapped on obstacles, and the detailed dynamics of how it becomes trapped and freed dominate the overall electrophoretic behavior. The overall motion is very irregular.

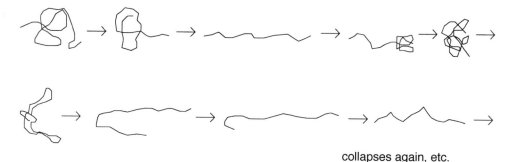

collapses again, etc.

**Figure 5.17**   DNA behavior in conventional gel electrophoresis as visualized by fluorescence microscopy of individual stained DNA molecules.

There may be DNA sizes where unhooking from obstacles is especially difficult at particular field strengths. This would explain the minimum in ordinary electrophoretic mobility seen macroscopically for certain DNA size ranges at high field strengths in dense gels.

When DNA molecules undergoing PFG are viewed by fluorescence microscopy, additional unexpected behavior is seen (Fig. 5.18). When the field is rotated by 90°, DNA molecules respond to the new direction not by motions of their ends but by herniation at several internal sites. This produces a series of kinks which start to move in the direction of the new field. These kinks grow and compete. Eventually one dominates, and this becomes the leading edge of the moving DNA. At some subsequent point the hairpin structure presumably unravels, and the DNA attains full elongation again. Note that this picture is quite at odds with the reptation model where the DNA is supposed to remain in its tube, except for motions at the ends. It suggests that a tube model is not at all appropriate for DNA in a gel.

The second experimental approach that revealed unexpected complexities of DNA behavior in gels was measurement of bulk electrophoretic orientation by linear dichroism (LD). Here the absorbance of polarized UV light by DNA in gels in the presence of an electrical field was measured. The base pairs of DNA are the dominant absorber of near-UV light (wavelengths around 260 nm). As shown in Figure 5.19, the base pairs preferentially absorb light polarized in the plane of the bases. DNA tends to orient in an electrical field with the helix axis parallel to the field. The LD is defined as

$$LD = A_z - A_y$$

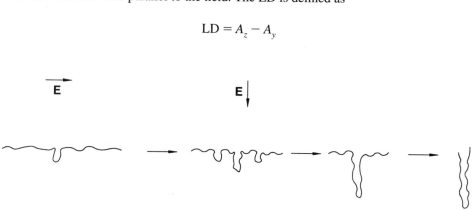

**Figure 5.18**   Fluorescence microscopic images of DNA in a gel after a 90° rotation of the electrical field direction. Shown from left to right are successive time points.

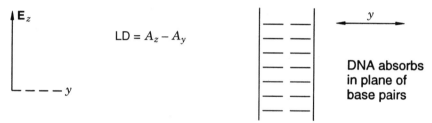

**Figure 5.19** Electric dichroism of DNA. An electrical field is applied in the $z$ direction $(E_z)$. Absorption of light polarized in the $z$ and $y$ directions is compared. $A_z$ and $A_y$ for DNA will be negative because the molecule orients along the $z$ axis but preferentially absorbs light in the planes of the base pairs which will be perpendicular to the $z$ axis.

where $A_z$ is the absorbance with polarizers set parallel to the field, and $A_y$ is the absorbance with perpendicular polarizers. This means that the net LD of oriented DNA will be negative, as shown in Figure 5.19. The magnitude of the LD is a measure of the net local orientation of the DNA helix axis.

The LD of DNA in the absence of an electrical field is zero because there is no net orientation. When the LD of DNA in a gel is monitored during the application of a square wave electrical field pulse, very surprising results are seen. What was expected is shown in Figure 5.20$a$: a monotonic increase in $-$LD until the orientation saturates; then a monotonic

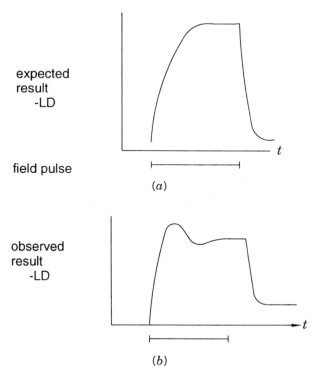

**Figure 5.20** Linear dichroism of DNA in a gel, as a result of a single applied electrical field pulse of sufficient intensity to cause a saturating level of orientation. *(a)* Expected result. *(b)* Observed result (from Nordén et al., 1991).

decrease in − LD soon reaching zero, after the field is removed. What is actually observed, under a fairly wide range of conditions, is shown in Figure 5.20b. Upon application of the field, the − LD increases, but it overshoots above the steady state orientation value and then goes through one or more oscillations before saturation occurs. When the field is turned off, most − LD is lost as expected, but a small amount takes a very long time to decay to zero. Furthermore the decay kinetics depend on the original field strength, even though the decay takes place in the absence of the field. This DNA behavior is very complex and difficult to explain by simple models.

The orientation of the agarose molecules that make up the gel itself can be examined by measuring the linear birefringence, which is just $n_z - n_y$, where $n$ is the refractive index and $z$ and $y$ indicate light polarization axes, as illustrated in Figure 5.19. Birefringence must be used to examine the gel rather than LD because the gel has only very weak near-UV absorbance. The birefringence is dominated by the gel, since it is present at much higher concentrations than the DNA. Without DNA, no field-dependent birefringence is seen at the field strengths used in these experiments. However, in the presence of DNA, the gel shows a rapid orientation after application of an electrical field pulse, and a much slower disorientation after the field is removed. This indicates that the DNA is interacting with the gel and remodeling its shape under the influence of the electrical field. Clearly simplistic explanations will not be adequate to explain DNA electrophoresis in gels.

When LD measurements are applied to monitor the effects of successive field pulses, even more complications emerge. These results are illustrated in Figure 5.21. Two patterns of behavior are seen when pulses are applied spaced by an interval $\Delta t$. When $\Delta t$ is comparable to the pulse time, and the second pulse is either in the same direction as the first or at 90° to the first, no overshoot or orientation oscillations are seen in response to the second pulse. In contrast, when $\Delta t$ is comparable to the pulse time, but the second pulse is oriented at 180°, or alternatively, when $\Delta t$ is very long relative to the pulse time, then the second pulse produces an overshoot and oscillations comparable to those produced by the first pulse. These observations have a number of profound implications. They suggest that the relaxed DNA coil is not equivalent to any moving state. The initial response of this relaxed configuration leads to hyperorientation, presumably because of the nature of the way the DNA becomes hooked on gel obstacles. Regaining the original relaxed configuration after a pulse can be very slow; it can take up to 30 minutes for 100 kb DNA, which is much longer than the times inferred for orientation and disorientation from macroscopic observations of PFG mobilities. Apparently inverted pulses act effectively to produce a relaxed-like state. Perhaps under these circumstances the DNA can largely retrace the paths it took when it became entangled with obstacles. Perpendicular pulses do not restore a relaxed configuration. One way to interpret the overshoot seen by LD is to postulate a phased response of the original population of relaxed molecules to the applied field. Once the DNAs have equilibrated in the field, a more complex set of orientations occurs which dephases subsequent responses.

To mimic macroscopic PFG more closely, LD measurements have been performed with continuous alternate pulsing. The time-averaged − LD was measured as a function of pulse time for different angles, as shown in Figure 5.22. It is apparent that relatively good, net local DNA orientation is seen with both long and short pulse times. However, with intermediate pulse times, much poorer net orientation is seen. This picture fits very well with the earliest ideas about a minimum in DNA orientation at interme-

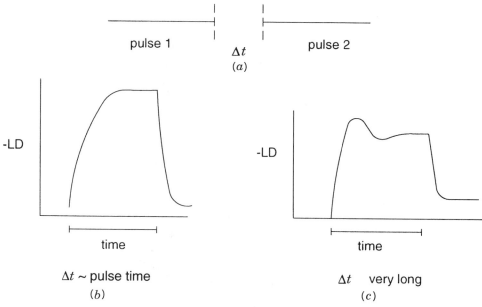

**Figure 5.21**  Dichroism of DNA in a gel produced by the second of two successive pulses, separated in time by $\Delta t$. *(a)* Time course of pulses. The remaining panels show LD in response to the second pulse. *(b)* Results of $\Delta t$ is comparable to the pulse time with the field direction of the second pulse either parallel to the first pulse or perpendicular to it. *(c)* Results if $\Delta t$ is very long compared to the pulse time, or if the second pulse is opposite in direction to the first pulse. (Adapted from Nordén et al., 1991.)

diate pulse times. However, the mechanisms that underlie these minima appear to be much more complex than originally envisioned. Note that the minimum in orientation is much more pronounced with 120° pulses than with 90° pulses. This correlates with the much more effective ability of 120° PFG to fractionate different size DNA molecules.

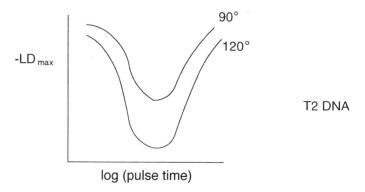

**Figure 5.22**  Linear dichroism seen for maximally oriented bacteriophage T2 DNA produced by continually alternating electrical field directions. (Adapted from Nordén et al., 1991.)

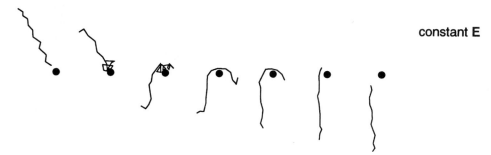

constant E

**Figure 5.23**   Behavior of DNA in gel electrophoresis as simulated by Monte Carlo calculations. Shown from left to right are successive time points. The black dot is an obstacle. The electrical field is vertical.

The third approach that has been used to improve our understanding of DNA electrophoresis in gels is computer simulations using Monte Carlo methods. Here the DNA is typically modeled as a set of charged masses attached by springs. The fluid is modeled by its contribution to the kinetic energy of the DNA through brownian motion. The gel is modeled as a set of rigid obstacles. Most early simulations were done in two dimensions to limit the amount of computer time required. This is unlikely to allow a realistic picture of either the gel or the DNA. However, more recent three-dimensional simulations are at least in qualitative agreement with the earlier two-dimensional results. Undoubtedly the model of the gel used in these simulations is unrealistically crude; the potential used to describe the interaction of the DNA with the gel is hopelessly oversimplified. The DNA model itself is not terribly accurate. Despite all these reservations, the picture of DNA behavior that emerges from the simulations is remarkably close to what is observed by microscopy. An example is shown in Figure 5.23. Collisions with obstacles dominate the motion of the DNA; hyperelongation of hooked structures, until they are released, apparently accounts for the overshoots seen in LD experiments.

Only a few simulations have yet been reported for PFG. These suggest that kinked structures play important roles. It seems logical that kinks should enhance the chances of DNA hooking onto obstacles (Fig. 5.24). Thus conditions that promote kinking might lead to minima in mobility. This notion, which is not yet proved, would at least be consistent with what is generally seen in macroscopic PFG experiments. The speed of most computers, and the efficiency of the algorithms used, has limited most simulations of electrophoresis to relatively short DNA chains at very dilute concentrations. Recently Yaneer Bar-Yam at Boston University has demonstrated orders of magnitude increases in simulation speed by implementing a molecular automaton approach on an intensely parallel computer architecture. This increase in the power of computer simulations should allow a wide variety of experimental conditions to be modeled much more efficiently.

**Figure 5.24**   Expected behavior of kinked DNA in electrophoresis.

Such increases in computation speed are needed because two additional variations of PFG show considerable promise for enhanced size resolution, but each of these introduces additional complications into both the experiments and any attempts to simulate them.

## DNA TRAPPING ELECTROPHORESIS

The technique of DNA trapping electrophoresis was devised by Levi Ulanovsky and Walter Gilbert as an approach to improving the resolution of DNA sequencing gels. These gels examine single-stranded DNA in crosslinked polyacrylamide under the denaturing conditions of 7 M urea and elevated temperatures. Typical behavior of DNA under such conditions is shown in Figure 5.25. A monotonic decrease in mobility with increasing DNA size is seen until at some threshold, usually around 1 kb, a limiting mobility is reached, and all larger molecules move through the gel at the same velocity. This presumably reflects complete orientation, just as we have argued for double-stranded DNA in agarose. To circumvent the loss in resolution at large DNA sizes, a globular protein was attached to one end of the DNA, as shown schematically in Figure 5.26. The actual protein used was streptavidin because of the ease of placing it specifically on the end-biotinylated target DNA. Streptavidin is a 50 kDa tetramer that can have a diameter of 40Å. This is considerably fatter than the 25-Å diameter of the DNA double helix. Note that streptavidin has such a stable tertiary and quaternary structure that it remains as a folded globular tetramer even under the harsh, denaturing conditions of DNA sequencing gel electrophoresis.

The larger size of the streptavidin, and its lack of charge under the electrophoretic conditions used, should ensure that the tagged end exists predominantly as the DNA tail. Periodically the head of the DNA chain will enter gel pores too large for the bulky tail to penetrate. This will lead to enhanced trapping, since the entire DNA will have to back out of the pore in order for net motion to occur. Above a certain DNA size, thermal energies may be insufficient to allow such backtracking, and once trapped, the tethered DNA might remain so indefinitely. A typical experimental result is shown in Figure 5.25.

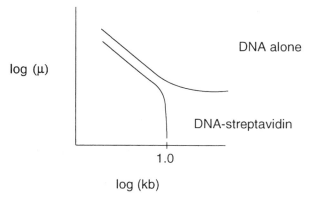

**Figure 5.25** DNA trapping electrophoresis. Shown is the dependence of the electrophoretic mobility of single-stranded DNA on size for ordinary DNA and DNA end-labeled with streptavidin (see Fig. 5.26). (Adapted from Ulanovsky et al., 1990.)

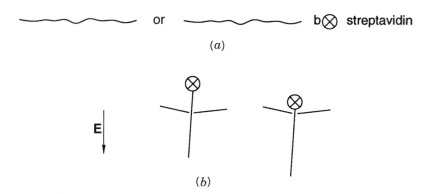

(a)

(b)

**Figure 5.26** Effect of a bulky end label on DNA mobility in gels. *(a)* Comparison of DNA with DNA terminally labeled with streptavidin. *(b)* Trapping of DNA in a pore because of the bulkiness of the streptavidin.

Instead of a limiting mobility, tagged DNA shows a zone of super high resolution until, at a key size, the mobility drops to zero. This is all in accord with the simple view illustrated in Figure 5.23.

If the picture just described is accurate, one can make the further prediction that FIGE will assist the escape of the trapped DNA. These results are shown in Figure 5.27. The three curves in this figure show the size dependence of DNA mobility under three different FIGE conditions with DNAs tethered to streptavidin. Very slow FIGE leads to a reappearance of the limiting mobility. Apparently under these conditions all trapped molecules can be rescued. Very rapid FIGE fails to remove any trapping effects.

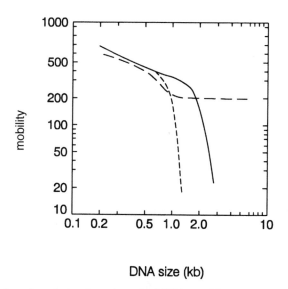

DNA size (kb)

**Figure 5.27** DNA trapping electrophoresis under FIGE conditions. Shown is FIGE mobility for three different sets of conditions, Forward/reverse (ms): 800/200 (long dashed line), 80/20 (solid line), 8/2 (short dashed line). In each case the sample was a single-stranded DNA terminally labeled with a streptavidin. (Adapted from Ulanovsky et al., 1990.)

Apparently these pulses are too short to allow the trapped molecules to escape. Intermediate FIGE conditions produce a very impressive increase in the useful separation range of the gel. Overall, DNA trapping electrophoresis appears to be a very interesting idea worth further elaboration and pursuit. It underscores the fact that trapping, and not sieving, appears to be the dominant effect underlying DNA electrophoretic size separations.

## SECONDARY PULSED FIELD GEL ELECTROPHORESIS (SPFG)

The history of SPFG is amusing. Tai Yong Zhang, a technician then working with us in the early 1990s, was instructed to perform the experiment shown schematically in Figure 5.28. The notion was to use a series of rapid field alternations to perturb the motion of the head of an oriented DNA chain moving in response to a slowly varying field. The result should be to fold the chain into a zigzag pattern, which ought to affect its subsequent orientation kinetics markedly when the primary field direction is switched. Zhang, whose English was quite imperfect at the time, misunderstood the original instructions and did the experiment shown schematically in Figure 5.29*a*. He applied periodic, short intense pulses, along the direction of net DNA motion. At about the same time Jan Noolandi's group did a similar experiment, shown in Figure 5.29*b*. They applied short, intense pulses opposite to the direction of net DNA motion. The results of both experiments, now called SPFG, are quite similar. The overall rate of DNA motion is dramatically increased. In addition, under some SPFG conditions, greatly improved resolution is seen, and larger molecules can be handled than is possible with conventional PFG alone under comparable field strengths and primary pulse times. Some examples are shown in Figure 5.30.

While the actual mechanism for the success of SPFG is unknown, one suggested mechanism shown in Figure 5.31 seems quite reasonable. DNA in PFG spends most of its time trapped on obstacles. As long as the field direction remains constant, the applied field will tend to keep the DNA trapped, as illustrated directly in trapping electrophoresis.

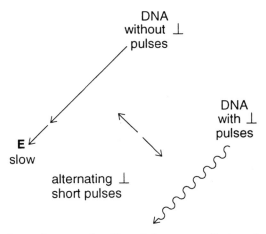

**Figure 5.28**    Untried experiment on the effect of short perpendicular pulses on PFG mobility.

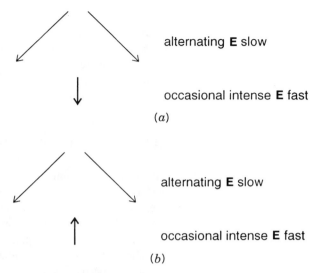

**Figure 5.29**   Electrical field configurations actually used in SPFG. *(a)* In our own work, *(b)* in the work of Jan Noolandi and collaborators.

Any tangential field, such as applied in either implementation of SPFG, will tend to bias the configuration of the trapped DNA, making it easier to slip off the obstacle in response to the primary field. At present it is difficult to find optimal SPFG conditions because of the large number of experimental variables involved, and the fact that these variables are highly interactive (Table 5.1). Perhaps the vastly improved rates of simulations of DNA electrophoresis will soon be applied to increase our understanding of SPFG.

## ENTRY OF DNAs INTO GELS

An additional, surprising aspect of SPFG was revealed when the behavior of molecules as large as (presumably) intact human DNAs was observed. It was found that the conditions required for gel entry were far more stringent than the conditions needed to move DNA, once the DNA was inside the running gel matrix. This may reflect a trivial artifact, that the DNA must be broken in order to enter the matrix, or it could be revealing an intriguing aspect of DNA behavior. Note that DNA samples for PFG or SPFG are all made in situ in agarose, as shown schematically in Figure 5.32. The sample agarose is a low-gelling temperature variety that has different pore sizes than the running gel. In addition the sample gel (0.5%) usually has half the agarose concentration of the running gel (1.0 to 1.1%). However, what is probably more significant is that once the sample cells are lysed, and the chromosomal DNAs freed of bound protein and any other cellular constituents, they find themselves in free solution inside a chamber much bigger than the gel pores. When the electrical field is turned on, the DNA coils in free solution are swept to one side of the chamber where they may encounter the gel in quite a different state than they experience once they have threaded its way into the first series of gel pores.

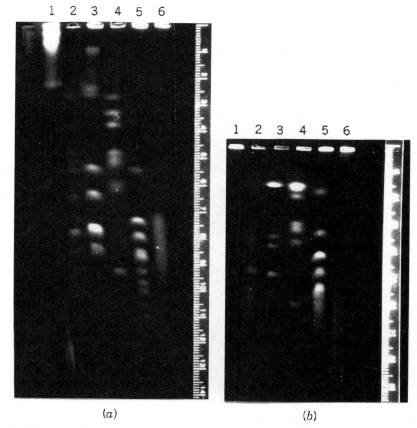

(a)                                              (b)

**Figure 5.30** Examples of the effect of secondary pulses on PFG behavior. Separation of chromosomal DNA molecules in the size range between 50 kb and 5.8 Mb by SPFG. Samples are (lane 1) *S. pombe,* (lane 2) Pichia 1A which consists of a mixture of *P. scolyti* and *P. mississipiensis,* (lane 3) *P. scolyti,* (lane 4) *P. mississippiensis,* (lane 5) *S. cerevisiae,* and (lane 6) lambda concatemers. Separation on a 1% agarose gel using *(a)* a pulse program as follows: (i) 4800 s, 2.2 volts/cm primary pulses with 1:15 s (1 second pulse every 15 seconds), 6 volts/cm secondary pulses for 12 h, (ii) 2400 s, 2.8 volts/cm primary pulses with 1:15 s, 6 volts, cm secondary pulses for 12 h, (iii) 240 s, 6 volts/cm primary pulses with 1:15 s; 10 volts/cm secondary pulses for 21 h, and (iv) 120 s, 6 volts/cm primary pulses with 1:15 s, 10 volts/cm secondary pulses for 10 h. *(b)* The same primary pulse program shown in *(a)* but without secondary pulses. (From Zhang et al., 1991.)

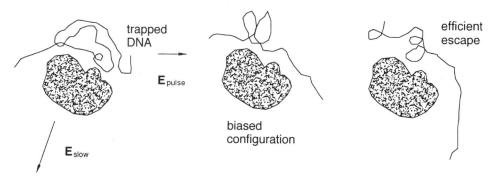

**Figure 5.31** Schematic picture of how secondary pulses may accelerate PFG by releasing DNAs from obstacles.

**TABLE 5.1   Interactive Parameters in Secondary Pulsed Field Electrophoresis (SPFG)**

| Electrical Parameters | Other Parameters[a] |
|---|---|
| Primary field strength | Agarose gel concentration |
| Primary pulse time | Particular type of agarose (low melting, high |
| Angle of alternation of primary fields | endoosmosis, etc.) |
| Secondary field strength | Ionic strength |
| Direction of secondary field | Temperature |
| Secondary pulse time | Specific ions present (e.g., acetate versus borate) |
| Phase between secondary and primary fields | |
| Homogeneity of primary field | |
| Homogeneity of secondary field | |

[a]Known to affect PFG and thus assumed to affect SPFG as well.

One possible consequence of the way in which large DNA is made and the obstacles it must overcome to enter a gel is shown schematically in Figure 5.33. Vaughan Jones, a mathematical topologist at the University of California, Berkeley, in thinking about large DNA molecules, suggested in 1989 that above a critical DNA size the conformation of these chains would always contain at least one potential knot. Much earlier, Maxim Frank-Kamenetskii, in thinking about DNA cyclization reactions, had very similar thoughts (Frank-Kamenetskii, 1997). Knot formation poses no problems inside cells where topoisomerases exist in abundance that can tie and untie knots at will. Similarly it is no problem for an isolated DNA molecule in solution. However, as DNA enters a gel, any knots it contains are prime targets for entanglement on obstacles. Despite the fact that this has occurred, the remainder of the DNA molecule will tend to be pulled further into the gel by the influence of the electrical field. This will result in tightening the knot, and it may lead to permanent trapping of the DNA at the site of entry into the gel. Secondary pulses could help to alleviate this problem by continually untrapping the knotted DNA from the obstacle until it has, by chance, assumed an unknotted configuration that allows it to fully enter the gel and move unimpeded. This idea has been explored recently by Jean Viovy who has demonstrated, using a clever multidimensional version of PFG, that DNA molecules do become irreversibly trapped when they are forced to run in agarose above a certain field strength (Viovy et al., 1992). Viovy argues from considerations of the forces involved that the effect of secondary pulses may be to prevent knots from tightening rather than to untie them once formed. What is needed is a clever way to test this intriguing but unproved suggestion.

insert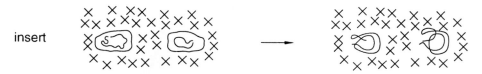

**Figure 5.32**   Preparation of large DNA molecules by in-gel cell lysis and deproteinization.

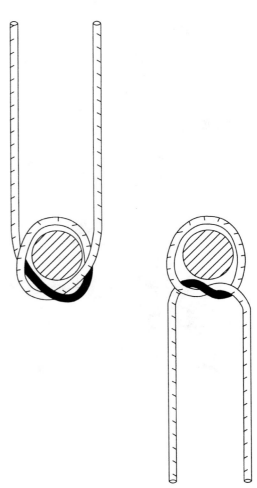

**Figure 5.33** How DNA knotting may prevent gel entry. Jean Viovy has obtained evidence that is consistent with the notion that knots are responsible for irreversible trapping of DNA within the gels. But there is still no direct proof for this idea (Viovy et al., 1992).

---

**BOX 5.2**
**PRACTICAL ASPECTS OF DNA ELECTROPHORESIS**

The ideal PFG fractionations would be rapid, high-yield resolution separations and allow larger quantities of sample to be purified to increase the ease and sensitivity of subsequent analyses. This ideal is not achievable because both speed and sample size generally lead to a loss in resolution, for reasons that are not fully understood.

The quality of PFG separations strongly deteriorates as DNA concentration is raised. Samples should be run at the lowest concentrations that allow proper visualization or other necessary analysis of the data or utilization of the separated samples. Examples are presented in Panel A.

*(continued)*

**BOX 5.2** *(Continued)*

**Panel A. Effect of DNA Concentration on Apparent Electrophoretic Mobility (Adapted from Doggett et al. 1992)**

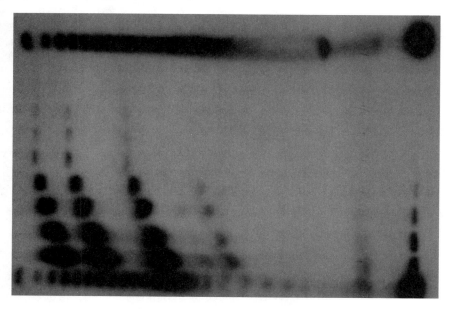

In ordinary agarose electrophoresis of DNA, most band broadening occurs not by diffusion but by an electrical-field-dependent process called *dispersion.* The factors that produce this dispersion remain to be clarified (Yarmola et al., 1996).

In PFG separations in all generally used apparatus, a trade-off must be made between band sharpness (resolution) and how straight the lanes are. The use of field gradients, an inherent property of the first PFG instrument designs, leads to band sharpening. However, no successful design of a PFG apparatus has been demonstrated that produces straight lanes and still allows gradients to be present to sharpen the bands. Examples of the trade-offs before resolution and straightness are shown in Panel B.

**Panel B. Effect of a Gradient in Electrical Field on the Apparent Sharpness of Bands**

The basic mechanism by which gradients lead to band sharpening is shown in Panel C:

*(continued)*

**BOX 5.2** *(Continued)*

CF 14
Decreasing concentration
→

Amount loaded    1/16 1/8 1/4 1/2 3/4 *I* *I* *I*

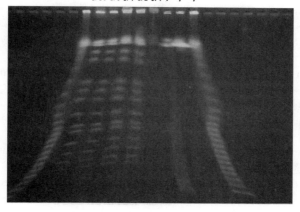

1/16 1/8 1/4 1/2 3/4

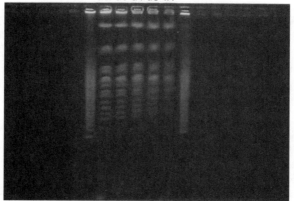

## Panel C. Rationale of the Effect of Electrical Field Gradients

Because molecules at the back of the zone are moving faster than those at the front, they will catch up until limits on zone thickness are reached that broaden the zone either by dispersion or by concentration-dependent effects.

Band sharpening can also be achieved by other methods. For example, a gradient of increasing gel concentration will mimic the effect of a gradient of decreasing electrical field strengths.

## SOURCES AND ADDITIONAL READINGS

Birren, B., and Lai, E., eds. 1993. *Pulsed Field Gel Electrophoresis*. San Diego: Academic Press.

Birren, B., and Lai, E. 1994. Rapid pulsed field separation of DNA molecules up to 250 kb. *Nucleic Acids Research* 22: 5366–5370.

Bustamante, C., Vesenka, J., Tang, C. L., Rees, W., Guthold, M., and Keller, R. 1992. Circular DNA molecules imaged in air by scanning force microscopy. *Biochemistry* 31: 22–26.

Cantor, C. R., and Schimmel, P. R. 1980. *Biophysical Chemistry*. San Francisco: W. H. Freeman, ch. 24.

Carlsson, C., and Larsson, A. 1996. Simulations of the overshoot in the build-up of orientation of long DNA during gel electrophoresis based on a distribution of oscillation times. *Electrophoresis* 17: 1425–1435.

Desruisseaux, C., and Slater, G. W. 1996. Pulsed-field trapping electrophoresis: A computer simulation study. *Electrophoresis* 17: 623–632.

Doggett, N. A., Smith, C. L., and Cantor, C. R. 1992. The effect of DNA concentration on mobility in pulsed-field gel electrophoresis. *Nucleic Acids Research* 20: 859–864.

Duke, T. A. J., Austin, R. H., Cox, E. C., and Chan, S. S. 1996. Pulsed-field electrophoresis in microlithographic arrays. *Electrophoresis* 17: 1075–1079.

Frank-Kamenetskii, M. F. 1997. *Unraveling DNA*. Reading MA: Addison-Wesley.

Gurrieri, S., Smith, S. B., Wells, K. S., Johnson, I. D., and Bustamante, C. 1996. Real-time imaging of the reorientation mechanisms of YOYO-labelled DNA molecules during 90 degrees and 120 degrees pulsed field gel electrophoresis. *Nucleic Acids Research* 24: 4759–4767.

Monaco, A. P., ed. 1995. *Pulsed Field Gel Electrophoresis: A Practical Approach*. New York: Oxford University Press.

Nordén, B., Elvingson, C., Jonsson, M. and Åkerman, B. 1991. Microscopic behavior of DNA during electrophoresis: electrophoretic orientation. *Quarterly Review of Biophysics* 24: 103–164.

Schwartz, D., and Cantor, C. R. 1984. Separation of yeast chromosome-sized DNAs by pulsed field gradient gel electrophoresis. *Cell* 37: 67–75.

Smith, M. A., and Bar-Yam, Y. 1993. Cellular automaton simulation of pulsed field gel electrophoresis. *Electrophoresis* 14: 337–343.

Turmel, C., Brassard, E., Slater, G. W., and Noolandi, J. 1990. Molecular detrapping and band narrowing with high frequency modulation of pulsed field electrophoresis. *Nucleic Acids Research* 18: 569–575.

Ulanovsky, L., Drouin, G., and Gilbert, W. 1990. DNA trapping electrophoresis. *Nature* 343: 190–192.

Viovy, J. L., Miomandre, F., Miquel, M. C., Caron, F., and Sor, F. 1992. Irreversible trapping of DNA during crossed-field gel electrophoresis. *Electrophoresis* 13: 1–6.

Whitcomb, R. W., and Holzwarth, G. 1990. On the movement and alignment of DNA during 120 degrees pulsed-field gel electrophoresis. *Nucleic Acids Research* 18: 6331–6338.

Yarmola, E., Sokoloff, H., and Chrambach, A. 1996. The relative contributions of dispersion and diffusion to band spreading (resolution) in gel electrophoresis. *Electrophoresis* 17: 1416–1419.

Zhang, T.-Y., Smith, C. L., and Cantor, C. R. 1991. Secondary pulsed field gel electrophoresis: a new method for faster separation of larger DNA molecules. *Nucleic Acids Research* 19: 1291–1296.

# 6 Genetic Analysis

## WHY WE NEED GENETICS

Despite the great power of molecular biology to examine the information coded for by DNA, we have to know where on the DNA to look to find information of relevance to particular phenomena. Some genes have very complex phenotypic consequences. We may never have the ability to look at a DNA sequence and infer, directly, that it regulates some aspect of facial features or mathematical reasoning ability. Genetic analysis offers a totally independent approach to determining the location of genes responsible for inherited traits. In this chapter we will explore the power of this approach and also some of its particular limitations. Genetics, in the human, will mostly serve as a low resolution method for localizing genes, but it is the only method available if the only prior information about a gene is some hypothesis about its function, or some presumed phenotype that results from a particular DNA sequence (allele) of that gene.

## BASIC STRATEGY FOR GENETIC ANALYSIS IN THE HUMAN: LINKAGE MAPPING

In most organisms, genetics is carried out by breeding specific pairs of parents and examining the characteristics of their offspring. Clearly this approach is not practical in the human. Even leaving aside the tremendous ethical issues such an approach would raise, the small size of our families, and the long lifespan of our species, make genetic experimentation all but impossible. Instead, what must be done is to perform retrospective analyses of inheritance in families. Statistical analysis of the pattern of inheritance is used instead of direct genetic manipulation to test hypotheses about the genetic mechanism underlying a particular trait.

Human beings and other mammals are diploid organisms. They contain two copies of each homologous chromosome. Normally an offspring receives one of its homologs from each parent (Fig. 6.1). This is accomplished by a specialized cell division process, meiosis, which occurs during the formation of sperm and eggs, a process known as gametogenesis. Sperm cells and egg cells (ova) are haploid; both contain only a single one of each of the homologous pairs of the parental chromosomes. When sperm and egg combine in fertilization, this reconstitutes a diploid cell which is a mosaic of halves of the genomes of its parents.

Linkage is the tendency for two observable genetic traits, called *markers,* to be coinherited if they lie near each other on the same chromosome. To be distinguishable genetically, a marker must occur in more than one form (e.g., eye color) in different members of the population. We call these forms alleles. Consider the case shown in Figure 6.1 where two detectable markers, one with alleles A or a and the other with alleles B or b, are being followed in a family. Markers A and B are both inherited from the father.

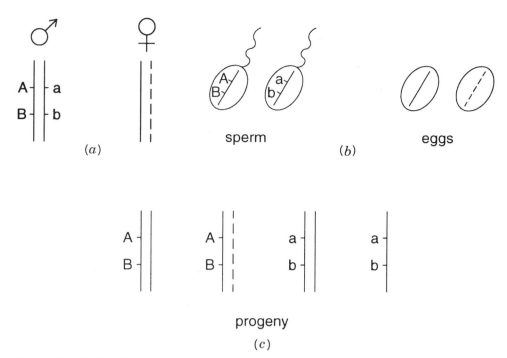

**Figure 6.1**  Basic scheme for chromosome segregation in the genetics of diploid organisms. *(a)* parental genome. *(b)* Possible gametes. *(c)* Possible offspring (all equally probable).

Markers on different chromosomes would, by chance, pass together to an offspring only 25% of the time, since there is a 50% chance of the offspring receiving each particular homologous chromosome. In contrast, markers on the same chromosome will always be coinherited, so long as the chromosome remains intact. Half of the offspring in this case would receive both allele A and allele B. This simple picture of inheritance is complicated by the process known as *meiotic recombination.* During the formation of sperm and eggs, at one stage in meiosis, homologous chromosomes pair up physically, and the DNA molecules within these chromosomes can form a four-ended structure known as a *Holliday junction.* This structure can be reverted to unpaired DNA duplexes in two different ways. Half the time the result is to produce a DNA molecule that consists of part of each parental DNA, as shown in Figure 6.2. When these rearranged chromosomes are passed to offspring, the result is that part of each parental homolog is inherited. This can separate two markers that were originally linked on one of the parental chromosomes.

Meiotic recombination in the human occurs at least once for each pair of homologous chromosomes.[1] If two markers A and B are 1 Mb apart, there is roughly a 1%

---

[1]It is interesting to speculate why meiotic recombination appears to be obligatory. In the absence of recombination, a chromosome that acquired a dominant lethal mutation would be lost to all future generations. Alleles that accidentally resided on this chromosome would also be lost. The resulting drift toward homozygosity could be harmful from an evolutionary standpoint. This drift is largely prevented by the requirement for meiotic recombination.

meiotic recombination during gametogenesis

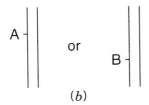

**Figure 6.2**  Minimalistic view of meiotic recombination in a single pair of homologous chromosomes. *(a)* Formation and resolution of a Holliday junction. *(b)* Typical offspring after a recombination event.

chance that they will be separated, and an offspring will be produced with A and b not B, and vice versa. Because human families are generally small, it becomes very difficult to detect recombination events for closely linked markers. In other words, the resolution of genetics in the human is quite limited. It would take families with 100 offspring to detect a single recombination event that would place two genes 1 Mb apart. Since such families are unavailable, one must pool data from different families in order to have a large enough population to provide a reasonable chance of seeing a rare recombination event. Such pooling makes the presumption that markers A and B in different families actually correspond to the same genes. For many reasons this is not always the case. For example, if the marker is the elimination of a complex metabolic pathway, it is possible that the phenotype might be similar regardless of which gene for one of the enzymes in the pathway was inactivated. Such multigenic complications may be quite common in human genetics, and they can severely impair our ability to use linkage analysis to find genes.

An additional complication is the restricted and uncontrolled variability of human DNA. In order for a marker to be useful, one must be able to distinguish which of the two parental homologous chromosomes has been inherited. If both chromosomes of a parent contain the same marker A, there is no way to tell from this marker alone which chromosome the offspring received. We say that this marker is uninformative. If both parents have the same set of allele pairs, A and a, as shown in Figure 6.3, the child can have allele pairs A, A or a, a or A, a. Once again we have no way of discerning from this marker alone which chromosome came from which parent. To carry out human genetics effectively, it is necessary to have large numbers of markers which are informative. That is, at any particular region, we would like to be able to distinguish easily among the four homologous chromosomes originally carried by the two parents.

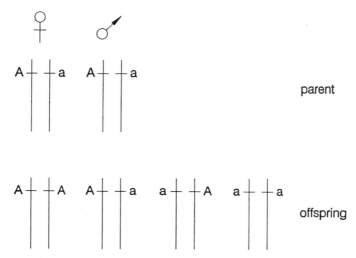

**Figure 6.3** Examples of cases where particular combinations of alleles in parents and offspring confound attempts to trace genetic events uniquely.

---

**BOX 6.1**
**MEIOSIS**

The process of meiosis and the recombination events that occur during meiosis, are actually much more complicated than the simple picture given in the text. Figure 6.4 illustrates this process for the relatively simple case of a cell with only two different chromosomes. The parental cell is diploid; it contains two copies of each homologous chromosome. When this cell duplicates its genome, in preparation for cell division, it becomes tetraploid. At this point each chromosome exists as a pair of sister chromatids. During the first meiotic cell division, the homologous chromosomes line up and associate, as shown in Figure 6.4. The resulting structures are pairs of homologous chromosomes, each containing two paired sister chromatids (identical copies of each other). Thus, altogether, four DNA molecules (eight DNA strands!) are involved in close proximity. In order for these molecules to separate, there appears to be an obligatory recombination event that involves at least one chromatid from each pair of the homologous chromosomes. As shown in Figure 6.4, this results in a reciprocal exchange of DNA between two chromatids, while, in the simple case illustrated, the other two chromatids remain unaltered.

The homologous chromosomes segregate to separate daughter cells during the last stages of the first meiotic division. When this is completed, the result is two diploid cells as in ordinary mitotic division. However, what is profoundly different about these cells is that each contains paired homologous chromosomes that now consist of some regions where both copies of the homolog arose by duplication of a single one of the initial homologous pairs. Because of recombination, one member of each chromosome pair at this point contains some material that is a mosaic of the two original parental homologs. In the second meiotic cell division, each the pair of homologous chromosomes is segregated into a separate daughter cell that goes on, eventually, to become a haploid gamete. Thus, as shown in Figure 6.4, some gametes receive chromosomes that are unaltered versions of the original parental material. Others receive chromosomes that have been rearranged by meiotic recombination. The true picture is even more complex than this, and we will return to it later.

**BOX 6.1** *(Continued)*

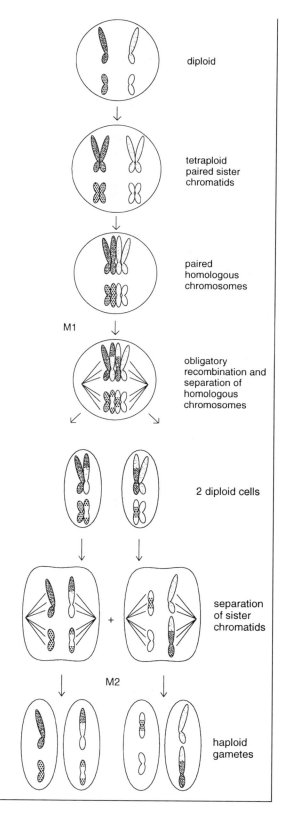

diploid

tetraploid
paired sister
chromatids

paired
homologous
chromosomes

M1

obligatory
recombination and
separation of
homologous
chromosomes

2 diploid cells

separation
of sister
chromatids

+

M2

haploid
gametes

**Figure 6.4** Some of the steps in meiosis, illustrated for a hypothetical cell with only two types of chromosomes.

## A GLOSSARY OF GENETIC TERMS

The science of genetics emerged and matured before DNA was known to be the genetic material. Many terms and concepts were coined without a knowledge of their chemical basis. Here we define some of the commonly used genetic terms needed to follow arguments in human genetics, and we provide an indication of their molecular basis. The first few of these notions are illustrated schematically in Figure 6.5.

**Gene.** A commonly used term for a unit of genetic information sufficient to determine an observable trait. In more contemporary terms, most genes determine the sequence and structure of a particular protein molecule. However, genes may code for untranslated RNAs like tRNA or rRNA, or they may code for more than one different protein when complex mRNA processing occurs.

**Locus.** A site on a chromosome, that is, site on a DNA molecule.

**Allele.** A variant at a particular locus, namely a DNA sequence variant (simple or complex) at the particular locus of interest.

**Homozygosity.** The two homologous chromosomes in a diploid cell have the same allele at the locus of interest.

**Heterozygosity.** The two homologous chromosomes in a diploid cell have different alleles at the locus of interest.

**Independent segregation.** The notion that each of the parental homologous chromosomes in a diploid cell has an equal chance of being passed to a particular daughter cell or offspring. Thus, considering only two chromosomes in Figure 6.6, there will be four possible types of gametes (neglecting recombination), and each of these will occur at equal frequencies.

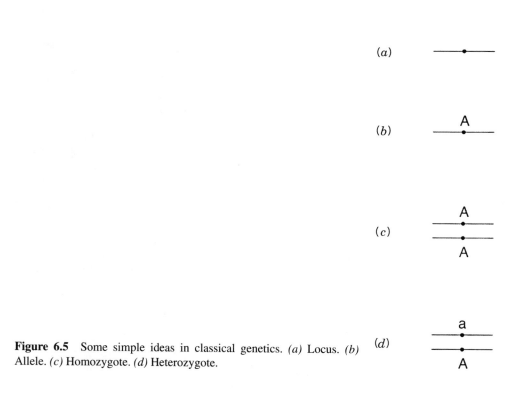

**Figure 6.5**   Some simple ideas in classical genetics. *(a)* Locus. *(b)* Allele. *(c)* Homozygote. *(d)* Heterozygote.

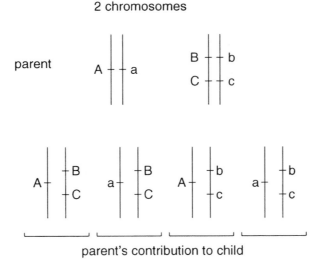

**Figure 6.6**    Independent segregation of alleles on two different chromosomes. Shown are parental genomes and possible gametes arising from these genomes.

**Phenotype.** What one sees, namely the complete set of observable inherited characteristics of an organism, or the particular set of characteristics controlled by the locus or loci of interest.

**Genotype.** What one gets, namely the genetic constitution, which is the particular set of alleles inherited by the organism as a whole, or for the particular locus or loci of interest.

**Haplotype.** The actual pattern of alleles on one chromosome, that is, on a single DNA molecule.

**Dominant.** An allele that produces an identical phenotype if it is present on one homologous chromosome, or on both. True simple dominant traits are rare in the human; one example may be Huntington's disease where individuals carrying one or two disease alleles appear to have indistinguishable disease phenotypes.

**Recessive.** An allele that produces a particular phenotype only if it is present on both homologous chromosomes. Obviously dominant and recessive are oversimplifications that ignore molecular mechanisms and the possibility of more than two alleles at a locus. In principle, with three alleles, A, a, and à, there are six possible genotypes: AA, Aa, Aà, aa, aà, and àà, and each of these might yield a distinguishable phenotype if we use precise enough definitions and measurements.

**Penetrance.** The chance that an allele yields the expected phenotype. This term covers a multitude of sins and oversimplifications. In simple genetic systems the penetrance should normally be 1.0. In outbred species like the human, there can be unknown genetic variants that interact with the gene product at the locus in question and produce different phenotypes in different individuals. Purely at the DNA level we expect all phenotypes to be distinguishable and to show complete penetrance because the phenotype at the level of DNA is identical to the genotype. An exception occurs if the phenotype requires the expression of a particular allele in an environmental context that might not be experienced by all individuals. For example, an allele that determined a serious adult emotional illness might require certain physical or mental stress during the development of the nervous system in order to actually result in phenotypically detectable disease.

**Phase.**  The distribution of particular alleles on homologous chromosomes. Suppose that alleles B and b can occur at one locus, alleles C and c at another. An individual with the phenotype BbCc can have two possible genotypes, as shown in Figure 6.7*a*. These are arbitrarily called cis and trans in analogy with chemical isomers. The two possible phases in this simple case can be distinguished easily by performing a genetic cross. An individual with phenotype BbCc is mated with an individual like bbcc (or BBCC). As shown in Figure 6.7*b*, this allows an unambiguous phase determination. In humans, where we cannot control mating, we need to find family members whose particular inheritance pattern allows an unambiguous determination of the phase of markers to be determined. This is not always possible, and one must frequently make unprovable assumptions about recombination (or the lack of it) in the family in order to assess the phase.

**Linkage.**  A statement that alleles at two loci, say B and C, tend to be coinherited. If linkage is complete they will be coinherited 100% of the time. In molecular terms, these alleles occur on loci that lie near to each other on the same DNA molecule.

**Linkage group.**  A set of loci that in general tend to be inherited together. In effect, a linkage group is a chromosome, a set of genes on a single DNA molecule.

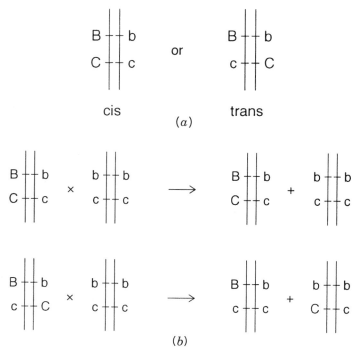

**Figure 6.7**  Phase of alleles at two loci on the same chromosome. *(a)* Definition of two possible phases, arbitrarily called cis and trans. *(b)* Determination of the phase in a particular individual by performing a genetic cross with an appropriately informative partner.

**Unlinked loci.** Two loci that show random cosegregation. Ordinarily, if an allele at one locus, A, is inherited, there is a 50% chance that an allele, B, at another unlinked locus will also be inherited. Loci on different (nonhomologous) chromosomes behave as unlinked. So do loci very far apart on the same large chromosome.

**Meiotic recombination.** A process, already described at the chromosomal level in Box 6.1, that makes the description of genetic inheritance much more complex. However, it is also the major event that allows us to map the location of genes. Simply put, meiotic recombination is the breakage and reunion of DNA molecules prior to their segregation to daughter cells during the process of gamete formation. Because of meiotic recombination, alleles at linked loci may occasionally fail to show co-segregation (Fig. 6.8*a*). In the human genome, meiotic recombination occurs at a frequency of about 1% for two loci spaced 1 Mb apart. The actual frequency observed between two loci is called $\Theta$. We define a 1% recombination frequency as 1 cM (centiMorgan after the geneticist Morgan). Most human chromosomes are more than 100 cM in length; thus genes at opposite ends of these chromosomes behave as unlinked. Note that recombination is an obligatory event for each pair of homologs in meiosis. This implies that smaller chromosomes may have overall higher average levels of recombination.

**Genetic map.** An ordered set of loci with relative spacing (and order) determined from measured recombination frequencies. The simplest map construction would be based on pairwise recombination frequencies (Fig. 6.8*b*). However, more accurate maps can be made by considering the inheritance of multiple loci simultaneously. This allows for multiple crossovers in recombination, multiple DNA breaks and joins, to be considered. Frequently genetic maps are displayed as statements about the relative likelihood of the order shown. For each pair of adjacent loci, what is presented is the relative odds in favor of the order shown, say ABCD, as opposed to an inverted order,

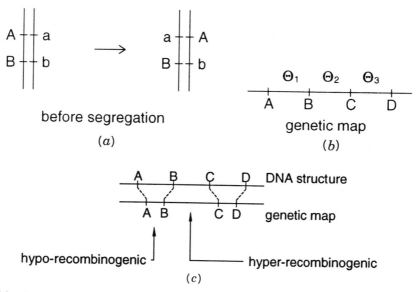

**Figure 6.8**   Genetic maps constructed from recombination frequencies between pairs of adjacent loci. *(a)* Effect of recombination on two adjacent loci before segregation. *(b)* Genetic map constructed from recombination frequencies. *(c)* Typical relationship between a genetic map and the underlying structure of the DNA, a physical map.

ACBD, given the available recombination data. In practice, the current human genetic map has an average spacing between markers of around 2 cM. The ultimate map we are likely to achieve in the absence of significant improvements in our analytical methods is about a 1 cM map (see Fig. 6.29c).

## RELATIONSHIP BETWEEN THE PHYSICAL AND THE GENETIC MAPS

The order of loci must be the same on the genetic map and on its corresponding physical map. Both must ultimately reflect the DNA sequence of the chromosome. However, the genetic map is based on recombination frequencies, while the physical map is based directly on measurements of DNA structure. The meiotic recombination frequency is not uniform along most chromosomes. Regions are observed where recombination is much more frequent than average, hot spots, and where recombination is much less frequent than average, cold spots. Such regions lead to considerable metrical distortion when genetic and physical maps are laid side by side (Fig. 6.8c). We know very little about the properties of recombination hot spots. Attempts to narrow them down to short specific DNA sequence elements have not yet succeeded in mammalian samples. However, we do know that recombination hot spots can, themselves, be inherited alleles. This can lead to serious complications in trying to interpret genetic maps directly in terms of relative physical distances along the DNA.

Two other complications can lead to errors in the construction of genetic maps, and confuse the issue when these maps are compared directly with physical maps. If not properly identified and accounted for, both can lead to inconsistent evidence for gene order. Multiple crossovers can occur between a pair of DNAs strand during meiotic recombination. One schematic example of this is given in Figure 6.9, and the consequences for puta-

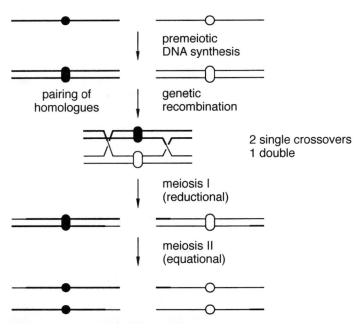

**Figure 6.9** A more detailed view of meiotic recombination showing the association between pairs of homologous chromosomes after DNA synthesis, and the possibility of multiple crossovers before segregation.

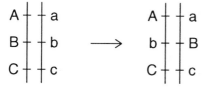

multiple crossovers

**Figure 6.10** Effect of a double crossover on the pattern of inheritance at three loci.

tive gene order are illustrated in Figure 6.10. Obviously multiple crossovers between very close loci are improbable, in general, but they are observed, particularly in regions where recombination hotspots abound. Such a double crossover makes more distant loci appear closer than they should.

The second genetic complication is gene conversion. Here information from one homologous chromosome is copied onto the corresponding region of another homolog. A schematic mechanism is given in Figure 6.11, and the consequences for gene order are shown in Figure 6.12. Gene conversion appears to be relatively frequent in yeast. Evidence for human or mammalian gene conversion is much more spotty, but it is generally believed that this process does play a significant role. Gene conversion can make a nearby marker appear to be far away, as shown in Figure 6.12. The exact biological functions of gene conversion remain to be clarified. It potentially forms a mechanism for very rapid evolution, since it allows a change in one copy of a gene to be spread among identical or nearly identical copies. Gene conversion is also believed to play a role in the sorting out of homologous chromosome pairs prior to crossing over, as described in Box 6.2.

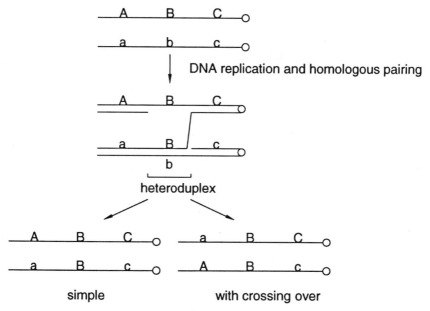

**Figure 6.11** An example of gene conversion, an event that destroys the usual 1 : 1 segregation pattern seen in typical Mendelian inheritance.

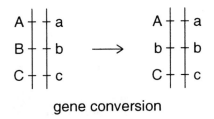

gene conversion

**Figure 6.12**   Effect of a gene conversion event on the pattern of inheritance at three loci.

---

**BOX 6.2**
**TWO-STAGE MODEL OF RECOMBINATION**

For recombination to occur, homologous chromosomes must pair up with their DNA sequences aligned. Studies in yeast have led to a rather complex model for this process. The complications occur because DNA sequence similarities or identities occur not only between the pairs of homologs but also between different chromosomes as a result of dispersed repeated DNA sequences (see Chapter 14) or dispersed gene families with similar or identical members. The recombination apparatus in bacteria, yeast, and presumably all higher organisms has the ability to catalyze sequence similarity searches. In these, a DNA duplex is nicked, and a single-stranded region is exposed and covered with protein. This is then used to scan duplex DNA by processes we still do not understand very well (see discussion of *rec*A protein in Chapter 14). When a close sequence match is found, the single strand can invade the corresponding duplex and displace its equivalent sequence there. Depending on what happens next, this can result either in a gene conversion event in which information is copied from one homolog to the other, a Holliday structure, which may eventually lead to strand rearrangement, or in simple displacement of the invading strand and restoration of the original DNA molecules.

Figure 6.13 illustrates some of the stages of these processes in a hypothetical example of a cell with two different homologous pairs of chromosomes. In meiosis each chromosome starts as a pair of identical sister chromatids linked at the centromere. Initial strand exchange (shown as dotted lines between the chromosomes) occurs both between homologs and between nonhomologs (Fig. 6.13*a*). Some gene conversion events may result at this stage. As the system is driven toward increasing amounts of strand exchange (in a manner we do not know), it is clearly much more likely that homologous pairing dominates (Fig. 6.13*b*). Finally, after suitable alignment is reached (again, judged by mechanisms that we have no current knowledge about), crossing-over events occur (Fig. 6.13*c*), and the homologs segregate to daughter cells (Fig. 6.13*d*). Each chromosome at that point consists of a pair of sister chromatids that are no longer identical because of the different gene conversion and crossing-over events they have experienced.

*(continued)*

**BOX 6.2** *(Continued)*

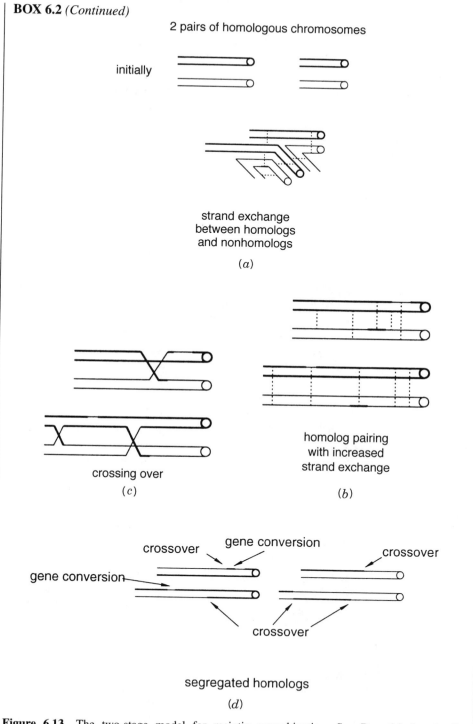

2 pairs of homologous chromosomes

initially

strand exchange
between homologs
and nonhomologs

*(a)*

homolog pairing
with increased
strand exchange

*(b)*

crossing over

*(c)*

crossover    gene conversion                    crossover

gene conversion

crossover

segregated homologs

*(d)*

**Figure 6.13**  The two-stage model for meiotic recombination. See Box 6.2 for details. (Adapted from Roeder, 1992.)

To distinguish single and multiple crossover events, and gene conversion, large numbers of highly informative loci are very helpful. Several features of genetic loci make them particularly powerful for mapping. Ideally one has many different alleles, and these occur at reasonable frequencies in the population. Where this occurs, there is a very good chance that all four parental homologs are distinguishable at the locus because they each carry different alleles. A major thrust in human genetics has been the systematic collection of such highly informative loci. This will be discussed in a later section. It is also helpful to have many offspring in families under study and to have many generations of family members. Because it is so difficult to satisfy these conditions, human genetics is often rendered rather impotent compared with the genetics of more easily manipulatable experimental organisms.

## POWER OF MOUSE GENETICS

Mice are among the smallest common mammals. They are relatively inexpensive to maintain, have large numbers of offspring, and mature quickly enough to allow many generations to be examined. However, this alone does not explain why mice form such a powerful genetic system. Several reasons exist that make the mouse the preeminent model for mammalian genetics. First, so much is already known about so many mouse genes, that an extensive genetic map already exists. Second, many highly inbred strains of laboratory mice exist. These tend to be homozygous at most alleles, and thus a description of their phenotype and genotype is relatively simple. Also gene transfer and knockout technology exist for mice. What is particularly useful is that rather distant inbred strains of mice can be interbred to give at least some fertile offspring. This property dramatically simplifies the construction of high-resolution genetic maps. Several different sets of inbred strains can be used. One of the earlier choices was *Mus musculus,* the common laboratory mouse, and *Mus spretus,* a distant cousin.

Figure 6.14 illustrates how crosses between *M. musculus* and *M. spretus* generate very useful genetic information. Because these mice are so different, the F1 offspring of a direct cross tend to be heterozygous at almost any locus examined. It turns out that the males that result from such a cross are sterile, but the females are fertile. When an F1 *spretus* × *musculus* female is bred with an *M. musculus* (a procedure called a *backcross*), in most regions of the genome the progeny either resemble wild type *M. musculus* homozygotes, or F1 *musculus* × *spretus* heterozygotes. However, every time a recombination event occurs, there is a switch between the homozygous pattern and the heterozygous pattern (Fig. 6.14b). Given the dense set of genetic markers available in these organisms, the location of many recombination events can be determined in each set of experimental animals. The result is that one develops and refines genetic maps extremely rapidly.

## WEAKNESS OF HUMAN GENETICS

Humans, from a genetic standpoint, are a stark contrast with mice. We have a relatively long generation time; it precludes the simultaneous availability of large numbers of generations. Our families are small; most are far too small for effective genetics. Crosses cannot be controlled, and there are no inbred strains, only the occasional result of very limited inbreeding in particular cultures that promote such practices as marriages between

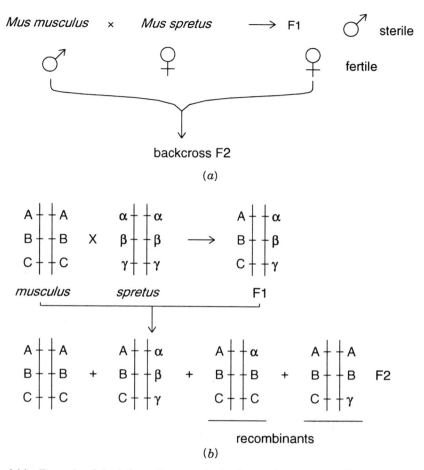

**Figure 6.14**   Example of the informativeness of a back cross between two distant mouse species *M. musculus* and *M. spretus*. *(a)* Design of the back cross. *(b)* Alleles seen for three loci in the parents and the first (F1) and typical second (F2) generation of offspring, with and without recombination.

cousins. For all of these reasons, the design and execution of prospective genetic experiments is impossible in the human. Instead, one must do retrospective genetic analysis on existing families. Ideally these will consist of family units where at lest three generations are accessible for study. Results from many different families usually must be pooled in order to have enough individuals segregating a trait of interest to allow a statistically significant test of hypotheses about the model for its inheritance. Usually that test is to ask if the trait is linked to any other known trait in the genome. This is a very tedious task, as we will illustrate. However, large numbers of ongoing studies use this approach because it is the only effective way we have to find a human gene location if the only available information we have is a disease phenotype. If there are animal models for the trait in question, or if one has a hint about the functional defect, one can sometimes cut short a search of the entire genome by focusing on candidate genes. However, these genes still must be examined in linkage studies with human markers because the arrangement of genes in humans and model organisms, while similar (Fig. 2.24), is not identical.

There are several additional major weaknesses that compromise the power of human genetics. In many cases our ability to evaluate the phenotype is imprecise. For example, in inherited diseases it is not at all uncommon to have imprecise or even incorrect diagnoses. These result in the equivalent of a recombination event as far as genetics is concerned, and a few such errors can often destroy the chances that a search for a gene will be successful. A second common problem is missing or uncooperative family members. In such cases the family tree, called a *pedigree,* is incomplete, and phase or other information about the inheritance of a disease trait, or a potential nearby marker, is lost. Homozygosity in key individuals is another frequent problem. As illustrated earlier, this makes it impossible to determine which parental homolog in the region of interest has been inherited. As genetic markers become denser and more informative, this problem is becoming less severe, but it is by no means uncommon yet.

The final problem that frequently plagues human genetic studies is mispaternity. This is relatively easy to discover by using the highly informative genetic markers currently available. Usually the true parent is not identified; this results in a missing family member for genetic studies.

## LINKAGE ANALYSIS IGNORING RECOMBINATION

The statistical analysis of linked inheritance is the tool used for almost all genetic studies in the human. Here we introduce this approach, and the Bayesian statistics used to provide a quantitative evaluation of the pattern of inheritance, assuming for the moment that recombination does not occur. The result will be a test of whether two loci are linked, but there will be no information about how far away on the genetic map these linked loci are. The treatment in this and the two following sections follows closely a previous exposition by Eric Lander and David Botstein (1986).

Consider the simple family shown in Figure 6.15. This is in fact the simplest case that can be used to illustrate the basic features of linkage analysis. We deal with two loci with two alleles each: A and a, D and d. We assume that phenotypic analysis allows all the possible independent genotypes at these two loci to be distinguished. Thus all individuals can be typed as AA, Aa, or aa and as DD, Dd, or dd. In linkage studies we ask whether particular individuals tend to inherit alleles at the two loci independently or in common. Our simple family has two parents, one heterozygous at the two loci and one homozygous at both loci. There are two offspring; both are heterozygous at both loci. The issue at hand is to assess the statistical significance of these data to reveal whether or not the two loci at linked; that is, whether they are on the same chromosome.

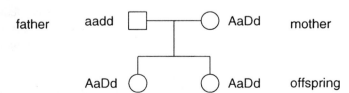

**Figure 6.15**   A typical family used to test the notion that two genetic loci A/a and D/d might be linked. (Adapted from Lander and Botstein, 1986). Circles show females, squares show males.

Suppose that the two loci are on different chromosomes. We then know the genotypes of the mother and the father unambiguously as shown in Figure 6.16. Independent segregation of alleles on different chromosomes leads to four possible genotypes for the offspring of these parents. A priori the probability of occurrence of each of these phenotypes should be equal, so each should occur with an expected frequency of $\frac{1}{4}$. Thus the a priori probability that the family in question should have both children with the same particular phenotype, AaDd is $\frac{1}{4} \times \frac{1}{4} = \frac{1}{16}$. We need to compare this probability, calculated from the hypothesis that the loci are unlinked, with the probability of seeing the same result if the loci are linked.

If the two loci are linked, they are on the same chromosome. In this case the genome of the homozygous father can be described unambiguously, but there are two possible phases for the mother. These are shown in Figure 6.17a. Unless we have some a priori knowledge about the mother's genotype, we must assume a priori that these two possible phases are equally probable. Thus there is a $\frac{1}{2}$ chance that the mother is cis and $\frac{1}{2}$ that she is trans. Since we are not allowing for the possibility of recombination, if the mother is trans, the probability that she would have an AaDd daughter is zero, since both the A and D alleles have to come from the mother in the family shown in Figure 6.15, and this would be impossible in the trans configuration where the A and D alleles are on different homologs.

If the mother is cis for the two loci, there are two possible genotypes for any offspring. These are shown in Figure 6.17b. In the absence of any intervening factors, the a priori probability of observing these genotypes should be equal. Thus the chance of seeing a child with the genotype AaDd under these circumstances is $\frac{1}{2}$. The chance of a family with two children both of whom are AaDd will be $\frac{1}{2} \times \frac{1}{2} = \frac{1}{4}$. However, since we have no a priori knowledge about the phase of the mother, we must average the chances of seeing the expected offspring across both possible phases. The overall odds of the observed family inheritance pattern is then $\frac{1}{2}(0) + \frac{1}{2}(\frac{1}{4}) = \frac{1}{8}$ if the two loci are linked.

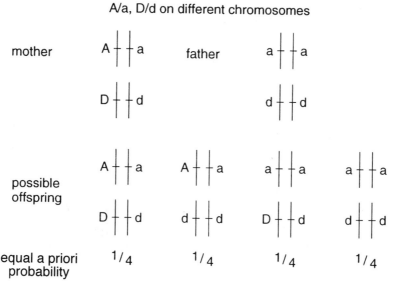

**Figure 6.16**  Genotypes of the parents and expected offspring if the two loci and unlinked, and there is no recombination.

A/a, D/d on same chromosome

two possible phases for mother

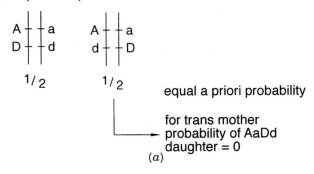

*(a)*

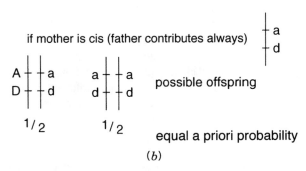

*(b)*

**Figure 6.17** Inheritance patterns with linkage, but no recombination. *(a)* Possible maternal genotypes if the two loci are linked. *(b)* Possible offspring if the mother is cis. For the example considered, other offspring genotypes will be seen only if the loci are not linked.

The odds ratio is a test of the likelihood that our hypothesis that linkage exists is correct. This is the ratio of the calculated probabilities with and without linkage:

$$\text{odds ratio} = \frac{P_\text{linked}}{P_\text{unlinked}} = \frac{\frac{1}{8}}{\frac{1}{16}} = 2$$

In human genetics we frequently know the phase of an individual because this is available from data on other generations or other family members. In this case the odds ratio becomes

$$\text{odds ratio} = \frac{P_\text{linked}}{P_\text{unlinked}} = \frac{\frac{1}{4}}{\frac{1}{16}} = 4$$

The greater power of linkage analysis with known phase is readily apparent.

With or without known phase, the single family shown in Figure 6.15 provides a small amount of statistical evidence in favor of linkage. To strengthen (or contradict) this evidence, we need to pool data from many such families. The overall odds ratio then becomes

$$\frac{P_\text{linked}}{P_\text{unlinked}} = \text{odds}_1 \times \text{odds}_2 \times \text{odds}_3 \ldots$$

where the subscripts refer to different families under study. It is convenient mathematically to deal with a sum rather than a product of such data, and this is accomplished by taking the logarithm of both sides of the above equation. Since

$$\log(A \times B \times C \ . \ . \ .) = \log(A) + \log(B) + \log(C) + \ . \ . \ .$$

the result is

$$\text{LOD} = \log\left(\frac{P_{\text{linked}}}{P_{\text{unlinked}}}\right) = \log(\text{odds}_1) + \log(\text{odds}_2) + \log(\text{odds}_3) \ . \ . \ .$$

This is called a *LOD* (rhymes with cod) *score.* The LOD score is calculated from the data seen with a particular family or set of families. Some feeling for the number of individuals that must be examined for a LOD score to be statistically significant can be captured by the expected LOD score calculated for a particular inheritance model, called an *ELOD.* However, the inheritance model we use has to include the possibility of recombination to be realistic enough to represent actual data.

## LINKAGE ANALYSIS WITH RECOMBINATION

Consider a pair of markers at loci that appear to be linked by available data. There are three possible cases to deal with

1. The markers are unlinked, but random segregation gives the appearance of linkage.
2. The markers are really linked.
3. The markers are linked, but recombination has disguised this linkage.

We will deal with two markers as in the previous case. Here, however, it simplifies matters if one of these is a locus where D is an allele of a disease gene that we are trying to find. A is an allele at another locus, and we are interested in testing the hypothesis that in a particular family A and D are linked. The chance that a recombination event occurs between the two loci in each meiosis is an unknown variable $\Theta$. We need to calculate the odds in favor of linkage, for data from a particular family, as a function of $\Theta$. Actual $\text{LOD}(\Theta)$ calculations are complex. To illustrate the considerations that go into such calculations, we will calculate the expected contribution of a single individual observed to inherit the disease allele D to the overall LOD score. This contribution is called the expected LOD or $\text{ELOD}(\Theta)$.

We will analyze the case where a parent is AaDd. Usually we are dealing with a relatively rare disease, and the other parent does not have the D allele. We assume for simplicity that the healthy parent also either has the a allele or some other allele that we can distinguish from A and a. We look only at offspring that are detected to carry the disease allele D. If the two loci are unlinked, the offspring inherit pairs of two different chromosomes carrying A or a and D or d at random, as shown in Figure 6.18a. We look only at offspring carrying D; thus there is a 0.5 probability that such an offspring will also inherit A.

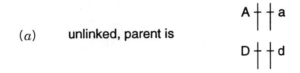

(a)    unlinked, parent is

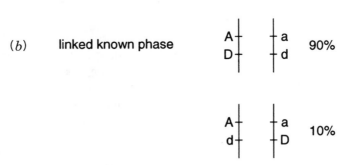

(b)    linked known phase

**Figure 6.18**    Analysis of the inheritance of a disease allele D and a possible linked allele A. *(a)* Possible parental contributions to an offspring if no linkage occurs. *(b)* Possible parental contributions to an offspring if the loci are linked but the recombination frequency between them is 0.1.

We will consider the case where two loci are linked and the phase in the parent is known to be cis (Fig. 6.18*b*). What we want to calculate is the effect of observing one child of this parent on the odds in favor of linkage of A and D. Suppose that $\Theta$ is 0.1. Such a 10% chance of recombination corresponds to an average distance of 10 Mb in the genome. This is near the maximum distance across which linkage is visible in the analysis of only two loci at once. We can calculate the chance of two outcomes:

1.  Probability that a child with D inherits AD from the parent is 0.9.
2.  Probability that a child with D inherits aD from the parent is 0.1.

The contribution of case 1 to the expected LOD score is

$$\log\left(\frac{0.9}{0.5}\right)$$

which is the ratio of the odds of seeing A and D with linkage versus without linkage. The contribution of case 2 to the expected LOD score is

$$\log\left(\frac{0.1}{0.5}\right)$$

which is the ratio of the odds seeing aD with linkage to the odds of seeing aD without linkage.

Thus the average contribution from one child to the ELOD is the sum of these two cases weighted by their expected frequency. Since recombination across 10 cM occurs only 10% of the time,

$$\text{ELOD}(0.1) = 0.9 \log \left(\frac{0.9}{0.5}\right) + 0.1 \log\left(\frac{0.1}{0.5}\right)$$

$$\text{ELOD}(0.1) = + 0.23 - 0.07 = 0.16$$

Thus observation of cosegregation of A and D adds to the probability of linkage, while observation of separation of A and D subtracts from the evidence for linkage.

What we need to do is develop the tools to assess the statistical significance of a particular ELOD score. Since some markers will appear to cosegregate by chance in any study with a relatively small number of affected individuals, there is always a chance of seeing a significantly positive LOD score, simply because of the random fluctuations. A near consensus in human genetics is that an observed LOD of 3.0 or higher is required before the probability of purely accidental linkage can be reduced to the point where few errors are made. For the example just described, the number of individuals segregating D with unambiguous pedigrees that would have to be combined to generate a LOD score of 3.0 can be estimated as 3/0.16 = 18. For common inherited diseases this is not a problem, but for very rare diseases it may be extremely difficult to find 18 genetically informative individuals for a particular marker with an unambiguous diagnosis.

Note that several constraints apply to the linkage analysis described above. One must have access to a parent with known phase between A and D. The marker A to be tested for linkage must have useful heterozygosity. The diagnosis of D must be unambiguous in all the individuals tested. Note that failing to diagnose an individual who is carrying D (a false negative) does not hurt the analysis, since in this case the individual and the parent are not scored. However, misclassifying an individual as carrying D instead of d (a false positive) causes serious problems because it will weaken the evidence about which alleles at other loci are cosegregating with D.


## INTERVAL MAPPING

Once a genetic map is available for a region of interest, the process of linkage analysis can be made more powerful by examining several markers simultaneously. We will consider the simplest possible case, illustrated in Figure 6.19a. As in the previous discussion of simple linkage analysis, we will calculate the average contribution of the LOD score from a single, informative individual inheriting a disease allele D. We wish to test a region of the genome containing two linked loci with markers A and B to see if the disease allele D lies between them or is unlinked. (Here we ignore the case that it might be linked to A and B but lie outside them rather than between them.)

Suppose that the loci containing A and B are 20 cM apart. This is a reasonable model for how human genetic maps are used in average regions of the genome. $\Theta_{AB} = 0.2$. First we calculate the possible contributions from a parent carrying D to a child, also carrying D, if there is no linkage between A and B with D (Fig. 6.19b). Since A and B are on the same chromosome, D, if unlinked, they must lie on a different chromosome. Assuming that the parent is heterozygous and informative at all these loci, there are four possible contributions from the parent to the child (Fig. 6.19c).

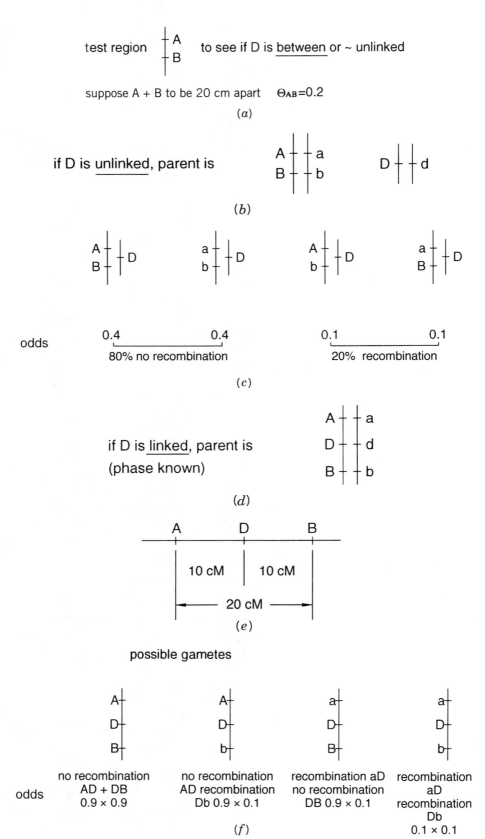

test region A | B to see if D is between or ~ unlinked

suppose A + B to be 20 cm apart   $\Theta_{AB}=0.2$

*(a)*

if D is <u>unlinked</u>, parent is

A — a
B — b        D — d

*(b)*

A
B — D       a
b — D       A
b — D       a
B — D

odds    0.4          0.4        0.1          0.1
        80% no recombination    20% recombination

*(c)*

if D is <u>linked</u>, parent is
(phase known)

A — a
D — d
B — b

*(d)*

A        D        B

10 cM | 10 cM

← 20 cM →

*(e)*

possible gametes

A        A        a        a
D        D        D        D
B        b        B        b

odds    no recombination    no recombination    recombination aD    recombination
        AD + DB             AD recombination    no recombination    aD
        0.9 × 0.9           Db 0.9 × 0.1        DB 0.9 × 0.1        recombination
                                                                    Db
                                                                    0.1 × 0.1

*(f)*

If no recombination between A and B occurs (80% probability for markers 20 cM apart), the child will either inherit ABD (0.4 odds) or abD (0.4 odds). If recombination between A and B occurs (20% probability), the child will inherit either AbD (0.1 odds) or aBD (0.1 odds).

If D is linked and located between A and B, assuming the phase of the parent is known, the two homologous chromosomes of the parent carry alleles ADB and adb, as shown in Figure 6.19d. In principle, D may lie anywhere between A and B and the actual position of D is a variable that must be included in the calculations. Here we will consider the simple case where D lies midway between A and B. Assuming that the recombination frequency is uniform in this region of the chromosome, we can then place D 10 cM from A and 10 cM from B (Fig. 6.19e). There are four possible sets of alleles that can be passed from this parent to a child who inherits D (Fig. 6.19f). These are as follows:

*ADB:* resulting from no recombination between A and D, and no recombination between D and B (odds are 0.9 × 0.9).

*ADb:* resulting from no recombination between A and D but recombination has occurred between D and B (odds are 0.9 × 0.1).

*aDB:* recombination has occurred between A and D, but no recombination has occurred between D abd B (odds are 0.1 × 0.9).

*aDb (a double crossover event):* recombination has occurred both between A and D and between D and B (odds are 0.1 × 0.1).

Thus the same four possible genotypes can arise either with or without linkage. However, the odds of particular genotypes vary considerably in the two cases. For the four possible offspring:

| Alleles | ADB | ADb | aDB | aDb |
|---|---|---|---|---|
| Odds (linked/unlinked) | 0.81/0.4 | 0.09/0.1 | 0.09/0.1 | 0.01/0.4 |

The ELOD for a single statistically representative child can be calculated from these results by realizing that if there is linkage, the probabilities of seeing the four patterns of alleles are 0.81, 0.09, 0.09, and 0.01, respectively. Thus the ELOD is given by

$$\text{ELOD}(0.2) = 0.81 \log\left(\frac{0.81}{0.4}\right) + 0.09 \log\left(\frac{0.09}{0.1}\right) + 0.09 \log\left(\frac{0.09}{0.1}\right)$$
$$+ 0.01 \log\left(\frac{0.01}{0.4}\right)$$

$$\text{ELOD}(0.2) = +0.25 - 0.004 - 0.004 - 0.02 = 0.23$$

**Figure 6.19**  Interval mapping to test the hypothesis that a disease allele D is located equidistant between two linked markers A and B, separated by 20 cM. *(a)* Map of the test region. *(b)* Parental chromosomes if D is unlinked to A and B. *(c)* Possible chromosomes inherited by an offspring carrying the disease allele in the absence of linkage. *(d)* Parental chromosomes if D lies between A and B. *(e)* Map location assumed for D for the example calculated in the text. *(f)* Possible parental contributions to an offspring inheriting the disease allele.

Note that this ELOD is larger in the case of interval mapping than in the simple case of linkage analysis we considered earlier. The number of informative individuals that would have to be examined to achieve a LOD score of 3 would be $3/0.23 = 14$.

## FINDING GENES BY GENETIC MAPPING

What is done, in practice, is to repeat the kinds of calculations previously described with all possible values of $\Theta$ as a variable using actual genotype data from real families. For simple linkage analysis the sorts of results obtained are shown schematically in Figure 6.20. These yield the expected LOD score as a function of $\Theta$. The critical results are the maximum LOD value and the confidence limits on possible values of $\Theta$. With interval mapping, the results are more complex, but the basic kind of information obtained is similar, as shown by the example in Figure 6.21. For details, see Ott (1991) and Lalouel and White (1966).

In a typical case, no a priori information exists about the putative location of a gene of interest. To have a reasonable chance of finding it, one must test the hypothesis that it lies near (or between) any of about 150 informative markers. This will subdivide the genome into intervals spaced about 20 cM apart. Each marker must be tested with a sufficient number of informative individuals to achieve a LOD score of 3.0 or higher if that particular marker is linked to the gene. With present technology this search is often carried out one marker and one individual at a time. It is easy to estimate that around 150 markers $\times$ 40 to 60 individuals (parents and offspring) must be tested in ideal cases where parental phase is known and markers are very informative. If the analysis is carried out by ordinary Southern blotting (Chapter 3) of DNA bands 6000 to 12,000 gel electrophoresis lanes have to be examined by hybridization to afford a reasonable chance of finding a gene, and this is an ideal case! Schemes have recently been described that can reduce the workload by an order of magnitude through the use of pools of samples (Churchill et al., 1993; see also Chapter 9 for examples of the power of pooling).

If a LOD score of 3.0 or greater is achieved, there is a reasonable chance that the correct location of the disease gene of interest has been found. What is usually done is to celebrate, publish a preliminary report, and fend off overoptimistic members of the press or families segregating the disease of interest who confuse the first sighting of a gene location with the identification of the actual disease gene itself. Knowing the location of a disease gene does provide improved diagnostics for the disease but, initially, only in those families where the phase of the disease allele and nearby markers is known.

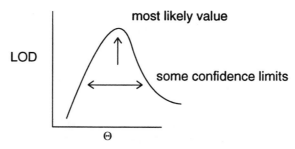

**Figure 6.20** LOD score for linkage of two genes, with a particular recombination frequency, that would be seen in a typical set of family inheritance data.

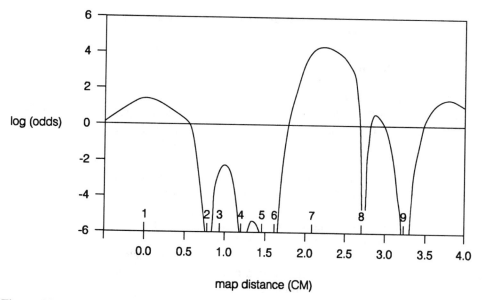

**Figure 6.21**  Example of interval mapping data. Shown is the expected LOD score [log(odds)] as a function of the possible location of a gene within the interval mapped by two known linked genes. (Adapted from Leppert et al., 1987).

Furthermore, at a 10 cM distance, the amount of recombination between the marker and the disease allele in each meiosis is still 10%, so the accuracy of any genetic testing is quite limited. More accurate approaches are outlined in Box 6.3 and Box 6.4.

---

**BOX 6.3**
**MULTIPOINT MAPPING**

More accurate genetic maps can be constructed by considering all the loci simultaneously rather than just dealing with pairs of loci. In this case what one establishes, primarily, is the order of the loci and the relative odds in favor of that order based on the sum of all the available data. In principle, one can write down all possible genetic maps and calculate the relative likelihood of each being correct in the context of the available data. In practice, it is usually quite tedious to do this. Instead, as shown in Figure 6.22, one usually plots the most likely map, and gives the relative odds that the order of each successive pair of markers is reversed from the true order.

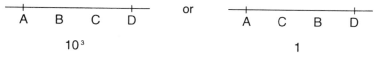

**Figure 6.22**  Typical map data by multipoint analysis. Shown are the relative odds in favor of two orderings of the markers A, B, C, and D.

The next goals are to strengthen the evidence for linkage and narrow the putative location of the gene. Additional examples of affected individuals can be examined using only the closest known markers. If this increases the LOD score, there is little doubt that the gene location has been correctly identified. Once the interval containing the gene is known, one can look for additional markers in the region of interest. Various methods to find polymorphic markers in selected DNA regions will be described later. These methods are quite powerful so long as the region is actually polymorphic in the population. Note that once the approximate gene location is found, the markers used to refine that location need not be informative in all patients in the sample. What is key is to find particular individuals who demonstrate recombination between the disease gene and nearby markers. Until linkage was established such individuals actually weakened the search because there was no way of knowing a priori that they were recombinants, and thus, as shown in earlier examples, they subtracted from the expected LOD score. Once the gene is known to be nearby, such individuals can be recognized as recombinants and properly scored as shown by the example in Figure 6.23. Just two informative individuals with recombination events defined by their haplotypes (patterns of alleles on a single chromosome) are sufficient to pinpoint the location of the disease gene, barring the unlikely occurrence of a gene conversion or double crossover.

## MOVING FROM WEAK LINKAGE CLOSER TO A GENE

Failure to find a linked marker in an initial test does not mean that no marker is linked to the gene. A disease gene must lie somewhere in the genome. A possibility is that the model for inheritance used in the linkage study was wrong. One must consider dominant and recessive inheritance as well more complex cases where multiple alleles or even multiple genes are involved. It is very tempting in cases where the maximum LOD score obtained is less than 3.0 to review individual families contributing to the LOD score and ask if the score can be improved by dropping some of the families. This implicitly challenges the diagnosis in these families or presumes that the disease is heterogeneous—that it is influenced by other factors in addition to the particular gene in question.

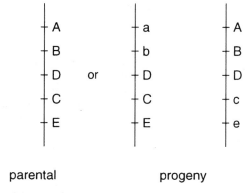

**Figure 6.23**  Examples of two recombinant genotypes seen from a parent with known phase. Once the disease allele D is known to lie in this region, the genotype of the two recombinants restricts the possible location of the disease gene to between markers b and c.

This is a very dangerous practice, since if one starts with a sufficient number of families, it will almost always to possible to achieve an alluring LOD score by selectively choosing among them. Clearly the appropriate statistical tests must be employed to discount the resulting LOD score against such selective manipulation of the data. The real issue is not whether one can increase a LOD score by dropping a family with a negative contribution. The issue is whether the magnitude of the increase in LOD is sufficient to justify the additional parameterization implicit in dropping this family. A much safer procedure is to collect more families and try additional markers near the ones that have already shown a hint of linkage if not yet compelling evidence for linkage. When this has been done, some LODs of 2.0 eventually have produced the desired gene; others have faded into oblivion.

Eventually genetic linkage studies may narrow down the location of a gene to a 2 cM region. However, in such an interval of the genome, there may be a single gene or more than 80. It is very difficult to use conventional linkage analysis to narrow the location further. The available families are likely to have only a limited number of recombination events in the region of interest because they represent just a few generations, which means any recombinations seen must have occurred recently. A 2 cM localization means that already 50 informative meioses have been found. It is usually not efficient to keep gathering more families at this point, although it is efficient to keep trying to find additional informative markers, since these can narrow down the location of any recombination events.

## LINKAGE DISEQUILIBRIUM

In fortunate cases, a variant on linkage analysis can be used to home in on the likely location of a disease gene once it has been assigned to a mapped region of a chromosome. Suppose that most affected individuals have the same disease allele D. This is the case, for example, with the Huntington's disease individuals who live near Lake Maricaibo in Venezuela; it is also the case with individuals affected with sickle cell disease, and with most individuals of northern European descent afflicted with severe cystic fibrosis. In such cases it is possible that the disease is the result of a founder effect: all affected individuals have inherited the same disease allele-carrying chromosome from a common progenitor. (When no evidence for a single disease allele exists, but phenotypic variation in the disease is evident, one can try to subtype the disease by severity, age of onset, particular symptoms, and test the presumption that, for this subtype, a founder effect may exist.)

If a disease allele arose once by mutation on a single chromosome, it will be created in the context of a particular haplotype (Fig. 6.24). The chromosome that first carries the disease will have a particular set of polymorphic markers. It will have a particular genetic background. As this chromosome is passed through many generations of offspring, it

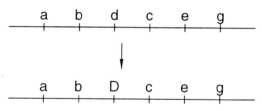

**Figure 6.24**  Generation of a disease allele by a mutation on a founder haplotype sets the stage for linkage disequilibrium.

will suffer frequent meiotic recombination events. These will tend to blur the memory of the original haplotype of the chromosome; they will average out the original genetic background with the general distribution of markers in the human population. However, those markers very close to the disease gene will tend, more likely than average, to retain the haplotype of the original chromosome because, as the distance to the disease gene shrinks, it becomes less likely that recombination events will have occurred in this particular location.

Humans are an outbred population. Most alleles were established when the species was established, and a sufficient number of generations has passed since then that frequent recombination events have occurred between any pair of neighboring loci resolved on our genetic maps. For this reason the distribution of particular haplotypes in neighboring loci in the population (as opposed to particular families) should be close to random. Consider the case shown in Figure 6.25, for two neighboring loci with two alleles each. Within the population, the frequencies $X$ of the alleles at a particular locus must sum to 1.0.

$$X_a + X_A = 1.0 \quad X_b + X_B = 1.0$$

The frequencies of particular haplotypes, $f$, should be given by simple binomial statistics:

$$f_{AB} = X_A X_B \quad f_{Ab} = X_A X_b \quad f_{aB} = X_a X_B \quad f_{ab} = X_a X_b$$

Deviations from these results, measured, for example, as $f_{AB}$ observed $- f_{AB}$ calculated, are evidence for linkage disequilibrium, and they indicate that the individuals examined are not a random sample of the population. Note, however, that deviation of allele frequencies from those expected by binomial statistics may have other causes besides genetic linkage. Deviations can reflect improper sampling of the population, or they can reflect actual functional association between specific alleles. The latter process could occur, for example, if the protein products of the two genes in question actually interacted biochemically. (For further discussion see Ott, 1991.)

To search for a gene by linkage disequilibrium, one does not examine families segregating a disease allele D. Instead, one looks across a broad spectrum of the population for unrelated individuals who have the disease allele D. If evidence for linkage disequilibrium is found, it reflects recombinations along the chromosome all the way back in time to the original founder. Since this may extend back hundreds of years, more than ten generations may be involved, and thus the number of recombination events viewed will be much greater than possible with any contemporary family. In the case of linkage disequilibrium, we expect to see the general results shown in Figure 6.26. There will be a gradient of increasing deviation from equilibrium as the neighborhood of the disease gene is reached because of the diminishing likelihood of recombination events occurring in an ever-shrinking region.

any 2 allele sets

**Figure 6.25**  Possible haplotypes in a two-allele system used to examine whether loci are at equilibrium.

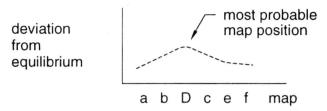

**Figure 6.26** Gradient of linkage disequilibrium seen near a disease allele in a case where a founder effect occurred.

## COMPLICATIONS IN LINKAGE DISEQUILIBRIUM AND GENETIC MAPS IN GENERAL

The human genome is a potential minefield of uncharted genetic events, hidden rearrangements, new mutations, and genetic heterogeneity. Failure to see linkage disequilibrium near a gene does not mean that the gene is far away. Two of the most plausible potential complications are the existence of more than one founder or the existence of a significant fraction of alleles in the population that have arisen by new mutations. For example, in the case of dominant lethal diseases (those in which, nominally, the affected individuals have no offspring), one must expect that most disease alleles will be new mutations. Multiple founders can occur in distinct geographical populations, and they can be tested for by subdividing the linkage disequilibrium analysis accordingly. However, our increasingly mobile population, at least in developed countries, will make such analyses increasingly difficult.

Two other reasonable explanations for a failure to see linkage disequilibrium near a disease gene of interest are shown in Figure 6.27. The first of these is the possible presence of recombination hot spots. If the recombination pattern in the region of interest is punctate, then an even gradient of linkage disequilibrium will not be seen. Instead, markers that lie within a pair of hot spots will appear to be in disequilibrium, while those that lie on opposite sides of a hot spot will appear to have equilibrated. The occurrence of any disequilibrium in the region is presumptive evidence that a disease gene is there, since this is the basis for selection of the particular set of individuals to be examined. However, the complex pattern of allele statistics in the region will make it difficult to narrow in on the location of the disease gene.

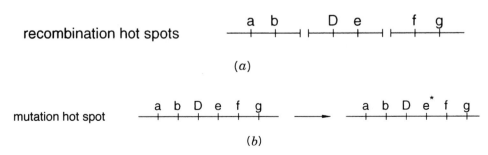

**Figure 6.27** complications that can obscure evidence for linkage disequilibrium. *(a)* Recombination hotspots near the disease gene. *(b)* Mutation hot spots near the disease gene.

A second potential source of confusion is the presence of mutation hotspots. These are quite common in the human genome. For example, the sequence CpG is quite mutagenic in those regions of the genome where the C is methylated, as discussed in Chapter 1. When mutation hotspots are present, these alleles appear to have equilibrated with their neighbors, while more distant pairs of alleles may still show deviations from equilibrium. As in the case of recombination hotspots, disequilibrium indicates that one has not sampled the population randomly. This is presumptive evidence for a disease gene nearby, but mutation hotspots weaken the power of the disequilibrium approach to actually focus in on the location of the desired gene.

## DISTORTIONS IN THE GENETIC MAP

We have already discussed briefly the occurrence of recombination hot spots and their deleterious effect on attempts to find genes by linkage disequilibrium. Some hot spots are inherited; in the mouse Major Histocompatibility Complex (MHC), a set of genes that regulates immune response, a hot spot allele has been found that raises the local frequency of recombination by a hundredfold. All of the recombination events caused by this hot spot have been mapped within the second intron of the $E^b$ gene, 4.3 kb in size. While we are not sure what has caused this hot spot, the region has been sequenced, and one peculiarity is the occurrence of four sequences with 9/11 bases equal to a consensus sequence TGGAAATCCCC. Such sequences have also been found in regions associated with other recombination hot spots.

The genetic map of the human, and other organisms is not uniform. Recombination is generally higher near the telomers and lower near centromeres. The map is strikingly different in males and females—that is, meiosis in males and females appears to display a very different pattern of recombination hot spots. A typical example is shown for a selected region of human chromosome 1 in Figure 6.28. Note that some regions that have short genetic distances in the female have long distances in the male, and vice versa. Genetic linkage analysis is more powerful in regions where recombination is prevalent because, the more recombinants per Mb, the more finely the genetic data will serve to subdivide the region. In general, genetic maps based on female meioses are considerably longer than those based on male meioses. This is summarized in Table 6.1. A frequent practice is to pool data and show a sex-averaged genetic map. It is not very clear that this is a reasonable thing to do. Instead, it would seem that once a region of interest has been selected, meioses should be chosen from either the female or the male depending on which set produces the most expanded and informative map of the region. At present it does not appear that most workers pay much attention to this.

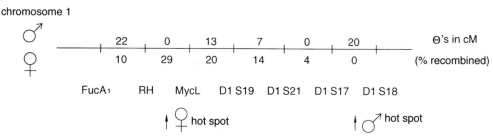

**Figure 6.28**  Comparison of low-resolution genetic maps in female and male meiosis. Shown is a portion of the map of human chromosome 1.

**TABLE 6.1    Genetic and Physical Map Lengths of the Human Chromosomes**

| Chromosome[a] | Physical Size (Mb) | Genetic Size | | | | | |
|---|---|---|---|---|---|---|---|
| | | Sex Averaged | cM/Mb | Female | cM/Mb | Male | cM/Mb |
| 1 | 263 | 292.7 | 1.11 | 358.2 | 1.36 | 220.3 | 0.84 |
| 2 | 255 | 277.0 | 1.09 | 324.8 | 1.27 | 210.6 | 0.83 |
| 3 | 214 | 233.0 | 1.09 | 269.3 | 1.26 | 182.6 | 0.85 |
| 4 | 203 | 212.2 | 1.05 | 270.7 | 1.33 | 157.2 | 0.77 |
| 5 | 194 | 197.6 | 1.02 | 242.1 | 1.25 | 147.2 | 0.76 |
| 6 | 183 | 201.1 | 1.10 | 265.0 | 1.45 | 135.2 | 0.74 |
| 7 | 171 | 184.0 | 1.08 | 187.2 | 1.09 | 178.1 | 1.04 |
| 8 | 155 | 166.4 | 1.07 | 221.0 | 1.43 | 113.1 | 0.73 |
| 9 | 145 | 166.5 | 1.15 | 194.5 | 1.34 | 138.5 | 0.96 |
| 10 | 144 | 181.7 | 1.26 | 209.7 | 1.46 | 146.1 | 1.01 |
| 11 | 144 | 156.1 | 1.08 | 180.0 | 1.25 | 121.9 | 0.85 |
| 12 | 143 | 169.1 | 1.18 | 211.8 | 1.48 | 126.2 | 0.88 |
| 13q | 98 | 117.5 | 1.20 | 132.3 | 1.35 | 97.2 | 0.99 |
| 14q | 93 | 128.6 | 1.38 | 154.4 | 1.66 | 103.6 | 1.11 |
| 15q | 89 | 110.2 | 1.24 | 131.4 | 1.48 | 91.7 | 1.03 |
| 16 | 98 | 130.8 | 1.33 | 169.1 | 1.73 | 98.5 | 1.01 |
| 17 | 92 | 128.7 | 1.40 | 145.4 | 1.58 | 104.0 | 1.13 |
| 18 | 85 | 123.8 | 1.46 | 151.3 | 1.78 | 92.7 | 1.09 |
| 19 | 67 | 109.9 | 1.64 | 115.0 | 1.72 | 98.0 | 1.46 |
| 20 | 72 | 96.5 | 1.34 | 120.3 | 1.67 | 73.3 | 1.02 |
| 21q | 39 | 59.6 | 1.53 | 70.6 | 1.81 | 46.8 | 1.20 |
| 22q | 43 | 58.1 | 1.35 | 74.7 | 1.74 | 46.9 | 1.09 |
| X | 164 | 198.1 | 1.21 | 198.1 | 1.21 | | |

*Source:* Adapted from Dib et al. (1996).

[a]Only the long arm is shown for five chromosomes in which the short arm consists largely of tandemly repeating ribosomal DNA. The approximate full length of these chromosomes are given in parentheses.

## CURRENT STATE OF THE HUMAN GENETIC MAP

Several major efforts to make genetic maps of the human genome have occurred during the past few years, and recent emphasis has been on merging these efforts to forge consensus maps. A few examples of the status of these maps several years ago are given in Figure 6.29a and b. These are sex-averaged maps because they have a larger density of markers. The ideal framework map will have markers spaced uniformly at distances around 3 to 5 cM, since this is the most efficient sort of map to use to try to find additional markers or disease genes with current technology. On this basis the map shown for chromosome 21 is quite mature, while the map for chromosome 20, a much less studied chromosome, is still relatively immature. Chromosome 21 is completely covered (except for the rDNA-containing short arm) with relatively uniform markers; in contrast, chromosome 20 has several regions where the genetic resolution is much worse than 5 cM. The current status of the maps can only be summarized highly schematically, as shown in Figure 6.29c. Here 2335 positions defined by 5264 markers are plotted. On average, this is almost 1 position per cM.

## 20

## 21

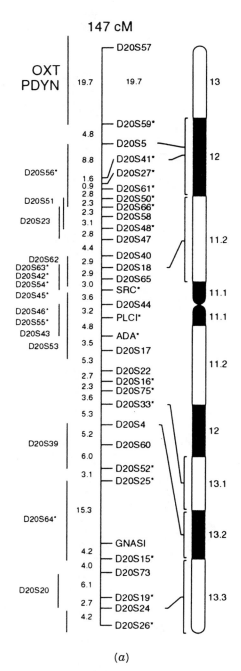

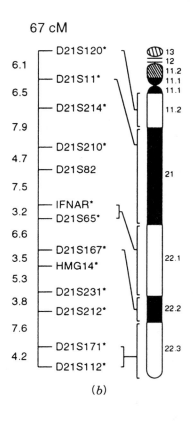

(a)

(b)

**Figure 6.29** Genetic maps of human chromosomes. *(a,b)* Status of sex-averaged maps of chromosomes 20 and 21 several years ago. *(c)* Schematic summary of current genetic map (from Dib et al., 1996).

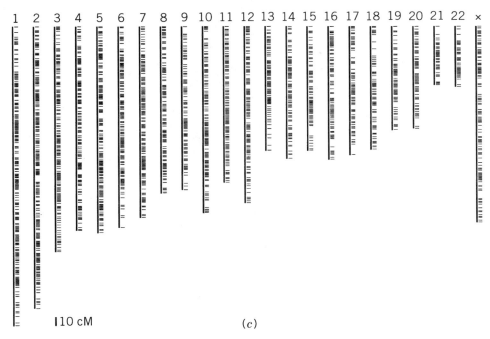

110 cM                                        (*c*)

**Figure 6.29**    *(Continued)*

The physical lengths of the human chromosomes are estimated to range from 50 Mb for chromosome 21, the smallest, to 263 Mb for chromosome 1, the largest. Table 6.1 summarizes these values and compares them with the genetic lengths, seen separately from male and female meioses. Several interesting generalizations emerge from an inspection of Table 6.1. The average recombination frequency per unit length (cM per Mb) varies over a broad range from 0.73 to 1.46 for male meioses and 1.09 to 1.81 for female meioses. Smaller chromosomes tend to have proportionally greater genetic lengths, but this effect is by no means uniform. Recombination along the X chromosome (seen only in females) is markedly suppressed compared with autosomal recombination.

Leaving the details of the genetic map aside, the current version is an extremely useful tool for finding genes on almost every chromosome. As this map is used on a broader range of individuals, we should start to be able to pinpoint potential recombination hot spots and explore whether these are common throughout the human population or whether some or all of them are allelic variants. The present map is already a landmark accomplishment in human biology.

## GENETICS IN THE PSEUDOAUTOSOMAL REGION

Meiotic recombination in the male produces a special situation. The male has only one X and one Y chromosome. These must segregate properly to daughter cells. They must pair with each other in meiotic metaphase, and by analogy with the autosomes, one expects that meiotic recombination will be an obligatory event in proper chromosome segregation.

ptel ———▶

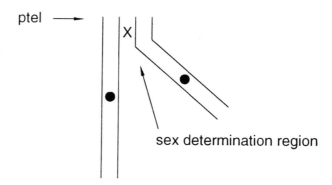

sex determination region

**Figure 6.30**  Pairing between the short arms of the X and Y chromosomes in male meiosis.

The problem this raises is that the X and Y are very different in size, and much of these two chromosomes appear to share little homology. How then does recombination occur between them? It turns out that there are at least two regions of the Y chromosome that share close enough of the X to allow recombination. These are called *pseudoautosomal regions,* for reasons that will become apparent momentarily.

The major psuedoautosomal region of the Y chromosome is located at the tip of the short arm. It is about 2.5 Mb in length and corresponds closely in DNA sequence with the 2.5 Mb short arm terminus of the X chromosome. These two regions are observed to pair up during meiosis, and recombination presumably must occur in this region (Fig. 6.30). If we imagine an average of 0.5 crossovers per cell division, this is a very high recombination rate indeed compared to a typical autosomal region.

Since the Y chromosome confers a male phenotype, somewhere on this chromosome must lie a gene or genes responsible for male sex determination. We know that this region lies below the pseudoautomal boundary, a place about 2.5 Mb from the short term telomere. Below this boundary, genes appear to be sex linked because, by definition, these genes must not be able to separate away from the sex determination region in meiosis. A genetic map of the pseudoautomal region of the Y chromosome is shown in Figure 6.31. There is a gradient of recombination probability across the region. Near the p telomere, all genes will show 50% recombination with the sex-determining region because of the obligatory recombination event during meiosis. Thus these genes appear to be autosomal, even though they are located on a sex chromosome, because like genes on autosomes they

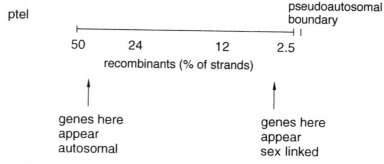

**Figure 6.31**  Genetic map of the pseudoautosomal region of the X and Y chromosomes.

have a 50% probability of being inherited with each sex. As one nears the pseudoautosomal boundary, the recombination frequency of genes with the sex determination region approaches a more normal value, and these genes appear to be almost completely sex linked.

Recently data have been obtained that indicate that a second significant pseudoautosomal region may lie at the extreme ends of the long arms of the X and Y chromosomes. About 400 kb of DNA in these regions appears to consist of homologous sequences, and a 2% recombination frequency in male meioses between two highly informative loci in these regions has been observed. The significance of DNA pairing and exchange in this region for the overall mechanism of male meiosis is not yet known. It is also of interest to examine whether in female meiosis any or all of the X chromosome regions that are homologous to the pseudoautosomal region of the Y chromosome show anomalous recombination frequencies.

---

**BOX 6.4**
**MAPPING FUNCTIONS: ACCOUNTING**
**FOR MULTIPLE RECOMBINATIONS**

If two markers are not close, there is a significant chance that multiple DNA crossovers may occur between them in a particular meiosis. What one observes experimentally is the net probability of recombination. A more accurate measure of genetic distance will be the average number of crossovers that has occurred between the two markers. We need to correct for the occurrence of multiple crossovers in order to compute the expected number of crossovers from the observed recombination frequency. This is done by using a mapping function.

The various possible recombination events for zero, one, and two crossovers are illustrated schematically in Figure 6.32. In each case we are interested in correlating the observed number of recombinations between two distant markers and the actual average number of crossovers among the DNA strands. In all the examples discussed below it is important to realize that any crossovers that occur between sister chromatids (identical copies of the parental homologs) have no effect on the final numerical results. The simplest case occurs where there are no crossovers between the markers; clearly in this case there is no recombination between the markers. Next, consider the case where there is a single crossover between two different homologs (Fig. 6.32b). The net result is a 0.5 probability of recombination because half of the sister chromatids will have been involved in the crossover and half will not have been.

When two crossovers occur between the markers, the results are much more complex. Three different cases are illustrated in Figure 6.32c. Two single crossovers can occur, each between a different set of sister chromatids. The net result, shown in the figure, is that all the gametes show recombination between the markers; the recombination frequency is 1.0. There are four discrete ways in which these crossovers can occur. Alternatively, the two crossovers may occur between the same set of sister chromatids. This is a double-crossover event. The net result is no observed recombination between the distant markers. There are four discrete ways in which a double crossover can occur. Note that the net result of the two general cases we have considered thus far is 0.5 recombinant per crossover.

*(continued)*

**BOX 6.4** *(Continued)*

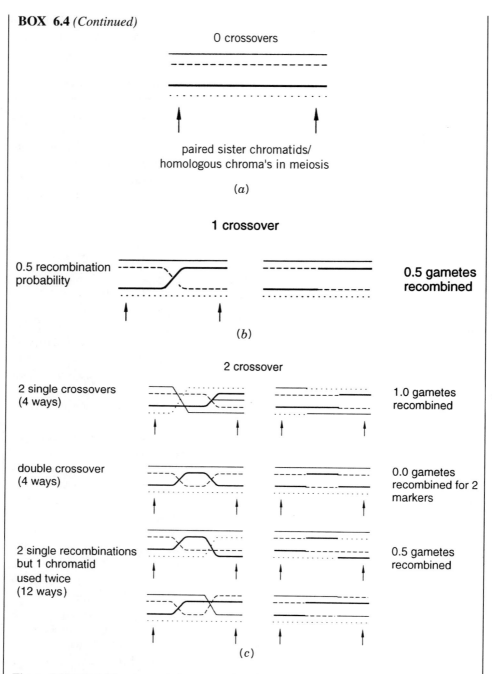

0 crossovers

paired sister chromatids/
homologous chroma's in meiosis

*(a)*

**1 crossover**

0.5 recombination
probability

0.5 gametes
recombined

*(b)*

2 crossover

2 single crossovers
(4 ways)

1.0 gametes
recombined

double crossover
(4 ways)

0.0 gametes
recombined for 2
markers

2 single recombinations
but 1 chromatid
used twice
(12 ways)

0.5 gametes
recombined

*(c)*

**Figure 6.32** Possible crossovers between pairs of genetic markers and the resulting meiotic recombination frequencies that would be observed between two markers (vertical arrows) flanking the region. *(a)* No crossovers. *(b)* One crossover. *(c)* Various ways in which two crossovers can occur.

*(continued)*

**BOX 6.4** (*Continued*)

The final case we need to consider for two crossovers are those in which three of the four paired sister chromatids are involved; two are used once and one is used twice. There are 12 possible ways in which this can occur. Two representative examples are shown in Figure 6.32c. Each results in half of the DNA strands showing a recombination event between distant markers, and half showing no evidence for such an event. Thus, on average, this case yields an observed recombination frequency of 0.5. This is also the average for all the cases we have considered except the case where no crossovers have occurred at all. It turns out that it is possible to generalize this argument to any number of crossovers. The observed recombination frequency, $\Theta_{obs}$, is

$$\Theta_{obs} = 0.5 \, (1 - P_0)$$

where $P_0$ is the fraction of meioses in which no crossovers occur between a particular pair of markers.

It is a reasonable approximation to suppose that the number of crossovers between two markers will be given by a Poisson distribution, where $\mu$ represents the mean number of crossovers that take place in an interval the size of the spacing between the markers. The frequency of $n$ crossovers predicted by the Poisson distribution is

$$P_n = \frac{\mu^n \exp(-\mu)}{n!}$$

and the frequency of zero crossovers is just $P_0 = \exp(-\mu)$. Using this, we can rewrite

$$\Theta_{obs} = 0.5 \, (1 - \exp(-\mu))$$

This can easily be rearranged to give $\mu$ as a function of $\Theta_{obs}$:

$$\mu = -\ln(1 - 2\,\Theta_{obs})$$

The parameter $\mu$ is the true measure of mapping distance corrected for multiple crossovers. It is the desired mapping function.

## WHY GENETICS NEEDS DNA ANALYSIS

In almost all of the preceding discussion, we assumed the ability to determine genotype uniquely from phenotype. We allowed that we could always find heterozygous markers when we needed them, and that there was never any ambiguity in determining the true genotype from the phenotype. This is an ideal situation never quite achieved in practice, but we can come very close to it by the use of DNA sequences as genetic markers.

The simplest DNA marker in common use is a two allele polymorphism—a single inherited base pair difference. This is shown schematically in Figure 6.33. From the DNA sequence it is possible to distinguish the two homozygotes from the heterozygote. Earlier, in Chapter 4, we demonstrated how this can be accomplished using allele-specific PCR.

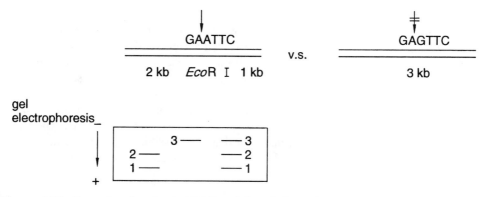

**Figure 6.33** Example of a simple RFLP and how it is analyzed by gel electrophoresis and Southern blotting. Such an allele can also be analyzed by PCR or allele-specific PCR, as described in Chapter 4.

An alternative approach, less general but with great historical precedent, is the examination of restriction fragment length polymorphisms (RFLPs). Such a case is illustrated in Figure 6.33. Where a single-base polymorphism adventitiously lies within the sequence recognized and cleaved by a restriction endonuclease, the polymorphic sequence results in a polymorphic cleavage pattern. All three possible genotypes are distinguishable from the pattern of DNA fragments seen in an appropriate double digest. If this is analyzed by Southern hybridization, it is helpful to have a DNA probe that samples both sides of the polymorphic restriction site, since this prevents confusion from other possible polymorphisms in the region of interest.

The difficulty with two-allele systems is that there are many cases where they will not be informative in a family linkage study, since too many of the family members will be homozygotes or other noninformative genotypes. These problems are rendered less serious when more alleles are available. For example, two single-site polymorphisms in a region combine to generate a four-allele system. If these occur at restriction sites, the alleles are both sites cut, one site cut, the other site cut, and no sites cut. Most of the time the resulting DNA lengths will all be distinguishable.

The ideal DNA marker will have a great many easily distinguished alleles. In practice, the most useful markers have turned out to be minisatellites or variable number tandem repeated sequences (VNTRs). An example is a block like $(AAAG)_n$ situated between two single-copy DNA sequences. This is analyzed by PCR from the two single-copy flanking regions or by hybridization using a probe from one of the single-copy regions (Figure 6.34). The alleles correspond to the number of repeats. There are a large number of possible alleles. VNTRs are quite prevalent in the human genome. Many of them have a large

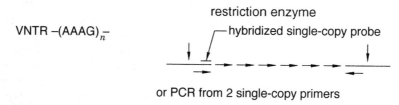

or PCR from 2 single-copy primers

**Figure 6.34** Use of PCR to analyze length variations in typical VNTR. In genetics such markers are extremely informative and can be easily found and analyzed.

number of alleles in the actual human population, and thus they are very powerful genetic markers. See, for example, Box 6.5. Most individuals are heterozygous for these alleles, and most parents have different alleles. Thus the particular homologous chromosomes inherited by an offspring can usually be determined unambiguously.

---

**BOX 6.5**
**A HIGHLY INFORMATIVE POLYMORPHIC MARKER**

A particularly elegant example of the power of VNTRs is a probe described by Alec Jeffries. This single-copy probe which detects a nearby minisatellite was originally called MS32 by Jeffries (MS means minisatellite). When its location was mapped on the genome, it was found to lie on chromosome 1 and was assigned by official designation D1S8. Here D refers to the fact that the marker is a DNA sequence, 1 means that it is on chromosome 1, S means that it is a single copy DNA sequence, and 8 means that this was the eighth such probe assigned to chromosome 1. The power of D1S8 is illustrated in Figure 6.35. The minisatellite contains a *Hin*f I cleavage site within each 29 base repeat. In addition some repeats contain an internal *Hae* II cleav-

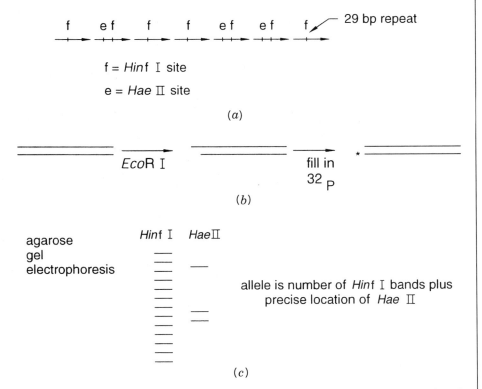

Figure 6.35  A highly informative genetic marker D1S8 that can be used for personal identity determinations. *(a)* Repeating DNA structure of the marker. *(b)* PCR production of probes to analyze the marker. *(c)* Typical results observed when the marker is analyzed after separate, partial *Hae* II and *Hin*f I digestions.

*(continued)*

**BOX  6.5** *(Continued)*

age site. When appropriately chosen PCR primers are used, one can amplify the region containing the repeat and radiolabel just one side of it. Then a partial digestion with the restriction enzyme *Hae* II or *Hin*f I generates a series of DNA bands whose sizes reflect all of the enzyme-cutting sites within the repeat. This reveals not only the number of repeats but also the locations of the specific *Hae* II sites. When this information is combined, it turns out that there are more than $10^{70}$ possible alleles of this sequence. Almost every member of the human population (exempting identical twins) would be expected to have a different genotype here; thus this probe is an ideal one not only for genetic analysis but also for personal identification.

By comparing the alleles of D1S8 in males and in sperm samples, the mutation rage of this VNTR has been estimated. It is extremely high, about $10^{-3}$ per meiosis or $10^5$ higher than the average rate expected within the human genome. The mechanism believed to be responsible for this very high mutation rate can be inferred from a detailed analysis of the mutant alleles. It turns out that almost all of the mutations arise from interallelic events, as shown in Figure 6.36. These include possible slippage of the DNA during DNA synthesis, and unequal sister chromatid exchange. Only 6% of the observed mutations arise from ordinary meiotic recombination events between homologous chromosomes.

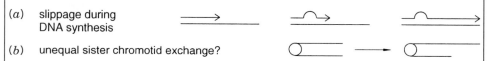

(*a*)   slippage during DNA synthesis

(*b*)   unequal sister chromotid exchange?

**Figure 6.36**   Recombination events that generate diversity in the marker D1S8. *(a)* Intra-allelic recombination or polymerase slippage, a very common event. *(b)* Inter-allelic recombination, a relatively rare event.

## DETECTION OF HOMOZYGOUS REGIONS

Because the human species is highly outbred, homozygous regions are rare. Such regions can be found, however, by traditional methods or by some of the fairly novel methods that will be described in Chapter 13. Homozygous regions are very useful both in the diagnosis of cancer and in certain types of genetic mapping. The significance of homozygous regions in cancer is shown in Figure 6.37. Oncogenes are ordinary cellular genes, or foreign genes that under appropriate circumstances can lead to uncontrolled cell growth, that is, to cancer. Quite a few oncogenes have been found to be recessive alleles, ordinarily silenced in the heterozygous

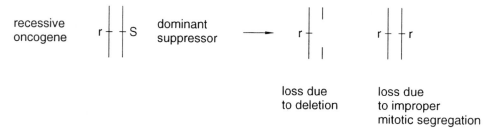

**Figure 6.37**   Generation of homozygous DNA regions in cancer.

state by the presence of a homologous dominant suppressor gene (called a tumor suppressor gene). Consider what happens, when that suppressor is lost either due to mutation, deletion, or improper mitotic segregation so that a daughter cell receives two copies of the homologous chromosome with the recessive oncogene on it. These are all rare events, but once they occur, the resulting cells have a growth advantage; there is a selection in their favor. In the case of a loss due to deletion, the resulting cells will be hemizygous in the region containing the oncogene. Strictly speaking, they will show only single alleles for all polymorphisms in this region, and thus a zone of apparent homozygosity will mark the location of the oncogene. In the case of improper mitotic segregation, the entire chromosome carrying the oncogene may be homozygous, and once this is discovered, it is useful for diagnostic purposes but less useful for finding the oncogene, since an entire chromosome is still a very large target to search.

A second application of homozygous regions is the genetic technique known as homozygosity mapping. This is illustrated in Figure 6.38. It is useful in those relatively rare

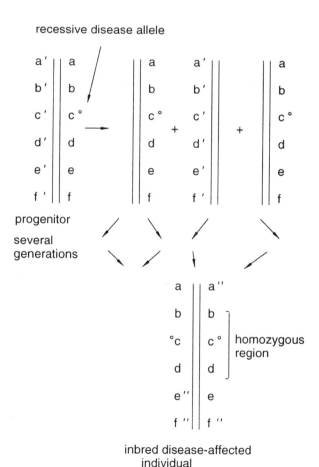

**Figure 6.38** Example of homozygosity mapping. Inbreeding within a family can reveal an ancestral disease allele by a pattern of homozygous alleles surrounding it.

---

**BOX 6.6**
**IMPRINTING**

For a small number of genes in the human and mouse, the phenotype caused by an allele depends on the parent who transmitted it. Some alleles are preferentially expressed from the gene coded for by the father's gamete (for example, insulinlike growth factor 2 and small ribonucleoprotein peptide n [Snrpn] in the mouse). Others are expressed preferentially from the mother's gamete (such as insulinlike growth factor 2 receptor and H19 in the mouse). These effects can be seen strikingly when one parent or the other contains a deletion in an imprinted region.

DNA methylation is believed to play the major role in marking which parent of origin will dominate. Generally methylation silences gene expression, so methylation of a region in a particular parent will silence the effect of that parent. The process of imprinting is complicated. Imprinting may not be felt the same in all tissues in which the gene is expressed. The imprint must be established in the gametes; it may be maintained during embryogenesis and in somatic adult tissues, but must be erased in the early germline so that a new imprint can be re-established in the gametes of the individual who is now ready to transmit it to the next generation. There are well-known examples of human diseases that show a parent of origin phenotype arising from differences in methylation. For example, Prader-Willi syndrome is used to describe the disease phenotype when a damaged or deleted human analog of the mouse Snrpn gene is inherited from the father, whereas Angelman syndrome describes the rather different phenotype when the deleted allele is inherited from the mother. For further reading see Razin and Cedar (1994).

---

cases where there is significant inbreeding in the human population. Consider a progenitor carrying a recessive disease allele. Several generations later come two individuals who have inherited this allele from the original progenitor mate and have offspring together. One or more offspring receive two copies of the disease allele, and as a result they are affected by the inherited disease. Because these offspring have inherited the allele from the same original chromosome, they are likely to share not only that allele but also nearby alleles. While the chromosome will have suffered several recombination events during the generations that have ensued between the progenitor and the affected individuals, most of the original chromosome has presumably remained intact. This approach will be useful in finding disease genes that segregate in inbred populations, and it has the advantage that only affected individuals need to be examined to find the zone of homozygosity that marks the approximate location of the disease gene.

A third application of homozygous regions is their frequent occurrence in the rather remarkable genetic phenomenon called imprinting. Here, as described in Box 6.6, it matters from which parent a gene is inherited.

## SOURCES AND ADDITIONAL READINGS

Churchill, G. A., Giovannoni, J. J., and Tanksley, S. D. 1993 Pooled-samplings makes high-resolution mapping practical with DNA markers. *Proceedings of the National Academy of Sciences USA* 90: 16–20.

Dib, C., Fauré, S., Fizames, C., Samson, D., et al. 1996. A comprehensive genetic map of the human genome based on 5,264 microsatellites. *Nature* 380: 152–154.

Dietrich, W. F., Miller, J., Steen, R., Merchant, M. A., et al. 1996. A comprehensive genetic map of the mouse genome. *Nature* 380: 149–152.

Donis-Keller, H., Green, P., Helms, C., et al. 1987. A genetic linkage map of the human genome. *Cell* 51: 319–337.

Lalouel, J.-M., and White, R. L. 1996. Analysis of genetic linkage. In Rimoin, D. L., Connor, J. M., and Pyeritz, R. E., eds. *Principles and Practice of Medical Genetics.* New York: Churchill Livingstone, pp. 111–125.

Lander, E., and Botstein, D. 1986. Mapping complex genetic traits in humans: New methods using a complete RFLP linkage map. *Cold Spring Harbor Symposium* 51: 1–15.

Lander, E., and Botstein, D. 1987. Homozygosity mapping: A way to map human recessive traits with the DNA of inbred children. *Science* 236: 1567–1570.

Leppart, M., Dobbs, M., Scambler, P. et al. 1987. The gene for familial polyposis coli maps to the long arm of chromosome 5. *Science* 238: 1411–1413.

Lindahl K. F. 1991. His and hers recombinational hotspots. *Trends in Genetics* 7: 273–276.

Nicholls, R. D., Knoll, J. H. M., Butler, M. G., Karam, S., and Lalande, M. 1989. Genetic imprinting suggested by maternal heterdisomy in non-delection Prader-Willi syndrome. *Nature* 342: 281-286.

Ott, J. 1991. *Analysis of Human Genetic Linkage*, rev. ed. Baltimore: John Hopkins University Press.

Razin, A. and Cedar, H. 1994. DNA methylation and genomic imprinting. *Cell* 77: 473–476.

Roeder, G. S. 1990. Chromosome synapsis and genetic recombination: their roles in meiotic chromosome segregation. *Trends in Genetics* 6: 385–389.

Shuler, G. D., Boguski, M. S., Stewart, E. A., Stein, L. D., et al. 1996. A gene map of the human genome. *Science* 275: 540–546.

Swain, J. L., Stewart, T. A., and Leber, P. 1987. Parental legacy determines methylation and expression of an autosomal transgene: A molecular mechanism for parental imprinting. *Cell* 50: 719-727.

# 7 Cytogenetics and Pseudogenetics

## WHY GENETICS IS INSUFFICIENT

The previous chapter showed the potential power of linkage analysis in locating human disease genes and other inherited traits. However, as demonstrated in the chapter, genetic analysis by current methods is extremely tedious and inefficient. Genetics is the court of last resort when no more direct way to locate a gene of interest is available. Frequently one already has a DNA marker believed to be of interest in the search for a particular gene, or just of interest as a potential tool for higher-resolution genome mapping. In such cases it is almost always useful to pinpoint the approximate location of this marker within the genome as rapidly and efficiently as possible. This is the realm of cytogenetic and pseudogenetic methods. While some of these have potentially very high resolution, most often they are used at relatively low resolution to provide the first evidence for the chromosomal and subchromosomal location of a DNA marker of interest. Unlike genetics, which can work with just a phenotype, most cytogenetics and pseudogenetics are best accomplished by direct hybridization using a cloned DNA probe or by PCR.

## SOMATIC CELL GENETICS

In Chapter 2 we described the availability of human-rodent hybrid cells that maintain the full complement of mouse or hamster chromosomes but lose most of their human complement. Such cells provide a resource for assigning the chromosomal location of DNA probes or PCR products. In the simplest case a panel of 24 cell lines can be used, each one containing only a single human chromosome. In practice, it is more efficient to use cell lines that contain pools of human chromosomes. Then simple binary logic (presence or absence of a signal) allows the chromosomal location of a probe to be inferred with a smaller number of hybridization experiments or PCR amplifications. In Chapter 9 some of the principles for constructing pools will be described. However, the general power of this approach is indicated by the example in Figure 7.1. The major disadvantage of using pools is that they are prone to errors if the probe in question is not a true single-copy DNA sequence but, in fact, derives from a gene family with representatives present on more than one human chromosome. In this case the apparent assignment derived from the use of pools may be totally erroneous. A second problem with the use of hybrid cells, in general, is the possibility that a given human DNA probe might cross-hybridize with rodent sequences. For this reason it is preferable to measure the hybridization of the probe not by a dot blot but by a Southern blot after a suitable restriction enzyme digestion. The reason for this is that rodents and humans are sufficiently diverged that although a similar DNA sequence may be present, there is a good chance that it will lie on a different-sized DNA restriction fragment. This may allow the true human-specific hybridization to be distinguished from the rodent cross-hybridization.

| Chromosome | Base 3 | ← Right digit → | | | ← Center digit → | | | ← Left digit → | | |
| | | Pool A = 0 | Pool B = 1 | Pool C = 2 | Pool D = 0 | Pool E = 1 | Pool F = 2 | Pool G = 0 | Pool H = 1 | Pool I = 2 |
|---|---|---|---|---|---|---|---|---|---|---|
| 1 | 001 | | + | | + | | | + | | |
| 2 | 002 | | | + | + | | | + | | |
| 3 | 010 | + | | | | + | | + | | |
| 4 | 011 | | + | | | + | | + | | |
| 5 | 012 | | | + | | + | | + | | |
| 6 | 020 | + | | | | | + | + | | |
| 7 | 021 | | + | | | | + | + | | |
| 8 | 022 | | | + | | | + | + | | |
| 9 | 100 | + | | | + | | | | + | |
| 10 | 101 | | + | | + | | | | + | |
| 11 | 102 | | | + | + | | | | + | |
| 12 | 110 | + | | | | + | | | + | |
| 13 | 111 | | + | | | + | | | + | |
| 14 | 112 | | | + | | + | | | + | |
| 15 | 120 | + | | | | | + | | + | |
| 16 | 121 | | + | | | | + | | + | |
| 17 | 122 | | | + | | | + | | + | |
| 18 | 200 | + | | | + | | | | | + |
| 19 | 201 | | + | | + | | | | | + |
| 20 | 202 | | | + | + | | | | | + |
| 21 | 210 | + | | | | + | | | | + |
| 22 | 211 | | + | | | + | | | | + |
| X (23) | 212 | | | + | | + | | | | + |
| Y (24) | 220 | + | | | | | + | | | + |

Examples:

| samples | pool analysis (left, center, right) | conclusions |
|---|---|---|
| mixture 5 + 8 | 0, 1 and 2, 2 | => 012, 022 = 5 + 8 |
| mixture 10 + 22 | 1 and 2, 0 and 1, 1 | => 101, 211 = 10 + 22 |
| | | or 111, 201 = 13 + 19 |

**Figure 7.1** Assignment of a cloned DNA probe to a chromosome by hybridization against a panel of cell lines containing limited sets of human chromosomes in a rodent background. This hypothetical example is based on a base three pooling (see Chapter 9) which is quite efficient compared with randomly selected pools. If the probe is derived from a single chromosome, it can be uniquely assigned. If it contains material that hybridizes to two or more chromosomes, in some cases all of the chromosomes involved can be identified, but in most cases the results will be ambiguous as shown by the example at the bottom of the figure.

An alternative way to circumvent the problem of rodent cross-hybridization in assigning DNA probes to chromosomes is to use flow-sorted human chromosomes, preferably chromosomes that have been sorted from human-rodent hybrid cells so that the purified human fraction has very little contamination from other human chromosomes (Chapter 2). This is a very powerful approach; its major limitation is the current scarcity of flow-sorted human chromosomal DNA.

## SUBCHROMOSOMAL MAPPING PANELS

Frequently it is possible to assemble sets of cell lines containing fragments of only one chromosome of interest. The occurrences of a DNA probe sequence in a subset of these cell lines can be used to assign the probe to a specific region of the chromosome. The accuracy of that assignment is determined by the accuracy with which the particular chromosome fragments contained in each cell line are known. This can vary quite considerably, but in general this approach is quite a powerful and popular one. A simple example is shown in Figure 7.2. Here three cell lines are used to divide a chromosome into three regions. The key aspect of these cell lines are breakpoints that determine which chromosome segments are present and which are absent.

A more complex example of cell lines suitable for regional assignment is given in Figure 7.3. Here the properties of a chromosome 21 mapping panel are illustrated. It is evident that some regions of the chromosome are divided by the use of this panel into very high resolution zones while others are defined much less precisely. In the case shown in Figure 7.3, human cell lines were assembled from available patient material: individuals with translocations or chromosomal deletions. Then these chromosomes were moved by cell fusion into rodent lines to simplify hybridization analyses. Because no systematic method was used to assemble the panel, the resulting breakpoint distribution is very uneven.

If a selectable marker exists near the end of a chromosome, it is possible to assemble a very orderly and convenient mapping panel. Assuming the selection can be applied in a hybrid cell, what is done is to irradiate the cell line lightly with X rays so as to cause an average of less than one break per chromosome, and then grow the cells in culture and allow broken chromosomes to heal. Since the selection applies to only one end of the human chromosome, there will be a tendency to keep chromosomes containing that end and lose pieces without that end. The result is a fine, ordered subdivision of the chromosome, as shown in Figure 7.4. This resource is exceptionally useful for mapping the location of cloned probes because the results are usually unambiguous. The pattern of hybridization should be a step function and the point where positive signal disappears indicates the most distal possible location of the probe.

A few caveats are in order concerning the use of cell lines that derive from broken or rearranged human chromosomes. Broken chromosome are always healed by fusion with other chromosomes (or by acquisition of telomeres). For example, cell lines described as 21q- are missing the long arm of chromosome 21. Frequently such cell lines have other human chromosomes present that have been quietly ignored or forgotten. This is not a problem when a single-copy probe is used, assuming that this probe is already known to

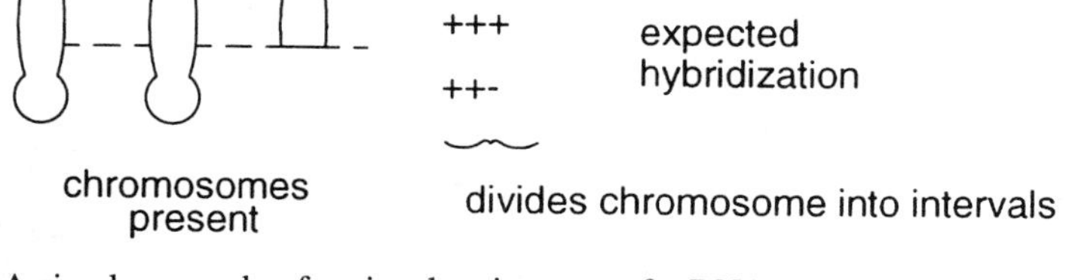

**Figure 7.2**   A simple example of regional assignment of a DNA sample by hybridization to three cell lines containing incomplete copies of the chromosome of interest. The two break points in the chromosomes are used to divide the chromosome into three intervals.

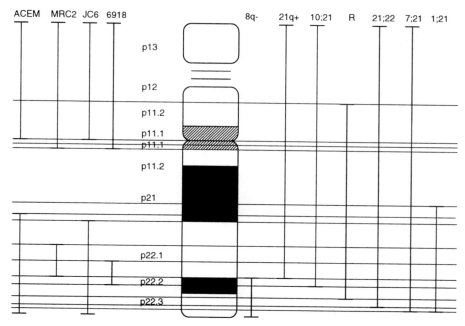

**Figure 7.3**   A mapping panel of hybrid cell lines used to assign probe locations on human chromosome 21. Vertical lines indicate the human chromosome content of each cell line. Note the unevenness of the resulting intervals (horizontal lines).

be located on the chromosome of interest, and the goal is just narrowing down its subchromosomal location. However, the problem is much more serious if the probe of interest comes from a gene family or contains repeated sequences. The actual chromosome content of a set of cell lines described as chromosome 21 hybrids is illustrated in Figure 7.5. Many of these lines, in fact, contain other human chromosomes. Cross-hybridization to rodent background is always a potentially serious additional complication when working with hybrid cells. For this reason it is best to do Southern blots or PCR analysis rather than dot blots, as we already explained earlier in the chapter.

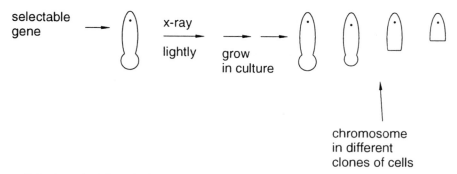

**Figure 7.4**   A method for selection of broken chromosomes that can be used to generate a much more even mapping panel than the one shown in Figure 7.3.

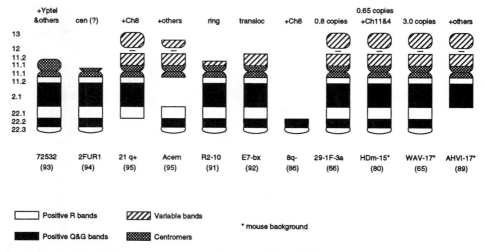

**Figure 7.5**   Actual chromosome content of some of the cell lines used to produce the mapping panel summarized in Figure 7.3. Note that many contain other human chromosomes besides number 21.

## RADIATION HYBRIDS

It is possible to make hybrid cells containing small fragments of human chromosomes. These cell lines represent, essentially, a way of cloning small continuous stretches of human DNA. The method was originally described by S. Goss and H. Harris, and later rediscovered and elaborated by D. Cox and R. Meyers (Cox et al., 1990). The basic technique for producing these cells, called *radiation hybrids,* is shown schematically in Figure 7.6. The starting point is a hybrid cell line containing only a single human chromosome in a Chinese hamster ovary (CHO) cell with a wild type hypoxanthine ribosyl transferase (hprt) gene. This cell line is subjected to a very high X-ray dose, 8000 rads, which is sufficient to break every chromosome into five pieces on average. After irradiation the fragments are allowed to heal, and the resulting cells are fused with an hprt⁻CHO line, one in which an inactive allele of the hprt gene is present. Cell clones that have received and retained hprt⁺ DNA are selected and maintained by growth on a particular set of conditions called *HAT medium,* which require an active hprt gene. Some of these clones are recipients of both CHO and human DNA. Note that no human-specific selection has been used to focus the choice of hybrid cells on any particular region of the human chromosome in question. Thus the population of hybrids should represent a random sampling of that chromosome if there is no intrinsic selection method operative.

Selection for an hprt⁺ phenotype ensures that all of the cell lines maintained will have been recipients of some foreign DNA. Those that have human DNA present can be detected by hybridization with human-specific repeated sequences. Then these lines are allowed to grow in the absence of any selection for human markers. This will lead to random loss of human sequences as time progresses. The cells are continually screened for the presence of human markers until 30% to 60% of the hybrids contain a particular marker of interest. At this point the radiation hybrids contain quite a few discrete fragments of human DNA integrated between hamster DNA segments.

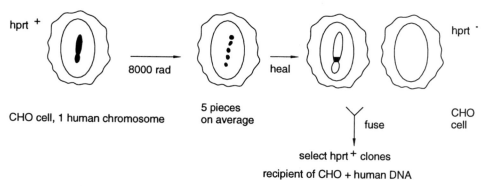

**Figure 7.6** Steps in the production of a set of radiation hybrid cell lines. Note that there is no selection for any of the human DNA fragments, so they will be lost progressively as the cells are grown.

They are now used to look at the statistics of co-retention of pairs of human DNA markers. The basic idea is illustrated below, for three markers on a chromosome:

$$\text{————————A —— B————————————————C——}$$

Even when the chromosome is fragmented into five pieces, there is a good chance that A and B will still be found on the same DNA piece; it is much more likely that C will be on a different piece. Thus cell lines that have lost A are also likely to have lost B, but they are less likely to have lost C. By establishing the presence or absence of particular markers, we gain information about the relative proximity of the markers because nearby markers tend to be co-retained.

Here we will examine the statistics of co-retention of marker pairs. There is a basic problem with this approach. If only two markers at a time are considered, it turns out that there are more variables than constraints, and approximations have to be used to analyze the results. With more than two simultaneous markers, the mathematics becomes more robust but only at a cost of much greater complexity. For didactic purposes we will analyze the two-marker case; anyone with a serious interest in pursuing this further should be aware that in practice, one must consider four or more markers at once to avoid the additional assumptions used later in this section. The parameters we must consider are illustrated below. The two markers A and B reside either on the same DNA fragment or on separate fragments as shown by the horizontal dashed lines. These fragments will have individual probabilities of retention, as indicated.

retention probability $P_{AB}$

retention probabilities $P_A$, $P_B$

The probability that a break has occurred between A and B in a particular irradiation treatment is given by the additional parameter $\phi$. All four parameters $P_A$, $P_B$, $P_{AB}$, and $\phi$ have values between 0 and 1.

What is done experimentally is to examine a set of hybrid cell clones and to ask in each whether markers A and B are present. The fraction of them that have both markers present is $f_{AB}$, the fractions with only A or only B are $f_A$ and $f_B$, respectively, and the fraction that have retained neither marker is $f_0$. These four observables can be related to the four parameters by the following equations:

$$f_{AB} = P_{AB}(1 - \phi) + \phi P_A P_B$$

The first term denotes cells that have maintained A and B on an unbroken piece of DNA; the second term considers those cells that have suffered a break between A and B but have subsequently retained both separate DNA pieces.

$$f_A = \phi P_A (1 - P_B)$$
$$f_B = \phi P_B (1 - P_A)$$

The top equation indicates cells that have had a break between A and B and kept only A; the bottom equation indicates cells with a similar break, but these have retained only B.

$$f_0 = (1 - P_{AB})(1 - \phi) + \phi(1 - P_A)(1 - P_B)$$

The first term indicates cells that originally contained A and B on a continuous DNA piece but subsequently lost this piece (probability $1 - P_{AB}$); the second term indicates cells that originally had a break between A and B and subsequently lost both of these DNA pieces.

The four types of cells represent all possibilities;

$$f_{AB} + f_A + f_B + f_0 = 1$$

This means that we have only three independent observables, but there are four independent unknown parameters. The simplest way around this dilemma would be to assume that the probability of retention of individual markers was equal, for example, that $P_A = P_B$. However, available experimental data contradict this. Presumably some markers lie in regions that because of chromosome structural elements like centromeres or telomers, or metabolically active genes, convey a relative advantage or disadvantage for retention of their region. The approximation actually used by Cox and Myers to allow analysis of their data is to assume that $P_A$ and $P_B$ are related to the total amount of A and B retained, respectively:

$$P_A = f_{AB} + f_A$$
$$P_B = f_{AB} + f_B$$

Upon inspection this assumption can be shown to be algebraically false, but in practice, it seems to work well enough to allow sensible analysis of existing experimental data.

The easiest way to use the two remaining independent observables is to perform the sum indicated below:

$$f_A + f_B = \phi P_A (1 - P_B) + \phi P_B (1 - P_A)$$

This can be rearranged to yield a value for $\phi$, which is the actual parameter that should be directly related to the inverse distance between the markers A and B:

$$\phi = \frac{f_A + f_B}{P_A + P_B - 2 P_A P_B}$$

Here the parameters $P_A$ and $P_B$ are evaluated from measured values of $f_{AB}, f_A,$ and $f_B$ as described above.

Note that in using radiation hybrids, one can score any DNA markers A and B whether they are just physical pieces of DNA or whether they are inherited traits that can somehow be detected in the hybrid cell lines. However, the resulting radiation hybrid map, based on values of $\phi$, is expected to be a physical map and not a genetic map. This is because $\phi$ is related to the probability of X-ray breakage, which is usually considered to be a property of the DNA itself and not in any way related to meiotic recombination events. In most studies to date, $\phi$ has been assumed to be relatively uniform. This means that in addition to marker order, estimates for DNA distances between markers were inferred from measured values of $\phi$.

An example of a relatively mature radiation hybrid map, for the long arm of human chromosome 21, is shown in Figure 7.7. The units of the map are in centirays (cR), where 1 cR is a distance that corresponds to 1% breakage at the X-radiation dosage used. The density of markers on this map is impressively high. However, the distances in this map are not as uniform as previously thought. As shown in Table 7.1, the radiation hybrid map is considerably elongated at both the telomeric and centromeric edges of the chromosome arm, relative to the true physical distances revealed by a restriction map of the chromosome. It is not known at present whether these discrepancies arose because of the oversimplifications used to construct the map (discussed above) or because, as has been shown in several studies, the probability of breakage or subsequent repair is nonuniform along the DNA. A radiation hybrid map now exists for the entire human genome. It is compared to the genetic map and to physical maps in Table 7.2.

## SINGLE-SPERM PCR

A human male will never have a million progeny, but he will have an essentially unlimited number of sperm cells. Each will display the particular meiotic recombination events that generated its haploid set of chromosomes. These cells are easily sampled, but in

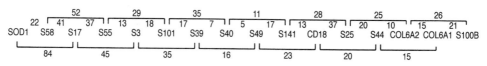

**Figure 7.7** A radiation hybrid map of part of human chromosome 21. The units of the map are in centirays (cR). The map shows the order of probes and the distance between each pair. The telomere is to the right, near S100B. Distances for adjacent markers are shown between the markers, and for nearest neighbor pairs above and below the map. (Adapted from Burmeister et al., 1994.)

**TABLE 7.1 Comparison of the Human Chromosome 21 *Not* I Genomic Map with the Radiation Hybrid Map**

| Marker Pairs | Physical Distance, kb[a] | cR[b] | kb/cR[a] |
|---|---|---|---|
| D21S16–D21S11, 1 | 5300 ± 2400 | 87 | 36 ± 16 |
| D21S11, 1–D21S12 | 7300 ± 2200 | 140 | 52 ± 16 |
| D21S12–SOD | 6065 ± 2495 | 66 | 92 ± 38 |
| SOD–D21S58 | 1490 ± 480 | 22 | 68 ± 22 |
| D21S58–D21S17 | 2310 ± 1040 | 39 | 59 ± 26 |
| D21S17–D21S55 | 3860 ± 1600 | 35 | 110 ± 46 |
| D21S55–D21S39 | 3700 ± 1350 | 43 | 86 ± 31 |
| D21S39–D21S141, 25 | 3690 ± 660 | 40 | 91 ± 16 |
| D21S151, 25–COL6A | 2330 ± 540 | 54 | 43 ± 10 |

*Source:* From Wang and Smith (1994).

[a]The uncertainty shown for each distance is the maximum possible range. The distance given is the mean of that range. The uncertainty shown for each ratio is the maximum range, considering only uncertainty in the placement of markers on the restriction map and ignoring any uncertainty in the radiation hybrid map.

[b]Taken from the data of Cox et al. (1990) and Burmeister et al. (1991). These data were obtained at 8000 rads.

**TABLE 7.2 Characteristics of the Human Radiation Hybrid Map**

| Chromosome | Physical Length (Mb) | Average Relative Metrics | |
|---|---|---|---|
| | | cM/cR[a] | kb/cR[a] |
| 1 | 263 | 0.20 | 197 |
| 2 | 255 | 0.21 | 225 |
| 3 | 214 | 0.27 | 233 |
| 4 | 203 | 0.28 | 256 |
| 5 | 194 | 0.29 | 272 |
| 6 | 183 | 0.24 | 243 |
| 7 | 171 | 0.23 | 229 |
| 8 | 155 | 0.26 | 271 |
| 9 | 145 | 0.38 | 305 |
| 10 | 144 | 0.26 | 253 |
| 11 | 144 | 0.30 | 270 |
| 12 | 143 | 0.21 | 234 |
| 13q | 98 | 0.22 | 179 |
| 14q | 93 | 0.34 | 208 |
| 15q | 89 | 0.36 | 203 |
| 16 | 98 | 0.43 | 201 |
| 17 | 92 | 0.23 | 147 |
| 18 | 85 | 0.30 | 172 |
| 19 | 67 | 0.20 | 110 |
| 20 | 72 | 0.30 | 191 |
| 21q | 39 | 0.31 | 151 |
| 22q | 43 | 0.95 | 185 |
| X | 164 | 0.31 | 231 |
| Genome | 3154 | 0.31 | 208 |

*Source:* Adapted from McCarthy (1996).

Note: The data were obtained at 3000 rads. Hence physical distances relative to cR are roughly 8/3 as large as the data shown in Table 7.1.

[a]cM/cR scales the genetic and radiation hybrid maps. kb/cR scales the physical and radiation hybrid maps.

order to analyze specific recombination events, it is necessary to be able to examine the markers *cosegregating in each single sperm cell.* Sperm cannot be biologically amplified except by fertilization with an oocyte, and the impracticality of such experiments in the human is all too apparent. Thus, until in vitro amplification became a reality, it was impossible to think of analyzing the DNA of single sperm. With PCR, however, this picture has changed dramatically.

A series of feasibility tests for single-sperm PCR was performed by Norman Arnheim and co-workers. First they asked whether single diploid human cells could be genotyped by PCR. They mixed cells from donors who had normal hemoglobin, homozygous for the $\beta_A$ gene with cells from donors who had sickle cell anemia, homozygous for the $\beta_S$ gene. Single cells were selected by micromanipulation and were tested by PCR amplification of the $\beta$ locus followed by hybridization of a dot blot with allele-specific oligonucleotide probes. The purpose of the experiment was to determine whether the pure alleles could be detected reliably, or whether there would be cross-contamination between the two types of cells. The results for 37 cells analyzed were 19 $\beta_A$, 12 $\beta_S$, 6 none, and 0 both. This was very encouraging.

The next test involved the analysis of single sperm from donors with two different LDL receptor alleles. For 80 sperm analyzed, the results were 22 allele 1, 21 allele 2, 1 both alleles, and the remainder neither allele. Thus the efficiency of the single-sperm analysis was less than with diploid cells, but the cross-contamination level was low enough to be easily dealt with. A final test was to look at two nonlinked two-allele systems. One was on chromosome 6 ($a_1$ or $a_2$), and the other on chromosome 19 ($b_1$ or $b_2$). Four types of gametes were expected in equal ratios. What was actually observed is

| $a_1b_1$ | $a_1b_2$ | $a_2b_1$ | $a_2b_2$ |
|----------|----------|----------|----------|
| 21       | 18       | 14       | 17       |

This is really quite close to what was expected statistically,

The final step was to do single-sperm linkage studies. In this case one does simultaneous PCR analyses on linked markers. Consider the example shown in Figure 7.8. Nothing needs to be known about the parental haplotypes to start with. PCR analysis automatically provides the parental phase. In the case shown in the figure, most of the sperm measured at loci a and b have either alleles $a_1$ and $b_2$ or $a_3$ and $b_1$. This indicates that these are the haplotypes of the two homologous chromosomes in the donor. Occasional recombinants are seen with alleles $a_1$ and $b_1$ or $a_3$ and $b_2$. The frequency at which these are observed is a true measure of the male meiotic recombination frequency between markers a and b. Since large numbers of sperm can be measured, one can determine this frequency very accurately. Perhaps more important, with large numbers of sperm, very rare meiotic recombination events can be detected, and thus very short genetic distances can be measured.

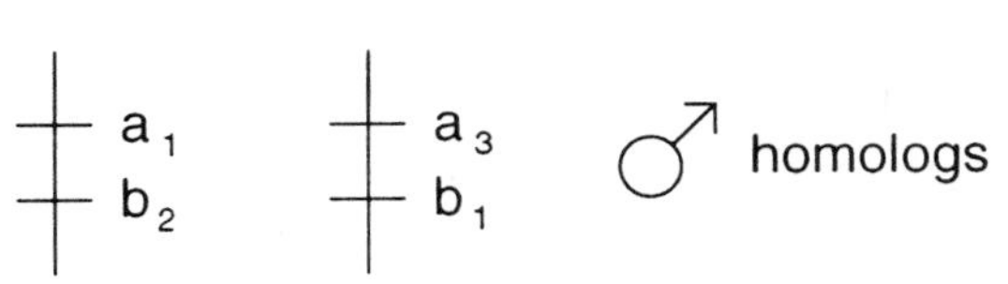

**Figure 7.8**  Two unrecombined homologous chromosomes that should predominate in the sperm expected from a hypothetical donor. Note that the observed pattern of alleles also determines the phase at these loci if it is not known in advance.

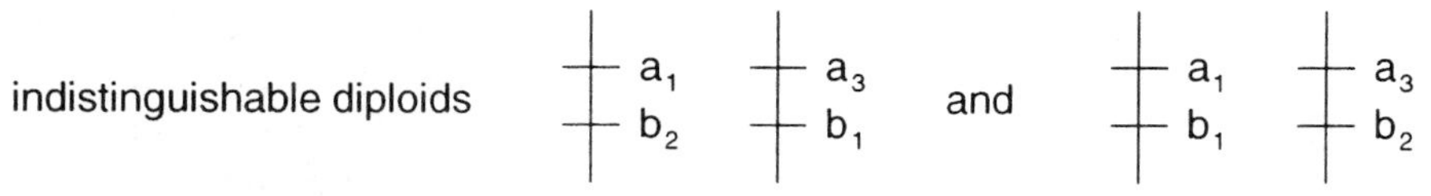

indistinguishable diploids $\quad \dfrac{a_1}{b_2} \quad \dfrac{a_3}{b_1} \quad$ and $\quad \dfrac{a_1}{b_1} \quad \dfrac{a_3}{b_2}$

**Figure 7.9**  The allele pairs present in single DNA molecules will determine the phase of two markers in a pair of homologous chromosomes.

The map that results from single-sperm PCR is a true genetic map because it is a direct measure of male meiotic recombination frequencies. However, it has a number of limitations. Only male meioses can be measured. Only DNA markers capable of efficient and unique PCR amplification can be used. Thus this genetic map cannot be used to locate genes on the basis of their phenotype. Not only must DNA be available, but enough of it must be sequenced to allow for the production of appropriate PCR primers. The phenotype is irrelevant, and in fact it is invisible. Perhaps the greatest limitation of direct single-sperm PCR is that the sperm are destroyed by the PCR reaction. In principle, all markers of interest on a particular sperm cell must be analyzed simultaneously, since one will never be able to recreate that precise sperm cell again. This is not very practical, since simultaneous multi-locus PCR with many sets of primers has proved to be very noisy. One solution is to first do random primed PCR (PEP), tagged random primed (T-PCR), or degenerate oligonucleotide primed (DOP)-PCR, as described in Chapter 4. The sample is saved, and aliquots are used for the subsequent analysis of specific loci, one at a time, in individual, separate ordinary PCR reactions.

A variation on single-sperm PCR is single DNA molecule genetics. Here one starts with single diploid cells, prepares samples of their DNA, and dilutes these samples until most aliquots have either a single DNA molecule or none at all. The rationale behind this tactic is that it allows determination of the phase of an individual without any genetic information about any other relatives. A frequent problem in clinical genetics is that only a single parent is available because the other is uncooperative, aspermic, or dead. Phase determination means distinguishing among the two cases shown in Figure 7.9. It is clear that simultaneous PCR analysis of the alleles present on particular individual DNA molecules will reveal the phase in a straightforward manner. This is a very important step in making genetic mapping more efficient.

## IN SITU HYBRIDIZATION

A number of techniques are available to allow the location of DNA sequences within cells or chromosomes to be visualized by hybridization with a labeled specific DNA probe. These are collectively called in situ hybridization. This term actually refers to any experiment in which an optical image of a sample (large or small) is superimposed on an image generated by detecting a specifically labeled nucleic acid component. In the current context we mean superimposing images of chromosomes or DNA molecules in the light microscope with images of the locations of labeled DNA probes. Radioactive labels were originally used, and the resulting probe location was determined by autoradiography superimposed on a photomicrograph of a chromosome. Fluorescent labels have almost totally supplanted radioisotopes in these techniques because of their higher spatial resolution, and because both the chromosome image and the specific hybridization image can be captured on the same film, eliminating the need for a separate autoradiographic development step. Short-lived radioisotopes like $^{32}$P have decay tracks that are too long for microscopic images, and considerable resolution is lost by imprecision in the location of the origin of a track. Longer-lived isotopes like $^3$H or $^{14}$C would have shorter tracks, but these would be less efficient to detect, and more seriously, one would have to wait unrealistically long periods of time before an image could be detected. The technique in widespread use today depends on several cycles of stoichiometric amplification such as streptavidin biotin amplification (see Chapter 3) to increase the sensitivity of fluorescent detection. It is called FISH, for fluorescent in situ hybridization.

Metaphase chromosomes, until recently, were the predominant samples used for DNA mapping by in situ hybridization. A schematic illustration is given in Figure 7.10. In typical protocols, cell division is stopped, and cells are arrested in metaphase by adding drugs such as colchicine; chromosomes are isolated, dropped from a specified height onto a microscope slide, fixed (partially denatured and covalently crosslinked), and aged. If this seems like quite a bit of magic, it is. The goal is to strike a proper balance between maintaining sufficient chromosome morphology to allow each chromosome to be recognized and measured, but disrupting enough chromosome structure to expose DNA sequences that can hybridize with the labeled DNA probe. Under such circumstances the hybridization reactions are usually very inefficient. As a result a probe with relatively high DNA complexity must be used to provide sufficient illumination of the target site. Typically one starts with probes containing a total of $10^4$ to $10^5$ base pairs of DNA from the site of interest, and these probes are broken into small pieces prior to annealing to facilitate the hybridization.

Recently improved protocols and better fluorescence microscopes have allowed the use of smaller DNA probes for in situ hybridization. For instance, one recent procedure called PRINS (primed in situ hybridization) can use very small probes. In this method the probes hybridized to long target DNAs are used as primers for a DNA polymerase extentions. During the DNA polymerase extentions, modified bases containing biotin or digoxygenin are incorporated. This allows subsequent signal amplification by methods described in Chapter 3. Thus even short cDNAs can be used.

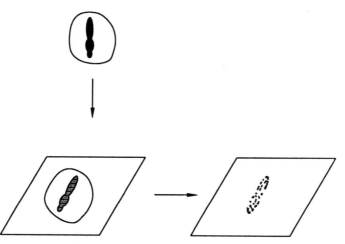

**Figure 7.10**  Schematic illustration of the preparation of metaphase chromosomes for in situ hybridization. See text for explanation.

In practice, it is usually much easier to first use the cDNA to find a corresponding cosmid clone, and then use that as a probe. In conventional FISH, 40 kb clones are used as probes. Complex probes like cosmids or YACs always have repeated human DNA sequences on them. It is necessary to eliminate the effect of these sequences; otherwise, the probe will hybridize all over the genome. An effective way to eliminate the complications caused by repeats is to fragment the probe into small pieces and prehybridize these with a great excess of $C_0t = 1$ DNA (see Chapter 3). A typical metaphase FISH result with a single-copy DNA probe is shown schematically in Figure 7.11. In an actual color image a bright pair of yellow fluorescent spots would be seen on a single red chromosome. The yellow comes from fluorescein conjugated to the DNA probe via streptavidin and biotin. The red comes from a DNA stain like DAPI used to mark the entire chromosome. The pair of yellow dots result from hybridization with each of the paired sister chromatids. In the simplest case one can determine the approximate chromosomal location by measuring the relative distance of the pair of spots from the ends of the chromosome. The identity of the particular chromosome is revealed in most cases by its size and the position of its centromeric constriction.

The interpretation of FISH results requires quantitative analysis of the image in the fluorescent microscope. It is not efficient to do this by photography and then processing of the image. Instead, direct on-line imaging methods are used. One possibility is to equip a standard fluorescence microscope with a charge couple device array (CCD) camera that can acquire and process the image as discrete pixels of information. The alternate approach is to use a scanning microscope like a confocal laser microscope that records the image's intensity as a function of position. In either case it is important to realize that the number of bits of information in a single microscope image is considerable; only a small amount of it actually finds its way into the final analyzed probe position. Either a large amount of mass storage must be devoted to archiving FISH images, or a procedure must be developed to allow some arbitration or reanalysis of any discrepancies in the data after the original raw images have been discarded.

Many enhancements have been described that allow FISH to provide more accurate chromosomal locations than the simple straightforward approach illustrated in Figure 7.11. Several examples are given in Figure 7.12. Chromosome banding provides a much more accurate way of identifying individual chromosomes and subchromosomal regions than simple measurements of size and centromere position. Each chromosome in the microscope is an individual—the amount of stretching and the nature of any distortion can vary considerably. Clearly, by superimposing banding on the emission from single-copy labeled probes, one not only provides a unique chromosome identifier but also local markers at the band locations that allow more accurate positioning of the single-copy probe.

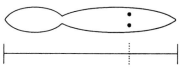

**Figure 7.11**   Appearance of a typical metaphase chromosome in FISH when a single fluorescein-labeled DNA probe is used along with a counterstain that lightly labels all DNA. Shown below is the coordinate system used to assign the map location of the probe.

map by relative position

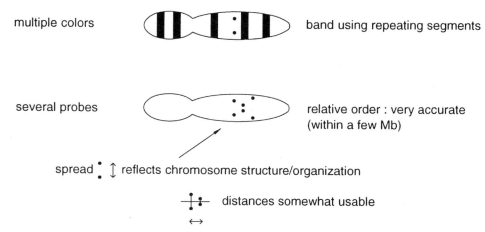

**Figure 7.12**    Some of the refinements possible in metaphase FISH. (*a*) Simultaneous use of a single-copy probe and a repeated DNA probe like the human *Alu* sequence which replicates the banding pattern seen in ordinary Giemsa staining. (*b*) Use of three different single-copy DNA probes.

Standard DNA banding stains are incompatible with the fluorescent procedures used for visualizing single-copy DNA. Fortunately it turns out that the major repeated DNA sequences in the human have preferential locations within the traditional Giemsa bands. Thus *Alu* repeats prefer light G bands, while L1 repeats prefer dark G bands. One can use two different colored fluorescent probes simultaneously, one with a single-copy sequence, the other with a cloned repeat, and the results, like those shown in Figure 7.12, represent a major improvement.

The accuracy gained by FISH over conventional radioactively labeled in situ hybridization is illustrated in Figure 7.13. The further increase in accuracy when FISH is used on top of a banded chromosome stain is also indicated. It is clear that the new procedures completely change the nature of the technique from a rather crude method of chromosome location to a highly precise mapping tool. It is possible to improve the resolution of FISH mapping even further by the simultaneous use of multiple single-copy DNA probes. For example, as shown in Figure 7.12, when metaphase chromosomes are hybridized with three nearby DNA segments, each labeled with a different color fluorophore, a distinct pattern of six ordered dots is seen. Each color is a pair of dots on the two sister chromatids. The order of the colors gives the order of the probes down to a resolution limit of about 1 to 2 Mb. The spread of the pair of dots relative to the long axis of the chromatids reflects details of chromosome structure and organization that we do not understand well today. It is a reproducible pattern, but our lack of knowledge about the detailed arrangement of packing of DNA in chromosomes compromises current abilities to turn this information into quantitative distance estimates between the probes. This is frustrating, but fortunately there is an easy solution, as described below. Ultimately, as we understand more about chromosome structure, and as high-resolution physical maps of DNA become available, FISH on metaphase chromosomes will undoubtedly turn out to be a rich source of information about the higher-order packing of chromatin within condensed chromosomes.

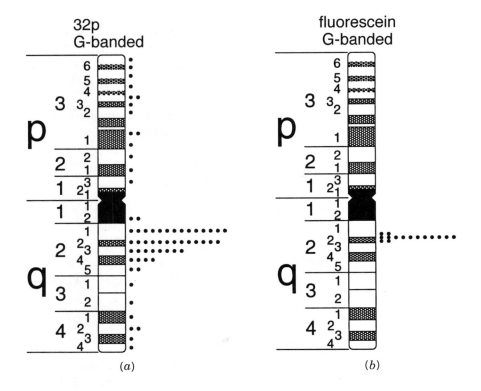

(a)

(b)

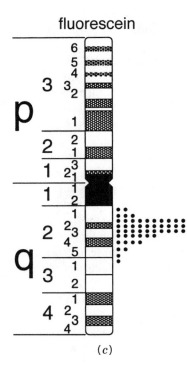

(c)

**Figure 7.13**  Examples of the accuracy of various in situ hybridization mapping techniques. In each case the probe is the blast1 gene. Dots show the apparent position of the probe on individual chromosomes. (a) $^{32}$P-labeled probe on Giemsa-banded chromosomes. (b) Fluorescein-labeled probe on fluorescently banded chromosomes. (c) Fluorescein-labeled probe on unbanded chromosomes. (Adapted from Lawrence et al., 1990.)

Today metaphase FISH provides an extremely effective way to assign a large number of DNA probes into bins along a chromosome of interest. An example from work on chromosome 11 is shown in Figure 7.14. Note, in this case, that the mapped, cloned probes are not distributed evenly along the chromosome; they tend to cluster very much in several of the light Giemsa bands. This is a commonly observed cloning bias, and it severely complicates some of the approaches, used to construct ordered libraries of clones, which will be described in Chapters 8 and 9.

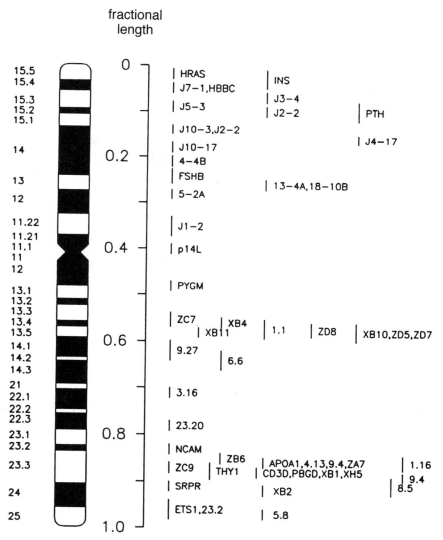

**Figure 7.14** Regional assignment of a set of chromosome 11 cosmids by FISH. Note the regional biases in the distribution of clones. (Adapted from Lichter et al., 1990b.)

## HIGH-RESOLUTION FISH

To increase the resolution of FISH over what is achievable with metaphase chromosomes, it is necessary to use more extended chromosome preparations. In most cases this results in sufficient disruption of chromosome morphology that no apriori assignment of a single probes to particular chromosomes or regions is possible. Instead, what is done is to determine the relative order and distances among a set of closely spaced DNA probes by simultaneous multicolor FISH, as we have already described for metaphase chromosomes. The different approaches reflect mostly variations in the particular preparation of extended chromosomes used.

Several different methods exist for systematic preparation of partially decondensed chromosomes. One approach is to catch cells at the pro-metaphase stage, before chromosomes have become completely condensed. A second approach to fertilize a hamster oocyte with a human sperm. The result of this attempted interspecies cross is called a humster. It does not develop, but within the hamster nucleus the human chromosomes, which are highly condensed in the sperm head, become partially decondensed; in this state they are very convenient targets for hybridization. A third approach is to use conditions that lead to premature chromosome condensation.

The most extreme versions of FISH use highly extended DNA samples. One way to accomplish this is to look at interphase nuclei. Under these conditions the chromatin is mostly in the form of 30 nm fibers (Chapter 2). One can make a crude estimate of the length of such fiber expected for a given length of DNA as follows. The volume of 10 bp of DNA double helix is given by $\pi r^2 d$ where $r$ is the radius and $d$ is the pitch. Evaluating these as roughly 10 Å and 34 Å, respectively, yields $10^4$ Å per 10 bp or $10^3$ Å per bp. The volume of a micron of 30-nm filament is $\pi r^2 d$, where $r$ is 150 Å and $d$ is 1 $\mu = 10^4$ Å. This is $8 \times 10^8$ Å$^3$, and roughly half of it is DNA. Thus one predicts that a micron of 30-nm filament will contain on average about 0.4 Mb of DNA. This estimate is not in bad agreement with what is observed experimentally (Box 7.1). Since the resolution of the fluorescence microscope is about 0.25 $\mu$, the ultimate resolution of FISH based on interphase chromatin should be around 0.1 Mb.

Even higher resolution is possible if the DNA is extended further in methods called *fiber FISH*. One way to do this is the technique known as a Weigant halo. Here, as shown schematically in Figure 7.15, nuclei are prepared and then treated so that the DNA is deproteinized and exploded from the nucleus. Since naked DNA is about 3 Å per base pair, such samples should show an extension of $3 \times 10^3$ Å per kb, which is 0.3 $\mu$ per kb. Thus the ultimate resolution of FISH under these circumstances could approach 1000 base pairs. To take advantage of the high resolution afforded by extended DNA samples, one must use multicolored probes to distinguish their order, and then estimate the distance between the different colors. The probes themselves will occupy a significant distance along the length of the DNA, as shown in Figure 17.16.

Interphase chromatin, or naked DNA, in contrast to metaphase chromosomes, is not a unique structure, and it is not rigid (Fig. 7.17*a*). To estimate the true distance between markers from the apparent separation of two probes in the microscope, one must correct for the fact that the DNA between the markers is not straight. The problem becomes ever more severe as the distance between the markers increases (Fig. 7.17*b*). No single molecule measurement will suffice because there is no way of knowing what the unseen DNA configuration is. Instead it is necessary to average the results over observations on many molecules, using a model for the expected chain configuration of the DNA.

**Figure 7.15** A Weigant halo, which is produced when naked DNA is allowed to explode out of a nucleus with portions of the nuclear matrix still intact.

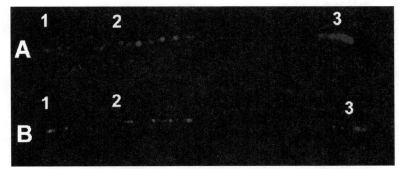

**Figure 7.16** An example of the sorts of results obtainable with current procedures for interphase FISH. (From Heiskanen et al., 1996.) Figure also appears in color insert.

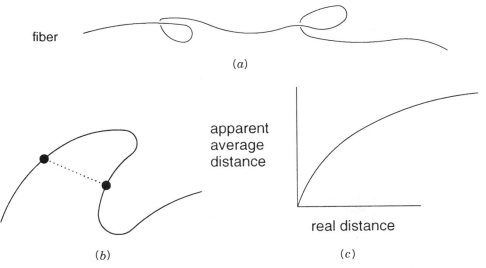

**Figure 7.17** Typical configuration of a chromatin fiber in interphase FISH. (*a*) Appearance of one region of a fiber. (*b*) Apparent versus real distance between two loci. (*c*) Expected dependence of the apparent distance on the true distance for a random walk model.

**BOX 7.1**
**QUANTITATIVE HIGH-RESOLUTION FISH**

The quantitative analysis of distances in high-resolution FISH has been pioneered by two groups headed by Barb Trask, currently at the University of Washington, and Jeanne Lawrence, at the University of Massachusetts in Amherst. Others have learned their methods and begun to practice them. A few representative analyses are shown in Figures 7.18 and 7.19. The distribution of measured distances between two fixed markers in individual samples of interphase chromatin varies over quite a wide range, as shown in Figure 7.18a. However, when these measurements are averaged and plotted as a function of known distance along the DNA, for relatively short distances a reasonably straight plot is observed, but for unknown reasons, in this study, it does not pass through the origin (Fig. 7.18b). In another study, data were analyzed two ways. The cumulative probability of molecular distances was plotted as a function of measured distance, and the resulting curves appeared to be well fit by a random walk model (Fig. 7.19a). Alternatively, the square of the measured distance was an approxi-

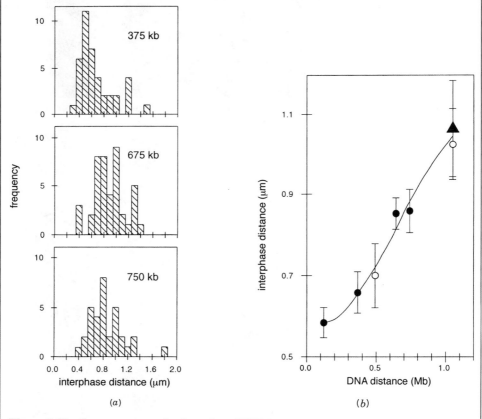

(a)                                   (b)

**Figure 7.18**   Some representative interphase FISH data. (a) Results for three pairs of probes: Histograms indicate the number of molecules seen with each apparent size. (c) Apparent DNA distance as a function of the true distance, for a larger set of probes. (Adapted from Lawrence et al., 1990.)

*(continued)*

**BOX 7.1** *(Continued)*

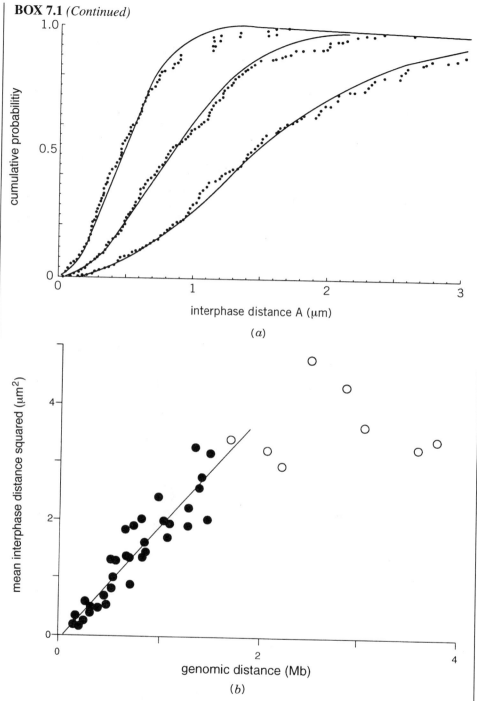

**Figure 7.19** Additional examples of interphase FISH data. (*a*) Distribution of apparent distances seen for three probes. What is plotted is the fraction of molecules with an observed distance less than or equal to a particular value, as a function of that value. The solid lines are the best fit of the data to a random walk model. (*b*) Plot of the square of the apparent distance as a function of the true distance for a large set of probes. It is clear that the random walk model works quite well for distances up to about 2 Mb. (Adapted from Van den Engh et al., 1992.)

*(continued)*

**BOX 7.1** *(Continued)*

mately linear function of the true distance, out to about 1.5 Mb (Fig. 7.19*b*). Thus these results support the use of a random walk model. The difficulty with the available results to date is that they do not agree on the scaling. Three studies are summarized below:

| INVESTIGATOR | DISTANCES MEASURED | REAL |
|---|---|---|
| Lawrence | 1 $\mu$ | 1 Mb |
| Trask | 1 $\mu$ | 0.5 Mb |
| Skare | 1 $\mu$ | 1.2 Mb or more |

It is not clear if these considerable differences are due to differences in the methods used to prepare the interphase chromatin or differences in the properties of interphase chromatin in different regions of the genome. Note that the scaling, at worst, is within a factor of three of what was estimated in the text from a very crude model for the structure of the 30-nm filament.

If this is taken to be a random walk, then the measured distance between markers should be proportional to the square root of the real distance between them (Fig. 7.17*c*; see also Box 7.1).

An example of the utility of interphase FISH is shown in Figure 7.20. Here interphase in situ hybridization was used to estimate the size of a gap that existed in the macrorestriction map of a region of human chromosome 4 near the tip of the short arm and now known to contain the gene for Huntington's disease. The gap was present because no clones or probes could be found between markers E4 and A252. In all other segments in the region, the accord between distances on the macrorestriction map and distances inferred from interphase in situ hybridization (using a scaling of 0.5 Mb per $\mu$) is quite good. This has allowed the conclusion that the gap in the physical map would have to be small to maintain this consistency.

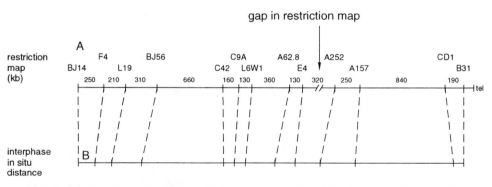

**Figure 7.20**    Comparison of a macrorestriction map of the tip of the short arm of human chromosome 4 with interphase FISH results for the same set of probes. (Adapted from Van den Engh et al., 1992.)

## CHROMOSOME  PAINTING

For most chromosomes a dense set of mapped markers now exists. An even larger set of clones is available and assigned to a particular chromosome but not yet mapped. These clones can be used to examine the state of the entire chromosome, either in metaphase or in less condensed states. This practice is called *chromosome painting*. Metaphase chromosome painting is a useful tool to see if a chromosome is intact or has rearranged in some way. A novel application of this is illustrated in Figure 7.21. Here the high degree of homology between chimp and human DNA sequences was used to examine the relationship between human chromosome 2 and its equivalent in the chimp. Probes from human chromosome 2 were used first on the human, where they indicated even and exhaustive coverage of the chromosome. Next the same set of probes was used on the chimp. Here two smaller chimp chromosomes were painted, numbers 12 and 13. Each of these is acrocentric, while human chromosome 2 is metacentric. These results make it clear that human chromosome 2 must have arisen by a Robertsonian (centromeric) fusion of the two smaller chimp chromosomes.

When interphase chromatin is painted, one can observe the cellular location of particular segments of DNA. The complexity of interphase chromatin makes it difficult to view more than small DNA regions simultaneously. This technique is still in its infancy, but it is already clear that it has the potential to provide an enormous amount of information of how DNA is organized in a functioning nucleus (Chandley et al., 1996, Seong et al., 1994).

A recently developed variation of chromosome painting shows considerable promise both in facilitating the search for gene involved in diseases like cancer and in assisting the development of improved clinical diagnostic tests for chromosomal disorders. In this technique, called *comparative genome hybridization* (CGH), total DNA from two samples to be compared is labeled using different specificity tags. Actual procedures use nick translation: For one sample a biotinylated dNTP is used; for the other a digoxigenin-labeled dNTP is used (see Chapter 3). These two samples are then allowed to hybridize simultaneously to a metaphase chromosome spread. The results are visualized by two-

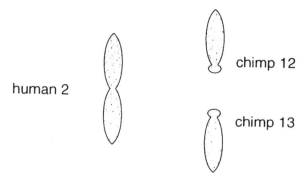

**Figure 7.21**  An example of metaphase chromosome painting, in which a human chromosome 2 probe is hybridized to the full set of human chromosomes, and separately to the full set of chimpanzee chromosomes. Shown are just those resulting chromosomes from both species that show significant hybridization.

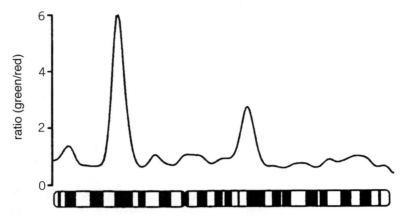

**Figure 7.22**    An example of comparative genome hybridization. Shown is the green to red color ratio seen along human chromosome 2 after competitive hybridization with DNA from two cell lines. The red cell line is normal human DNA. Three regions of DNA amplification are apparent in the green labeled cell line. The most prominent of these corresponds to the oncogene N-*myc* known to be amplified in the cell line used. (From Kallioniemi et al., 1992.)

color detection using fluorescein-labeled streptavidin or avidin to detect the biotin and rhodamine-labeled antidigoxigenin to detect the digoxigenin. The ratio of the two colors should indicate the relative amounts of the two probes hybridized. This should be the same if the initial probes have equal concentrations everywhere in the genome. However, if there are regions in one target sample that are amplified or deleted, the observed color ratio will shift. The results can be dramatic as shown by the example in Figure 7.22. The color ratio shifts provide an indication of the relative amounts of each amplification or deletion, and they also allow the locations of all such variations between two samples to be mapped in a single experiment.

## CHROMOSOME MICRODISSECTION

Frequently the search for a gene has focused down to a small region of a chromosome. The immediate task at hand is to obtain additional DNA probes for this region. These are needed to improve the local genetic map and also to assist the construction of physical maps. They may also be useful for screening cDNA libraries if there is any hint of the preferred tissue for expression of the gene of interest. The question is how to focus on a small region of the genome efficiently. One approach has been microdissection of that region from metaphase chromosomes and microcloning of the DNA that results. Two approaches to microdissection are shown schematically in Figure 7.23. In one, a fine glass needle is used to scratch out the desired chromosome region and transfer the material to a site where it can be collected. In the other approach, a laser is used to ablate all of the genome except for the small chromosome region of interest. In either case, the problem in early versions of this method was to develop efficient cloning schemes that could start from the very small amounts of DNA that could be collected by microdissection.

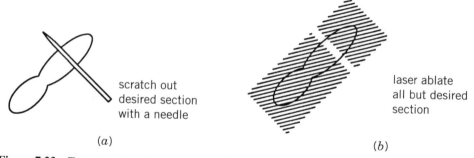

**Figure 7.23** Two examples of chromosome microdissection. (*a*) A fine needle is used to scratch a chromosome and transfer a section of it to another site where it can be further manipulated. (*b*) A laser is used to destroy the DNA in all but a preselected segment of a chromosome.

Hundreds of microdissection products would be combined and placed in liquid micro-drops suspended in oil (Fig. 7.24). Restriction enzymes, vectors, and other components needed for cloning were delivered by micromanipulators. The result was a technical *tour de force* that did deliver clones from the desired region but usually in very small numbers.

With the development of genome-wide PCR amplification methods (Chapter 4), the need for microcloning dissected chromosome samples is eliminated. Using these methods, it is relatively easy to amplify the dissected chromosome material by PCR until there is enough DNA to be handled by more conventional methods. Note that the microdissection method has several important potential uses beyond the hunt for specific disease genes. Most DNA libraries in any type of vector are biased, and some regions are under-represented or not represented at all. Microdissection offers an attractive way to compensate for cloning biases, especially when they are severe. The alternate approach, in common use today, is to combine clones from different types of libraries in the hope that biases will compensate for each other. This can be a very effective strategy, but it increases substantially the number of DNA samples that must be handled. Another potential use for microdissection will be to pull clones from a particular individual who may have a region of special interest, such as a suspected rearrangement. This will not always be an effective strategy, but it may well be necessary in cases where simpler approaches fail to yield a definitive picture of what has happened.

**Figure 7.24** Microcloning of DNA in a tiny droplet containing microdissected chromosomes.

## SOURCES AND ADDITIONAL READINGS

Boyle, A. L., Feltquite, D. M., Dracopoli, N. C., Housman, D. E., and Ward, D. C. 1992. Rapid physical mapping of cloned DNA on banded mouse chromosomes by fluorescence *in situ* hybridization. *Genomics* 12: 106–115.

Burmeister, M., Kim, S., Price, E. R., de Lange, T., Tantravahi, U., Meyers, R. M., and Cox, D. 1991. A map of the distal region of the long arm of human chromosome 21 constructed by radiation hybrid mapping and pulsed field gel electrophoresis. *Genomics* 9: 19–30.

Chandley, A. C., Speed, R. M., and Leitch, A. R. 1996. Different distributions of homologous chromosomes in adult human Sertoli cells and in lymphocytes signify nuclear differentiation. *Journal of Cell Science* 109: 773–776.

Cherif, D., Julier, C., Delattre, O., Derré, J., Lathrop, G. M., and Berger, R. 1990. Simultaneous localization of cosmids and chromosome R-banding by fluorescence microscopy: Application to regional mapping of human chromosome 11. *Proceedings of the National Academy of Sciences USA* 87: 6639–6643.

Cox, D. R., Burmeister, M., Price, E. R., Kim, S., and Meyers, R. 1990. Radiation hybrid mapping: A somatic cell genetic method for constructing high-resolution maps of mammalian cell chromosomes. *Science* 250: 245–251.

Ellis, N., and Goodfellow, P. N. 1989. The mammalian pseudoautosomal region. *Trends in Genetics* 5: 406–410.

Heiskanen M., Kallioniemi, O., and Palotie, A. 1996. Fiber-FISH: Experiences and a refined protocol. *Genetic Analysis (Biomolecular Engineering)* 12: 179–184.

Jeffreys, A. J., Neumann, R., and Wilson, V. 1990. Repeat unit sequence variation in minisatellites: A novel source of DNA polymorphism for studying variation and mutation by single molecule analysis. *Cell* 60: 473–485.

Kallioniemi, A., Kallioniemi, O. P., Sudar, D., Rutovitz, D., Gray, J. W., Waldman, F., and Pinkel, D. 1992. Comparative genomic hybridization for molecular cytogenetic analysis of solid tumors. *Science* 258: 818–821.

Lawrence, J., Singer, R. H., and McNeil, J. A. 1990. Interphase and metaphase resolution of different distances within the human dystrophin gene. *Science* 249: 928–932.

Li, H., Cui, A., and Arnheim, N. 1990. Direct electrophoresis detection of the allelic state of single DNA molecules in human sperm by using the polymerase chain reaction. *Proceedings of the National Academy of Sciences USA* 87: 4580–4584.

Li, H., Gyllensten, U. B., Cui, X., Saiki, R. K., Erlich, H. A., and Arnheim, N. 1988. Amplification and analysis of DNA sequences in single human sperm and diploid cells. *Nature* 335: 414–417.

Lichter, P., Ledbetter, S. A., Ledbetter, D. H., and Ward, D. C. 1990. Fluorescence in situ hybridization with *Alu* and L1 polymerase chain reaction probes for rapid characterization of human chromosomes in hybrid cell lines. *Proceedings of the National Academy of Sciences USA* 87: 6634–6638.

Lichter, P., Tang, C. C., Call, K., Hermanson, G., Evans, G. A., Housman, D., and Ward, D. C. 1990. High resolution mapping of human chromosome 11 by in situ hybridization with cosmid clones. *Science* 247: 64–69.

McCarthy, L. C. 1996. Whole genome radiation hybrid mapping. *Trends in Genetics* 12: 491–493.

Nilsson, M., Krejci, K., Koch, J., Kwiatkowski, M., Gustavsson, P., and Landegren, U. 1997. Padlock probes reveal single nucleotide differences, parent of origin and in situ distribution of centromeric sequences in human chromosomes 13 and 21. *Nature Genetics* 16: 252–255.

Ruano, G., Kidd, K. K., and Stephens, J. C. 1990. Haplotype of multiple polymorphisms resolved by enzymatic amplification of single DNA molecules. *Proceedings of the National Academy of Sciences USA* 87: 6296–6300.

Seong, D. C., Song, M. Y., Henske, E. P., Zimmerman, S. O., Champlin, R. E., Deisseroth, A. B., and Siciliano, M. J. 1994. Analysis of interphase cells for the Philadelphia translocation using painting probe made by Inter-*Alu*-polymerase chain reaction from a radiation hybrid. *Blood* 83: 2268–2273.

Silva, A. J., and White, R. 1988. Inheritance of allelic blueprints for methylation patterns. *Cell* 54: 145–152.

Stewart, E. A., McKusick, K. B., Aggarwal, A., et al. 1997. An STS-based radiation hybrid map of the human genome. *Genome Research* 7: 422–433.

Strong, S. J., Ohta, Y., Litman, G. W., and Amemiya, C. T. 1997. Marked improvement of PAC and BAC cloning is achieved using electroelution of pulsed-field gel-separated partial digests of genomic DNA. *Nucleic Acids Research* 25: 3959–3961.

Van de Engh, G., Sachs, R., and Trask, B. J. 1992. Estimating genomic distance from DNA sequence location in cell nuclei by a random walk model. *Science* 257: 1410–1412.

Wang, D., and Smith, C. L. 1994. Large-scale structure conservation along the entire long arm of human chromosome 21. *Genomics* 20: 441–451.

Wu, B-L., Milunsky, A., Nelson, D., Schmeckpeper. B, Porta, G., Schlessinger, D., and Skare, J. 1993. High-resolution mapping of probes near the X-linked lymphoproliferative disease (XLP) locus. *Genomics* 17: 163–170.

Yu, J., Hartz, J., Xu, Y., Gemmill, R. M., Korenberg, J. R., Patterson, D., and Kao, F.-T. 1992. Isolation, characterization, and regional mapping of microclones from a human chromosome 21 microdissection library. *American Journal of Human Genetics* 51: 263–272.

# 8  Physical Mapping

## WHY HIGH-RESOLUTION PHYSICAL MAPS ARE NEEDED

Physical maps are needed because ordinary human genetic maps are not detailed enough to allow the DNA that corresponds to particular genes to be isolated efficiently. Physical maps are also needed as the source for the DNA samples that can serve as the actual substrate for large-scale DNA sequencing projects. Genetic linkage mapping provides a set of ordered markers. In experimental organisms this set can be almost as dense as one wishes. In humans one is much more limited because of the inability to control breeding and produce large numbers of offspring. Distances that emerge from human genetic mapping efforts are vague because of the uneven distribution of meiotic recombination events across the genome. Cytogenetic mapping, until recently, was low resolution; it is improving now, and if it were simpler to automate, it could well provide a method that would supplant most others. Unfortunately, current conventional approaches to cytogenetic mapping seem difficult to automate, and they are slower than many other approaches that do not involve image analysis.

It is worth noting here that in some other organisms, cytogenetics is more powerful than in the human. For example, in Drosophila the presence of polytene salivary gland chromosomes provides a tool of remarkable power and simplicity. The greater extension and higher-resolution imaging of polytene chromosomes allows bands to be seen at a typical resolution of 50 kb (Fig. 8.1). This is more than 20 times the resolution in the best human metaphase FISH. Furthermore the large number of DNA copies in the metaphase salivary chromosomes means that microdissection and cloning are much more powerful here than they are in the human. It would be nice if there were a convenient way to place large continuous segments of human DNA into Drosophila and then use the power of the cytogenetics in this simple organism to map the human DNA inserts.

Radiation hybrids offer, in principle, a way to measure distances between markers accurately. However, in practice, the relative distances in radiation hybrid maps appear to be distorted (Chapter 7). In addition there are a considerable number of unknown features of these cells; for example, the types of unseen DNA rearrangements that may be present need to be characterized. Thus it is not yet clear that the use of radiation hybrids alone can produce a complete accurate map or yield a set of DNA samples worth subcloning and characterizing at higher resolution. Instead, currently other methods must be used to accomplish the three major goals in physical mapping:

1. Provide an ordered set of all of the DNA of a chromosome (or genome).
2. Provide accurate distances between a dense set of DNA markers.
3. Provide a set of DNA samples from which direct DNA sequencing of the chromosome or genome is possible. This is sometimes called a *sequence-ready map.*

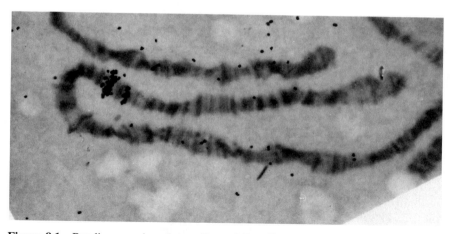

**Figure 8.1**  Banding seen in polytene Drosophila salivary gland chromosomes. Shown is part of chromosome 3 of for which a radiolabeled probe to the gene for heat-shock protein hsp70 has been hybridized. A strong and weak site are seen. (Figure kindly provided by Mary Lou Pardue.)

Most workers in the field focus their efforts on particular chromosomes or chromosome regions. Usually a variety of different methods are brought to bear on the problems of isolating clones and probes and using these to order each other and ultimately produce finished maps. However, some research efforts have produced complete genomic maps with a restricted number of approaches. Such samples are turned over to communities interested in particular regions or chromosomes that finish a high-resolution map of the regions of interest.

There are basically two kinds of physical maps commonly used in genome studies: restriction maps and ordered clone banks or ordered libraries. We will discuss the basic methodologies used in both approaches separately, and then, in Chapter 9, we will show how the two methods can be merged in more powerful second generation strategies. First we outline the relative advantages and disadvantages of each type of map.

## RESTRICTION MAPS

A typical restriction map is shown in Figure 8.2. It consists of an ordered set of DNA fragments that can be generated from a chromosome by cleavage with restriction enzymes individually or in pairs. Distances along the map are known as precisely as the lengths of the DNA fragments generated by the enzymes can be measured. In practice, the lengths are measured by electrophoresis in virtually all currently used methods; they are accurate to a single base pair up to DNAs around around 1 kb in sizes. Lengths can be measured with better than 1% accuracy for fragments up to 10 kb, and with a low percent of accuracy for fragments up to 1 Mb in size. Above this length, measurements today are still fairly qualitative, and it is always best to try to subdivide a target into pieces less than 1 Mb before any quantitative claims are made about its true total size.

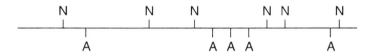

**Figure 8.2**   Typical section of a restriction map generated by digestion of genomic or cloned DNA with two enzymes with different recognition sites A, and N.

In an ideal restriction map each DNA fragment is pinned to markers on other maps. Note that if this is done with randomly chosen probes, the locations of these probes within each DNA fragment are generally unknown (Fig. 8.3). Probes that correspond to the ends of the DNA fragments are more useful, when they are available, because their position on the restriction map is known precisely. Originally probes consisted of unsequenced DNA segments. However, the power of PCR has increasingly favored the use of sequenced DNA segments.

A major advantage of a restriction map is that accurate lengths are known between sets of reference points, even at very early stages in the construction of the map. A second advantage is that most restriction mapping can be carried out using a top-down strategy that preserves an overview of the target and that reaches a nearly complete map relatively quickly. A third advantage of restriction mapping is that one is working with genomic DNA fragments rather than cloned DNA. Thus all of the potential artifacts that can arise from cloning procedures are avoided. Filling in the last few small pieces is always a chore in restriction mapping, but the overall map is a useful tool long before this is accomplished, and experience has shown that restriction maps can be accurately and completely constructed in reasonably short time periods.

In top-down mapping one successively divides a chromosome target into finer regions and orders these (Fig. 8.4). Usually a chromosome is selected by choosing a hybrid cell in which it is the only material of interest. There have been some concerns about the use of hybrid cells as the source of DNA for mapping projects. In a typical hybrid cell there is no compelling reason for the structure of most of the human DNA to remain intact. The biological selection that is used to retain the human chromosome is actually applicable to only a single gene on it. However, the available results, at least for chromosome 21, indicate that there are no significant differences in the order of DNA markers in a selected set of hybrid and human cell lines (see Fig. 8.49). In a way this is not surprising; even in a human cell line most of the human genome is silent, and if loss or rearrangement of DNA were facile under these circumstances, it should have been observed. Of course, for model organisms with small genomes, there is no need to resort to a hybrid cell at all—their genomes can be studied intact, or the chromosomes can be purified in bulk by PFG.

In a typical restriction mapping effort, any preexisting genetic map information can be used as a framework for constructing the physical map. Alternatively, the chromosome of interest can be divided into regions by cytogenetic methods or low-resolution FISH.

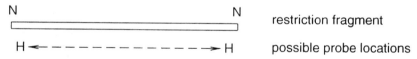

**Figure 8.3**   Ambiguity in the location of a hybridization probe on a DNA fragment.

*(a)*

**Chromosome**

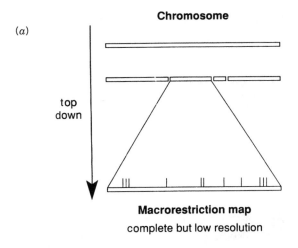

top
down

**Macrorestriction map**
complete but low resolution

*(b)*

**Linked library**
detailed but incomplete

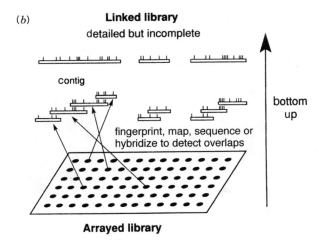

contig

bottom
up

fingerprint, map, sequence or
hybridize to detect overlaps

**Arrayed library**

**Figure 8.4**   Schematic illustration of methods used in physical mapping. (*a*) Top-down strategy used in restriction mapping. (*b*) Bottom-up strategy used in making an ordered library.

Large DNA fragments are produced by cutting the chromosome with restriction enzymes with very rare recognition sites. The fragments are separated by size and assigned to regions by hybridization with genetically or cytogenetically mapped DNA probes. Then the fragments are assembled into contiguous blocks, by methods that will be described later in this chapter. The result at this point is called a *macrorestriction map*. The fragments may average 1 Mb in size. For a simple genome this means that only 25 fragments will have to be ordered. This is relatively straightforward. For an intact human genome, the corresponding number is 3600 fragments. This is an unthinkable task unless the fragments are first assorted into individual chromosomes.

If a finer map is desired, it can be constructed by taking the ordered fragments one at a time, and dissecting these with more frequently cutting restriction nucleases. An advantage of this reductionist mapping approach is that the finer maps can be made only in those regions where there is sufficient interest to justify this much more arduous task.

The major disadvantage of most restriction mapping efforts is that they do not produce the DNA in a convenient, immortal form where it can be distributed or sequenced by available methods. One could try to clone the large DNA fragments that compose the macrorestriction map, and there has been some progress in developing the vectors and techniques needed to do this. One could also use PCR to generate segments of these large fragments (see Chapter 14). For a small genome, most of the macrorestriction fragments it contains can usually be separated and purified by a single PFG fractionation. An example is shown in Figure 8.5. In cases like this, one does really possess the

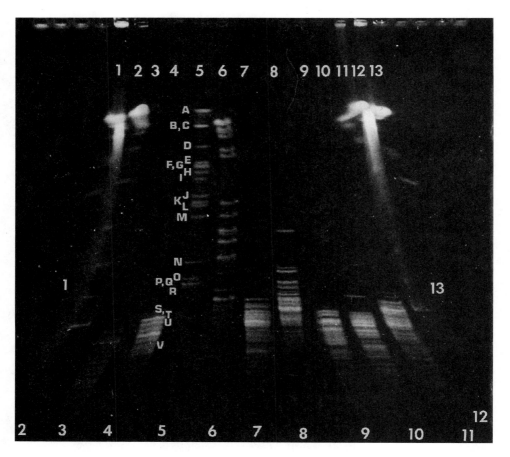

**Figure 8.5** Example of the fractionation of a restriction enzyme digest of an entire small genome by PFG. *Above: Not* I digest of the 4.6 Mb *E. coli* genome shown in lane 5; *Sfi* I digest in lane 6. Other lanes shown enzymes that cut too frequently to be useful for mapping. (Adapted from Smith et al., 1987.) *Left:* Structure of the ethidium cation used to stain the DNA fragments.

genome, but it is not in a form where it is easy to handle by most existing techniques. For a large genome, PFG does not have sufficient resolution to resolve individual macrorestriction fragments. If one starts instead with a hybrid cell, containing only a single human chromosome or chromosome fragment, most of the human macrorestriction fragments will be separable from one another. But they will still be contaminated, each by many other background fragments from the rodent host.

## ORDERED LIBRARIES

Most genomic libraries are made by partial digestion with a relatively frequently cutting restriction enzymes, size selection of the fragments to provide a fairly uniform set of DNA inserts, and then cloning these into a vector appropriate for the size range of interest. Because of the method by which they were produced, the cloned fragments are a nearly random set of DNA pieces. Within the library, a given small DNA region will be present on many different clones. These extend to varying degrees on both sides of the particular region (Fig. 8.6). Because the clones contain overlapping regions of the genome, it is possible to detect these overlaps by various fingerprinting methods that examine patterns of sequence on particular clones. The random nature of the cloned fragments means that many more clones exist than the minimum set necessary to cover the genome. In practice, the redundancy of the library is usually set at five- to tenfold in order to ensure that almost all regions of the genome will have been sampled at least once (as discussed in Chapter 2). From this vast library the goal is to assemble and to order the minimum set of clones that covers the genome in one contiguous block. This set is called the *tiling path*.

Clone libraries have usually been ordered by a bottom-up approach. Here individual clones are initially selected from the library at random. Usually the library is handled as an array of samples so that each clone has a unique location on a set of microtitre plates, and the chances of accidentally confusing two different clones can be minimized. The clone is fingerprinted, by hybridization, by restriction mapping, or by determining bits of DNA sequence. Eventually clones appear that share some or all of the same fingerprint pattern (Fig. 8.7). These are clearly overlapping, if not identical, and they are assembled into overlapping sets, called *contigs,* which is short for contiguous blocks. There are several obvious advantages to this approach. Since the DNA is handled as clones, the map is built up of immortal samples that are easily distributed and that are potentially suitable for direct sequencing. The maps are usually fairly high resolution when small clones are used, and some forms of fingerprinting provide very useful internal information about each clone.

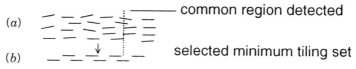

**Figure 8.6**   Example of a dense library of clones. (*a*) The large number of clones insures that a given DNA probe or region (vertical dashed line) will occur on quite a few different clones in the library. (*b*) The minimum tiling set is the smallest number of clones that can be selected to span the entire sample of DNA.

$(a)$

$(b)$

$(c)$

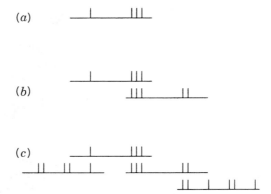

**Figure 8.7**  Example of a bottom-up fingerprinting strategy to order a dense set of clones. $(a)$ A clone is selected at random and fingerprinted. $(b)$ Two clones that share an overlapping fingerprint pattern are assembled into a contig. $(c)$ Longer contigs are assembled as more overlapping clones are found.

There are a number of disadvantages to bottom-up mapping. While this process is easy to automate, no overview of the chromosome or genome is provided by the fingerprinting and contig building. Additional experiments have to be done to place contigs on a lower-resolution framework map, and most approaches do not necessarily allow the orientation of each contig to be determined easily. A more serious limitation of pure bottom-up mapping strategies is that they do not reach completion. Even if the original library is a perfectly even representation of the genome, there is a statistical problem associated with the random clone picking used. After a while, most new clones that are picked will fall into contigs that are already saturated. No new information will be gained from the characterization of these new clones. As the map proceeds, a diminishingly smaller fraction of new clones will add any additional useful information. This problem becomes much more serious if the original library is an uneven sample of the chromosome or genome. The problem can be alleviated somewhat if mapped contigs are used to screen new clones prior to selection to try to discard those that cannot yield new information. An additional final problem with bottom-up maps is shown in Figure 8.8. Usually the overlap distance between two adjacent members of a contig is not known with much precision. Therefore distances on a typical bottom-up map are not well defined.

The number of samples or clones that must be handled in top-down or bottom-up mapping projects can be daunting. This number also scales linearly with the resolution desired. To gain some perspective on the problem, consider the task of mapping a 150-Mb chromosome.

or                                    or

**Figure 8.8**  Ambiguity in the degree of clone overlap resulting from most fingerprinting or clone ordering methods.

This is the average size of a human chromosome. The numbers of samples needed are shown below:

| RESOLUTION | RESTRICTION MAP | ORDERED LIBRARY ($5\times$ REDUNDANCY) |
|---|---|---|
| 1 Mb | 150 fragments | 750 clones |
| 0.1 Mb | 1500 fragments | 7500 clones |
| 0.01 Mb | 15,000 fragments | 75,000 clones |

With existing methods the current convenient range for constructing physical maps of entire human chromosomes allows a resolution of somewhere between 50 kb (for the most arduous bottom-up approaches attempted) to 1 Mb for much easier restriction mapping or large insert clone contig building.

The resolution desired in a map will determine the sorts of clones that are conveniently used in bottom-up approaches. Among the possibilities currently available are

| | |
|---|---|
| Bacteriophage lambda | 10 kb inserts |
| Cosmids | 40 kb inserts |
| P1 clones | 80 to 100 kb inserts |
| Bacterial artificial chromosomes (BACs) | 100 to 400 kb inserts |
| Yeast artificial chromosomes (YACs) | 100 to 1300 kb inserts |

The first three types of clones can be easily grown in large numbers of copies per cell. This greatly simplifies DNA preparation, hybridization, and other analytical procedures. The last two types of clones are usually handled as single copies per host cell, although some methods exist for amplifying them. Thus they are more difficult to work with, individually, but their larger insert size makes low-resolution mapping much more rapid and efficient.

## RESTRICTION NUCLEASE GENOMIC DIGESTS

Generating DNA fragments of the desired size range is critical for producing useful libraries and genomic restriction maps. If genomes were statistically random collections of the four bases, simple binomial statistics would allow us to estimate the average fragment length that would result from a given restriction nuclease recognition site in a total digest. For a genome where each of the four bases is equally represented, the probability of occurrence of a particular site of size $n$ is $4^{-n}$; therefore the average fragment length generated by that enzyme will be $4^n$. In practice, this becomes

| SITE SIZE ($N$) | AVERAGE FRAGMENT LENGTH (kb) |
|---|---|
| 4 | 1 |
| 6 | 4 |
| 8 | 64 |
| 10 | 1000 |

This tabulation indicates that enzymes with four or six base sites are convenient for the construction of partial digest small insert libraries. Enzymes with sites ten bases long would be the preferred choice for low-resolution macrorestriction mapping, but such enzymes are unknown. Enzymes with eight-base sites would be most useful for currently achievable large-insert cloning. More accurate schemes for predicting cutting frequencies are discussed in Box 8.1.

A list of the enzymes currently available that have relatively rare cutting sites in mammalian genomes is given in Table 8.1. Unfortunately, there are only a few known restriction enzymes with eight-base specificity, and most of these have sites that are not well-predicted by random statistics. A few enzymes are known with larger recognition sequences. In most cases there is not yet convincing evidence that the available preparations of these enzymes have low enough contamination with random nonspecific nucleases to allow them to be used for a complete digest to generate a discrete nonoverlapping set of large DNA fragments. Several enzymes exist that have sites so rare that they will not occur at all in natural genomes. To use these enzymes, one must employ strategies in which the sites are introduced into the genome, their location is determined, and then cutting at the site is used to generate fragments containing segments of the region where the site was inserted. Such strategies are potentially quite powerful, but they are still in their infancy.

The unfortunate conclusion, from experimental studies on the pattern of DNA fragment lengths generated by genomic restriction nuclease digestion, is that most currently available 8-base specific enzymes are not useful for generating fragments suitable for macrorestriction mapping. Either the average fragment length is too short, or the digestion is not complete, leading to an overly complicated set of reaction products. However, mammalian genomes are very poorly modeled by binomial statistics, and thus some enzymes, which statistically might be thought to be useless because they would generate fragments that are too small, in fact generate fragments in useful size ranges. As a specific example, consider the first two enzymes known with eight-base recognition sequences:

*Sfi* I        GGCCN^NNNNGGCC
*Not* I        GC^GGCCGC

Here the symbol N indicates that any of the four bases can occupy this site; the caret (^) indicates the cleavage site on the strand shown; there is a corresponding site on the second strand.

Human DNA is A–T rich; like that of other mammals it contains approximately 60% A+T. When this is factored into the predictions (Box 8.1), on the basis of base composition, alone, these enzymes would be expected to cut human DNA every 100 to 300 kb. In fact *Sfi* I digestion of the human genome yields DNA fragments that predominantly range in size from 100 to 300 kb. In contrast, *Not* I generates DNA fragments that average closer to 1 Mb in size. The size range generated by *Sfi* I would make it potentially useful for some applications, but the cleavage specificity of *Sfi* I leads to some unfortunate complications. This enzyme cuts within an unspecified sequence. Thus the fragments it generates cannot be directly cloned in a straightforward manner because the three base overhang generated by *Sfi* I is a mixture of 64 different sequences. Another problem, introduced by the location of the *Sfi* I cutting site, is that different sequences turn out to be cut at very different rates. This makes it difficult to achieve total digests efficiently, and as described later, it also makes it very difficult to use *Sfi* I in mapping strategies that depend on the analysis of partial digests.

## BOX 8.1
## PREDICTION OF RESTRICTION ENZYME-CUTTING FREQUENCIES

An accurate prediction of cutting frequencies requires an accurate statistical estimation of the probability of occurrence of the enzyme recognition site. To accomplish this, one must take into account two factors: First, the sample of interest, that is, the human genome, is unlikely to have a base composition that is precisely 50% G + C, 50% A + T. Second, the frequencies of particular dinucleotide sequences often vary quite substantially from that predicted by simple binomial statistics based on the base composition. A rigorous treatment should also take the mosaicism into account (see Chapters 2 and 15).

Base composition effects alone can be considered, for double-stranded DNA with a single variable: $X_{G+C} = 1 - X_{A+T}$, where $X$ is a mole fraction. Then the expected frequency of occurrence of a site with $n$ G's or C's and $m$ A's or T's is just

$$\left(\frac{1}{2}\right)^{n+m} (X_{G+C})^n (1 - X_{G+C})^m$$

To take base sequence into account, Markov statistics can be used. In Markov chain statistics the probability of the $n$th event in a series can be influenced by the specific outcome of the prior events such as the $(n-1)$th and $(n-2)$th events. Thus Markov chain statistics can take into account known frequency information about the occurrences of sequences of events. In other words, this kind of statistics is ideally suited for the analysis of sequences.

Suppose that the frequencies of particular dinucleotide sequences are known for the sample of interest. There are 16 possible dinucleotide sequences: The frequency $X_{A,C}$, for example, indicates the fraction of dinucleotides that has a AC sequence on one strand base paired to a complementary GT on the other. The sum of these frequencies is 1. Only 10 of the 16 dinucleotide sequences are distinct unless we consider the two strands separately. On each strand, we can relate dinucleotide and mononucleotide frequencies by four equations:

$$X_{A,C} + X_{A,A} + X_{A,T} + X_{A,G} = X_A$$

since base A must always be followed by some other base (the X's indicate mole fraction). The expected frequency of occurrence of a particular sequence string, based on these nearest-neighbor frequencies is just

$$X_{12} \prod_{i=2}^{n} \frac{X_{i,i+1}}{X_i}$$

where the product is taken successively over the mole fractions $X_{i,i+1}$, respectively, and $X_i$ of all successive dinucleotides $i, i+1$, and mononucleotides $i$ in the sequence of interest. Predictions done in this way are usually more accurate than predictions based solely on the base composition. Where methylation occurs that can block the cutting of certain restriction nucleases, the issue becomes much more complex, as discussed in the text.

**TABLE 8.1    Restriction Enzymes Useful for Genomic Mapping**

| Enzyme[a] | Recognition Site (5'–3')[b] | Source[a,c] |
|---|---|---|
| *Enzymes with Extended Recognition Sites* | | |
| I-*Sce* I | TAGGGATAA/CAGGGTAAT | B |
| *VDE* | TATSYATGYYGGTGY/ | O |
| | GGRGAARKMGKKAAWGAAWG | |
| I-*Ceu* I | TAACTATAACGGTCCTA/AGGTAGCGA | N |
| I-*Tli* I | GGTTCTTTATGCGGACAC/TGACGGCTTTATG | N |
| I-*Ppo* I | CTCTCTTAA/GGTAGC | P |
| *Enzymes with >6-bp Recognition Site* | | |
| *Pac* I | TTAAT/TAA | N |
| *Pme* I | GTTT/AAAC | N |
| *Swa* I | ATTT/AAAT | B |
| *Sse*83888t1 | CCTGCA/GC | T |
| *Enzymes with >6-bp Recognition that Cut in CpG Islands* | | |
| *Rsr* II (*Csp**) | CG/GWCCG | N, P* |
| *SgrA* I | CR/CCGGYG | B |
| *Not* I | GC/GGCCGC | B, N, S, P |
| *Srf* I | GCCC/GGGC | P, S |
| *Fse* | GGCCGGCC | B |
| *Sfi* I[d] | GGCCNNNN/NGGCC | B, N, P |
| *Asc* I | GG/CGCGCC | N |
| *Enzymes that cut in CpG Islands: Fragments Average >200 kb* | | |
| *Mlu* I | A/CGCGT | B, N, P, S |
| *Sal* I | G/TCGAC | N |
| *Nru* I | TCG/CGA | N |
| *Bss* HII | G/CGCGC | |
| *Sac* II | CCGC/GG | N |
| *Eag* I (*EcI* XI*, *Xma* III) | C/GGCCG | B*, N |
| *Enzymes that cut in CpG Islands: Fragments Average <200 kb* | | |
| *Nar* I | GG/CGCC | N, P |
| *Sma* I | CCC/GGG | N |
| *Xho* I | C/TCGAG | N |
| *Pvu* I | CGAT/CG | N |
| *Apa* I | GGGCC/C | N |
| *Enzymes with TAG in their Recognition Sequence* | | |
| *Avr* II (*Bln* I*) | C/CTAGG | N, T* |
| *Nhe* I | G/CTAGC | N, P, S |
| *Xba* I | T/CTAGA | B, N, P, S |
| *Spe* I | A/CTAGT | B, N, P, S, T |
| *Nhe* I | G/CTAGC | P |
| *Dra* I | TTT/AAA | B, P, S |
| *Ssp* I | AAT/ATT | P |

[a]Asterisk indicates preferred enzyme and source.

[b]R, A, or G; Y, C, or T; M, A, or C; K, G, or T; S, G, or C; W, A, or T; N, A, or C or G or T.

[c]B, Boehringer Mannheim; N, New England Biolabs; O, not commercially available; P, Promega; S, Stratagene; T, Takara.

[d]Two sites are needed in order for cleavage to occur at both of them.

Note that the recognition site for *Sfi* I contains no CpG's, while the site recognized by *Not* I contains 2 CpG's. In Chapter 1 we showed that the frequency of occurrence of this dinucleotide sequence is reduced in mammalian DNA to about 1/4 of the level expected statistically. On this basis *Not* I can be expected to behave like a ten-base specific enzyme rather than an eight-base specific enzyme; as a result the expected fragment sizes are predicted to lie around 1 Mb, in agreement with experimental results. However, this is an oversimplified argument for it ignores the effect of DNA methylation. Overall, in the human genome about 80% of the CpG sequences are methylated to 5-meC, and *Not* I (and most other restriction nucleases) is unable to cleave at sites that are methylated. If we factor this effect into the calculation, we can now predict that for random methylation only about 1/25 of the *Not* I sites have no $^m$C and thus are cleavable. This yields an average DNA fragment size of 25 Mb, making *Not* I all but useless for conventional macrorestriction mapping. Fortunately the distribution of CpG methylation is not random. In practice, it appears that about 90% of the *Not* I sites in the human genome are not methylated, and so they are accessible to cleavage by the enzyme. This explains why *Not* I can generate fragments that average about 1 Mb in size.

## HTF ISLANDS

The peculiar distribution of methylated CpG, which has such a dramatic effect on the frequency of *Not* I cutting sites, is a reflection of a more general statistical unevenness in mammalian genomes. This was first discovered when genomic digestions were carried out with a much less specific restriction enzyme, *Hpa* II. A typical genomic digest generated by this enzyme that recognizes the sequence CCGG is shown in Figure 8.9. A statistically random genome would be expected to give a roughly Gaussian distribution of fragment sizes. Instead, the two-phase distribution observed in practice is striking. *Hpa* II is inhibited by DNA methylation. It cannot cut the sequence C$^m$CGG. To as good approximation, the fragment sizes shown in Figure 8.9 can be fit by assuming that the genome is divided into regions where no methylation occurs and regions where most of the CpG's are methylated. The large fragments in the latter regions are what were expected from an *Hpa* II digest. The small fragments were unexpected. They were named *Hpa* II tiny fragments (HTFs), and the regions that contain them have been called *HTF islands*. As we have learned more about the properties of these regions, many researchers have preferred to call them *CpG islands,* but for others, the original term has stuck.

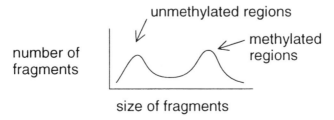

**Figure 8.9** Distribution of the sizes of the DNA fragments generated by a complete *Hpa* II digest of a typical mammalian genome.

HTF islands have a number of very interesting properties. They tend to be located near genes, most often at the 5'-edge of genes. They are very rich in G+C. The frequency of CpG is as expected from binomial statistics—that is, there is no suppression of CpG in these regions and no elevation of TpG (which results from $^{m}$CpG mutagenesis as described in Chapter 1.) In HTF islands the CpG sequences are unmethylated. These results are self-consistent: If there is no methylation in these regions, there should be no progressive loss of CpG by mutation, and thus the frequency of this sequence should just reflect the local C+G content. More than 90% of the known *Not* I sites appear to be located in HTF islands; this is understandable. To produce a cleavable *Not* I site requires two nearby unmethylated CpG's. This is an event most unlikely to occur in the bulk of the genome where CpG's are both very rare and methylated.

Many HTF islands have been studied by sequencing the DNA flanking *Not* I sites. This is relatively easy to do because, as we will discuss later in the chapter, there are straightforward ways to clone DNA that contains a cutting site for this enzyme (or almost any other enzyme for that matter). Two representative human DNA sequences flanking *Not* I sites on chromosome 21 are described in Figure 8.10, which shows the local base composition as a function of distance from the *Not* I site. The first of these examples is extraordinarily G + C rich throughout a 600–800 base region. The second example shows a transition from an HTF island to more ordinary genomic DNA; the *Not* I site is near the edge of the island. The distribution of CpG, GpC, and TpG sequences in the first of these examples is plotted as a function of position in Figure 8.11. It is evident that GpC's and GpC's are extraordinarily prevalent; more significantly, their prevalence is roughly equivalent, showing the lack of any significant CpG suppression.

## ORDERING RESTRICTION FRAGMENTS

The simplest way to view top-down mapping is by projecting a low-resolution map with ill-defined distances onto a higher-resolution map with more carefully defined distances. Suppose that a genetic or cytogenetic map already exists for a chromosome of interest, and this is actually the case today for almost all regions of the human genome, at least at 2 Mb resolution (Chapter 6). Each genetically mapped or cytogenetically mapped DNA marker can be radiolabeled and used as a hybridization probe to identify the corresponding large DNA fragments that it resides on, as shown in Figure 8.12. If DNA fragments can be generated that are comparable in size to the density of available DNA markers, then most of the construction of a restriction map would be accomplished with a relatively small number of direct DNA hybridizations. In reality, for almost all regions of the genome, the probes available today are not dense enough to order a complete set of restriction fragments. Instead one must utilize procedures that allow the restriction map to be extended outward from the position of known fragments into neighboring regions. A second problem is that the DNA fragment sizes generated by total digestion with available rare-cutting enzymes are quite diverse. Thus, although the average fragment size seen with *Not* I is about 1 Mb, many fragments are 3 Mb in size, and many are 0.2 Mb or smaller. Available DNA probes are very likely to be found that recognize the 3-Mb fragments; it is much less likely that any preexisting probes will correspond to the 0.2-Mb fragments at typical probe densities of 1 per 1 to 2 Mb.

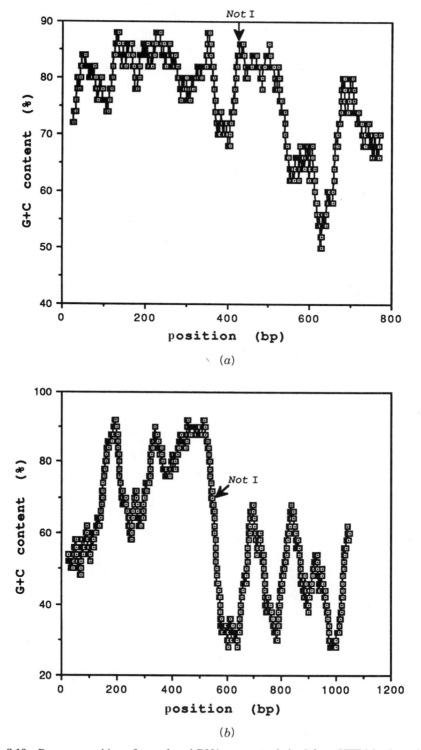

**Figure 8.10** Base composition of two cloned DNA segments derived from HTF islands on human chromosome 21. Plotted is the local average base composition as a function of the position within the clone. A *Not* I site that allowed the selective cloning of these DNA pieces is also indicated. (*a*) Clone centered in an HTF island. (*b*) Clone at one edge of an HTF island. (Adapted from Zhu et al., 1993.)

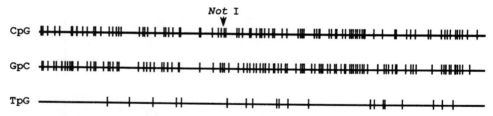

**Figure 8.11** Distribution of three dinucleotide sequences within the clone described in Figure 8.10*a*.

To carry out macrorestriction mapping projects efficiently, a number of needs must be met that allow one to circumvent the general problems raised above:

1. Probes must be isolated that are not preferentially located on large restriction fragments.
2. DNA must be cut less frequently to generate large fragments in any desired area of the genome.
3. Isolated probes must correspond to fragments of interest, those fragments where probes are needed to complete a map.
4. Neighboring fragments must be unequivocally identified.
5. Any tiny fragments generated by chance, by enzymes that on average yield very large fragments, must not be ignored.

Methods now exist that deal with all of these problems reasonably efficiently. Most will be described in this chapter; a few will be deferred to Chapter 14 where we deal with some of the specialized methods that have been developed to manipulate particular DNA sequences.

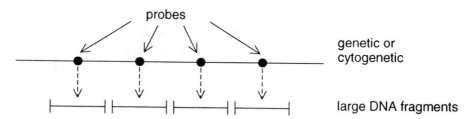

**Figure 8.12** The genetic or cytogenetic map provides a set of anchor points to place selected large DNA fragments in order.

## IDENTIFYING THE DNA FRAGMENTS GENERATED BY A RARE-CUTTING RESTRICTION ENZYME

The restriction digestion pattern generated by any enzyme that yields pieces large enough to be useful in genomic or chromosomal restriction mapping must be analyzed by PFG. Since this separation technique has a limited window of high resolution, under any fixed experimental conditions it is usually necessary to fractionate the digest by PFG using a range of three or four different pulse times: 30-second pulses for fragments 0 to 200 kb, 60-second pulses for fragments 200 to 1000 kb, 1000-second pulses at lower field strengths for 1 to 3 Mb pieces, and 3600-second pulses or secondary PFG for fragments larger than this.

The entire distribution of DNA fragments can be visualized by staining the PFG fractionated material in the gel with ethidium bromide. This dye binds in between every other base pair (total stoichiometry 0.5 dye per base pair), and shows a more than 25-fold fluorescence enhancement when bound. Ethidium bromide appears to have no significant specificity for any particular base compositions or base sequences. It is sensitive enough for all routine use. Recently a number of other dyes have been reported that may offer the promise of higher-sensitivity DNA detection than ethidium bromide. However, the interactions of these dyes with very large DNA molecules have not yet been fully characterized.

If the genome under analysis is a relatively simple one, ethidium staining allows all the pieces to be visualized, as we showed earlier in Figure 8.5. Since the amount of ethidium bound is proportional to the size of the DNA fragment, the resulting distribution of fluorescence intensity in a result like Figure 8.5 gives the weight average distribution of DNA. Small fragments are very hard to see. In general, a monotonic increase in fragment intensity is expected as fragment size increases. Deviations from this pattern indicate multiplets: size fractions that contain two or more unresolved DNA pieces, or heterogeneity; DNA fragments present in substoichiometric amounts because they arise from a contaminant in the sample, from a restriction site that has been only partially cleaved, or from DNA partially degraded (by nuclease or even by electrophoresis itself). These complications aside, a quantitative analysis of the pattern of ethidium staining will usually allow an accurate analysis of the number of DNA fragments in the genome. When these sizes are summed, an estimate for the total genome size is produced. Indeed, when ethidium staining is carried out very carefully, the stained intensity of DNA bands is not just a monotonic function of their size; it is very close to a linear function of their size up to a few Mb. Thus from the relative staining intensity alone it is sometimes possible to make reliable estimates of DNA sizes, even when proper size standards, for some reason, are not present or useable.

With complex genomes only a smear of ethidium staining intensity is generally seen. At the highest obtainable PFG resolution, the entire human and mouse genome actually show a very discrete and reproducible banding pattern in ethidium-stained gels of DNA digested with *Not* I or other rare-cutting restriction enzymes (Fig. 8.13). The significance of this pattern has never been explained—but it does allow DNA from different species to be identified. In those unusual cases where a major repeating sequence has a rare-cutting site in it, a bright DNA band will sometimes be seen above the background ethidium smear.

To develop a restriction map of a chromosome from a complex genome, like the human genome, it is best to start with that chromosome in a hybrid cell. Even better, for the larger mammalian chromosomes, are hybrid cells that contain only a 10 to 50 Mb fragment of the chromosome of interest. There are two major advantages for starting with such a hybrid cell. First, the chromosome of interest will almost certainly have a unique genotype. Even if multiple copies are present, these are likely to be identical. In effect the sample is homozygous. This eliminates any confusion that would otherwise result if two different polymorphic structures of a region were merged into the same map.

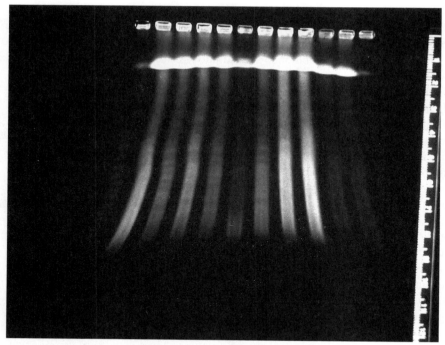

**Figure 8.13**   Ethidum-stained PFG fractionation of *Not* I digested human and mouse DNA from different cell lines. Fragments resolved in this gel range in size from 50 kb to 1 Mb. The distinct pattern of banding seen depends on the particular species and enzyme used.

An example of such a useful cell line is WAV17 which has two to three identical copies of human chromosome 21 in a mouse background.

The second advantage of using a hybrid cell is that it is possible to view most or all of the human component above the background of DNA from the rodent. This is accomplished by hybridizing a blot of a PFG-fractionated restriction enzyme digest of the cellular DNA with various human-specific repeating DNA sequences. For example, the most common human-specific repeat is the *Alu* sequence. This will be described in much greater detail in Chapter 14. Here, however, it is sufficient to note that this sequence occurs on average about once every 3 kb in some regions of the human genome and about once every 10 kb in others. Probes for *Alu* exist that show little significant cross-hybridization with rodent DNA. These can then be used as a human-specific stain to detect the presence of human restriction fragments in a mouse or hamster cell. An example of such an analysis is shown in Figure 8.14.

The human *Alu* sequence should be seen on almost every large fragment of human DNA if our genome were well fit by a statistically random model. This can be shown by using a simple Poisson model for the distribution of *Alu*'s. From the Poisson distribution (Chapter 6) we can estimate that the probability of an *Alu* not occurring on a fragment of interest will be

$$P(\text{no } Alu) = e^{(-\text{average fragment size}/Alu \text{ spacing})}$$

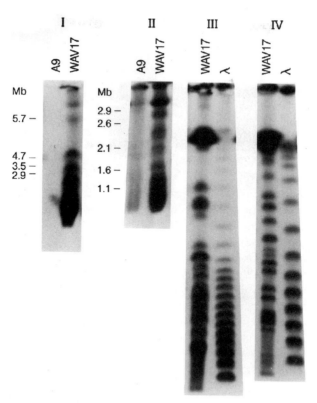

**Figure 8.14** PFG fractionation of *Not* I-digested DNA from a mouse cell line that contains chromosome 21 as the only human component. After electrophoresis the gel was blotted, and the blot was hybridized with the highly repeated human-specific *Alu* sequence. Shown in some of the panels are size standards used to estimate the length of particular human DNA fragments. (Taken from Sainz et al., 1992.)

The smallest fragments of much interest in human macrorestriction mapping are around 50 kb; the least *Alu*-dense regions of the genome have a 10-kb spacing between *Alu*'s; thus in these regions the chance that a 50-kb fragment has no *Alu* is exp($-5$). The chance, at random, that a 1 Mb fragment should lack an *Alu* is infinitesimal. We have actually characterized the distribution of *Alu*'s on more than 50 *Not* I fragments of human chromosome 21. In practice, all but one contain *Alu* as detected by hybridization experiments. The sole exception, however, is a 2.3 Mb fragment. The probability of seeing such an event at random is $e^{-230}$ assuming it derives from an *Alu*-poor region of the chromosome. Clearly we will have to refine our statistical picture of human DNA sequences quite considerably as more cases like this surface.

When human DNA fragments are detected by hybridization with human-specific repeated sequences the resulting distribution of intensity should still be generally proportional to fragment size. Note, however, that considerable variations around this mean will occur because repeated sequences tend to cluster and because small fragments with small numbers of repeats on average will show typical small number fluctuations. One way to generate labeling intensity that is independent of fragment size is to label the ends.

This can be done with enzymes like polynucleotide kinase that place a phosphate on the 5′-end of a DNA chain, or it can be done by filling in any inset 3′-ends of duplex DNA with DNA polymerase and radiolabeled dppN's. This procedure is readily applied within agarose gels, and it has been useful in the analysis of small genomes. For mammalian genomes the procedure is not useful because there is no specific way to label just the ends of the human DNA fragments in a rodent hybrid cell.

## MAPPING IN CASES WHERE FRAGMENT LENGTHS CAN BE MEASURED DIRECTLY

The classical approach to constructing restriction maps of small DNAs like plasmids and viruses is illustrated in Figure 8.15. Two or more restriction enzymes are used separately and in double digests to fragment the DNA of interest. The sizes of all pieces seen in an ethidium-stained gel are measured. Usually the pattern of sizes allows alignment of the different cutting sites in a single map. This procedure is clearly not a rigorous one, and it breaks down severely once the maps become complex, or when many similar-sized fragments are involved. In principle, each fragment could be isolated by electrophoretic fractionation, radiolabeled, and used as a probe. This would allow all overlapping fragments from digests with other enzymes to be unambiguously identified. In practice, however, it is usually easier to employ partial digestion strategies with end-labeled probes, as we will describe later for macrorestriction mapping.

   With mammalian DNAs, multiple restriction enzyme digest mapping is much less effective. Although one can determine the size of the fragments generated by each enzyme, by using repeated sequence hybridizations, there is an annoying tendency for many of the restriction enzymes that yield large DNA fragments to cut in the same regions. This is because most such enzymes prefer HTF islands. Thus the usefulness of double digestion in most regions of mammalian genomes is far less than illustrated by the example in Figure 8.15. A more serious problem is that with large numbers of fragments in single-enzyme digests, double digests become hopelessly complicated to analyze. With mammalian DNA, in contrast to DNA from simple genomes, one cannot easily access each purified fragment because it is contaminated by other human or rodent fragments. PCR can help circumvent this problem, as we will demonstrate in Chapter 14. In general, though, one must rely on hybridization with single-copy probes in order to simplify the pattern of DNA fragments to the point where it can be analyzed. The example in Figure 8.15 shows clearly that fragments from the end of large DNA pieces are particularly useful hybridization probes. The figure also indicates that with only a limited set of probes from the region, double digests are frequently impossible to analyze because many of the fragments in these digests will not correspond to any of the available probes. This discussion should make it clear that new strategies had to be developed to simplify the construction of macrorestriction maps of segments of complex genomes.

**Figure 8.15**  Schematic illustration of the double digestion procedure used to assemble simple restriction maps.

## GENERATION OF LARGER DNA FRAGMENT SIZES

Some of the problems illustrated by the example in Figure 8.15 would be alleviated if there were a systematic way to generate large DNA fragments in a region of interest. One general approach for doing this will be discussed here; others, like the RARE method, are deferred to Chapter 14.

It has already been mentioned that most restriction enzymes are inhibited by DNA methylation. One can take advantage of this to increase the specificity of certain restriction enzymes by methylating a subset of their cutting sites. Methylation can also be used as a general way to promote partial digests by using a methylase that recognizes all the cutting sites but does not allow the reaction to go to completion. Here, however, we deal with methylation reactions that are carried to completion. Consider the DNA sequence shown below. It contains a *Not* I site flanked by additional G–C pairs shown in boldface:

$$\text{\textbf{CG}CGGCCGC\textbf{G}} \quad \rightarrow \quad {}^{\text{m}}\text{CGCGGC}{}^{\text{m}}\text{CGCG}$$

$$\text{\textbf{GC}GCCGGCG\textbf{C}} \quad\quad\quad \text{GCG}{}^{\text{m}}\text{CCGGCG}{}^{\text{m}}\text{C}$$

Roughly one-quarter of all *Not* I sites will have an extra C at their 5′-end; an additional quarter will have an extra G at their 3′-end as shown in the above example. These extra residues generate a recognition site for the *Fnu*D II methylase that converts CGCG to ${}^{\text{m}}$CGCG. Since some of the methylation is within the *Not* I cutting site, this inhibits any subsequent attempts to cleave the site with *Not* I. Thus by methylation one can inactivate about half of all the *Not* I cleavage sites and double the average fragment size generated by this enzyme. Many variations on this theme exist.

Methylation also plays a key role in a whole set of potential schemes for site-selective cleavage of DNA employing the unusual restriction endonuclease *Dpn* I. This enzyme recognizes the sequence GATC, but it requires that the A be methylated in order for cleavage to occur. The preferred substrate is

$$\text{G}{}^{\text{m}}\text{ATC}$$

$$\text{CT}{}^{\text{m}}\text{AG}$$

A major complication is that the monomethyl derivative is also cut, but much more slowly. Hence, in the schemes described below, it is essential to expose the DNA substrate to *Dpn* I for the minimum time needed to cleave at the desired sites before the background of undesired additional cleavages becomes overwhelming.

*Dpn* I is converted to an infrequently cutting enzyme by starting with unmethylated target DNA, which *Dpn* I cannot cleave at all, and selectively introducing methyl groups by treatment with methylases with recognition sequences that overlap part of the *Dpn* I site. The simplest example, shown below, employs the *Taq* I methylase. This enzyme recognizes the sequence TCGA and converts it to TCG${}^{\text{m}}$A. If two such sequences lie adjacent to each other in a genome, the result, once both are methylated is

$$\text{TCG}{}^{\text{m}}\text{ATCG}{}^{\text{m}}\text{A}$$

$$^{\text{m}}\text{AGCT}{}^{\text{m}}\text{AGCT}$$

Comparison of this sequence with the *Dpn* I recognition site shown above indicates that a *Dpn* I cleavage site has been generated. An eight-base sequence is required to specify this site by the procedure we have used. Since it contains two CpG's, it will be a very rare site in mammalian genomes. Thirty to 40 variations on this theme exist, some generating sites as large as 16 base pairs. Some of these sites are predicted to occur less than once, on average, in most genomes. However, one difficulty with these schemes, is the lack of availability of many of the necessary methylases in sufficiently nuclease-free form to generate the very large DNA fragments specified by such large recognition sites. At present the exploitation of *Dpn* I remains an extremely attractive method that is not yet generally practiced because of some of these unsolved experimental limitations.

## LINKING CLONES

In this and the next few sections we discuss several of the methods that allow the order of restriction fragments to be determined, de novo, without access to other mapping information. By far the most robust of these, in many respects, is the use of specialized clones called *linking clones*. These clones contain the same rare-cutting sites that were used to produce macrorestriction fragments from the sample to be mapped. As shown in Figure 8.16, because these clones have an internal rare-cutting site, they must overlap two adjacent large DNA fragments. (The sequence properties of two authentic human linking clones were illustrated in Figs. 8.10 and 8.11.) When used as hybridization probes, the linking clones should identify two fragments that are adjacent in the genome. This will occur unless the rare-cutting site is so close to one edge of the linking clone that not enough single-copy material remains beyond the site to hybridize effectively. To eliminate this possibility, it is useful to work with more than one set of linking clones, constructed in such a way as to have different distributions of DNA flanking the rare-cutting site. A convenient way to do this is to start with separate total digest libraries made with different restriction nucleases such as *Eco*R I and *Hin*d III. In principle, there is sufficient information in linking clone hybridizations that if a complete linking library were available, its use in hybridizations would produce data that would allow the entire set of restriction fragments to be ordered.

Two of the methods that have been used to prepare libraries of linking clones are summarized in Figure 8.17. In each case one starts with a small insert genomic library contained into a circular vector that is lacking any sites for the restriction enzyme of interest. If such a library does not preexist, it is easily constructed by subcloning an existing library into such a vector. DNA from the entire library is purified and digested with the rare-cutting restriction nuclease. Only those clones containing a genomic insert that includes a site for this enzyme will be linearized; the rest remain circular. Two different methods can be used to select out the clones that have been linearized. One approach is to use trapping electrophoresis: ordinary or PFG electrophoresis at high fields such that linear molecules migrate into the gel effectively while circles remain trapped in the sample

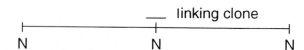

**Figure 8.16**    A linking clone spans two adjacent large DNA fragments.

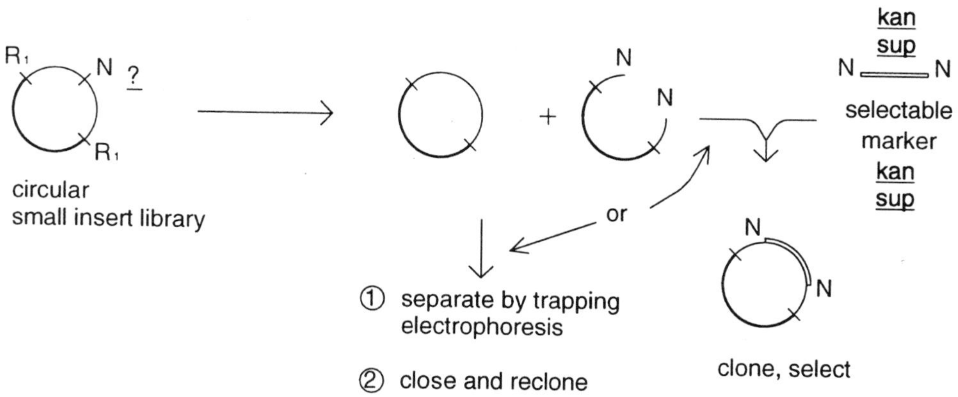

**Figure 8.17** Two procedures used to make libraries of linking clones. Both require starting with a library of the desired chromosome (or other target) in a circular vector.

well or sample plug. In this way the linking clones are selectively captured. They can be religated at low concentration to avoid the production of chimeras, and then re-introduced into a bacterial host.

The alternative approach, after the rare-site-containing clones have been linearized, is to ligate a selectable marker into the rare-cutting site. Both suppressor tRNA genes and kanamycin resistance genes have been successfully used for this purpose. The library is then used to transform a bacterial host in the presence of the selection. The only clones that should survive are the linking clones. Both of the methods described work reasonably well. However, the procedures are not perfect, and it is important, after a linking library has been made, to screen out artifactual clones. These may include such annoyances as chimeras of rare cutter fragments, clones that contain a rare cutter site that is not actually cut in the parental genome because it is methylated, and clones containing small individual rare cutter fragments. One way to screen for useful clones is to use them as probes in Southern hybridizations with genomic DNA singly and doubly digested with the restriction enzyme used for constructing the library such as *Eco*R I or *Hin*d III, and the rare-cutting enzyme. Proper linking clones will show single bands or no bands in the absence of rare cutter cleavage, and double bands once the rare site has been cut.

Once linking clones are available, they have many other useful applications besides the ordering of macrorestriction fragments. Linking clones serve as effective bridges between two different kinds of restriction nuclease digests (or other types of DNA samples) because they come from a known position, the ends, of large DNA fragments. As a second example, one can cut linking clones at their internal rare cutter site and isolate the two fragments. The resulting samples are called *half-linking clones*. They are useful for the analysis of the internal structure of large DNA fragments, as shown by the schematic experiment illustrated in Figure 8.18. Here a *Not* I half-linking clone is used as an indirect end label by hybridizing it to a blot of a PFG-fractionated partial *Eco*R I digest of a *Not* I fragment. The sizes of the DNA pieces seen in the partial digest provide the location of the internal *Eco*R I sites. This type of analysis was originally developed for smaller DNA pieces by Smith and Birnstiel. It is a very powerful approach. Half-linking clones are also very useful for the analysis of partial digests with enzymes that cut infrequently, as we will demonstrate later.

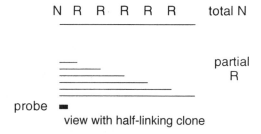

**Figure 8.18** Use of a half-linking clone as an indirect end label to reveal the pattern of internal restriction sites in a partial digest. This approach to restriction mapping, called the Smith-Birnstiel method after the researchers who first described it, remains the most powerful way to accumulate large amounts of restriction map data rapidly.

Another potential use of linking clones will be to order complete digest libraries of cloned large DNA fragments, once we are able to produce such libraries. The difficulty today is that too large a percentage of the fragments generated from mammalian DNA by an enzyme like *Not* I exceed the capacities of current large insert cloning vectors. However, an example of the potential power of this approach can be seen for an enzyme like *Eag* I that recognizes the sequence CCGCCG, which is the internal 6 base pairs of the *Not* I recognition sequence. *Eag* I cuts genomic DNA into fragments that average 200 to 300 kb in size. These fragments can be cloned into conventional YAC vectors once the vectors are equipped with the proper cloning sites. The result of probing such a library with *Eag* I linking clones is shown in Figure 8.19. In principle, there is enough information in the two sets of clones, linking clones and YACs, to completely order both sets of samples, and the method has two advantages. The materials used are only a tiny bit larger than the minimum possible tiling path for any ordered library, and distances along this path are known with great precision. In practice, there is no example of a library that has been fully ordered in this way.

One limitation with currently used linking clones must be realized. These clones tend to come from very G + C rich regions because this is where the sites of most rare-cutting enzymes are preferentially located. These regions make subsequent PCR analyses fairly difficult because, in current PCR protocols, very G + C rich primers and templates do not work especially well. Secondary structure in the single strands of these materials is presumably the major cause of the problems, but a generally effective solution to these problems is not yet in hand.

**Figure 8.19** Linking clones and large DNA fragments generated by the same enzyme could be assembled, in principle, into an ordered library with almost the minimum possible tiling length.

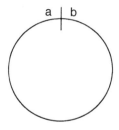

**Figure 8.20**  Basic notion of a jumping clone: Two discontinuous pieces of the genome (*a*) and (*b*), but related by some map or fragment information, are assembled into the same clone.

## JUMPING LIBRARIES

Jumping clones offer a way of dealing with two discontinuous pieces of a chromosome. The basic notion of a jumping clone is shown in Figure 8.20. It is an ordinary small insert clone except that two distinct and distant pieces of the original DNA target are brought together and fused in the process of cloning. A simple case of a jumping clone would be one that contained the two ends of a large DNA fragment, but all of the internal DNA of the fragment had been excised out. The way in which such clones can be made is shown in Figure 8.21.

A genomic digest with a rare-cutting enzyme is diluted to very low concentration and ligated in the presence of a vector containing a selectable marker. The goal is to cyclize the large fragments around vector sequences. At sufficiently low concentrations of target, it becomes very unlikely that any intermolecular ligation of target fragments will occur. One can use excess vector, provided that it is dephosphorylated so that vector–vector ligation is not possible. When successful, ligations at low concentrations can produce very large DNA circles. These are then digested with a restriction enzyme that has much more frequent sites in the target DNA but no sites in the vector. This can be arranged by suitable design of the vector.

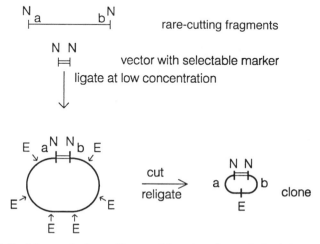

**Figure 8.21**  Method for producing a library of jumping clones derived from the ends of large DNA fragments.

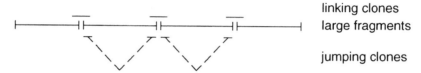

**Figure 8.22**   Potential utility of jumping clones that span large DNA fragments.

The result is a mixture of relatively small DNA pieces; only those near the junction of the two ends of the original large DNA fragments contain vector sequences. A second ligation is now carried out, again at very low concentration. This circularizes all of the DNAs in the sample. When these are reintroduced into *E. coli,* under conditions where selectable markers on the vector are required for growth, a jumping library results.

There are potentially powerful uses for jumping libraries that span adjacent sites of infrequently cutting restriction enzymes. These are illustrated in Figure 8.22. Note that each of the jumping clones will overlap two linking clones. If one systematically examines which linking clones share DNA sequences with jumping clones, and vice versa, the result will be to order all the linking clones and all the jumping clones. This could be done by hybridization, PCR, or by direct DNA sequencing near the rare-cutting sites. In the latter case DNA sequence comparisons will reveal clone overlaps. Such an approach, called *sequence-tagged rare restriction sites* (STARs), becomes increasingly attractive as the ease and throughput of automated DNA sequencing improves.

A more general form of jumping library is shown schematically in Figure 8.23. Here a genome is partially digested by a relatively frequent cutting restriction nuclease. The resulting, very complex, mixture is fractionated by PFG, and a very narrow size range of material is selected. If this is centered about 500 kb, then the sample contains all contiguous blocks of 500-kb DNA in the genome. This material is used to make a jumping library in the same general manner as described above, by circularizing it around a vector at low concentration, excising internal material, and recircularizing. The resulting library is a size-selected jumping library, consisting, in principle, of all discontinuous sets of short DNA sequences spaced 500 kb apart in the genome. The major disadvantage of this library is that it is very complex. However, it is also very useful, as shown in Figure 8.24*a*. Suppose that one has a marker in a region of interest, and one would like another marker spaced approximately 500 kb away. The original marker is used to screen a 500-

(*a*)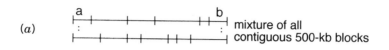

(*b*)     all short sequences spaced 900 kb in genome

**Figure 8.23**   DNA preparation (*a*) used to generate a more general jumping library (*b*) consisting of a very dense sampling of genomic fragments of a fixed length.

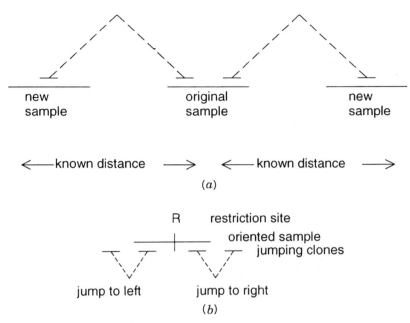

**Figure 8.24**  Example of the use of a general jumping library to move rapidly from one site in a genome to distantly spaced sites. (*a*) Half-jumping clones provide new probes, at known distances from the starting probe. (*b*) Information about the direction of the jump can be preserved if the map orientation of the original probe is known.

kb jumping library. This will result in the identification of clones that can be cut into half-jumping probes that flank the original marker by 500 kb on either side. It is possible to tell which jump has occurred in which direction if the original marker is oriented with respect to some other markers in the genome and if it contains a convenient internal restriction site (Fig. 8.24*b*). Selection of jumping clones by using portions of the original marker will let information about its orientation be preserved after the jump. The major limitation in the use of this otherwise very powerful approach is that it is relatively hard to make long jumps because it is difficult to ligate such large DNA circles efficiently.

## PARTIAL DIGESTION

In many regions of the human genome, it is difficult to find any restriction enzyme that consistently gives DNA fragments larger than 200 kb. In most regions, genomic restriction fragments are rarely more than a few Mb in size. This is quite inefficient for low-resolution mapping, since PFG is capable of resolving DNA fragments up to 5 or 7 Mb in size. To take advantage of the very large size range of PFG, it is often extremely useful to do partial digests with enzymes that cut the genome infrequently. However, there is a basic problem in trying to analyze the result of such a digest to generate a unique map of the region. This problem is illustrated in Figure 8.25. If a single probe is used to examine the digest by hybridization after PFG fractionation, the probe will detect a mixture of DNA fragments that extends from its original location in both directions. It is not straightforward to analyze this mixture of products and deduce the map that generated it.

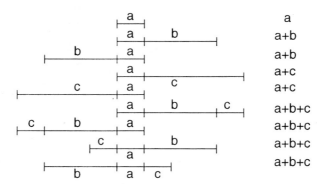

**Figure 8.25**   Ambiguity in interpreting the fragment pattern seen in a partial restriction nuclease digest, when hybridized with only a single probe.

If the enzyme cut all sites with equal kinetics, one could probably use likelihood or maximum entropy arguments to select the most likely fragment orderings consistent with a given digestion pattern. The difficulty is that many restriction enzymes appear to have preferred cutting sites. In the case of *Sfi* I these are generated by the peculiarities of the interrupted recognition sequence. In the case of *Not* I and many other enzymes, hot spots for cutting appear to occur in regions where a number of sites are clustered very closely in the genome (Fig. 8.26). Since the possibility of a hot spot in a particular region is hard to rule out, a priori, one plausible candidate for a restriction map that fits the partial digestion data is a null map in which all partially cleaved sites lie to one side of the probe and a fully cut hot spot lies to the other side (Fig. 8.27). While such a map is statistically implausible in a random genome, it becomes quite reasonable once the possibility of hot spots is allowed.

There are three special cases of partial digest patterns that can be analyzed without the complications we have just raised above. The first of these occurs when the region is flanked by a site that is cut in every molecule under examination. There are two ways to generate this site (which is, in effect, a known hot spot). Suppose that the probe used to analyze the digest is very close to the end of a chromosome (Fig. 8.28). Then the null map is the correct map. Using a telomeric probe is equivalent to the Smith-Birnstiel mapping approach we discussed earlier for smaller DNA fragments. It is worth noting that even if the probe is not at the very end of the mapped region, any prior knowledge that it is close to the end will greatly simplify the analysis of partial digest data. A second case of such a site occurs if we can integrate into a chromosome the recognition sequence of a very rare cutting enzyme, such that this will be the only site of its kind in the region of interest.

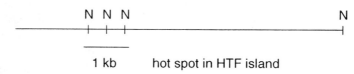

**Figure 8.26**   Example of a restriction enzyme hot spot that can confound attempts to use partial digest mapping methods.

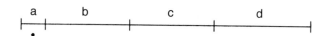

**Figure 8.27**  Null ordering of fragments seen in a partial digest.

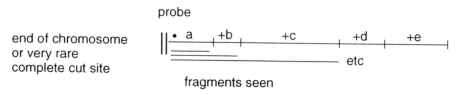

**Figure 8.28**  Assembling a restriction map from partial digest data when a totally digested site (or a chromosome end) is known to be present in the region.

Then we cleave at the very rare site completely and partially digest with an enzyme that has somewhat more frequent cleavage sites. The power of this approach as a rapid mapping method is considerable. However, it is dependent on the ability to drive the very rare cleavage to completion. As we have indicated before, in many of the existing schemes for very rare cleavage, total digestion cannot be guaranteed. Thierry and Dujon in Paris have recently shown that sites for the nuclease I-*Sce* I (Table 8.1) can be inserted at convenient densities in the yeast genome and that the enzyme cuts completely enough to make the strategy we have just described very effective.

A second case occurs that allows analysis of partial digestion data if the region of interest is next to a very large DNA fragment. Then, as shown in Figure 8.29, all DNA fragments seen in the partial digest that have a size less than the very large fragment can be ordered so long as the digest is probed by hybridization with a probe that is located on the first small piece next to the very large fragment. Again, as in the case of telomeric probes, relaxing this constraint a bit still enables considerable map information to be inferred from the digest.

The third case where partial digests can be analyzed occurs when one has two probes known, a priori, to lie on adjacent DNA fragments. This is precisely the case in hand, for example, when two half-linking clones are used as probes of a partial digest made by using the same rare site present on the clones. In this case, as shown in Figure 8.30, those fragment sizes seen with one probe and not the other must extend in the direction of that probe. Bands seen with both probes are generally not that informative. Thus linking clones play a key role in the efficient analysis of partial digests.

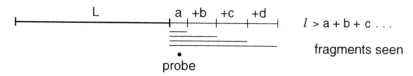

**Figure 8.29**  Assembling a restriction map from partial digest data adjacent to a very large restriction fragment.

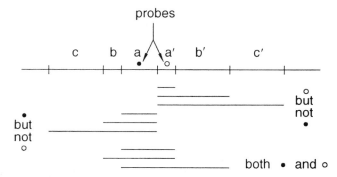

**Figure 8.30**  Assembling a restriction map when two hybridization probes are available and are known to lie on adjacent DNA fragments. Such probes are available, for example, as the halves of a linking clone.

The power of the approaches that employ partial digests is that a probe in one location can be used to reach out and obtain map data across considerable distances, where no other probes may yet exist. In order to carry out these experiments, however, two requirements must be met. First, reliable length standards and good high-resolution PFG fractionations are essential. The data afforded by a partial digest is all inherent in the lengths of the DNA bands seen. For example, if one used a probe known to reside on a 400-kb band, and found in a partial 1000-kb band, this is evidence that a 600-kb band neighbors the 400-kb band. One could then try to isolate specific probes from the 600-kb region to try to prove this assignment. However, if the 1000-kb band was mis-sized, and it was really only 900 kb, when one went to find probes in the 600-kb region of a gel, this would be the wrong region. The second constraint for effective partial digest analysis is very sensitive hybridization protocols. The yields of DNA pieces in partials can easily be only 1 to 10%. Detecting these requires hybridizations that are 10 to 100 times as sensitive as those needed for ordinary single-copy DNA targets. Autoradiographic exposures of one to two weeks are not uncommon in the analysis of partial digests with infrequently cutting enzymes.

## EXPLOITING DNA POLYMORPHISMS TO ASSIST MAPPING

Suppose that two different DNA probes detect a *Not* I fragment 800 kb long. How can we tell if they lie on the same fragment, or if it just a coincidence and they derive from two different fragments with sizes too similar to resolve? One approach is to cut the *Not* I digest with a second relatively rare-cutting enzyme. If the band seen by one probe shortens, and the band seen by the other does not, we know two different fragments are involved. If both bands shorten after the second digestion, the result is ambiguous, unless two different size bands are seen by the two probes and the sum of their sizes is greater than the size of the band originally seen in the *Not* I digest.

A more reliable approach is to try the two different probes on DNA isolated from a series of different cell lines. In practice, eight cell lines with very different characteristics usually suffices. When this is done with a single DNA probe, usually one or more of the lines shows a significant size difference or cutting difference from the others (Fig. 8.31). There are many potential origins for this polymorphism. Mammalian genomes are rampant with tandem repeats, and these differ in size substantially from individual to individual.

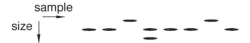

**Figure 8.31**   Example of the polymorphism in large restriction fragments seen with a single-copy DNA probe when a number of different cell lines are compared.

Most rare-cutting sites are methylation sensitive, and especially in tissue culture cell lines, a quite heterogeneous pattern of methylation frequently occurs. There are also frequent genetic polymorphisms at rare-cutting enzyme sites: These RFLPs arise because the sites contain CpGs that are potential mutation hotspots. For the purposes of map construction, the source of the polymorphism is almost irrelevant. The basic idea is that if two different probes share the same pattern of polymorphisms across a series of cell lines, whatever its cause, they almost certainly must derive from the same DNA fragment. One caveat to this statement must be noted. Overloading a sample used for PFG analysis will slow down all bands. If one cell line is overloaded, it will look like the apparent size of a particular *Not* I band has increased, but in practice, this is an artifact. To avoid this problem, it is best to work at very low DNA concentrations, especially when PFG is used to analyze polymorphisms. Further details are given in Chapter 5.

Polymorphism patterns can also help to link up adjacent fragments. Consider the example shown in Figure 8.32. Here one has probes for two DNA fragments that happen to share a common polymorphic site. In a typical case the site is cut in some cell lines, partially cut in others, and not cut at all in others. The pattern of bands seen by the two

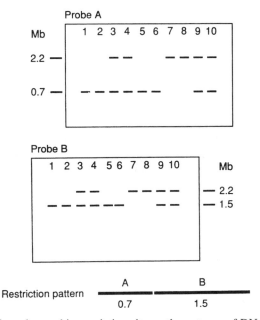

**Figure 8.32**   Effect of a polymorphic restriction site on the patterns of DNA fragments seen with two probes that flank this site (from Oliva et al., 1991). Hybridization of DNA from different cell lines (lanes 1–10) with two putatively linked probes (A and B) leads to the detection of different and common fragment sizes but identical polymorphism patterns.

probes will be correlated across a set of cell lines in a precise way. Furthermore one should see a band in some cell lines that is the sum of the two fragment lengths seen in others. This provides evidence that the two probes are, in fact, seeing neighboring bands. More complex cases exist where there are multiple polymorphisms in a small region. These become very hard to analyze. In fact such regions may not be that rare; there is some evidence from a few of the restriction maps that have been made thus far that poly-morphism, at the level of macrorestriction maps is a patchy phenomenon with regions rel-atively homogeneous across a set of cell lines alternating with regions that are much more variable. Methylation differences are probably at the heart of this phenomenon, but whether it has any biological significance or is just an artifact of in vitro tissue culture conditions is not yet known.

## PLACING SMALL FRAGMENTS ON MAPS

In most macrorestriction maps that have been constructed to date, occasional very small restriction fragments have been missed. These fragments add little value to the map, and they inevitably show up in subsequent work when a region is mapped more finely or con-verted to an ordered clone bank. However, from a purely aesthetic standpoint, it is unde-sirable to leave gaps in a map. The real question is how to fill these gaps with minimum effort. One approach that has been reasonably successful, is to end label DNA fragments from a complete digest with an infrequently cutting enzyme and use PFG with short short pulses or conventional electrophoresis to detect the presence of any small fragments. Once these are identified, a simple way to place them on the restriction map is to use them as a probe in hybridization against a partial digest. As shown by Figure 8.33, the small fragment itself is invisible in such a digest. However, when it is fused to fragments on either side, the sizes of these pieces will be detected in the partial digest, and in almost all cases these sizes will be sufficiently characteristic to place the small fragment uniquely on the map.

An alternative approach to identifying small fragments is to use PCR. Splints are li-gated onto the sticky ends generated by the restriction enzyme digestion. Primers specific for the splints are then used for PCR amplification. Given the limited size range of typical PCR amplifications, no macrorestriction fragments will be amplified. The PCR product will consist of only small restriction fragments if any of these were generated by the di-gest. These can be fractionated by size and individually used to probe a partial digest, as described above, in order to determine the position of the small fragments on the macrorestriction map.

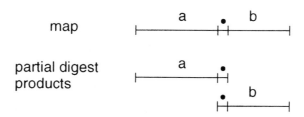

**Figure 8.33**  Placing a small DNA fragment on a macrorestriction map.

## REACHING THE ENDS OF THE PHYSICAL MAP: CLONING TELOMERES

Ten years ago there was a tremendous flurry of activity in cloning mammalian telomere-associated sequences. This was motivated by several factors. The simple sequence repeat at the very end of mammalian chromosomes had been determined, but little was known about what kinds of DNA sequences lay immediately proximal to this. Such sequences should be of considerable biological interest, since they ought to play a role in chromosome function, and they might help determine the identity of different chromosomes. Clones near telomeres will be extremely useful probes for partial digests, as we have already described. Finally telomeres are, by definition, the ends of all linear chromosome maps, and until they are anchored to the remainder of a genetic or physical map, one cannot say that that map is truly complete.

True telomeres would not be expected to be present in plasmid, bacteriophage, or cosmid libraries, and in fact it has been all but impossible to find them there. There are at least two explanations for this. First, the structure of the simple sequence telomere repeat contains a hairpin rather than a normal duplex or single-stranded overhang (Chapter 2). This hairpin is quite stable under the conditions used for ordinary DNA cloning. It will not ligate under these conditions, and thus one would expect that the telomeres would fail to be cloned. Second, simple tandem repeats are not very stable in typical cloning vectors; they are easily lost by recombination; the 10- to 30-kb simple human telomeric repeat would almost certainly not be stable in typical bacterial cloning strains. The sequences distal to the simple repeat are themselves rich in medium-size tandem repeats, and whole blocks of such sequences are repeated. Thus these regions are also likely to be quite unstable in *E. coli.* For all these reasons it is not surprising that early attempts to find telomeres in conventional libraries failed.

To circumvent the problems discussed above and find mammalian telomeric DNA, about six research groups simultaneously developed procedures for selectively cloning these sequences in the yeast, *S. cerevisiae.* The standard cloning vector used in yeast is a yeast artificial chromosome (YAC; see Box 8.2). We and others have reasoned that because telomeres are such key elements in chromosome function, and because the properties of telomeres are so well conserved through most species, there was a chance that a mammalian telomere can function in yeast even if its structure is not identical to the yeast telomere and associated sequences. To test this idea, we used half YACs as cloning vectors. To survive in yeast as a stable species, the half YAC would have to become ligated to another telomere-containing DNA fragment.

We did not expect that the process of cloning telomeres would be very efficient. To enhance the chances for a human telomere to be successfully cloned, we developed the procedure shown in Figure 8.34. Total genomic human DNA was digested with the restriction enzyme *Eco*R I. The reaction mixture was diluted, and ligase was added. All normal *Eco*R I DNA fragments have two sticky ends generated by the original restriction enzyme digest. These are capable of ligation, and at the reduced concentrations the primary ligation products become intramolecular circles (just as in the case of the jumping library construction). The only fragments that do not lose their sticky ends by ligation are telomeres, which have only one sticky end and cannot circularize, and very short *Eco*R I fragments, which cannot bend enough to form a circle. By this ligation procedure most of the nontelomeric restriction fragments can be selectively eliminated; in practice, about a $10^4$ enrichment for telomeric restriction fragments is produced.

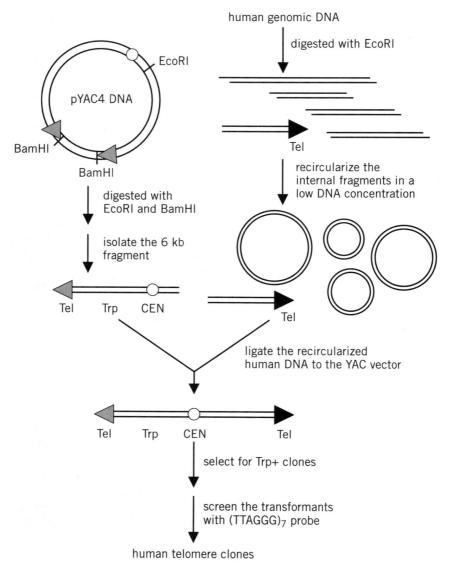

**Figure 8.34** Strategy used to pre-enrich human DNA for telomeric fragments and then to clone these fragments. (From Cheng et al., 1989.)

The resulting mixture was then ligated to an *Eco*R I-digested YAC vector. DNA was transfected into yeast, and colonies that grew in the presence of the selectable marker carried on the half YAC arm were screened for the presence of human repeated sequences and human telomere simple repeats. The first clone found in this way proved to be an authentic human telomere. Its structure is shown in Figure 8.35. The human origin of the clone is guaranteed by the presence of a human *Alu* repeat, and a short stretch of human-specific telomere simple sequence. Most of the simple sequence that originally had to be present at the authentic human telomere has been lost, and some of it has been replaced

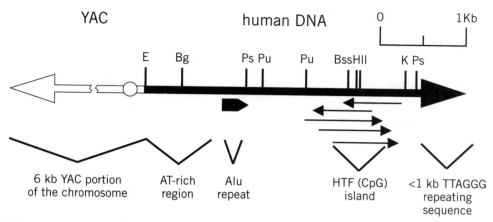

**Figure 8.35** Structure of a cloned human telomere, as isolated in yeast. Clone yHT1 is a chimeric minichromosome containing human DNA (filled arrow) and yeast DNA (open arrow). Arrowheads represent regions of telomeric simple repeats. The open circle represents a yeast centromere. Locations of *Eco*R I(E), *Bgl* II (Bg), Pst I (Ps), *Pvu* II (Pv), and *Kpn* I (K) restriction sites are shown above the map. The position of an *Alu* repetitive element is indicated by a solid arrowhead below the map, which points in the direction of the poly(dA) stretch. (From Cheng et al., 1989.)

---

**BOX 8.2**
**CLONING LARGE DNA FRAGMENTS AS**
**ARTIFICIAL CHROMOSOMES**

Both yeast and bacterial large fragment cloning systems are developed. Yeast artificial chromosomes (YACs) are constructed using small plasmids that are grown in *E. coli,* as shown in Figure 8.36. The YAC vectors have telomere sequences at each of their ends, a centromere sequence, and usually a DNA replication origin and selectable marker in each arm of the vector. Large fragments are cloned into a multicloning site occurring between the two arms of the YAC. Initially both arms of the YAC were cloned into the same plasmid. This meant that the plasmid was cleaved both at the cloning site and at a site between the converging telomeric sequences. Subsequently the vector arms were divided between two plasmids. The vector arms are ligated to high molecular DNA that has been partially digested with a restriction enzyme like *Eco*R I. The recombinant DNAs are introduced into *S. cerevisiae* cells chemically treated so that they could take up DNA. YACs up to about 1 Mb have been created.

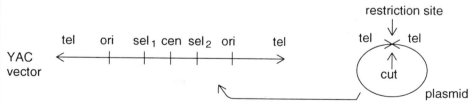

**Figure 8.36** General design of yeast artificial chromosome (YAC) cloning vectors and how they are propagated in bacteria: tel, telomere, ori, replication origin, cen, centromere, sel1 and sel2, selectable markers.

*(continued)*

**BOX 8.2** *(Continued)*

Even though transformation frequencies were very low (hundreds per mg of input DNA), eventually most of the human genome was cloned in this manner. The YAC libraries have facilitated the cloning of many human genes. YACs are not as easy to manipulate as bacterial clones, and many are chimeric, containing DNA from more than one region of the human genome, or contain interstitial deletions.

The large insert bacterial cloning systems are based on well-characterized autonomously replicating extrachromosomal DNA elements of *E. coli*. P1 artificial chromosomes (PACs) use vector sequences for P1 bacteriophage (Box 2.3), whereas bacterial artificial chromosomes (BACs) are based on the fertility factor (F-factor, Figure 8.37). P1 bacteriophage replicates as an extrachromosomal low-copy plasmid and as a high-copy lytic phage. Induction of the lytic replicon in PACs ensures that large amounts of recombinant PAC DNA can be synthesized. Although both the P1 bacteriophage and F-factors genomes are 100 kb in size, only a small portion of these genome codes for essential functions and can be deleted for cloning. Up to about 2.5 Mb of *E. coli* DNA has been known to be stably maintained in F-plasmids. Initially PACs were developed so that efficient transfection systems could be used to introduce recombinant DNA into cells. This limited the recombinant DNA size to ~100 kb so that it could be packaged into bacteriophage particles. Later recombinant PAC DNAs up to about 250 kb were introduced into cells using electroporation. Unlike YACs which are linear DNA molecular, PACs and BACs must be circular to be stable in *E. coli*. The efficiency of circulation in vitro by DNA ligase decreases with the size of the molecule. Hence BAC and PAC systems were developed that took advantage of a P1 coded site-specific recombination enzyme, namely cre recombinance. This enzyme promotes recombination between two loxP sites. Cre-promoted recombination circularizes a linear DNA fragment containing loxP sites at both ends. Genes specifying proteins such as the cre enzyme are moved to the host chromosome to minimize the BAC or PAC vector sequences and to allow for independent gene expression. Bacterial-based BAC or PAC cloning systems are easier to manipulate than YAC systems, appear to contain much fewer rearrangements and allow for easy manipulation of clone DNA.

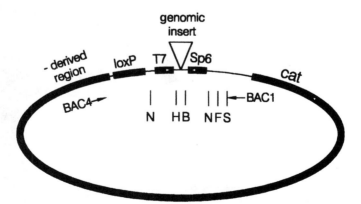

**Figure 8.37**   A typical bacterial artificial chromosome (BAC) cloning vector. It contains a loxP sequence to promote circularization, and bacteriophage T7 and Sp6 promoters to allow strand-specific transcription of the cloned insert. Also shown inside the circle are restriction enzyme cleavage sites and useful PCR primers.

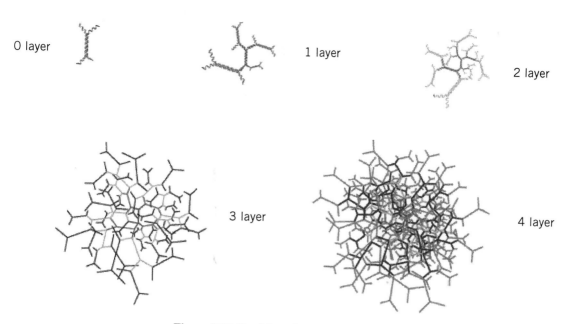

0 layer

1 layer

2 layer

3 layer

4 layer

**Figure 3.27**  Dendrimer layer growth.

**Figure 7.16**  An example of the sorts of results obtainable with current procedures for interphase FISH. (From Heiskanen et al., 1996.)

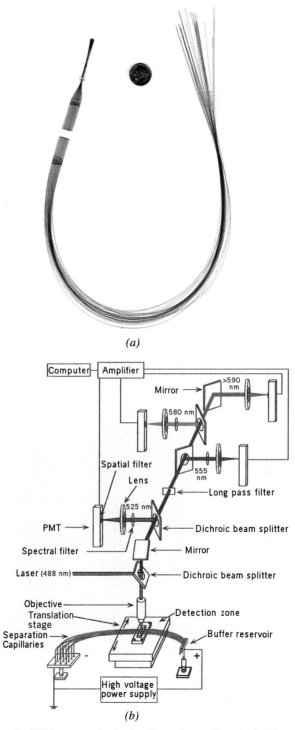

**Figure 10.19** Apparatus for DNA sequencing by capillary electrophoresis. *(a)* An array of gel-filled capillaries used for DNA sequencing. *(b)* On-line detection by confocal scanning fluorescence microscopy. Figure provided by Richard Mathies.

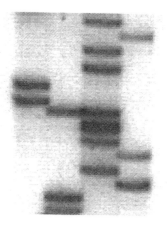

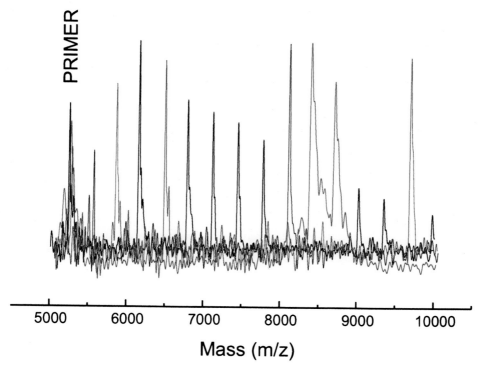

**Figure 10.26** Sequencing a compression region in the beta globin gene by gel electrophoresis *(top)* and MALDI-TOF mass spectrometry *(bottom)*. Note in the gel data a poorly resolved set of G's topped by what appears to be a C, G heterozygote. The true sequence, GGGGC, is obvious in the mass spectrum. Provided by Andi Braun.

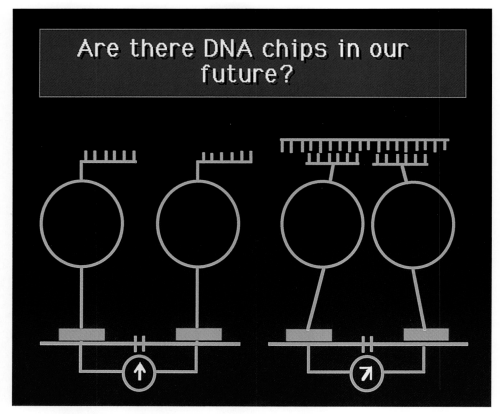

**Figure 12.24** Possible future direct reading oligonucleotide hybridization chip.

by yeast telomeric simple sequence repeats. This indicates that the human simple se-quence is unstable in yeast but that some feature of it is eventually recognized by the yeast telomerase, which then converts the end of the clone into a yeastlike telomere.

The particular clone that we first isolated turned out to be useful in mapping a few Mb of DNA at the long-arm telomeres of human chromosomes 4 and 21. However, it could only be used in a hybrid cell line because no where on this clone could any single-copy human DNA be found. Since this early work, others have systematically cloned and char-acterized single-copy probes from a number of human telomeres. These findings promise to be useful also for clinical diagnostics, since DNA rearrangements at telomeres are not uncommon in diseases like cancer.

## OPTICAL MAPPING

Recently an optical method for restriction mapping has been described that could speed the process considerably (Kai et al., 1995; Samad et al., 1995). This method has success-fully been applied to bacteriphage and YAC clones and to natural yeast chromosomes. Recently it has been semi-automated and extended to even larger DNAs. In optical map-ping DNA molecules are elongated by gentle flow as they are fixed by capture onto poly-L-lysine derivatized glass surfaces. The restriction enzyme cleavage is used to fragment the fixed molecules. A small portion of the stretched chain relaxes at each cleavage site. This leaves a gap that is visible by fluorescence microscopy after staining the DNA sam-ples. The contour length of each fragment seen indicates its size. However, a very signifi-cant feature of this method is that the pattern of organization of the fragments is main-tained by the initial fixation, and thus the order of the fragments is immediately known. Because this is a single-molecule method, it can deal effectively with any sample hetero-geneity. If sufficient numbers of molecules are examined, a complete map of each class of species in the sample should be revealed.

## BOTTOM-UP LIBRARY ORDERING

The conventional approach to constructing an ordered library of clones is to fingerprint a dense set of samples and look for clones that share overlapping properties. A key variable in designing such strategies is the minimum fraction of the clones that must be in common in order for their overlap to be detectable. The smaller the overlap required, the fewer clones needed to produce an ordered set, and the faster the process proceeds (Fig. 8.38).

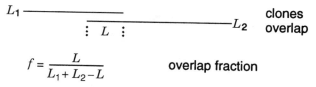

**Figure 8.38** The degree of overlap, $f$, of two clones of length $L1$ and $L2$ will determine the resolu-tion of the fingerprinting procedure needed to identify them.

Typical methods usually require a five- to tenfold redundant set of clones to ensure that there is a good chance of sampling the entire target at sufficient density to allow overlap detection. A simplistic estimate based on this redundancy would indicate that overlaps between clones should be perfectly scored at the 80 to 90% overlap level and perhaps half detected at the 40 to 50% overlap level. The key thing is to avoid false positives: predicting an overlap where none exists in fact. This is a serious error because it will result in the assignment of clones to incorrect regions of the target. A number of different methods have been used to detect overlaps in past studies. In all of these a key requirement is to have some sort of statistical way of determining the most likely set of overlaps in cases where there are ambiguities or potential inconsistencies. One very effective approach for evaluating overlap data is summarized in Box 8.3.

The earliest clone fingerprinting methods used restriction fragment sizes seen in single and double digests with 6-base specific enzymes like *Eco*R I and *Hind* III. This approach was first used to order bacteriophage lambda clones of *S. cerevisiae*. It was actually very difficult because of inaccuracies in sizing DNA fragments in the 1 to 10 kb range, particularly when results obtained on different gels electrophoresed on different days had to be compared. More accurate sizing is possible with smaller DNA fragments, but the number of such fragments that result from digestion of a bacteriophage or larger clone with a 4-base specific restriction enzyme is too large to allow all of them to be separated cleanly. One way around this problem is to use an end-labeling strategy as shown in Figure 8.39. This was first developed to order a cosmid library from the nematode *C. elegans*. First the clone is digested with a 6-base specific enzyme. The ends of the resulting fragments clone are labeled. Then a second digest is done with a 4-base specific enzyme. This results in DNA pieces that can be analyzed with single-base resolution of DNA sequencing gels, but their number is restricted to a manageable set.

An alternative approach for clone fingerprinting by restriction enzyme digestion is illustrated in Figure 8.40. This procedure was first used by Kohara in constructing an *E. coli* library in bacteriophage lambda clones. Indirect end labeling from probes in the vector sequence was used to determine the positions of restriction sites seen in separate partial digests, each generated with one of eight different restriction enzymes. The key advantage of this approach was that all eight digests were analyzed in a single gel electrophoresis. Under these circumstances, even though the DNA size information might be

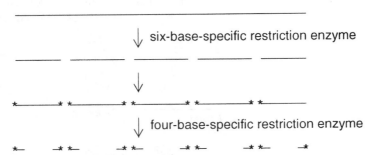

**Figure 8.39**  End-labeling procedure (asterisk) to produce a set of discrete DNA sizes which serve as a fingerprint.

**BOX 8.3**
## STATISTICAL EVALUATION OF CLONE OVERLAPS

A very effective method that can be used for clone analysis is an adaptation of the procedure originally developed by Branscomb et al. (1996) for contig building by restriction fragment fingerprinting. However, the method can also be applied to hybridization based fingerprinting like the *S. pombe* example discussed in Chapter 9. Clones are considered in pairs; each pair either overlaps ($O$), with a fraction of overlap $f$, defined in Figure 8.38, or does not overlap ($N$). Each piece of fingerprinting data available is tested for its support of the hypothesis that two clones overlap with a fraction $f$. The data can be a concordant or a discordant hybridization result with a particular probe or the presence or absence of particular size restriction fragments or restriction site order. Let $q$ represent the fraction of all of the clones that do not have a particular restriction fragment size or order (or alternatively that do not hybridize to a certain probe). Then $1 - q$ is the fraction that would be expected to show that fragment size (or hybridize to a certain probe) due to chance alone. We consider six cases in all: $P(+ +)$ means that both clones are detected, $P(+ -)$ means that only one clone is detected (scored positive), and $P(- -)$ means that neither clone is detected. Simple statistical considerations have been used to develop the following equations:

$$P(+ +, N) = (1 - q)^2$$

$$P(+ -, N) = 2q(1 - q)$$

$$P(- -, N) = q^2$$

$$P(+ +, O) = 1 - P(+ -, O) - P(- -, O)$$

$$P(+ -, O) = 2(1 - qq^{-f})q$$

$$P(- -, O) = q^2 q^{-f}$$

The first three equations are obvious because they derive from simple binomial statistics for two uncorrelated clones. The remaining terms are not rigorous, but they do model the statistics of coincident detection of partially overlapping clones approximately. For example, the sixth equation, if the overlap, $f$, is zero, becomes equal to the third equation. If the overlap is 1, $P(- -, O) = q$ which is reasonable, since in effect one now is dealing with only a single distinct clone. The fifth equation, in the limit $f = 0$, is equal to the second equation. In the limit $f = 1$, the fifth equation is zero which is the desired behavior because identical clones cannot be expected to show discordant behavior. The fourth equation, $P(+ +, O)$ also observes the correct limits. These equations have been used successfully in the construction of ordered libraries by fingerprinting and by hybridization. A more exact treatment must take into account the relative lengths of the probes and the clones in hybridization assays. For example, if two clones partially overlap, a probe larger than the clones has a higher chance of detecting both than a probe smaller than the clones.

*(continued)*

**BOX 8.3** *(Continued)*

To use the above equations one must evaluate $P(- -, O)$ and $P(+ -, O)$ before $P(+ +, O)$ can be calculated. When these results are considered for the entire set of restriction digests or hybridization results seen, we can compute the likelihood, $L$, of any particular overlap value, $f$, for each pair of clones as

$$L(f) = \prod_n \frac{P(X_n, O)}{P(X_n, N)}$$

where the outcome for the comparison of the two clones with the $n$th test is $x_n$, and the product is taken over the entire set of tests. The calculation is very time intensive because all pairs of probes must be considered separately for all probes and all overlap values. The procedure can be simplified by using results with longer clones to arrange the shorter clones into bins (see Fig. 9.4). Then overlap calculations need to be carried out only within bins and between the edges of neighboring bins.

imprecise, the relative order of the different restriction sites would be known with great accuracy. This order information was the major source of data used to fingerprint the clones.

Some of these approaches can be improved upon by using the power of current automated four-color DNA sequence readers (see Chapter 10). An example is shown in Figure 8.41. Here restriction fragments are filled in by end labeling in order to reduce the number of fragments seen to a manageable level, as just described. In this case, however, instead of just labeling with a single color (radioactive phosphate), a restriction site is filled in with a mixture of four different colored dpppNs. By chosing a restriction enzyme that cuts outside of its recognition sequence, as shown in the figure, one develops a very informative fingerprint of the end of the restriction fragments labeled. Now, instead of knowing that two clones share a common length restriction fragment, one learns that they share this length, plus they share a particular terminal sequence that will only occur, by chance, in 1/16 of the sites cleaved by the particular enzyme used. Thus the chances of distinguishing true overlaps, from accidental similarities, become greatly enhanced.

The example just described shows the power of obtaining "color" information about a clone; that is, more information beyond just fragment sizes. Another way to do this is shown schematically in Figure 8.42. Here each clone is digested with several restriction enzymes, the fragments are separated by electrophoresis, and the resulting gel is blotted and hybridized with several different interspersed repeating sequence probes. These probes provide a signature that goes beyond pure size measurements and indicates those sizes that contain particular repeats. This adds considerable information to each analysis, and it makes it much less likely that coincident similarities will be scored as false positives. The kind of clarity with which this approach allows overlaps to be viewed is shown by the example in Figure 8.43.

Several other powerful variations of clone fingerprinting have been described, and some of these are now being tested intensively. If repeats can be nulled out, then the use of clones or sets of clones as hybridization probes against other clones or sets of clones becomes an effective fingerprinting method. This approach will be described in detail in

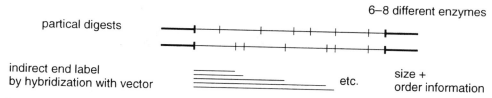

**Figure 8.40** Fingerprinting clones by Smith-Birnstiel restriction mapping. An indirect end label is used to probe the pattern of fragments seen in a set of different, separate partial digests.

the next chapter. An extreme version is to use individual or mixtures of arbitrary, short oligonucleotides as hybridization probes to fingerprint individual clones. In the limit of this approach, one would actually determine the DNA sequence of all of the clones by repeated hybridization experiments. The powers and limitations of oligonucleotide hybridization fingerprinting will be discussed in Chapter 12. Here it is sufficient to note that this approach has worked well in the construction of ordered libraries.

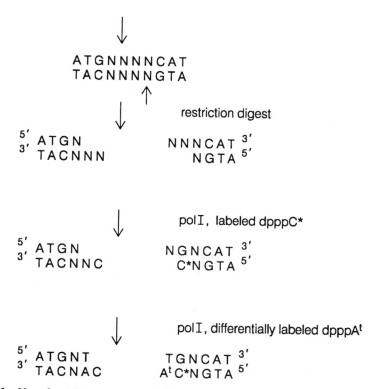

**Figure 8.41** Use of restriction enzymes with imperfectly defined cutting sites to label different restriction sites with characteristic colors. This greatly increases the informativeness of the fingerprint pattern generated by the sizes of these fragments.

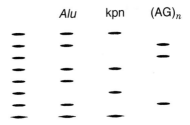

total restriction digest

**Figure 8.42** Fingerprinting a clone by hybridization with different repeated DNA sequences. See Chapter 14 for a description of these sequences.

hybridization pattern

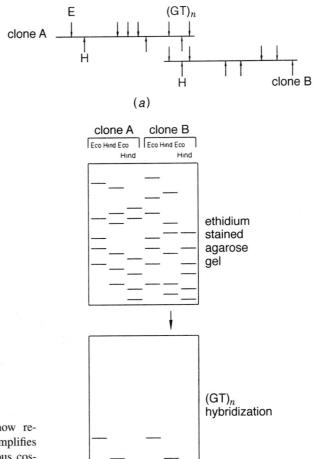

(a)

**Figure 8.43** An example of how repeated sequence hybridization simplifies the identification of two contiguous cosmid clones. (a) Two overlapping clones. (b) Restriction fragments and Southern blot. (Adapted from Stallings et al., 1990.)

(b)

# MEASUREMENTS OF PROGRESS IN BUILDING ORDERED LIBRARIES

The process of assembling contigs by fingerprinting clones can be treated in relatively straightforward mathematical ways. One makes the key assumption that the genome is being sampled uniformly, and sets, as a parameter, the degree of overlap between two clones necessary to constitute positive evidence that they are contiguous. Lander and Waterman have modeled this process of clone ordering. The sorts of results they obtained are shown in Figure 8.44. It is assumed that clones are fingerprinted one at a time, and the number of clones assembled into contigs of two or more clones is plotted as a function of the number of clones fingerprinted. At early times in the project, there are almost no contigs because the odds of picking overlapping clones, chosen at random, are small. Eventually overlaps start to build up, but most contigs contain just two clones. These begin to coalesce into larger contigs as the genome is sampled deeper and deeper. However, the effectiveness of the contig building begins to saturate long before all clones are assembled into a single contig. This saturation is partly determined by the lack of completeness of the library; if any regions are not represented at all, contigs cannot be built across them. The saturation is also a function of the effectiveness of overlap detection; due to chance, some clones that actually are contiguous may not overlap enough to be counted as a positive score. Several ongoing programs in contig building have been evaluated by the Lander-Waterman approach. Actual progress on these projects is in remarkably good agreement with predictions.

Eventually the pure bottom-up approach must be abandoned if a complete ordered library is desired. The point at which a switch in strategy is profitable is said to be somewhere between 60% and 90% coverage, when almost all progress in typical bottom-up mapping stops. The early stages of bottom-up mapping are very efficient. DNA preparations, fingerprinting, and data analysis have all been completely automated for some of the schemes we have described. Contig assembly is also done by computer software. Once the saturation point is reached, a typical project will still have hundreds or thousands of separate contigs. The challenge is to close the gaps between them in an efficient way. Several different approaches are useful at this stage. The contigs can be ordered by FISH localization of individual clone representatives from each contig. Once one knows that two contigs are very close to each other, frequently overlap data that were marginal

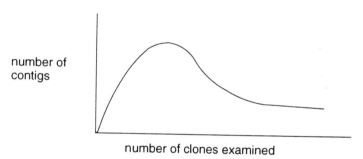

**Figure 8.44**    Progress in a pure bottom-up clone-ordering strategy, as calculated from the Lander-Waterman model. Plotted is the number of contigs as a function of randomly chosen clones examined.

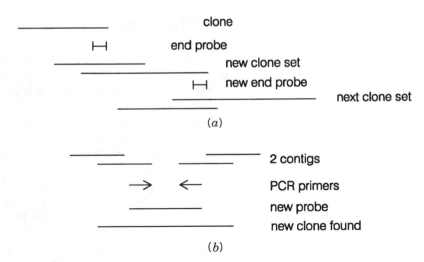

**Figure 8.45**    Two strategies for finishing the construction of contig maps. (*a*) Walking by probing existing or new libraries of clones with the ends of existing contigs. (*b*) Attempting to PCR across the gaps between two contigs suspected of being adjacent on the map (generally hinted at from other data such as FISH results).

before can now be used to fuse the contigs. The easiest way to fill major gaps where they are suspected is to switch to another library. Here regional assignment of clones from that library (or microdissection, Chapter 7) can help to focus on clones most likely to lie in regions where contiguity is not yet established.

A generally useful endgame strategy is to use existing contigs to screen a library of clones and subtract out those that have already been found. This greatly improves the odds of finding new, useful clones, once additional random picking from the remainder is reinitiated. Perhaps the single most useful method, once a dense set of contigs exists, is walking (Fig. 8.45*a*). Here one takes clones from the ends of existing tiling path contigs and uses them to screen libraries. Both the original library and totally new libraries can be used. The goal is to identify new clones that allow the contig to be extended. It is often particularly useful to change from one type of library to another in the walking process. Frequently a gap will exist because the sequence within it is not cloneable, say in cosmids, but it may be easily cloneable in YACs, and vice versa. Multiplex walking methods have been described that allow the simultaneous walking from many contig ends.

A final useful endgame strategy is to sequence the ends of contigs. Sequence information is much more robust than any other kind of fingerprinting. Even if two clones overlap by as few as 15 base pairs, sequence information can determine that they actually overlap. Sequence information at the ends of contigs can also be used to design PCR primers that face outward from the contigs (Fig. 8.45*b*). These primers can be used to test systematically whether two contigs suspected of being located near enough to each other are actually within a few kb apart. This technique turns out to be extremely powerful, in practice, because in actual projects, thus far, many of the hardest to close gaps turn out to be very small, and PCR can be carried out across them.

## SURVEY OF RESTRICTION MAP AND ORDERED LIBRARY CONSTRUCTION

Complete macrorestriction maps have been produced for a number of prokaryotic genomes, some simpler eukaryotic genomes, and sections of complex genomes. The first of these maps, a *Not* I map of *E. coli,* is shown in Figure 8.46. The most complex of all these maps, that for human chromosome 21q, is shown in Figure 8.47. A number of features of this map are of interest. Note that small *Not* I fragments and large *Not* I fragments tend to cluster. This must eflect wide oscillations in the density of HTF islands along the chromosome, since *Not* I sites occur almost exclusive in these islands.

The *Not* I map of human chromosome 21 was actually executed, not in a single cell line but in a set of eight cell lines. Polymorphisms among these lines were helpful in establishing the map as described earlier in the chapter. The full pattern of polymorphisms is illustrated in Figure 8.48. While the extent of polymorphism is considerable, almost all of it is consistent with varying degrees of methylation in the cell lines studied. There is little or no compelling evidence for major shifts in the lengths of DNA between existing *Not* I sites. Most important, there is no evidence that any significant amounts of DNA have been rearranged or lost in these cell lines.

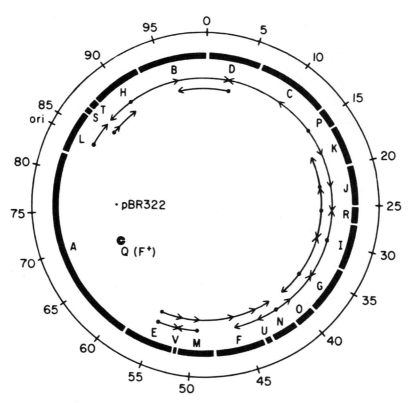

**Figure 8.46**  *Not* I restriction map of *E. coli.* (Adapted from Smith et al., 1987.)

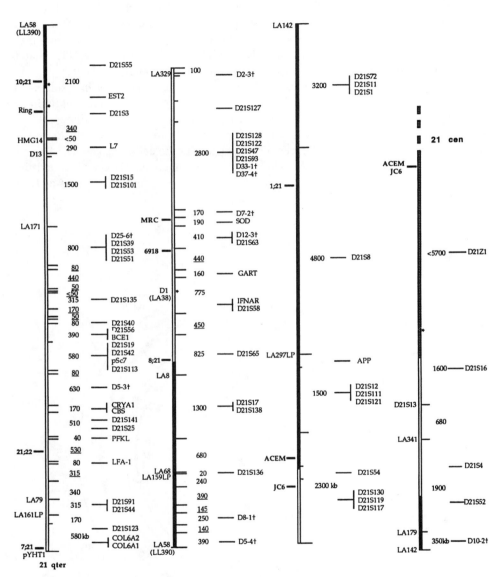

**Figure 8.47**  *Not* I restriction map of the long arm of human chromosome 21. (Taken from Wang and Smith, 1994.)

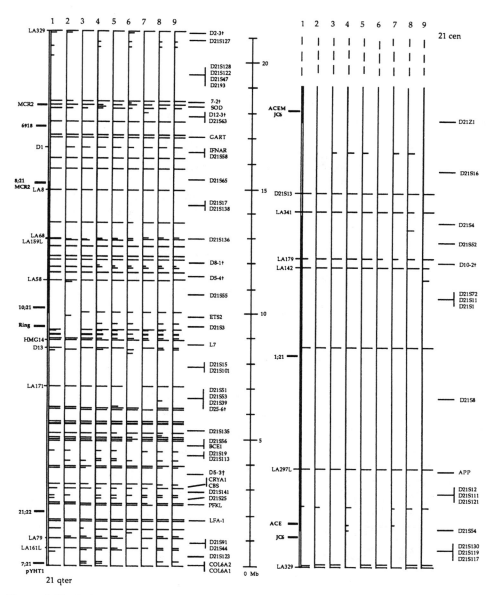

**Figure 8.48** Polymorphisms seen in the *Not* I map of human chromosome 21q in nine different cell lines (lanes 1 to 9). (Taken from Wang and Smith, 1994.)

A number of successful projects have been reported that have produced complete, or almost complete ordered clone libraries. The first of these was the ordered bacteriophage lambda library covering the *E. coli* genome. Other model organisms now mapped include the yeasts *S. cerevisiae*, and *S. pombe*, and the nematode *C. elegans*. Extensive map data also exists for Drosophila and for the human genome. A relatively complete YAC map covering the informative part of the Y chromosome has been reported, and a complete YAC maps exist that cover most human chromosomes. Extensive cosmid ordering projects on chromosomes 16 and 19 are virtually complete. Gaps not covered in cosmids are mostly covered in YACs or BACs.

An example of some of the data used to construct the chromosome YAC 21 map is shown in Figure 8.49. It is apparent that at the present stage some of the overlap evidence would be strengthened by interpolating results from additional clones, and some YACs used show evidence of rearrangements that are potential sources of error. Indeed, when the YAC contig for chromosome 21 is compared with the *Not* I restriction map, several YACs appear to be assigned to the wrong locations on the chromosome (Fig. 8.50). This is almost certainly partly the result of YAC chimeras which can seriously confuse clone ordering (see Chapter 9). Other discrepancies appear to result from the use of several probes with confused identities. Nevertheless, a remarkable amount of information and a goodly number of useful clones are now available for this chromosome.

A complete YAC map and three complete cosmid maps are available for the yeast *S. pombe*. The tiling path YACs from this map are shown in Figure 8.51, alongside the *Not* I restriction map of this organism and a sketch of the genetic map. This view, which presents a very simple looking map, hides the complex process that actually went into the construction of the map. Figure 8.52 illustrates the actual YAC clones

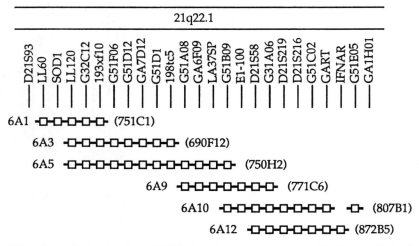

**Figure 8.49** A contiguous section of YACs from human chromosome 21. The contig is about 2.3 Mb long; 18 probes (STSs) were needed to assemble it. Note that several of the YACs appear to have internal deletions.

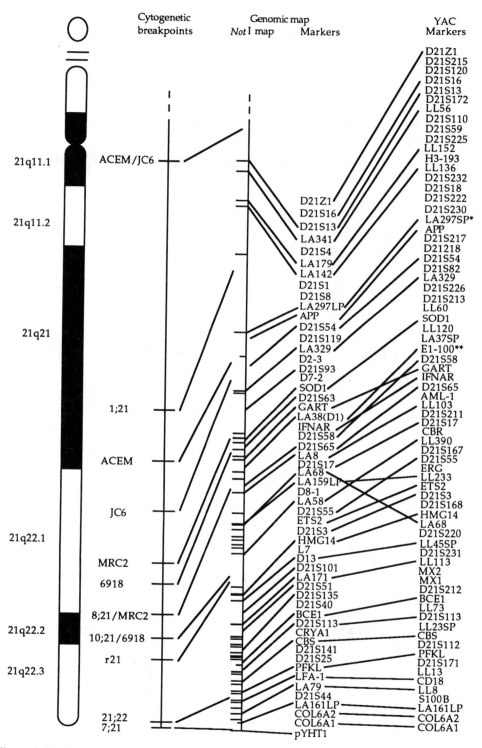

**Figure 8.50**  Comparison of marker order in the *Not* I restriction map of human chromosome 21 and the chromosome 21 YAC contig map. (Taken from Wang and Smith, 1994.)

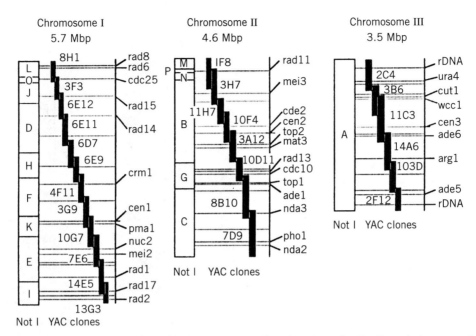

**Figure 8.51** Three maps of the fission yeast *S. pombe*. Plotted are the *Not* I restriction map, the 26-clone tiling set of a YAC contig map, and markers from the genetic map. Dotted lines indicate genetic markers and cosmids, which were hybridized to *Not* I digests of *S. pombe* and to YACs. (Taken from Maier et al., 1992.)

and probes studied along the way to map completion, and the selection of a simple tiling set. The large number of samples required, even for a simple organism, can barely be displayed as a legible figure. This should make it clear that any mapping project, with contemporary technology, is not to be undertaken lightly. The cosmid maps of *S. pombe* are even more complex and hard to display visually. Some details about the procedures that were used to construct one of these maps will be given in Chapter 9.

An issue that still leads to considerable debate is when to end a mapping project. How important is it to close the last gap, that is, to confirm the relative order within a contig to beyond any doubt? The simplest way to deal with this question is to recall the purpose of maps. We need them to access the genome, both for biological studies and for eventual DNA sequencing. A map that is 70% complete has seen only the beginning of the effort required to make a fully finished map—but it already provides access to 70% of the chromosome. A 90% map is frankly, for most purposes, almost as useful as a fully completely map, unless one is so unfortunate as to need clones or sequence data in some of the regions that are still in small fragments or contigs. In general, the usefulness of mapping projects grows very rapidly in the early stages and then begins to increase much more slowly as the maps near completion. It is important to consider this in deciding how much effort should be devoted to fitting in the last contig, as opposed to breaking out into new, uncharted territory on another chromosome or in another genome.

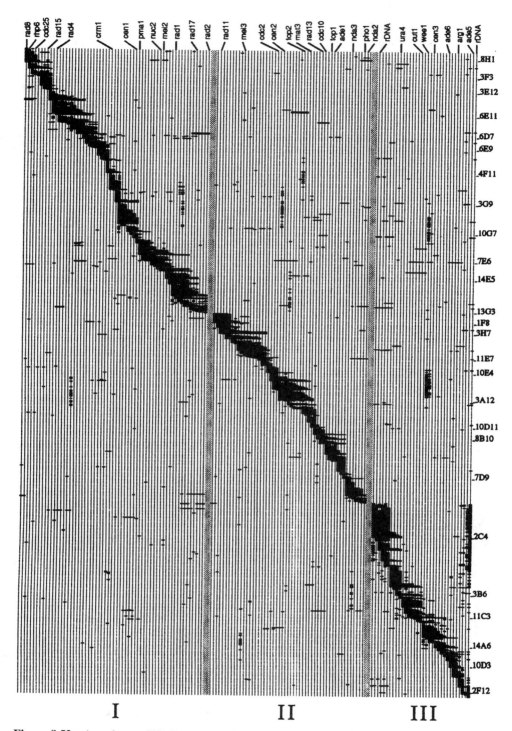

Probe labels (top, left to right): rad8, rhp6, cdc25, rad15, rad4, crm1, cen1, pma1, nuc2, mei2, rad1, rad17, rad2, rad11, mei3, cdc2, cen2, top2, mat3, rad13, cdc10, top1, ade1, nda3, pho1, nda2, rDNA, ura4, cut1, wee1, cen3, ade6, arg1, ade5, rDNA

YAC clone labels (right side, top to bottom): 8H1, 3F3, 3E12, 6E11, 6D7, 6E9, 4F11, 3G9, 10G7, 7E6, 14E5, 13G3, 1F8, 3H7, 11E7, 10E4, 3A12, 10D11, 8B10, 7D9, 2C4, 3B6, 11C3, 14A6, 10D3, 2F12

I   II   III

**Figure 8.52** Actual sets of YACs and probes needed to generate the YAC tiling set in Figure 8.51. YAC clones are shown on the vertical axis, where a subset of 26 clones spanning the entire genome is indicated. Probes are drawn on the horizontal axis; some of the genetic markers used are identified. Vertical gray bars separate the three chromosomes. Positive signal outside the contructed contigs indicates the locations of repeats. (Taken from Maier et al., 1992.)

283

## SOURCES AND ADDITIONAL READINGS

Branscomb, E., Slezak, T., Pae, R., Galas, D., Carrano, A. V., and Watermann, M. 1990. Optimizing restriction fragment fingerprinting methods for ordering large genomic libraries. *Genomics* 8: 351–366.

Cai, W., Aburatani, H., Stanton, V. P., Housman, D. E., Wang, Y. K., and Schwartz, D. C. 1995. Ordered restriction endonuclease maps of yeast artificial chromosomes created by optical mapping on surfaces. *Proceedings of the National Academy of Sciences USA* 92: 5164–5168.

Cheng, J. F., Smith, C. L., and Cantor, C. R. 1989. Isolation and characterization of a human telomere. *Nucleic Acids Research* 17: 6109.

Huang, M. E., Chuat, J. C., Thierry, A., Dujon, B., and Galibert, F. 1994. Construction of a cosmid contig and of an *Eco*R I restriction map of yeast chromosome X. *DNA Sequence* 4: 293–300.

Ioannou, P. A., Amemiya, C. T., Garnes, J., Kroisel, P. M., Shizya, H., Chen, H., Batzer, M. A., and de Jong, P. J. 1994. A new bacteriophage p1-derived vector for the propagation of large human DNA fragments. *Nature Genetics* 6: 84–89.

Jasin, M. 1996. Genetic manipulation of genomes with rare-cutting endonucleases. *Trends in Genetics* 12: 224–228.

Lennon, G. G., and Lehrach, H. 1991. Hybridization analyses of arrayed cDNA libraries. *Trends in Genetics* 7: 314–317.

Maier, E., Hoheisel, J. D., McCarthy, L., Mott, R., Grigoriev, A. V., Monaco, A. P., Larin, Z., and Lehrach, H. 1992. Complete coverage of the Schizosaccharomyces pombe genome in yeast artificial chromosomes. *Nature Genetics* 1: 273–277.

Oliva, R., Lawrance, S. K., Wue, T., and Smith, C. L. 1991. Chromosomes: Molecular studies. In Dulbecco, R., et al., eds., *Encyclopedia of Human Biology.* San Diego: Academic Press, pp. 475–488.

Palazzolo, M. J., Sawyer, S. A., Martin, C. H., Smoller, D. A., and Hartl, D. L. 1991. Optimized strategies for sequence-tagged-site selection in genome mapping. *Proceedings of the National Academy of Sciences USA* 88: 8034–8038.

Sainz, J., Pevny, L., Wu, Y., Cantor, C. R., and Smith, C. L. 1992. Distribution of interspersed repeats (*Alu* and *Kpn*) on *Not* I restriction fragments of human chromosome 21. *Proceedings of the National Academy of Sciences USA* 89: 1080–1085.

Samad, A., Huff, E. J., Cai, W., and Schwartz, D. 1995. Optical mapping: A novel, single-molecule approach to genomic analysis. *Genome Research* 5: 1–4.

Smith, C. L., Econome, J. G., Schutt, A., Klco, S., and Cantor, C. R. 1987. A physical map of the *Escherichia coli* K12 genome. *Science* 236: 1448–1453.

Stallings, R. L., Torney, D. C., Hildebrand, C. E., Longmire, J. L., Deaven, L. L., Jett, J. H., Doggett, N. A., and Moyzis, R. K. 1990. Physical mapping of human chromosomes by repetitive sequence fingerprinting. *Proceedings of the National Academy of Sciences USA* 87: 6218–6222.

Sternberg, N. 1992. Cloning high molecular weight DNA fragments by bacteriophage p1 system. *Genetic Analysis: Techniques and Applications* 7: 126–132.

Stubbs, L. 1992. Long-range walking techniques in positional cloning strategies. *Mammalian Genome* 3: 127–142.

Wang, D., Fang, H., Cantor, C. R., and Smith, C. L. 1992. A contiguous *Not* I restriction map of band q22.3 of human chromosome 21. *Proceedings of the National Academy of Sciences USA* 89: 3222–3226.

Wang, D., and Smith, C. L. 1994. Large scale structure conservation along the entire long arm of human chromosome 21. *Genomics* 20: 441–451.

Zhu, Y., Cantor, C. R., and Smith, C. L. 1993. DNA sequence analysis of human chromosome 21 *Not* I linking clones. *Genomics* 18: 199–205.

# 9  Enhanced Methods for Physical Mapping

## WHY BETTER MAPPING METHODS ARE NEEDED

In Chapter 8 we described the original top-down and bottom-up approaches that have led to the construction of a fair number of macrorestriction maps and ordered libraries. These methods are quite laborious, and it would be difficult to replicate them on very large numbers of mammalian genomes. New methods will be needed that are much more powerful if we are ever to be able to explore the full range of evolutionary diversity and the full range of human diversity by genome analysis. It seems fairly clear that in the future we will want the ability to go into any individual genome and obtain samples suitable for sequencing of large contiguous blocks of DNA. At least from the present perspective, this could require prior mapping studies to prepare the samples needed for subsequent sequencing. The key will be to develop approaches that allow a rapid focus on a particular region of interest and then a rapid collection and ordering of samples suitable for direct sequencing. It would be especially desirable if methods could eventually be developed that focus directly on map differences between individuals or species. Many future studies will doubtlessly be interested only in differences between two otherwise fairly homologous samples. Today techniques for effective differential mapping are unknown, and we will largely focus instead on methods for making direct mapping approaches much more efficient.

## LARGER YEAST ARTIFICIAL CHROMOSOMES (YACs)

YACs are a major tool currently used for making ordered libraries of large insert clones. The basic design and generation of YACs was described in Chapter 8. A major issue with YACs has been the size of the DNA insert. The first YAC libraries made had average insert sizes of 200 to 300 kb. This is a vast improvement over cosmid clones when used in schemes for rapid walking. However, since the first libraries, the sizes of YACs have continued to grow steadily. At least two improvements in YAC design have assisted this. Early on, in YAC development it was noted that mammalian DNA was not always rich in sequences that could serve serendipitously as yeast replication origins (autonomously replicating sequences). The original YAC vectors had only a single origin in one arm of the YAC vector. Requiring that this origin replicate the entire chromosome places potentially severe kinetic constraints on the viability of the chromosome. This problem can be alleviated considerably by building authentic YAC origins into both vector arms.

A second technique for increasing the size of YAC inserts has been to size-fractionate the DNA to be cloned both before and after ligation of YAC vector arms. The ligation is normally carried out in a melted agarose sample. Under these conditions Mb DNA is quite susceptible to shear breakage, which increases as the square of the length of the DNA. Any DNA that is fragmented by shear breakage or nuclease contamination during the ligation procedure, and also any contaminating vector arms, will be eliminated by the second size-fractionation. This is important because, otherwise, large numbers of vector arms will contaminate the true YACs. Since these carry the selectable markers, and can recombine with yeast chromosomal DNA, they lead to a high background of useless clones. Several groups have reported the construction of YAC libraries with average insert sizes of 500 to 700 kb and even larger. However, the greatest success has been seen with a continuing effort at Genethon to make larger and larger YACs. This has resulted in a series of libraries with average insert sizes of 700 kb, 1.1 Mb, 1.3 Mb, and 1.4 Mb. The largest insert libraries resulted from an extensive effort at Genethon. These had average insert sizes in excess of 1 Mb. The protocols for producing these megaYACs do not seem to be reproducible at this stage. Instead, by having the same team concentrate on the repeated construction of YAC libraries, the quality of these libraries appeared to improve on average, for unknown reasons as the team gained more experience. All of these libraries are made from PFG-fractionated *Eco*R I partial digests of genomic DNA. Usually DNA is transformed into yeast by electroporation.

The major problem with megaYACs (and most YAC libraries for that matter) are rearranged clones. These include chimeric clones, clones with deletions, and clones with insertions of yeast DNA. These are illustrated in Figure 9.1. Deletions and yeast insertions make it difficult to use the YACs directly as DNA sources for finer mapping or sequencing. However, such clones are still useful for the kinds of mapping strategies we will describe later in this chapter. Chimeric clones are more of a problem because they can lead to serious errors in mapping if they are not detected. The chimeric clones appear to contain two or more disconnected genomic regions. In some YAC libraries more than 50% of the clones are chimeras.

There are two potential origins for the occurrence of chimeras. Some may arise during ligation, especially if the insert DNA is at too high a concentration relative to the amount of YAC arms present. Co-ligation can be reduced substantially by more complex cloning strategies than used in early YAC library construction (Wada et al., 1994).

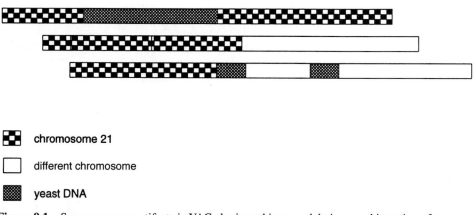

chromosome 21

different chromosome

yeast DNA

**Figure 9.1**   Some common artifacts in YAC cloning: chimeras, deletions, and insertion of yeast sequences.

For example, suppose that genomic DNA is partially digested with *Mbo* I in agarose, which produces fragments ending in

$$5'\text{———}$$
$$3'\text{———CTAG}$$

The single-stranded ends of the resulting sample are then partially filled in treatment with Klenow fragment DNA polymerase and dpppA and dpppG only. The resulting genomic DNA fragments will then have ends like

$$5'\text{———GA}$$
$$3'\text{———CTAG}$$

and thus they cannot be ligated to each other. In parallel, the YAC cloning vector is digested to completion with *Bam*H I to generate telomeres (see Box 8.1) and then digested with *Sal* I to yield fragments that end in

$$5'\text{———G}$$
$$3'\text{———CAGCT}$$

The ends of these are then partially filled in by treatment with Klenow DNA polymerase in the presence of only dpppT and dpppC. This yields fragments ending in

$$5'\text{———-GTC}$$
$$3'\text{———-CAGCT}$$

Now the vector arms produced cannot ligate to each other, but they are still capable of ligating to the genomic DNA fragments prepared as described above.

A second source of chimeras will arise from recombination. In preparing yeast for DNA transformation, usually a small fraction of the cells is rendered competent to pick up DNA, and it is not at all uncommon for these cells to pick up several YACs. Usually the YACs will separate in subsequent cell divisions. Occasionally a cell will stably maintain two different YACs. However, since mitotic recombination is very prevalent in yeast, two different YACs can recombine at shared DNA sequences, and as a result two chimeric daughters are produced. If each of these retains a centromere, they will usually segregate to separate daughter cells, each of which will now maintain a different single chimeric YAC (Fig. 9.2). Evidence for such recombination between two YACs has been obtained in at least one case where the two original inserts corresponded to DNA of known sequence, and thus the site of the recombination event could be identified. Alternatively, a dicentric YAC and an acentric YAC could be produced by a recombination event. In this case the latter clone will be lost, and the former may break unless one of the centromeres becomes inactivated.

Human DNA is likely to be a very favorable recombination target in yeast because of the large amount of interspersed repeated DNA sequences. This can also lead to instabilities within a single YAC which may lose part of its insert by an intramolecular recombination event. Just how prevalent these rearrangements are in particular libraries is not

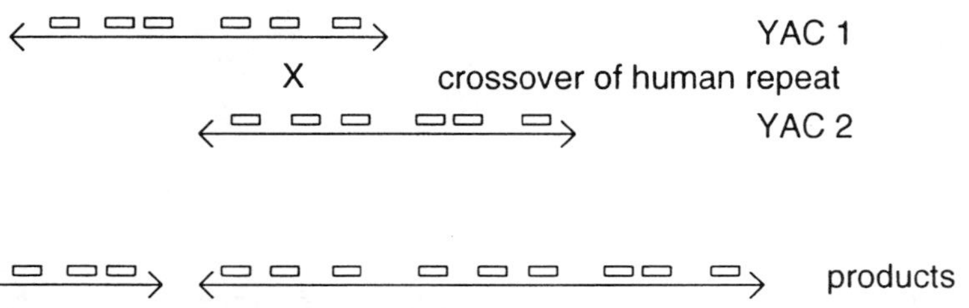

**Figure 9.2**   Recombination between human repeated sequences (shown as boxes) as a mechanism for the production of chimeric YACs.

known, but there are reasons to think that these are serious problems. The yeast strains used for almost all current YAC library construction have not had their recombination functions disabled. When a recombination deficient yeast host was used, a dramatic decrease in the fraction of chimeras was observed (Ling et al., 1993). Additional indirect evidence that recombination and not co-cloning is the major cause of chimeric YACs comes from observations on libraries made from hybrid cell lines. In most of these cases, the amount of chimerism is very low. This presumably arises because the human DNA is much more dilute in these samples, and human-rodent recombination is much less efficient because most repeated sequences are not well conserved between the two species.

When YACs are prepared directly from DNA obtained from flow-sorted chromosomes, the frequency of chimeras is also quite low. This is partly a result of the low DNA concentrations, which will diminish coligation and recombination events. However, the decreased chimera frequency also is likely to reflect the fact that in these samples all of the DNA preparation and manipulation was carried out in agarose. Agarose will reduce the number of DNA fragments with broken, unligatable ends, which are highly recombinogenic in yeast.

## HOW FAR CAN YACs GO?

Larger insert clones facilitate mapping projects in several ways. The number of clones one needs to order to fill the minimum tiling path is reduced. This greatly simplifies the process of clone ordering. If one has a fingerprinting method that requires a certain absolute amount DNA for demonstrating an overlap, the larger the clone the smaller a fraction of the clone this amount represents. Finally larger clones can easily be used to order clones one-half to one-third their size. Thus, as we will describe later, by having a tiered set of samples, the whole process of ordering them is greatly facilitated. The limit of the tier is determined by the largest stable and reliable clones that are available.

The chromosomes of *S. cerevisiae* range in size from 250 kb to 2.4 Mb. Thus the largest current YACs are in the midsize range of yeast chromosomes. It is not clear whether there is a size limit to yeast chromosomes. One issue already mentioned is the frequency of replication origins in the insert DNA. Early studies with artificial chromosomes in yeast indicated that stability, measured as retention over many generations of growth, actually increased sharply as a function of size, but these studies were not carried out up to the size range of current megaYACs. At some point the amount of foreign DNA that any organism can tolerate becomes limited by its competition for binding of key cellular enzymes or regulatory proteins. Where this limit occurs in yeast is unknown.

Some hints that *S. cerevisiae* can tolerate really large amounts of foreign DNA are available from studies with amplifiable YACs. The basic scheme behind such cloning vectors is shown in Figure 9.3*a*. Yeast centromeres can be inactivated if transcription occurs across them. To take advantage of this effect, a YAC vector arm has been designed with a strong, regulatable promoter extending toward a centromere. This vector is used to transform DNA into yeast in the ordinary way. The yeast is allowed to grow in the presence of selectable markers on the vector arms, first in the absence of transcription from the regulatable promoter. Then the promoter is activated, and growth is continued in the presence of selection. What happens is that with centromeres inactivated, the YACs segregate unevenly into daughter cells (Fig. 9.3*b*). Those daughters that receive many copies of the YACs have a selective advantage; those that receive very few copies are killed by the selection. The result is to increase, progressively, the average number of YACs per viable cell. This process continues up to the point where there are 10 to 20 copies of each YAC per cell. At this point the YAC DNA is 20 to 40% of the entire yeast DNA. This technique appears to be very promising, for it produces cells that are much easier to screen by hybridization than ordinary single-copy YACs. However, it has not yet been applied to libraries of megaYACs.

One potential improvement over standard YAC cloning methods might come from the use of *S. pombe* rather than *S. cerevisiae*. The former yeast has about the same genome size as the latter. However, *S. pombe* has only three chromosomes that range in size from 3.6 to 5.8 Mb. Based solely on this observation, it seems reasonable to speculate that *S. pombe* ought to be able to accommodate large YACs if there were a way to get them into the cell. One potential complication is that the centromeres of *S. pombe* and *S. cerevisiae* are very different in size. *S. cerevisiae* has a functional centromere that covers only a few hundred base pairs.

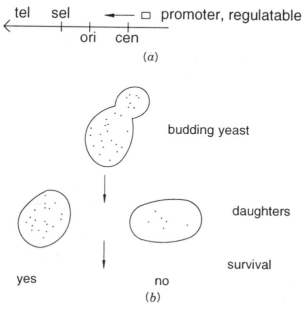

**Figure 9.3** Amplifiable YACs. (*a*) Vector used that allows regulation of centromere function. (*b*) Random segregation in mitosis leads to selective survival of cells with large numbers of YACs (dots).

In contrast, while the irreducible minimum centromere of *S. pombe* is unknown, past experience suggests that it could well be in excess of 100 kb. This would make the construction of cloning vectors for use in *S. pombe* very cumbersome.

An alternative cloning system for large DNA where considerable recent progress has been made uses bacterial artificial chromosomes or BACs (see Box 8.2). This system employs single-copy *E. coli* F-plasmids as vectors for DNA inserts. Natural F-factors can be a megabase in size. Thus BACs ought to have capacities in this size range if the resulting DNAs can be transfected into *E. coli* efficiently. The first BACs were rather small with inserts mostly in the 100- to 200-kb size range. However, recently larger BACs, with sizes from greater than 300 kb, have been reported. It remains to be seen how much this range can be enhanced by further modification and protocol optimization. BACs have the intrinsic advantage that the background *E. coli* DNA is only a third that of background yeast DNA. It is also relatively easy to purify the BAC DNA away from the host genomic DNA. Powerful bacterial genetic procedures can be used to manipulate the BAC sequences in vivo, and it is fair to say that comparable procedures can be used to manipulate YACs within their host cells. A key feature for both systems is that we now know the entire DNA sequence of both *E. coli* and *S. cerevisiae*. Procedures are likely to be developed and in place soon for direct DNA sequencing of large insert clones in one or both of these organisms. By knowing all of the host DNA sequence ahead of time, it will be possible to design sequencing or PCR primers in an intelligent, directed way. Any accidental host sequence that results from primer errors or homology will be immediately apparent once the putative clone sequence is compared with that of its host genome. Another bacterial large DNA cloning system that is in widespread use is the P1 artificial chromosome (PACs), described in Box 2.3.

## VECTOR OBSOLESCENCE

Based on past experience with ordinary recombinant DNA procedures over the past decade, the highly desirable vector of today is an inefficient, undesirable vector tomorrow. There is no way we can predict what the optimal vectors will be like five years from now; what bells and whistles they must have to facilitate the rapid mapping and sequencing procedures that will then be in use. To demonstrate how cloudy our crystal ball is in this respect, within five years the development of rapid methods to screen clones for possible functions is very likely. Just what form these screens will take, and what requirements they will impose on cloning vectors, are entirely unknown.

Imagine that tomorrow a vast improvement in some cloning vector has been achieved. All of the current map data and samples do not use this vector. How will we transfer the enormous number of samples used in genomic mapping from yesterday's obsolete vectors to the new ones? Certainly it will not be efficient to do this clone by clone. New strategies are badly needed that allow flexibility in the handling of samples to allow mass recloning or rescreening of entire ordered or partially ordered libraries to retain useful order information but equip the clones with the newly desired features. It is fair to say that today, while the problem is recognized, creative solutions to it are still lacking. We will either need clever selection, very effective automation for large numbers of separate samples, or very effective multiplexing that will allow many samples to be handled together and then sorted out in some very simple way afterward.

One consequence of the virtual certainty of vector obsolescence is that it is desirable to minimize the numbers of samples archived for storage and subsequent redistribution. Instead, it seems more efficient to develop procedures that will allow desired clones to be pulled easily from whatever new libraries are made. The advantage of PCR-based approaches is obvious in this regard. These approaches require storing only DNA sequence information that allows primers to be made whenever they are needed to assay for a given sequence in a sensitive way. Whatever library a desired DNA sequence is in, it should then be possible to find it in a relatively quick and inexpensive way, by PCR assays on pools of clones or hybridization assays on arrays of clones from that library, as we will describe later in this chapter. In this way no large numbers of samples need be stored for long time periods nor for mass distribution. Only pools of samples will have to be archived.

## HYBRID MAPPING STRATEGIES: CROSS-CONNECTIONS BETWEEN LIBRARIES

In any physical mapping project, sooner or later there is the need to handle a number of different types of samples. These include cell lines, radiation hybrids, and large restriction fragments for regional assignments and gaining an overview, as well as various clone libraries such as megaYACs, YACs, P1, and cosmid clones that actually form the eventual basis for DNA sequencing. Past projects have tended to concentrate on ordering at most a few of these samples across the chromosome, and then they resorted to using other types of samples in selected regions where these were needed to address particular problems. Based on these experiences, it now seems evident that much of the labor in handling all of these types of samples is preparing a dense set of labeled DNA probes or PCR primers (e.g., STSs *vide infra*) that are needed for most fingerprinting or mapping activities.

Once one has such probes or primers, an attractive scheme for ordering them is shown in Figure 9.4. The labeled probes or primers are used to interrogate, simultaneously, all samples that are of potential interest for the chromosome of question. If a dense enough set of probes or primers exists, and if the clone libraries are highly redundant, we will show that the result of the interrogation should be to order all the samples of interest in parallel. Ordering by any of the methods currently at our disposal involves finding overlap information. The larger the number of different samples used in the overlap procedure, the more likely one or more will cover the key region needed to form an informative overlap. This approach is neither top down or bottom up; in most respects it combines the best features of the two extreme views of map construction. They key issue is how to implement such a strategy in an efficient way. There are three basic issues: how to handle the probes, how to handle the samples, and how to do the interrogation.

Tens of thousands of DNA samples are involved in most large-scale mapping efforts. These cannot be handled routinely as individual liquid DNA preparations. One viable approach is to make dense arrays of DNA spots on filters. As an example consider the filter shown schematically in Figure 9.5. It consists of $2 \times 10^4$ individual cosmids. This must be prepared by an automated arraying device (Box 9.1). The key fact is that once the samples are prepared, the device can make as many copies of the filter as needed with relatively little additional effort. If each cosmid has an average of $4 \times 10^4$ base pairs of DNA, the array represents a total of $8 \times 10^8$ base pairs. This is a fivefold redundant coverage of

cross connections between DNA samples

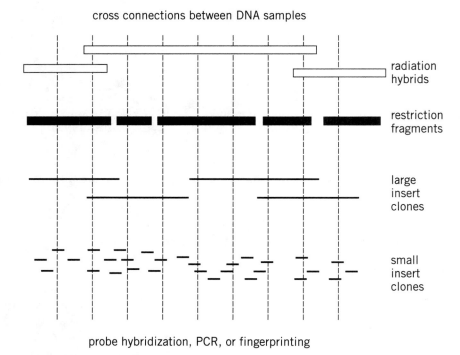

probe hybridization, PCR, or fingerprinting

**Figure 9.4**   Mapping by making cross-connections between a set of different DNA samples. The key variables are the density of available probes and the density or coverage of available samples.

an average, 150 Mb, human chromosome. It is sufficient for most mapping exercises that one can contemplate. For example, hybridization of a labeled probe to the filter will identify any cosmids that contain corresponding DNA sequences. All the cosmids can be examined in a single experiment, and if the signal to noise in the hybridization is good, the resulting data should be fairly unequivocal. We assume that any repeated DNA sequences in the probe will be competed out by methods described earlier in the book.

The alternative approach to making arrays is to make pools of samples. This has the potential advantage that the DNA is handled in homogeneous solution, which facilitates some screening procedures like PCR. It also has the advantage that a relatively small number of pools can replace a very large number of individual samples or spots on an array. Procedures for constructing these pools intelligently will be a major theme of this chapter. However, regardless of how they are made, a disadvantage of pools is that a single interrogation will not usually identify unique clone targets identified by a probe. Instead, one usually has to perform several successive probings or PCRs in order to determine which elements of particular pools were responsible for positive signals generated by the probe.

Hundreds to thousands of probes or primers are used in a large-scale mapping effort. The complexity of these samples is almost as great as that of the clones themselves. In the past most probes or primer pairs have been handled individually. Probes have consisted of small DNA clones, cosmids, YACs, large DNA fragments, or radiation hybrids. It is necessary to compete out any repeated DNA in these probes to prevent a background level of hybridization that would be totally unacceptable. PCR tricks abound that can be used to

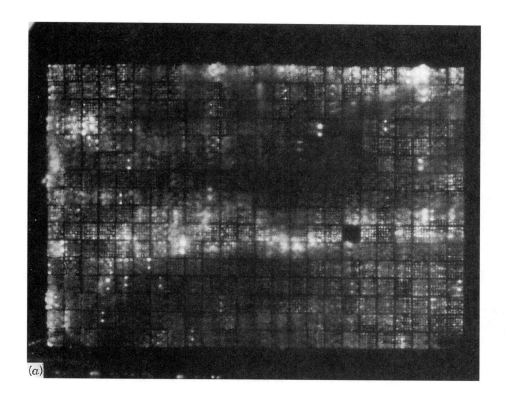

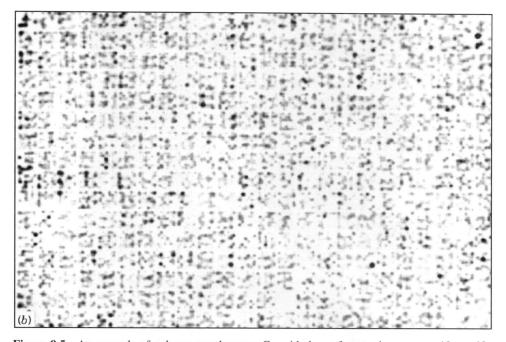

**Figure 9.5**  An example of a dense sample array. Cosmid clones from a chromosome 19-specific library were (*a*) arrayed by 36-fold compression from 384-well microtitre plates and (*b*) probed by hybridization with a randomly primed pool of five cosmids. (Unpublished work of A. Copeland, J. Pesavento, R. Mariella, and D. Masquelier of LLNL. Photographs kindly provided by Elbert Branscomb.)

## BOX 9.1
## AUTOMATED MANIPULATION OF SAMPLE ARRAYS

A considerable background of experience exists in automated handling of liquid samples contained in microtitre plates. While it is by no means clear that this is the optimal format for mapping and sequencing automation, the availability of laboratory robots already capable of manipulating these plates has resulted in most workers adopting this format. A typical microtitre plate contains 96 wells in an 8 by 12 format, which is about $3 \times 5$ cm in size (Fig. 9.6 bottom). Each well holds about 100 $\mu$l (liquid) of sample. Higher-density plates have recently become available: An 18 by 24 sample plate, the same size as the standard plate with proportionally smaller sample wells, seems easily adapted to current instruments and protocols (Fig. 9.6 top). A fourfold higher-density 36 by 48 sample plate has been made, but it is not yet clear that many existing robots have sufficient mechanical accuracy and existing detection systems sufficient sensitivity to allow this plate to be adopted immediately.

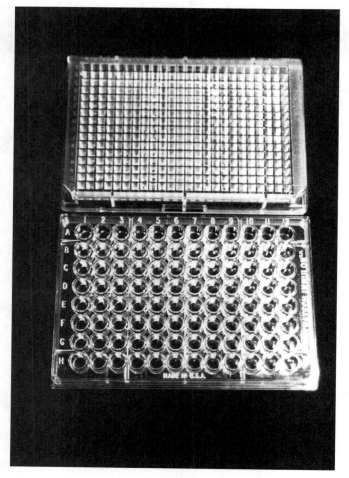

**Figure 9.6**    Typical microtitre plates: 384-well plate (top), 96-well plate (bottom).

*(continued)*

**BOX 9.1** *(Continued)*

Liquid-handling robots such as the Beckman Biomec, the Hewlett-Packard Orca chemical robot, among others, can use microtitre plates singly or several at a time. More complex sets of plates can be handled by storage and retrieval from vertical racks called microtitre plate hotels. A number of custom-made robots have also been built to handle specialized aspects of microtitre plate manipulation efficiently, such as plate duplication, custom sample pooling, and array making from plates. All of these instruments share a number of common design features. Plate wells can be filled or sampled singly with individual pipetting devices, addressable in the *x-y* plane. Rows and columns can be sampled or fed by multiple-headed pipetors. Entire plates can be filled in a single step by 96-head pipetors. This is done, for instance, when all the wells must be filled with the same sample medium for cell growth or the same buffer solution for PCR.

Most standard biochemical and microbiological manipulations can be carried out in the wells of the microtitre plates. A sterile atmosphere can be provided in various ways to allow colony inoculation, growth, and monitoring by absorbance. Temperature control allows incubation or PCR. Solid state DNA preparations such as agarose plug preparations of large DNA, or immobilized magnetic microbead-based preparations of plasmid DNA or PCR samples are all easily adapted to a microtitre plate format. The result is that hundreds to thousands of samples can be prepared at once. Standard liquid phase preparations of DNA are much more difficult to automate in the microtitre plate format because they usually require centrifugation. While centrifuges have been built that handle microtitre plates, loading and unloading them is tedious.

For microbiological samples, automated colony pickers have been built that start with a conventional array of clones or plaques in an ordinary petri dish (or a rectangular dish for more accurate mechanical positioning), optically detect the colonies, pick them one at a time by poking with a sharp object, and transfer them to a rectilinear array in microtitre plates. The rate-limiting step in most automated handling of bacteria or yeast colonies appears to be sterilization of the sharp object used for picking, which must be done after each step to avoid cross-contamination. With liquid handling, cross-contamination can also be a problem in many applications. Here one has the choice of extensive rinsing of the pipet tips between each sample, which is time-consuming, or the use of disposable pipet tips, which is very costly. As we gain more experience with this type of automation, more clever designs are sure to emerge that improve the throughput by parallelizing some of the steps. A typical example would be to have multiple sample tips or pipetors so that some are being sterilized or rinsed off line while others are being used. At present, most of the robots that have been developed to aid mapping and sequencing are effective but often painfully slow.

Dense sample arrays on filters are usually made by using offset printing to compress microtitre plate arrays. An example is shown in Figure 9.7. Here a 96-pin tool is used to sample a 96-well microtitre plate of DNA samples, and stamp its image onto a filter, for subsequent hybridization. Because the pin tips are much smaller than the microtitre plate wells, it is possible to intersperse the impressions of many plates to place all of these within the same area originally occupied by a single plate. With presently available robots, usually the maximum compression attained is 16-fold (a $4 \times 4$ array of 96 well images). This leads to a $3 \times 5$ cm filter area with about 1600 samples. Thus a $10 \times 10$ cm filter can hold about $10^4$ samples, more than enough for most current

*(continued)*

**BOX 9.1** (*Continued*)

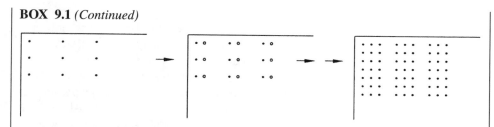

**Figure 9.7**   Making a dense array of samples by offset spotting of a more dilute array.

applications. Most dense arrays that have been made thus far for mapping projects are random. That is, no information is available about any of the DNA samples at the time the array is made. As the map develops, the array becomes much more informative, but the *x, y* indexes of each sample have only historical significance.

Once a map has been completed, it is convenient to reconfigure the array of samples so that they are placed in the actual order dictated by the map. While this is not absolutely necessary, it does allow for visual inspection of a hybridization to be instantly interpretable in many cases. There is no difficulty in instructing a robot to reconfigure an array. This procedure needs to be done only once, however slow the process, and then replicas of the new configuration can be made rapidly. The great motivation for achieving more compressed arrays is sample storage. Many clones and most DNA samples are stored at low temperature. Large libraries of samples can rapidly saturate all available temperature-controlled laboratory storage space, especially if a large number of replicas is made for subsequent distribution.

select out human material from hybrid cell lines or to reduce the complexity of a probe to desirable levels. Some of these techniques were described in Chapter 4; others will be dealt with in Chapter 14. It is worth noting that randomly chosen oligonucleotides often make very effective probes for fingerprinting. For example, a randomly chosen 10-mer should detect 1 out of every 25 cosmid clones. A key issue that we will address later in this chapter is whether it is possible to increase the efficiency of handling probes by pooling them instead of using them individually.

## SCREENING BY PCR VERSUS HYBRIDIZATION

A key variable in contemporary mapping efforts is whether the connections between probes and samples are made by hybridization or by PCR. The two techniques have compensating disadvantages and advantages, and the choice of which to use will depend on the nature of the samples available and the complexity of the mapping project. We will consider these differences at the level of the target, the probe, and the organization of the project.

The sensitivity of PCR at detecting small amounts of target is unquestionably greater than hybridization. The advantage of being able to use smaller samples is that with presently available methods, it is much easier to automate the preparation of relatively small DNA samples. It is also cheaper to make such samples. When pools of targets are used, the greater sensitivity of PCR allows more complex pools with larger numbers of samples to be used. What matters for detection is the concentration of the particular target

DNA that will be detected by one method or the other. With PCR this concentration can be almost arbitrarily low, so long as there are not contaminants that will give an unacceptable PCR background.

Hybridization has the advantage when a large number of physically discrete samples (or pools) must be examined simultaneously. We have already described the use of dense arrays of samples on filters to process in parallel large numbers of targets in hybridization against a single probe. These arrays can easily contain $10^4$ samples. In comparison, typical PCR reactions must be done in individually isolated liquid samples. Microtitre plates handle around $10^2$ of these. Devices have been built that can do thousands of simultaneous PCRs, but these are large; with objects on such a scale one could easily handle $10^5$ samples at once by hybridization.

At the level of the probe, PCR is more demanding than hybridization. Hybridization probes can be made by PCR or by radiolabeling any DNA sample available. Typical PCR analyses require a knowledge of the DNA sequence of the target. In contrast, hybridization requires only a sample of the DNA that corresponds to one element of the target. PCR primers require custom synthesis, which is still expensive and time-consuming although recent progress in automated synthesis has lowered the unit cost of these materials considerably. PCR with large numbers of different primers is not very convenient because in most current protocols the PCR reactions must be done individually or at most in small pools. Pooling of probes (or using very complex probes) is a powerful way to speed up mapping, as we will illustrate later in this chapter. However, in PCR, primers are difficult to pool. The reason is that with increasing numbers of PCR primers, the possible set of reactions rises as the square of the number of primers: Each primer, in principle, could amplify with any other primer if a suitable target were present. Thus the expected background will increase as the square of the number of primers. The desired signal will increase only linearly with the number of primers. Clearly this rapidly becomes a losing proposition. In contrast, it is relatively easier to use many hybridization probes simultaneously, since here the background will increase only linearly with the number of probes.

Both PCR and hybridization schemes lend themselves to large scale organized projects, but the implications and mechanics are very different. With hybridization, filter replicas of an array can be sent out to a set of distant users. However, the power of the array increases, the more about it one knows. Therefore, for an array to have optimal impact, it is highly desirable that all results of probe hybridizations against it be compiled in one centrally accessible location. In practice, this lends itself to schemes in place in Europe where the array hybridizations are actually done in central locations, and data are compiled there. Someone who wishes to interrogate an array with a probe, mails in that probe to a central site.

With PCR screening, the key information is the sequences of the DNA primers. This information can easily be compiled and stored on a centrally accessible database. Users simply have to access this database, and either make the primers needed or obtain them from others who have already made them. This allows a large number of individual laboratories to use the map, as it develops, and to participate in the mapping without any kind of elaborate distribution scheme for samples and without centralized experimental facilities. The PCR-based screening approach has thus far been more popular in the United States.

In the long run it would be nice to have a hybrid approach that blends the advantages of both PCR and hybridization. One way to think about this would be to use small primers or repeated sequence primers to reduce the complexity of complex probes by sampling either useful portions of pools of probes or large insert clones such as YACs. This would greatly increase the efficiency of PCR for handling complex pools and also decrease the cost of making large numbers of custom, specific PCR primers. What is still needed for an optimal method is spatially resolved PCR. The idea is to have a method that would allow one to probe a filter array of samples directly by PCR without having to set up a separate PCR reaction for each sample. However, in order for this to be efficient, the PCR products from each element of the array have to be kept from mixing. While some progress at developing in situ PCR has been reported, it is not yet clear that this methodology is generally applicable for mapping. One nice approach is to do PCR inside permeabilized cells, and then the PCR products are retained inside the cells. However, this procedure, thus far, cannot be carried out to high levels of amplification (Teo and Shaunak, 1995).

## TIERED SETS OF SAMPLES

In cross-connecting different sets of DNA samples and libraries, it is helpful to have dense sets of targets that span a range of sizes in an orderly way. The rationale behind this statement is illustrated in Figure 9.8. The figure shows an attempt to connect two sets of total digest fragments or clones by cross-hybridization or some other kind of complementary fingerprinting. When large fragment *J* is used as a probe, it detects three smaller fragments *a, b,* and *c* but does not indicate their order. When the smaller fragments *a, b,* and *c* are used as probes, *b* detects only *J*, which means that *b* is the central small fragment; *a* and *b* each detect additional flanking larger fragments, so they must be external. A generalization of this argument indicates that it is more efficient if one has access to a progression of samples where average sizes diminish roughly by factors of three. Less sharp size decreases will mean an unnecessarily large number of different sets of samples. Larger size decreases will mean that ordering of each smaller set will be too ambiguous.

The same kinds of arguments are applicable in more complex cases where overlapping sets of fragments or clones exist. Consider the example shown in Figure 9.9. The objective is to subdivide a 500 kb YAC into cosmid clones so that these can be used as more convenient sources of DNA for finding polymorphisms or for sequencing. The traditional approach to this problem would have been to subclone the purified YAC DNA into cosmids. However, this involves a tremendous amount of work. The more modern approach is to start with a YAC contig flanking the clone of interest. This contig will automatically arise in the context of producing an ordered YAC library. With a fivefold redundant array of YACs, the contig will have typically five members in this region. Therefore, if these members are used separately for hybridization or fingerprinting, they will divide the region into intervals that average about 100 kb in size. Each interval will serve as a bin to assign cosmids to the region. Since the intervals are less than three times the cosmid in-

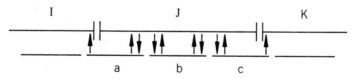

**Figure 9.8**  Clone ordering when tiered sets of clones are available.

sert size, the ordering should be reasonably effective. However, the key point in the over-all strategy is that one need not make the YAC contig in advance. Cross-connecting the various sets of samples in parallel will eventually provide all of the information needed to order all of them.

The example shown in Figure 9.9 is a somewhat difficult case because the size step taken, from 500 kb clones to 40 kb clones, is a big one. It would be better to have one or two tiers of samples in between; say 250 kb YACs and 100 kb P1 clones. This will compensate quite a bit for the inevitable irregularities in the distribution and coverage of each of the libraries in particular regions. An analogy that may not be totally farfetched is that the intermediate tiers of samples can help strengthen the ordering process in the same manner as intermediate levels in a neural net can enhance its performance (see Chapter 15).

YAC and cosmid libraries involve large numbers of clones, and it would be very ineffi-cient to handle such samples one at a time. We have already shown that cosmids can be handled very efficiently as dense arrays of DNAs on filters. YACs are less easily handled in this manner. The amount of specific sample DNA is much less, since typical YACs are single-copy clones in a 13-Mb genome background, while cosmids are multicopy in a 4.8-Mb background. Thus hybridization screening of arrays of YAC clones has not always been very successful. It would greatly help if there were an effective way to purify the YAC DNA away from the yeast genomic DNA. Automated procedures already have been developed to purify total DNA from many YAC-bearing strains at once (Box 9.1). Now what is needed is a simple automatable method for plucking the YACs out of this mixture. An alternative to purification would be YAC amplification. As described earlier in this chapter, this is possible but not yet widely used, and the amount of amplification is still only modest. Perhaps the most effective method currently available for increasing our sensitivity of working with YAC DNA is PCR. In Chapter 14 we will illustrate how human-specific PCR, based on repeating sequence primers like *Alu*'s, can be used to see just the human DNA in a complex sample. Almost every YAC is expected to contain mul-tiple *Alu* repeats. It is possible to do hundreds of PCR reactions simultaneously with com-monly available thermal cyclers, and ten to perhaps a hundred times larger number of samples is manageable with equipment that has been designed and built specially for this purpose.

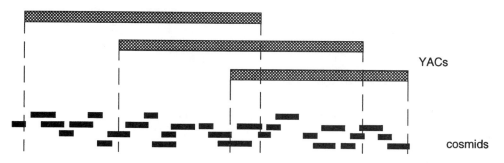

hybridization pattern arranges cosmids in (ordered) bins

**Figure 9.9** Division of a YAC into a contig of cosmids by taking advantage of other YACs known to be contiguous.

## SIMPLE POOLING STRATEGIES FOR FINDING A CLONE OF INTEREST

A major problem in genome analysis is to find in a library the clones that correspond to a probe of interest. This is typically the situation one faces in trying to find clones in a region marked by a specific probe that is suspected to be near a gene of interest. With cosmid clones, one screens a filter array, as illustrated earlier. With YAC clones, filter arrays have not worked well in many hands, and instead, PCR is used. But with today's methods it is inconvenient and expensive to analyze the YACs in a large library, individually, by PCR. For example, a single coverage 300-kb human genomic YAC library is $10^4$ clones. Fivefold coverage would require $5 \times 10^4$ clones. This number of PCR reactions is still daunting. However, if the objective is to find a small number of clones in the library that overlap a single probe, there are more efficient schemes for searching the library. These involve pooling clones and doing PCR reactions on the pools instead of on individual clones.

One of the simplest and most straightforward YAC pooling schemes involves three tiers of samples (Fig. 9.10). The YAC library is normally distributed in 96-well microtitre plates (Fig. 9.6$a$). Thus 100 plates would be required to hold $10^4$ clones. Pools are made from the clones on each plate. This can be done by sampling them individually or by using multiple headed pipetors or other tools as described in Box 9.1. Each plate pool contains 96 samples. The plate pools are combined ten at a time to make super pools. First, ten super pools are screened by PCR. This takes 10 PCR reactions (or 50 if a fivefold redundant library is used). Each positive superpool is then screened by subsequent, separate PCR reactions of each of the ten plate pools it contains. In turn, each plate pool that shows a positive PCR is divided into 12 column pools (of 9 YACs each) and, separately, 9 row pools (of 12 YACs each), and a separate PCR analysis is done on each of these samples. In most cases this should result in a unique row-column positive combination that serves to identify the single positive YAC that has been responsible for the entire tier of PCR amplifications. Each positive clone found in this manner will require around 41 PCR

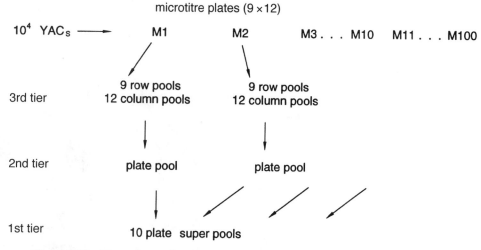

**Figure 9.10**   Three-tier pooling strategy for finding a clone of interest in a YAC library.

reactions. This is a vast improvement over the $10^4$ reactions required if YACs are examined one at a time. For a fivefold redundant library, if the super pools are kept the same size, then the total number of PCRs needed will be 81 per positive clone. This is still quite reasonable.

## SEQUENCE-SPECIFIC TAGS

The power of the simple pooling approach just described had a strong effect on early strategies developed for genome mapping. In reality the power is greatest when individual clones are sought, and it diminishes considerably when the goal is to order a whole library. However, the attractiveness of PCR-based screening has led to much consideration about the kinds of PCR primers that would be suitable for genome analysis. Since PCR ordinarily requires known sequences, the issue really becomes what kinds of DNA sequences should be used for finding genes or for genome mapping. A number of different types of approaches are currently being used; these are described below.

> *STS.* This is short for a sequence tagged site. The original notion was that any arbitrary bit of known DNA sequence could be used as a probe if it successfully generated useful PCR primers. One early notion was to take all of the existing polymorphic genetic probes and retro-fit these as STSs by determining a partial DNA sequence, and then developing useful PCR primers based on this sequence. This approach was never executed on a large scale, which is probably good in retrospect because recently developed genetic markers are far more useful for mapping than earlier ones, and the process for finding these automatically includes the development of unique sequence tags suitable for PCR.

> *STAR.* This stands for sequence tagged rare restriction site. The great utility of such probes has already been described in considering strategies for efficient restriction mapping or ordering total digest libraries. The appeal of STARs is that they allow precise placement of probes on physical maps, even at rather early stages in map construction. The disadvantage of STARs, mentioned earlier, is that many cleavage sites for rare restriction nucleases turn out to be very G + C rich, and thus PCR in these regions is more difficult to perform. For any kind of probe, clone ordering is most efficient if the probes come from the very ends of DNA fragments or clone inserts. As shown in Figure 9.11, this allows PCR to be done by using primers in the vector arms in addition to primers in the insert. Thus the number of unique primers that must be made for each probe is halved. STARs in total digest libraries naturally come from the ends of DNA fragments, and thus they promote efficient mapping strategies.

**Figure 9.11**   Use of vector primers in PCR to amplify the ends of clones like YACs.

*STP or STRP.* These abbreviations refer to sequence tagged polymorphism or polymorphic sequence tag. A very simple notion is involved here. If the PCR tag is a polymorphic sequence, then the genetic and physical maps can be directly aligned at the position of this tag. This allows the genetic and physical maps to be built in parallel. A possible limitation here is that some of the most useful genetic probes are tandemly repeating sequences, and a certain subset of these, usually very simple repeats like $(AC)_n$, tend to give extra unwanted amplification products in typical PCR protocols. However, it seems possible to find slightly more complex repeats, like $(AAAC)_n$, that are equally useful as genetic probes but show fewer PCR artifacts.

*EST.* This stands for expressed sequence tag. It could really refer to any piece of coding DNA sequence for which PCR primers have been established. However, in practice, EST almost always refers to a segment of the DNA sequence of a cDNA. These samples are usually obtained by starting with an existing cDNA library, choosing clones at random, sequencing as much of them as can be done in a single pass, and then using this sequence information to place the clone on a physical map (through somatic cell genetics or FISH). There are many advantages to such sequences as probes. One knows that a gene is involved. Therefore the region of the chromosome is of potential interest. The bit of DNA sequence obtained may be interesting: It may match something already in the database or be interpretable in some way (see Chapter 15). In general, the kinds of PCR artifacts observed with STPs, STRPs, and STARs are much less likely to occur with ESTs.

Despite their considerable appeal there are a number of potential problems in dealing with ESTs as mapping reagents. As shown in Figure 9.12, cDNAs are discontinuous samples of genomic DNA. They will typically span many exons. This can be very confusing. If the EST crosses a large intron, the probe will show PCR amplification, but genomic DNA or a YAC clone will not. A common strategy for EST production uses largely untranslated DNA sequence at the 3'-ends of the cDNA clones. It is relatively easy to clone these regions, and they are more polymorphic than the internal coding region. Furthermore cDNAs from gene families will tend to have rather different 3'-untranslated regions, and thus one will avoid some of the problems otherwise encountered with multiple positive PCR reactions from members of a gene family. These 3'-end sequences will also tend to contain only a single exon in front of the untranslated region. However, all of these advantages carry a price: The 3'-end sequence is less interesting and interpretable than the remainder of the cDNA.

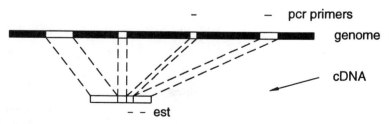

**Figure 9.12**  A potential problem with ESTs as mapping reagents is that an EST can cross one or more introns.

Eugene Sverdlov and his colleagues in Moscow have developed an efficient procedure for preparing chromosome-specific cDNA libraries. Their three-step procedure is an extension of simpler procedures that were tried earlier by others. This procedure uses an initial *Alu*-primed PCR reaction to make a cDNA copy of the hnRNA produced in a hybrid cell containing just the chromosome of interest. The resulting DNA is equipped with an oligo-G tail, and then a first round PCR is carried out using an oligo-C containing primer and an *Alu* primer. Then a second round of PCR is done with a nested *Alu* primer. The PCR primers are also designed so that the first round introduces one restriction site and the second round another. The resulting products are then directionally cloned into a vector requiring both sites. In studies to date, Sverdlov and his coworkers have found that this scheme produces a diverse set of highly enriched human cDNAs. Because these come from *Alu*s in hnRNA, they will contain introns, and this gives them potential advantages as mapping probes when compared with conventional cDNAs. As with the 3'-cDNAs discussed above, cDNA from hnRNAs will be more effective than ordinary cDNAs in dealing with gene families and in avoiding cross-hybridization with conserved exonic sequences in rodent-human cell hybrids.

## POOLING IN MAPPING STRATEGIES

All of the above methods for screening individual samples in a complex library are fine for finding genes. However, they are very inefficient in ordering a whole library. The schematic result of a successful screen for a gene-containing clone in an arrayed library is shown in Figure 9.13*a*. A single positive clone is detected, presumably containing the sample of interest. However, from the viewpoint of information retrieval, this result is very weak. A sample array is potentially an extremely informative source of information about the order of the samples it contains. A single positive hybridization extracts the minimum possible amount of information from the array and requires a time-consuming

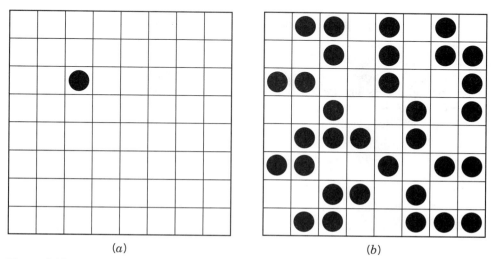

(*a*)                    (*b*)

**Figure 9.13**   Hybridization to arrays. (*a*) Typical pattern seen when screening a library for a particular gene. (*b*) Ideal pattern of hybridization in interrogating a dense sample array.

experiment. From the viewpoint of information theory, a much better designed interrogation of the array would produce the result shown in Figure 9.13*b*. In this ideal black-and-white case, roughly half the clones would be scored positive and half negative if we are considering only positive and negative hybridization results. When the amount of signal can be quantitated, much more information is potentially available from a single test of the array through all the gray levels seen for each individual clone.

The practical issue is how to take advantage of the power of arrays, or pools of samples, in a maximally efficient way so that all of the clones are ordered accurately with a minimum number of experiments. The answer will depend critically on the nature of the errors involved in interrogating arrays or pools. We start by considering fairly ideal cases. A detailed analysis of the effects of errors in real applications is still not available, but the power of these approaches is so great that it appears that a reasonable error rate can be tolerated.

The goal is to design probe or sample pools and arrays that allow roughly half of the targets to be scored positive in each hybridization or PCR. There are a number of different ways to try to accomplish this. One general approach is to increase the amount of different DNA sequences in the probes or the targets. The other general approach is to increase the fraction of potential target DNA that will be complementary to probe sequences. In principle, both approaches can be combined. A very simple strategy is to use complex probes. For example, purified large DNA fragments can be used as hybridization probes. DNA from hybrid cell lines or radiation hybrids can be used as probes. In some or all of these cases, it is helpful to employ human-specific PCR amplification so that the probe contains sufficient concentrations of the sequences that are actually serving to hybridize with specific samples in the target.

The logic behind the use of large DNA probes is that they automatically contain continuity information, and sets of target clones detected in such a hybridization should lie in the same region of the genome. It is far more efficient, for example, to assign an array of clones to chromosome regions by hybridizing with DNA purified from those regions, than it is to assign the regional location of clones one at a time by hybridizing to a panel of cell lines. The key technical advances that makes these new strategies possible are PCR amplification of desired sequences and suppression hybridization of undesired repeated sequences.

An alternative to large DNA probes is simple sequence probes. Oligonucleotides of lengths 10 to 12 will hybridize with a significant fraction of arrayed cosmids or YACs. Alternatively, one could use simple sequence PCR primers, as we discussed in Chapter 4. There is no reason why probe molecules must be used individually. Instead, one could make pools of probes and use this pool directly in hybridization. By constructing different sets of pools, it is possible, after the fact, to sort out which members of which pools were responsible for positive hybridization. This turns out to be more efficient in principle. Any sequences can be used in pools. One approach is to select arbitrary, nonoverlapping sets of single-copy sequences. Another approach is to build up pools from mixtures of short oligonucleotides.

Another basic strategy for increasing the efficiency of mapping is to use pools of samples. This is a necessary part of PCR screening methods, but there is no reason why it also could not be used for hybridization analyses. We will describe some of the principles that are involved in pooling strategies in the next few sections. These principles apply equally whether pools of probes or pools of samples are involved. Basically it would seem that one should be able to combine simultaneously sample pooling and probe pooling to increase the throughput of experiments even further. However, the most efficient ways to do this have not yet been worked out.

A caveat for all pooling approaches concerns the threshold between background noise or cross-hybridization and true positive hybridization signals. Theoretical, noise-free strategies are not likely to be viable with real biological samples. Instead of striving for the most ambitious and efficient possible pooling strategy, it is prudent to use a more overdetermined approach and sacrifice a bit of efficiency for a margin of safety. It is important to keep in mind that a pooling approach does not need to be perfect. Once potential clones of interest are found or mapped, additional experiments can always be done to confirm their location. The key idea is to get the map approximately right with a minimum number of experiments. Then the actual work involved in confirming the pooling is finite and can be done regionally by groups interested in particular locales.

## PROBE POOLING IN *S. POMBE* MAPPING

An example of the power of pooling is illustrated by results obtained in ordering a cosmid library of the yeast *S. pombe*. This organism was chosen for a model mapping project because a low-resolution restriction map was already available and because of the interest in this organism as a potential target for genomic sequencing. An arrayed cosmid library was available. This was first screened by hybridization with each of the three *S. pombe* chromosomes purified by PFG. Typical results are shown in Figure 9.14. It is readily apparent that most cosmids fall clearly on one chromosome by their hybridization. However there are significant variations in the amount of DNA present in each spot of the array. Thus a considerable amount of numerical manipulation of the data is required to correct for DNA sample variation, and for differences in the signals seen in successive hybridizations. When this is done, it is possible to assign, uniquely, more than 85% of the clones to a single chromosome based on only three hybridizations.

The next step is to make a regional assignment of the clones. Here purified restriction fragments are labeled and used as hybridization probes with the cosmid array. An example is shown in Figure 9.15. Note that it is inefficient to use only a single restriction fragment at a time. For example, once one knows the chromosomal location of the cosmids, one can mix restriction fragments from different chromosomes and use them simultaneously with little chance of introducing errors. After a small number of such experiments, one has most of the cosmids assigned to a well-defined region.

To fingerprint the cosmid array further, and begin to link up cosmid contigs, arbitrary mixtures of single-copy DNA probes were used. These were generated by the method shown in Figure 9.16. A FIGE separation of a total restriction enzyme digest of *S. pombe* genomic DNA was sliced into fractions. Because of the very high resolution of FIGE in the size range of interest, these slices should essentially contain nonoverlapping DNA sequences. Each slice was then used as a hybridization probe. As an additional fingerprinting tool, mixtures of any available *S. pombe* cloned single-copy sequences were made and used as probes. For all of these data to be analyzed, it is essential to consider the quantitative hybridization signal, and correct it both for background and for differences in the amount of DNA in each clone, the day-to-day variations in labeling, and overall hybridization efficiency.

The hybridization profile of each of the cosmid clones with the various probes used is an indication of where the clone is located. A likelihood analysis was developed to match

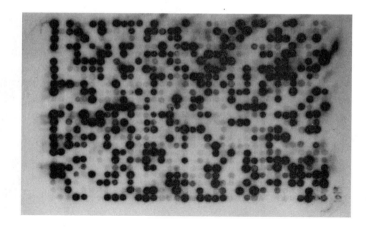

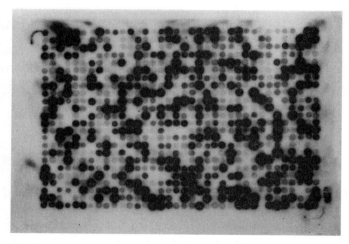

**Figure 9.14**    Hybridization of a cosmid array of *S. pombe* clones with intact *S. pombe* chromosomes I (*a*) and II (*b*).

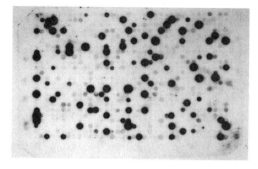

***S. pombe Sfi* I fragments K and L**

**Figure 9.15**    Hybridization of the same cosmid array shown in Figure 9.14 with two large restriction fragments purified from the *S. pombe* genome.

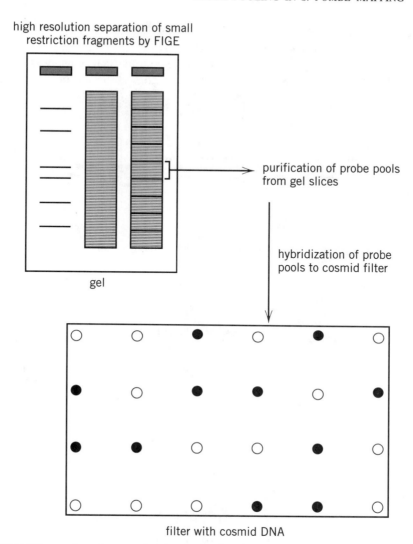

high resolution separation of small
restriction fragments by FIGE

purification of probe pools
from gel slices

hybridization of probe
pools to cosmid filter

gel

filter with cosmid DNA

**Figure 9.16** Preparation of nonoverlapping pools of probes by high-resolution FIGE separation of restriction enzyme-digested *S. pombe* genomic DNA.

up clones with similar hybridization profiles that indicate possible overlaps (Box 8.3). The basic logic behind this method is shown in Figure 9.17. Regardless of how pools of probes are made, actually overlapping clones will tend to show a concordant pattern of hybridization when many probe pools are examined. The likelihoods that reflect the concordancy of the patterns were then used in a series of different clone-ordering algorithms, including those developed at LLNL, LANL, a cluster analysis, a simulated annealing analysis, and the method sketched in Box 9.2. In general, we found that the different methods gave consistent results. Where inconsistencies were seen, these could often be resolved or rationalized by manual inspection of the data. Figure 9.18 shows the hybridization profile of three clones. This example was chosen because the interpretation of the profile is straightforward. The clones share hybridization with the same chromosome, the same restriction fragments, and most of the same complex probe mixtures. Thus they

must be located nearby, and in fact they form a contig. The clones show some hybridization differences when very simple probe mixtures are used, since they are not identical
clones. An example of the kinds of contigs that emerge from such an analysis is shown in
Figure 9.19. These are clearly large and redundant contigs of the sort one hopes to see in
a robust map.

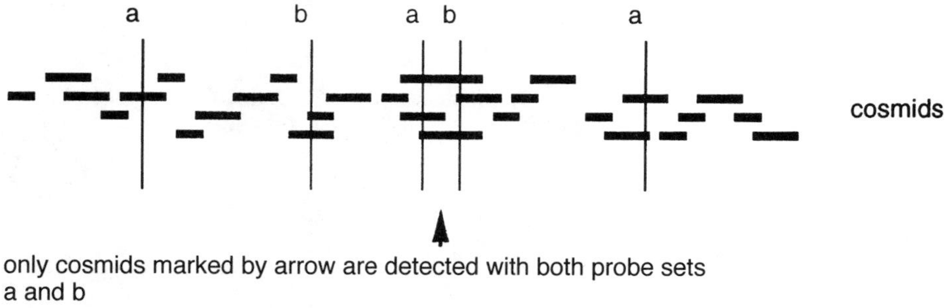

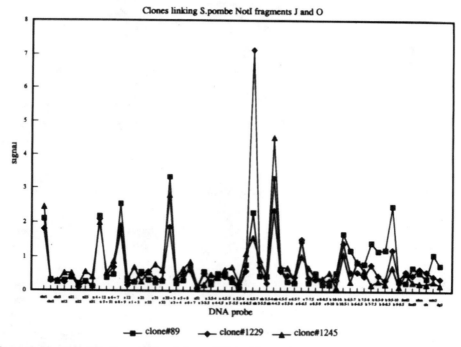

**Figure 9.17**    Even when complex pools of probes are used (e.g., pools *a* and *b*) overlapping clones
will still tend to show a concordant pattern of hybridization.

**Figure 9.18**    Hybridization profile of three different *S. pombe* cosmid clones, with a series of 63
different pools of probes.

**Figure 9.19**    *S. pombe* cosmid map constructed from the patterns of probe hybridization to the *S.
pombe* cosmid array. Cosmids are indicated by horizontal lines along the maps. Letters shown
above the map and fragment names (e.g., SHNF = *Sfi* I fragment H and *Not* I fragment F). Gaps in
contigs are indicated by a vertical bar at the right end. Positions of LTRs, 5S rDNAs, and markers
are indicated by *, #, and ×, respectively.

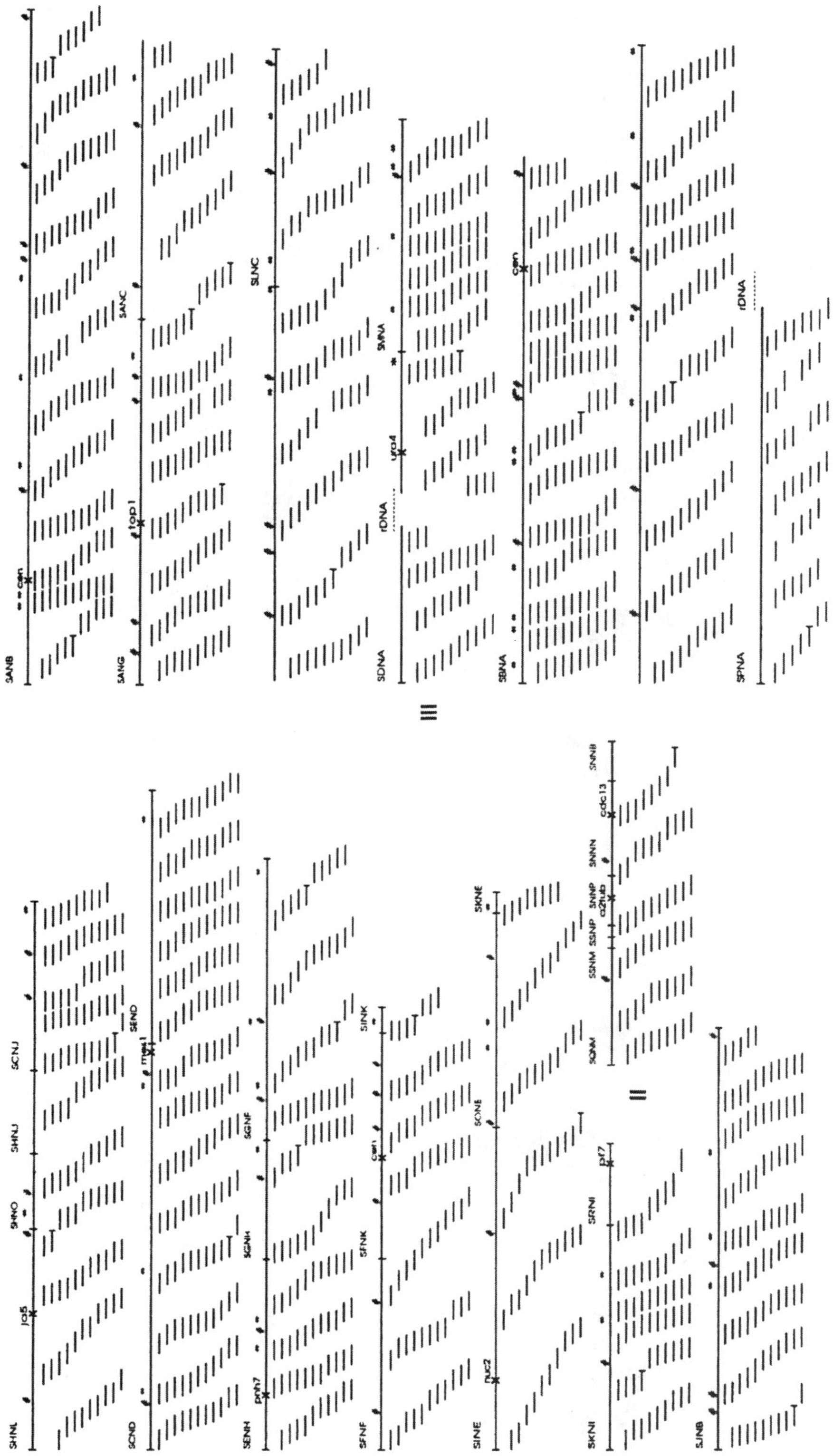

## BOX 9.2
## CONSTRUCTION OF AN ORDERED CLONE MAP

Once a set of likelihood estimates has been obtained for clone overlap, the goal is to assemble a map of these clones that represents the most probable order given the available overlap data. This is a computationally intensive task. A number of different algorithms have been developed for the purpose. They all examine the consistency of particular clone orders with the likelihood results. The available methods all appear to be less than perfectly rigorous, since they all deal with data only on clone pairs and not on higher clusters. Nevertheless, these methods are fairly successful at establishing good ordered clone maps.

Figure 9.20 shows the principle behind a method we used in the construction of a cosmid map of *S. pombe.* The objective in this limited example is to test the evidence in favor of particular schemes for ordering three clones A, B, and C. Various possible arrangements of these clone are written as the columns and rows of a matrix, each element of this matrix is represented by the maximum likelihood estimate that the clones $i$ and $j$ overlap by a fraction $f$: $L_{ij}(f)$. For each possible arrangement of clones, we calculated the weight of the matrix, $W_m$ defined as

$$W_m = \sum_{j>i} \sum_{i} |i - j| L_{ij}(f)$$

The result for a simple case is shown in the figure. The true map will have an arrangement of clones that gives a minimum weight or very close to this. The method is particularly effective in penalizing arrangements of clones where good evidence for overlap exists, and yet the clones are postulated to be nonoverlapping in the final assembled map.

| | A | B | C |
|---|---|---|---|
| A | x | $L_{AB}$ | $L_{AC}$ |
| B | | x | $L_{BC}$ |
| C | | | x |

$$W_m = L_{AB} + 2\,L_{AC} + L_{BC}$$

| | A | C | B |
|---|---|---|---|
| A | x | $L_{AC}$ | $L_{AB}$ |
| C | | x | $L_{BC}$ |
| B | | | x |

$$W_m = L_{AC} + 2\,L_{AB} + L_{BC}$$

**Figure 9.20** Example of an algorithm used to construct ordered clone maps from likelihood data. Details are given in Box 9.2.

The methods developed on *S. pombe* allowed a cosmid map to be completed to about the 98% stage in 1.5 person years of effort. Most of this effort was method or algorithm development, and we estimate that to repeat this process on a similar size genome would take only 3 person-months of effort. This is in stark contrast with earlier mapping approaches. Strict bottom-up fingerprinting methods, scaled to the size of the *S. pombe* genome, would require around 8 to 10 person-years of effort. Thus by the use of pooling and complex probes, we have gained more than an order of magnitude in mapping speed. The issue that remains to be tested is how well this kind of approach will do with mammalian samples where the effects of repeated DNA sequences will have to be eliminated. However, between the use of competition hybridization, which has been so successful in FISH, and selective PCR, we can be reasonably optimistic that probe pooling will be a generally applicable method.

## FALSE POSITIVES WITH SIMPLE POOLING SCHEMES

Row and column pools are very natural ideas for speeding the analysis of an array of samples. It is easy to implement these kinds of pools with simple tools and robots. However, they lead to a significant level of false positives when the density of positive samples in an array becomes large. Here we illustrate this problem in detail, since it will be an even more serious problem in more complex pooling schemes. Consider the simple example shown in Figure 9.21. Here two clones in the array hybridize with a probe (or show PCR amplification with the probe primers, in the case of YAC screening). If row and column pools are used for the analysis, rather than individual clones, four potentially positive clones are identified by the combination of two positive rows and two positive columns. Two are true positives; two are false positives. To decide among them, each isolated clone can be checked individually. In the case shown, only four additional hybridizations or PCR reactions would have to be done. This is a small addition to the number of tests required for screening the rows and columns. However, as the number of true positives grows linearly, the number of false positives grows quadratically. It soon becomes hopelessly inefficient to screen them all individually. An alternative approach, which is much more efficient, in the limit of high numbers of positive samples, is to construct alternate pools. For example, in the case shown in Figure 9.21, if one also included pools made along the diagonals, most true and false positives could be distinguished.

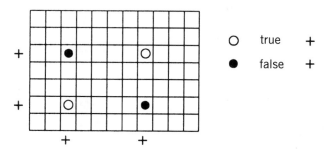

**Figure 9.21** An example of how false positives are generated when a pooled array is probed. Here row and sample pooling will reveal four apparent positive clones whenever only two real positives occur (unless the two true positives happen to fall on the same row or column).

## MORE GENERAL POOLING SCHEMES

A branch of mathematics called *sampling theory* is well developed and instructs us how to design pools effectively. In the most general case there are two significant variables: the number of dimensions used for the array and the pools, and the number of alternate pool configurations employed. In the case described in Figure 9.21, the array is two dimensional, and the pools are one dimensional. Rows and columns represent one configuration of the array. Diagonals, in essence, represent another configuration of the array because they would be rows and columns if the elements of the array were placed in a different order. Here we want to generalize these ideas. It is most important to realize that the dimensionality of an array or a pool is a mathematical statement about how we chose to describe it. It is not a statement about how the array is actually composed in space. An example is shown by the pooling scheme illustrated in Figure 9.22. Here plate pools are used in conjunction with vertical pools, made by combining each sample at a fixed *x-y* location on all the plates. The arrangement of plates appears to be three dimensional; the plate pools are two dimensional, but the vertical pools are only one dimensional.

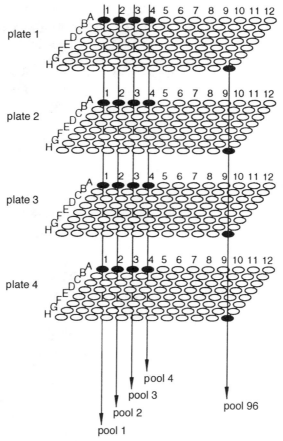

**Figure 9.22**  A pooling scheme that appears, superficially, to be three dimensional, but, in practice, is only two dimensional.

We consider first an $N$ element library, and assume that we have sufficient robotics to sample it in any way we wish. A two-dimensional square array is constructed by assigning to each element a location:

$$a_{ij} \qquad \text{where } i = 1, N^{1/2}; j = 1, N^{1/2}$$

If we pool rows and columns, each of these one-dimensional pools has $N^{1/2}$ components, and there are $2N^{1/2}$ different pools if we just consider rows and columns. The actual construction of a pool consists of setting one index constant, say $i = 3$, and then combining all samples that share that index.

The same $N$ element library can be treated as a three-dimensional cubic array. This is constructed, by analogy, with the two-dimensional array by assigning to each element a location:

$$a_{ijk} \qquad \text{where } i = 1, N^{1/3}; j = 1, N^{1/3}; k = 1, N^{1/3}$$

If we pool two-dimensional surfaces of this array, each of these pools has $N^{2/3}$ components. There are $3N^{1/3}$ different pools if we consider three orthogonal sets of planes. The actual process of constructing these pools consists of setting one index constant, say $j = 2$, and then combining all samples that share this index.

We can easily generalize the pooling scheme to higher dimensions. These become hard to depict visually, but there is no difficulty at handling them mathematically. To make a four-dimensional array of the library, we assign each of the $N$ clones an index:

$$a_{ijkl} \qquad \text{where } i = 1, N^{1/4}; j = 1, N^{1/4}; k = 1, N^{1/4}; l = 1, N^{1/4}$$

This array is what is actually called a hypercube. We can make cubic pools of samples from the array by setting one index constant, say $k = 4$, and then combining all samples that share this index. The result is $4N^{1/4}$ different pools, each with $N^{3/4}$ elements. The pools actually correspond to four orthogonal sets of sample cubes.

The process can be extended to five and higher dimensions as needed. Note that the result of increasing the dimensionality is to decrease, steadily, the number of different pools needed, at a cost of increasing the size of each pool. Therefore the usefulness of higher-dimension pooling will depend very much on experimental sensitivity. Can the true positives b distinguished among an increasingly higher level of background noise as the complexity of the pools grows?

The highest dimension possible for a pooling scheme is called a binary sieve. It is definitely the most efficient way to find a rare event in a complex sample so long as one is dealing with perfect data. In the examples discussed above, note that the range over which each sample index runs keeps dropping steadily, from $i = 1, N^{1/2}$ to $i = 1, N^{1/4}$ as the dimension of the array is increased from two to four. The most extreme case possible would allow each index only a single value; in this case there is really no array, the samples are just being numbered. A pooling scheme one step short of this extreme would be to allow each index to run over just two numbers. If we kept strictly to the above analogy we would say $i = 1, 2$; however, it is more useful to let the indices run from 0 to 1. Then we assign to a particular clone an index like $a_{101100}$. This is just a binary number (the equivalent decimal is 44 in this case). Pools are constructed, just as before, by selecting all clones with a fixed index, like $a_{ijk0mn}$.

With a binary sieve we are nominally restricted to libraries where $N$ is a power of 2: $N = 2^q$. Then the array is constructed by indexing

$$a_{ijklmn} \cdot \cdot \cdot \qquad \text{where } i, j, k, l, m, n, \ldots = 0, 1$$

Each of the pools made by setting one index to a fixed value will have $N/2$ elements. The number of pools will be $q = \log(N)/\log(2)$. This implicitly assumes that one scores each clone for 1 or 0 at each index, but not both. The notion in a pure binary sieve is that if the clone is not in one pool ($k = 1$), it must certainly be in the other ($k = 0$), and so there is no need to test them both. With real samples, one would almost certainly want to test both pools to avoid what are usually rather frequent false negatives. The size of each pool is enormous—it contains half of the clones in the library. However, the number of pools is very small. It cannot be further reduced without including additional sorts of information about the samples, such as intensity, color, or other measurable characteristics.

The binary sieve is constructed by numbering the samples with binary numbers, namely integers in base two. One can back off from the extreme example of the binary sieve by using indices in other bases. For example, with base three indices, the array is constructed as $a_{ijkl} \ldots$, where $i, j, k, l, \ldots = 0, 1, 2$. This results in a larger number of pools, each less complex than the pools used in the binary sieve. It is clear that to construct the actual pools used in binary sieves and related schemes would be quite complex if one had to do it by hand. However, it is relatively easy to instruct an $x$-$y$ robot to sample in the required manner. Specialized tools could probably be utilized to make the pooling process more rapid.

A numerical example will be helpful, here. Consider a library where $N = 2^{14} = 16,384$ clones. For a two-dimensional array, we need $2^7$ by $2^7$ or 128 by 128 clones. The row and column pools will have 128 elements each. There are 256 pools needed. In contrast, the binary array requires only 14 pools (28 if we want to protect against false negatives). Each pool, however, will have 8192 clones in it! Constructing these pools is not conceivable unless the procedure is fully automated.

As the dimensionality of pooling increases, there are trade-offs between the reduced number of pools and the increased complexity of the pools. The advantage of reduced pool number is obvious: Fewer PCR reactions or hybridizations will have to be done. A disadvantage of increased pool complexity, beyond background problems that we have already discussed, is the increasing number of false positives when the array has a high density of positive targets. For example, with a two-dimensional array, two positive clones $a_{36}$ and $a_{24}$ imply that false positives will appear at $a_{34}$ and $a_{26}$. In three dimensions, two positive clones $a_{826}$ and $a_{534}$ will generate false positives at $a_{824}$, $a_{836}$, $a_{834}$, $a_{536}$, $a_{524}$, and $a_{526}$. The number of false positives is smaller if some share a common orthogonal plane. In general for two positive clones there will be up to $2^n - 2$ false positives in an $n$-dimensional pool. By the time a binary sieve is reached, the false positives become totally impossible to handle. Thus the binary sieve will be useful only for finding very rare needles in very large haystacks. For realistic screening of libraries, much lower dimensionality pooling is needed.

The actual optimum dimension pooling scheme to use will depend on the number of clones in the array, the redundancy of the library (which will increase the rate of false positives) and the number of false positives that one is willing to tolerate, and then rescreen for individually or with an alternate array configuration. Figure 9.23 gives some

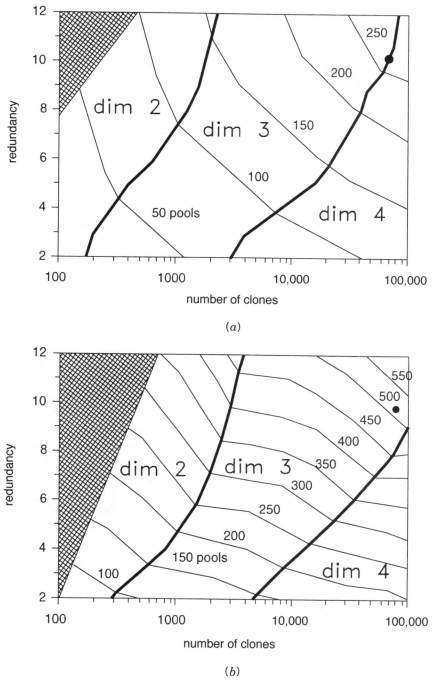

**Figure 9.23** Estimates of the ideal pool sizes and dimensions for screening libraries with different numbers of clones and different extents of redundancy. The actual characteristics of the CEPH YAC library are indicated by a dot. (*a*) Where one false positive is tolerable for each true positive. (*b*) Where only 0.01 false positive is tolerated per screen. Note that the shaded area represents situations where pooling strategies are not useful. (Taken from Barillot et al., 1991.)

calculated results for two realistic cases. These results cover cases where one false positive per true positive is acceptable, and where only 0.01 false positive per screen will be
seen. In general, the results indicate that for small numbers of clones, pooling is not efficient. The larger the library, the higher is the dimensionality of effective pools; however,
the higher the redundancy, the worse is the problem of false positives, and the lower is the
optimum dimensionality.

## ALTERNATE ARRAY CONFIGURATIONS

There is a generally applicable strategy to distinguish between true and false positives in
pools of arrayed clones. This strategy is applicable even where the density of true positives becomes very high. The basic principle behind the strategy is illustrated in Figure
9.24 for a two-dimensional array with two positive clones. Suppose that one has two versions of this array in which the positive clones happen to be at different locations, but the
relationship between the two versions is known. That is, the identity of each clone at each
position on both versions of the array is known. When row and column pools of the array
are tested, each configuration gives two positive rows and two positive columns, resulting
in four potentially positive clones in each case. However, when the actual identity of the
putative positive clones is examined, it turns out that the same true positives will occur in
both configurations, but usually the false positives will be different in the two configurations. Thus they can be eliminated.

The use of multiple configurations of the array is rather foolish and inefficient for
small arrays and small numbers of false positives. However, this process becomes very
efficient for large arrays with large numbers of false positives. Procedures exist called
*transformation matrices* or *Latin squares* that each show how to reconfigure an original
two-dimensional array into informative alternates. It is not obvious if efficient procedures
are known for arrays in higher dimensions. It is also not clear that any general reconfigu-
ration procedure is optimal, since the best scheme at a given point may depend on the
prior results. Suffice it to say that the use of several alternate configurations appears to be
a very powerful tool. This is illustrated by the example shown in Figure 9.25. Here five
positive clones are correctly identified using row and column pools from three configurations of a 7 × 7 array. This requires testing a total of 42 pools, which is only slightly less
than the 49 tests needed if the clones were examined individually. Again, however, the efficiency of the approach grows tremendously as the array size increases. Additional discussion of quantitative aspects of pooling strategies can be found in Bruno et al., (1995).

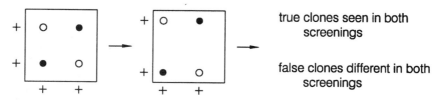

**Figure 9.24**   A simple example of how two different configurations of an array can be used to resolve the ambiguities caused by false positives.

1st configuration

$$\begin{bmatrix}
c^+_{0,0} & c^+_{0,1} & c^+_{0,2} & c^+_{0,3} & c^+_{0,4} & c^+_{0,5} & c^+_{0,6} \\
c^+_{1,0} & c^+_{1,1} & c^+_{1,2} & c^+_{1,3} & c^+_{1,4} & c^+_{1,5} & c^+_{1,6} \\
c^+_{2,0} & c^+_{2,1} & c^+_{2,2} & c^+_{2,3} & c^+_{2,4} & c^+_{2,5} & c^+_{2,6} \\
c^+_{3,0} & c^+_{3,1} & c^+_{3,2} & c^+_{3,3} & c^+_{3,4} & c^+_{3,5} & c^+_{3,6} \\
c^+_{4,0} & c^+_{4,1} & c^+_{4,2} & c^+_{4,3} & c^+_{4,4} & c^+_{4,5} & c^+_{4,6} \\
c_{5,0} & c_{5,1} & c_{5,2} & c_{5,3} & c_{5,4} & c_{5,5} & c_{5,6} \\
c_{6,0} & c_{6,1} & c_{6,2} & c_{6,3} & c_{6,4} & c_{6,5} & c_{6,6}
\end{bmatrix}$$

$\downarrow$ A

2nd configuration

$$\begin{bmatrix}
c^+_{0,0} & c_{6,2} & c_{5,4} & c^-_{4,6} & c_{3,1} & c^+_{2,3} & c^+_{1,5} \\
c_{1,6} & c_{0,1} & c_{6,3} & c_{5,5} & c_{4,0} & c_{3,2} & c_{2,4} \\
c^+_{2,5} & c_{1,0} & c_{0,2} & c^-_{6,4} & c_{5,6} & c^+_{4,1} & c^+_{3,3} \\
c_{3,4} & c_{2,6} & c_{1,1} & c_{0,3} & c_{6,5} & c_{5,0} & c_{4,2} \\
c^+_{4,3} & c_{3,5} & c_{2,0} & c^+_{1,2} & c_{0,4} & c^-_{6,6} & c^-_{5,1} \\
c_{5,2} & c_{4,4} & c_{2,6} & c_{2,1} & c_{1,3} & c_{0,5} & c_{6,0} \\
c_{6,1} & c_{5,3} & c_{4,5} & c_{3,0} & c_{2,2} & c_{1,4} & c_{0,6}
\end{bmatrix}$$

$\downarrow$ A²

3rd configuration

$$\begin{bmatrix}
c^+_{0,0} & c^-_{4,5} & c^-_{1,3} & c_{5,1} & c_{2,6} & c_{6,4} & c_{3,2} \\
c_{2,4} & c_{6,2} & c_{3,0} & c_{0,5} & c_{4,3} & c_{1,1} & c_{5,0} \\
c^+_{4,1} & c_{1,6} & c^-_{5,4} & c_{2,2} & c_{6,0} & c_{3,5} & c_{0,3} \\
c^-_{6,5} & c^+_{3,3} & c^-_{0,1} & c_{4,6} & c_{1,4} & c_{5,2} & c_{2,0} \\
c^+_{1,2} & c^-_{5,0} & c^+_{2,5} & c_{0,3} & c_{3,1} & c_{0,6} & c_{4,4} \\
c_{3,0} & c_{0,4} & c_{4,2} & c_{1,0} & c_{5,5} & c_{2,3} & c_{6,1} \\
c_{5,3} & c_{2,1} & c_{6,6} & c_{3,4} & c_{0,2} & c_{4,0} & c_{1,5}
\end{bmatrix}$$

**Figure 9.25** A more complex example of the use of multiple array configurations to distinguish true and false positives. In this case three configurations allowed the detection of five true positives (larger font characters at positions 0, 0; 1, 2; 2, 5; 3, 3; and 4, 1 in the first configuration) in a set of 49 clones. The +'s in the first configuration indicate positive rows and columns. In the second and third configuration the +'s indicate clones that could be positive; the −'s indicate those that are excluded by the results of the first and second configurations, respectively. (Taken from Barillot et al., 1991.)

## INNER PRODUCT MAPPING

The enormous attractiveness of large sample arrays as genome mapping tools has been made evident. What a pity it is with most contemporary methods there is no way to make these arrays systematically. Suppose that we had an arrayed library from a sample, and we wished to construct the equivalent array from a closely related sample (Fig. 9.26). The samples could be libraries from two different people, or from two closely related species. Constructing an array from the second sample that is parallel to the first sample is really making a map of the second sample, as efficiently as possible. We could do this one clone at a time, by testing each clone from the second library against the array of the first library. We could do this more rapidly by pooling clones from the second library. But ultimately we would have to systematically place each of the clones in the second library into its proper place in an array. This is an extremely tedious process. What we need, in the future, is a way of using the first array as a tool to order simultaneously all of the clones from the second array. In principle, it should be possible to use hybridization to do this; we just have to develop a strategy that works efficiently in practice.

One very attractive strategy for cross-correlating different sets of samples has recently been developed by Mark Perlin at Carnegie Mellon. The method has been validated by the construction of a YAC map for human chromosome 11. The basic idea behind the procedure, called *inner product mapping* (IPM) is shown in Figure 9.27. Radiation hybrids (RHs; see Chapter 7) can be analyzed relatively easily by PCR using STSs of known order on a chromosome. Because each RH encompasses a relatively large amount of human DNA, relatively few STS measurements suffice to produce a good RH map. RHs in turn provide an excellent source of DNA to identify YACs in corresponding regions. In practice, what is done is inter-*Alu* PCR (Chapter 14) both on each separate YAC and each

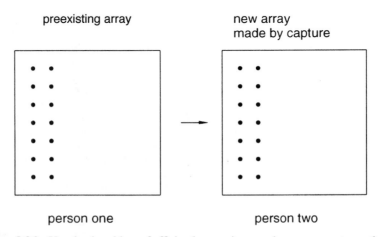

preexisting array                    new array
                                     made by capture

person one                           person two

**Figure 9.26**   Unsolved problem of efficiently mapping one dense array onto another.

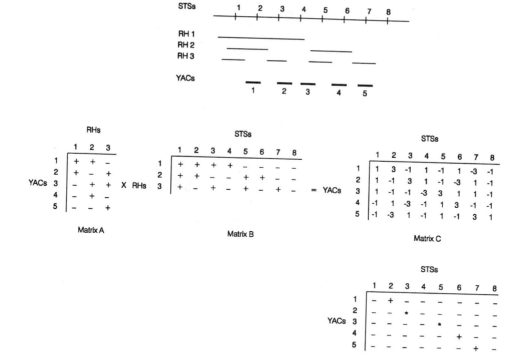

**Figure 9.27**  Inner product mapping (IPM). (See text for details.)

RH. An array is made of the YAC PCR products, and this is probed successively by hybridization with each RH PCR product. This assigns the YACs to RHs.

In previous STS mapping strategies, PCR with STS primers had to be used directly to analyze YACs in YAC pools. This is relatively inefficient for reasons that have been described earlier in this chapter. Instead, in IPM, the STS-YAC correspondences are built mathematically as shown in Figure 9.27. The inner product $C$ of two matrices $A$ and $B$ is computed as

$$C_{ij} = \sum_k A_{ik}B_{kj}$$

The YAC versus RH hybridization results can be scored as positive ($+1$) or negative ($-1$) as shown for matrix $A$. The RH versus STS PCR results can also be scored as positive ($+1$) or negative ($-1$) as shown in matrix $B$. The inner product matrix $C$ is computed element by element in a very simple fashion. It is a matrix that describes comparisons between STS's and YACs that reflects the separate RH results with each. The best estimate for each YAC-STS direct correspondence is the largest (most positive) element in each row of matrix $C$. This is shown as $+$ or $*$ in the simplified matrix $C'$ which indicates only the largest elements of C. The + symbols in C' indicate a YAC (column) that actually contains the indicated STS (row). The * symbols in C' indicate a YAC does not actually contain the STS (i.e., it would be scored negative in a direct PCR test) but must be located near this STS for the RH data to be self-consistent. This simple example indicates that IPM has more mapping power than direct STS interrogation of YACs even though the latter process would involve far more work.

The IPM mapping project of human chromosome 11 used 73 RHs, 1319 YACs, and 240 STSs to construct a YAC map. A total of 241 RH hybridizations of YAC clone arrays were required, and 240 STS interrogations of the 73 RHs were done with duplicate PCRs.

## SLICED PFG FRACTIONATIONS AS NATURAL POOLS OF SAMPLES

Previous considerations make clear that working with pools of probes or samples is often a big advantage. Sometimes this is also unavoidable when a region is too unstable or too toxic to clone in available vector systems. Cloning a region, even if it is stable, may also be too time-consuming or costly if one needs to examine the region in a large number of different samples. This would be the case where, for example, a region expanded in a set of different tumor samples is to be characterized. One way to circumvent the problem of subcloning a region is to find one or more slices of a PFG-fractionated restriction digest that contains the region. With enzymes like Not I that have rare recognition sites, generally 1 to 2 Mb regions will reside on at most a few fragments. Only a few probes from the region will suffice to identify these fragments. PFG separation conditions can then be optimized to produce these fragments in separation domains where size resolution is optimum. The resulting slice of separation gel will then contain, typically, about 2% of the total genome. For a 600-kb human DNA fragment, this slice will consist of 100 fragments, only one of which is the fragment of interest. This is probably too dilute to permit any kind of direct isolation or purification. But the slice can serve as an efficient sample for PCR amplifications that try to assign additional STSs or ESTs to the region. If the slice is examined from a digest of a single chromosome hybrid, it will contain only 1 or 2 human DNA fragments. Then, as shown in Chapter 14, PCR amplification based on human-specific repeating sequences can be used to produce numerous human-specific DNA probes from the region of interest.

## RESTRICTION LANDMARK GENOME SCANNING

An alternative method for systematically generating a dense array of samples from a genome has been developed. This method is called *restriction landmark genome scanning* (RGLS). It was originally conceived as a way of facilitating the construction of genetic maps by finding large numbers of useful polymorphic sequences. Thus it is a method set up to reveal differences between two genomes, and as such it fits the spirit of the kind of differential analysis that needs to be developed. RLGS, as currently practiced, is based, however, on genomic DNA rather than on cloned DNA. The basic idea behind RLGS is illustrated in Figure 9.28. A genome is digested with a rare-cutting restriction nuclease like Not I, and the ends of the fragments are labeled. This generates about 6000 labeled sites because there are about 3000 Not I sites in the genome, and each fragment will be

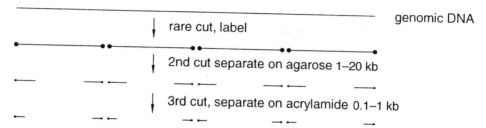

**Figure 9.28**   Steps in the preparation of DNA samples for restriction landmark genome scanning (RLGS).

labeled on both ends. The sample is then digested with a second restriction enzyme, one that cuts more frequently, say at a six base site. The resulting fragments are then fractionated by agarose gel electrophoresis in the size range of 1 to 20 kb. The agarose lane is excised and digested in situ with a third, more frequently cutting enzyme, one that recognizes a four-base sequence. The resulting small DNA fragments are now separated in a second electrophoretic dimension on polyacrylamide, which fractionates in the 0.1 to 1 kb size range. The result is a systematic pattern of thousands of spots, as shown in Figure 9.29*a*. Each spot reveals the distance between the original *Not* I site and the nearest site

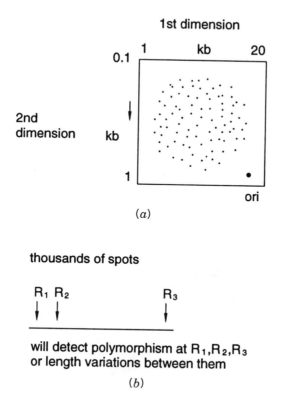

**Figure 9.29**   Example of the results seen by RLGS. (*a*) Two-dimensional electrophoretic separation of DNA fragments. (*b*) Sites where polymorphisms can be detected on a typical fragment.

for the second and third enzymes (Fig. 9.29*b*). Any polymorphisms in these distances, caused either by altered restriction sites or by DNA insertions or deletions, will appear as displaced spots in the two-dimensional fractionation. The method appears to be very powerful because so many spots can be resolved, and the patterns, at least for mouse DNA where the method was developed, are very reproducible.

## PROGNOSIS FOR THE FUTURE OF GENOME MAPPING

In the human, mouse, and other species officially sanctioned as part of the human genome project, genetic mapping is proceeding rapidly and effectively. Indeed the rate of progress appears to be better than originally projected. Dense sets of polymorphic genetic markers have been generated. These have served well to order megaYACs. Finer maps are still needed because of the difficulties in handling megaYACs and the need to break them down into smaller samples for subsequent manipulations. These finer maps, however, will be more easy to construct by using preexisting megaYAC contigs, just as the preexisting *S. pombe* restriction map allowed the design of efficient strategies to order an *S. pombe* cosmid library. BAC, PAC, or cosmid maps are still needed for current direct DNA sequencing technology. Direct sequencing from YACs or from genomic DNA is possible, as we will describe in Chapter 10, but it is not yet reliable enough to be routinely used in large-scale sequencing projects. As the genome project concentrates on DNA sequencing, the notion of a sequence-ready map has become important. Such a map consists of samples ready for DNA sequencing. Detailed order information on these samples could be known in advance, or it could be obtained in the process of DNA sequencing. See Chapter 11 for further discussion.

For species other than those already intensively studied, the best strategies will depend on the kinds of samples that are available. If radiation hybrids and mega clone libraries are made, these will obviously be valuable resources. If dense genetic maps can be made, the probes from these will order the megaYACs. If a genetic map is not feasible, FISH provides a readily accessible alternative. In other cases it may be possible to purify the chromosomes or fragments efficiently by flow sorting, improved microdissection, or other tricks to be described in Chapter 14.

The notion of having to make a map of a person for diagnostic purposes is still awesomely difficult. Mapping methods are complex and make major demands on both instrumentation and skilled personnel. New approaches will be needed before diagnostic mapping can be considered at all realistic. The use of radioactive $^{32}$P pervades most current mapping methods, and this is surely something to be avoided in a technique proposed for widescale clinical use. One area that may impact heavily on the prospects for diagnostic, mapping is the development of improved, sensitive, nonradioactive detection techniques. These will be described as we deal with DNA sequencing methods because it is here where these methods have first been used or tested.

# SOURCES AND ADDITIONAL READINGS

Allshire, R. C. 1995. Elements of chromosome structure and function in fission yeast. *Seminars in Cell Biology* 6: 55–64.

Ashworth, L. K., Hartman, M.-A., Burgin, M., Devlin, L., Carrano, A. V., and Batzer, M. A. 1995. Assembly of high-resolution bacterial artificial chromosome, P1-derived artificial chromosome, and cosmid contigs. *Analytical Biochemistry* 224: 565–571.

Barillot, E., Lacroix, B., and Cohen, D. 1991. Theoretical analysis of library screening using a *N*-dimensional pooling strategy. *Nucleic Acids Research* 19: 6241–6347.

Bruno, W. J., Knill, E., Balding, D. S., Bruce, D. C., Doggett, N. A., Sawhill, W. W., Stallings, R. L., Whittaker, C. C., and Torney, D. C. 1995. Efficient pooling designs for library screening. *Genomics* 26: 21–30.

Green, E. D., Riethman, H. C., Dutchik, J. E., and Olson, M. V. 1991. Detection and characterization of chimeric yeast artificial-chromosome clones. *Genomics* 11: 658–669.

Grigoriev, A., Mott, R., and Lehrach, H. 1994. An algorithm to detect chimeric clones and random noise in genomic mapping. *Genomics* 22: 482–486.

Grothues, D., Cantor, C. R., and Smith, C. L. 1994. Top-down construction of an ordered *Schizosaccharomyces pombe* cosmid library. *Proceedings of the National Academy of Sciences USA* 91: 4461–4465.

Hatada, I., Hayashizaki, Y., Hirotsune, S., Komatsubara, H., and Mukai, T. 1991. A genomic scanning method for higher organisms using restriction sitas landmarks. *Proceedings of the National Academy of Sciences USA* 88: 9523–9527.

Ling, L., Ma, N. S.-F., Smith, D. R., Miller, D. D., and Moir, D. T. 1993. Reduced occurrence of chimeric YACs in recombination-deficient hosts. *Nucleic Acids Research* 21: 6045–6046.

Mejia, J. E., and Monaco, A. P. 1997. Retrofitting vectors for *Escherichia coli*-based artificial chromosomes (PACs and BACs) with markers for transfection studies. *Genome Research* 7: 179–186.

Perlin, M., and Chakravarti, A. 1993. Efficient construction of high-resolution physical maps from yeast artificial chromosomes using radiation hybrids: Inner product mapping. *Genomics* 18: 283–289.

Perlin, M., Duggan, D., Davis, K., Farr, J., Findler, R., Higgins, M., Nowak, N., Evans, G., Qin, S., Zhang, J., Shows, T., James, M., and Richard, C. W. III. 1995. Rapid construction of integrated maps using inner product mapping: YAC coverage of human chromosome 11. *Genomics* 28: 315–327.

Smith, D. R., Smyth, A. P., and Moir, D. T. 1990. Amplification of large artificial chromosomes. *Proceedings of the National Academy of Sciences USA* 87: 8242–8246.

Teo, I., and Shaunak, S. 1995. Polymerase chain reaction in situ: An appraisal of an emerging technique. *Histochemical Journal* 27: 647–659.

Wada, M., Abe, K., Okumura, K., Taguchi, H., Kohno, K., Imamoto, F., Schlessinger, D., and Kuwano, M. 1994. Chimeric YACs were generated at unreduced rates in conditions that suppress coligation. *Nucleic Acids Research* 22: 1651–1654.

Whittaker, C. C., Mundt, M. O., Faber, V., Balding, D. J., Dougherty, R. L., Stallings, R. L., White, S. W., and Torney, D. C. 1993. Computations for mapping genomes with clones. *International Journal of Genome Research* 1: 195–226.

# 10 DNA Sequencing: Current Tactics

## WHY DETERMINE DNA SEQUENCE

A complete DNA sequence of a representative human genome is the major goal of the human genome project. Complete DNA sequences of other genomes are also sought. Why do we want or need this information? All descriptions of the organization of a genome, at lower resolution than the sequence, appear to offer little insight into genome function. Sometimes genes with common or related functions are clustered. This is particularly true in bacteria where the clustering allows polycistronic messages to ensure even production of a set of interactive gene products. However, in higher cells, related genes are not necessarily close together. For example, in humans, genes for alpha and beta globin chains are located on different chromosomes, even though it is desirable to produce their products in equal amounts because they associate to form a heterotetramer, (alpha)$_2$(beta)$_2$. The major purpose served by low-resolution maps is that they help us find things in the genome. We usually want to find genes in order to study or characterize their function. It is only at the level of the DNA sequence where we have any chance of drawing direct inferences about the function of a gene from its structure. Admittedly, our ability to do this today is still rather limited, as will be demonstrated in Chapter 15. However, from the rate of progress in our ability to interpret DNA sequences de novo in terms of plausible gene function, we can be reasonably optimistic that by the time the human genome is completely sequenced, coding regions will be identifiable with almost perfect accuracy, and most new genes will carry in their sequence immediately recognizable clues about function.

A second reason to have the DNA sequence of genomes is that it gives us direct access to the DNA molecules of these genomes via PCR. Using the sequence, it will almost always be possible to design primers that will amplify a small DNA target of interest, or to provide a probe that will uniquely allow effective screening of a library for a larger segment of DNA containing the region of interest. The key point is that once DNA sequence is available, clones do not have do be stored and distributed. DNA sequences also often allow us to search for similar genes in related organisms (or even more distant organisms) more efficiently than by using DNA probes of unknown sequence. For example, to find a mouse gene comparable to a human gene, one can try to use the human gene as a hybridization probe at reduced stringency (lower temperature, higher salt) against a mouse library or use the human gene to design PCR primers for probing the mouse genome. But, if one had both the relevant human and mouse DNA sequences available, a comparison among these might reveal consensus regions that are more highly conserved than average and thus better suited for hybridization or PCR to find corresponding genes in other species. This becomes increasingly important when searching for homologs of very distantly related proteins.

A continual debate in the human genome project is whether to determine the DNA sequence of the junk: DNA that as far as we can tell is noncoding. Sydney Brenner was quick to point out early in the project that this DNA is rightly called junk and not garbage because, like junk, this DNA has been retained, while garbage is discarded. Today, admittedly, we cannot interpret much from noncoding DNA sequences. But this does not mean they are nonfunctional. The fact that they remain in the genome argues for function, at least at the level of evolution. However, there are surely also functions for these sequences at the level of gene regulation, chromosome function, and perhaps properties we know nothing about today. The junk is certainly worth sequencing, but it will be best to do this later in the genome project when the cost of DNA sequencing has diminished. An analogy can be made between the genome project and the exploration of a new continent. At the time the interior of North America was first explored, a major target was river valleys because they were accessible and because they were commercially valuable. No one willingly spent much time in deserts or arctic slopes. However, most of our oil deposits are located far from river valleys, and if we had not pushed exploration of the continent to completion, we would never have found very valuable resources. It is probably this way also with the genome; when we finally make our way through the junk, systematically, there will be some unexpectedly valuable finds. We may not know enough today to realize they were valuable, even if we could find them.

## DESIGN OF DNA SEQUENCING PROJECTS

The first DNA sequence was determined in 1970 by Ray Wu at Cornell University. It consisted of the 12-base single-stranded overhang at each end of bacteriophage lambda DNA. The samples needed were readily in hand. Two investigators worked on the project for three years. Data handling and analysis did not present any unexpected or formidable problems. The major chore was developing techniques for actually determining the order of the bases. The method employed, selective addition of subsets of the four dpppN's, still has many attractive features, and we will revisit it several times in this and the next chapter.

Today, the complete DNA sequencing of 50-kb DNA targets, the size of the entire bacteriophage lambda, is a common task in specialized high-throughput sequencing laboratories. However, such projects are not yet routine in most laboratories that do DNA sequencing. The sequencing of targets 3 to 90 times larger has been accomplished in quite a few cases. Sequencing of continuous Mb blocks of human DNA is now becoming commonplace in quite a few research groups. These projects, even 50-kb projects, pose obstacles that were inconceivable at the dawn of DNA sequencing.

It is useful to divide discussion about DNA sequencing projects into tactics and strategy. Tactics is how the order of the bases on a single DNA sample is read and confirmed. Strategy, as illustrated in Figure 10.1, has a number of components. Presumably the target is selected in a rational manner, given the amount of effort that is actually required to complete a sequencing project. The upstream strategy is concerned with how the target is reduced to DNA samples suitable for application of the particular tactics selected. The tactics are then used, piece by piece, in as efficient and automated a way as possible. Then the downstream strategy consists in assembling the data into contiguous blocks of DNA sequence, filling any gaps, and correcting the inevitable errors that creep into all DNA sequence data.

Several caveats must be noted when thinking about DNA sequencing projects. Both the ideal tactics and strategy may depend on the types of targets. Effective strategies may

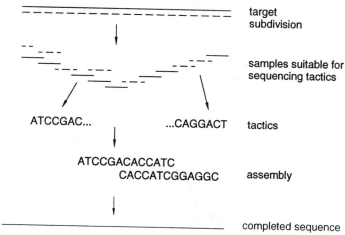

**Figure 10.1**   Design of a typical DNA sequencing project.

combine several types of targets and several types of tactics. The key variable to judge efficiency and cost is the throughput: the number of base pairs of DNA sequence generated per day for each individual working in the laboratory. With current methods, except at the largest and most efficient genome sequencing groups, personnel costs are the completely dominant expense; chemicals, enzymes, and instrument depreciation all pale in comparison with salaries. In a few very automated and experienced centers, reagents and supplies are now the dominant costs.

Three terms are useful in evaluating sequencing progress. Raw DNA sequence is the direct data read from an experimental curve or photograph with local error correction done, for example, a manual override to correct an ambiguous call by sequence reading software. Finished sequence is the assembled DNA sequence for the entire target, with error corrections made by comparing redundant samples. In general, the complete DNA sequence is read separately from both DNA strands. This is a major contributor to finding and correcting some of the most common kinds of errors. Sequencing redundancy is the ratio of the number of raw base pairs of sequence acquired to the number of base pairs of finished sequence determined. It is usually at least 2, because of the need just cited to examine both strands. In general, the redundancy is dependent on the strategy used, and it has often been as high as 10 in many of the relatively large DNA sequencing projects that have been accomplished to date.

## LADDER SEQUENCING TACTICS

Virtually all current de novo DNA sequencing methods are based on the ability to fractionate single-stranded DNA by gel electrophoresis in the presence of a denaturant with single base resolution. Information about the location of particular bases in the sequence is converted into a specific DNA fragment size. Then these fragments are separated and analyzed. The gels used are either polyacrylamide or variants on this matrix like Long Ranger™. The denaturant is usually 7 M urea. Its presence is required to eliminate most of the secondary structure that individual DNA strands can achieve by intramolecular base pairing, where this is allowed by the DNA sequence. It is possible, under ideal cases, to maintain single base resolution up to sizes of 1 kb. Some success has been reported with ever larger sizes by the use of gel-filled capillaries. The use of denaturing gels is an unfortunate aspect of current DNA sequencing. Since urea solutions are not stable to long-term storage, the gels must be cast within a few days of their use, and it is difficult to reuse most gels more than several times without a serious decrease in performance. In the two decades since Wu's first DNA sequencing, the ladder methods we will describe have produced more than 1,000 Mb of DNA sequence deposited in databases, and perhaps an equal amount or more that has not been published or deposited.

Two rather different approaches have been used to generate DNA sizes based on DNA sequence. We will describe how they are carried out starting with a single-stranded DNA template. Slightly more complex procedures are required if the original template is double stranded. The first of these methods, developed by Allan Maxam and Walter Gilbert, is shown in Figure 10.2. The ends of the DNA are distinguished by specifically labeling one of them. Usually this is done directly, and covalently, with a kinase that places a radiolabeled phosphate at the 5'-terminus of the template. There are other ways to label the 5'-end or 3'-end directly, and it is also possible to label either end indirectly, by hybridization with an appropriate complementary sequence. This requires that the end sequence be known; it usually is known, since the DNA template is cloned into a vector of known flanking sequence.

In Maxam-Gilbert sequencing base-specific or base-selective partial chemical cleavage is used to fragment the DNA. This is carried out under conditions where there is an average of only one cut per template molecule with each cleavage scheme employed. Thus a very broad range of fragment sizes is produced that reflects the entire sequence of the template. Four separate chemical fragmentation reactions are carried out; each one favors cleavage after a specific base. The fragments are fractionated, and the sizes of the labeled pieces are measured, usually in four parallel electrophoretic lanes. The DNA sequence can be read directly off the gel as indicated by the example in Figure 10.3. The pattern of bands seen is often called a ladder for reasons obvious from the figure. Note that in the Maxam-Gilbert approach, there are additional fragments produced that are not detected because they are not labeled, but they are present in the sample. For some alternate schemes of detecting DNA fragments for sequencing, like mass spectrometry, these additional pieces are undesirable.

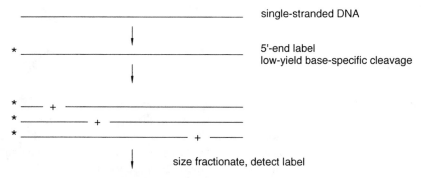

**Figure 10.2**   Maxam-Gilbert sequence technique: Preparation of end-labeled, size-fractionated DNA sample.

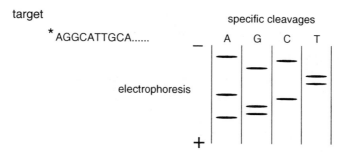

**Figure 10.3**   Typical Maxam-Gilbert sequencing ladder and its interpretation.

The second general approach to DNA fragmentation for ladder sequencing was developed by Frederick Sanger (Fig. 10.4). This is the approach in widespread use today, for a variety of reasons, including the ability to avoid the use of toxic chemicals and the ease of adapting it to four-color fluorescent detection. One starts with a single-stranded template. A primer is annealed to this template, near the 3'-end of the DNA to be sequenced. The primer must be long enough so that it binds only to one unique place on the template. This primer must correspond to known DNA sequence, either in the target or, more commonly, in the flanking vector sequence. A DNA polymerase is used to extend the primer in a sequence-specific manner along the template. However, the sequence extension is halted, in a base specific manner, by allowing the occasional uptake of chain terminators: dpppN analogs that cannot be further extended by the enzyme. Almost all current DNA sequencing uses dideoxy-pppN's as terminators. As shown in Figure 10.5, these derivatives lack the 3' OH needed to form the next phosphodiester bond. Four separate chain

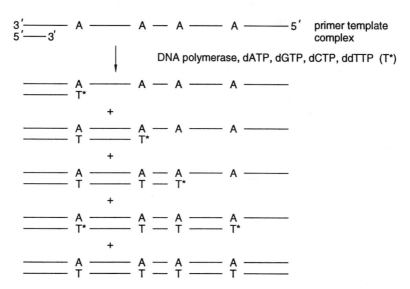

**Figure 10.4**   Sanger sequencing technique: Preparation of an end-labeled, size-fractionated DNA sample. The actual sequencing ladder will be virtually identical to that seen with the Maxam-Gilbert method.

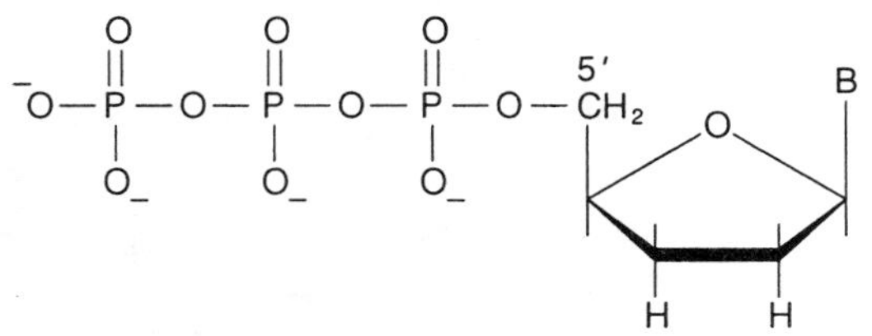

**Figure 10.5**   Structure of a dideoxynucleoside triphosphate terminator.

extension reactions are carried out—each one with a different terminator. Label can be introduced in several different ways: through the primer, the terminator, or internal dpppN's. The resulting mixture of DNA fragments is melted off the template and analyzed by gel electrophoresis.

## ISSUES IN LADDER SEQUENCING

The major goal is to maximize sequencing throughput. A second, significant goal is to minimize the number of sequencing errors. An important element of these goals is to be able to read the longest possible sequencing ladders, accurately. There are two significant variables in this. The resolution of the gel electrophoresis will determine how far the sequencing data can be read, if there are data to be read at all. Ultimately there are trade-offs between how fast the gel can be run, which also affects the throughput, how well certain artifacts can be eliminated, and how much sample must be applied. The more sample we have, the easier is the detection but, in general, the lower is the resolution. Large double-stranded DNAs show negligible diffusion during gel electrophoresis as described earlier in Chapter 5 (Yarmola et al., 1996). This is not the case for the smaller single-stranded DNAs used in sequencing where diffusion is a significant cause of band broadening. This motivates the use of higher fields where shorter running times can be achieved, hence minimizing the effects of diffusion. However, higher fields lead to greater joule heating. This increases the effects of thermal inhomogeneities which also lead to band broadening. The issues are complex because field strength also influences the shape of DNA in a gel and thus affects its diffusion coefficient. Other factors that affect band shape and thus resolution are the volume in which the sample is loaded, the volume sampled by the detector, and any inhomogeneities in gel concentration. For a thorough discussion of the effects of these variables, see Luckey et al. (1993), and Luckey and Smith (1993).

To a good approximation, the velocity, $v$, of DNA in denaturing acrylamide gel electrophoresis is proportional to $1/L$, where $L$ is the length of the molecule. In automated fluorescent detection (or the bottom wiper shown later in Fig. 10.10), the sample is examined at a constant distance from the starting point, $D$. The time it takes a fragment of a particular length to reach this distance is proportional to $D/v = DL$. Hence the spacing between two bands of length $L$ and $L - 1$ is $DL - D(L - 1) = D$. Thus the band spacing is independent of size, but it can be increased, more or less at will by using longer and longer running gels.

The second determinant of how far a ladder can be read is the uniformity of the sample fragment yield. It is important to realize that the larger the target is, the smaller the yield of each piece even if the distribution of fragments is absolutely uniform. Thus, with perfect cleavage, sequencing a 100-base piece of DNA will require only 10% the amount of sample that a 1-kb target requires. Put another way, for constant amounts of DNA sample loaded, the detection sensitivity will have to increase in proportion to the length of the DNA target. The relative yield of particular DNA fragments is affected by the choice of DNA polymerase, the nature of the terminators and primers used, the actual DNA template, and the reaction conditions. Much optimization has been required to produce reproducible runs of DNA sequence data that extend longer than 500 bases.

It is also important to realize that throughput is really the product of the number of lanes per gel and the speed of the electrophoresis. Speed can be controlled by the electrical field applied. In fact higher fields appear to improve electrophoretic performance. What limits the speed, once efficient cooling is provided to keep the running temperature of the gel constant, is the sensitivity of the detection scheme, if it is done on line. With off-line detection, the sensitivity is still important, not for speed, but for determining the number of lanes that can be used. The smaller the width of each lane, the more lanes one can place on a single gel but the smaller the amount of DNA one can actually load into each lane.

A major factor that affects the quality of DNA sequence data is the quality of the template DNA. When fluorescent labeling is used, great care must be taken not to introduce fluorescent contaminants into the DNA sample. A number of automated methods for DNA preparation routinely yield DNA suitable for sequencing. These methods are convenient because they are so standardized. A laboratory that tries to sequence DNA from many different types of sources will frequently encounter difficulties.

In early DNA sequencing, $^{32}$P was the label of choice, introduced from $\gamma$ [$^{32}$P] pppA via kinasing of the primer for Sanger sequencing or the strand to be cleaved for Maxam-Gilbert sequencing. This isotope has a short half-life which results in very high experimental sensitivity. However, $^{32}$P also has a relatively high energy beta particle, which causes an artifactual broadening of the thin fragment bands on DNA sequencing gels. Instead of $^{32}$P one can use the radioisotope $^{35}$S, as $\gamma$ thio-pppA. This still has a short half-life, but the decay is softer, leading to sharper bands. At first, DNA sequence data were obtained by using X-ray film in autoradiography to make an image of the sequencing gel. This can be read by hand, which is still done by some, perhaps with the help of devices and software to expedite transferring the data into a computer file. Alternatively, the film can be scanned and digitized by a device like a charge-couple device (CCD) camera. This then allows most of the data to be processed by image analysis software, with human intervention needed in difficult places. The accuracy of using film and some of the existing software does not appear to be as good as the fluorescent systems we will describe later.

A new approach to recording data from radioactive decay is the use of imaging plates. These consist of individual pixels that record local decays. After the plate is exposed, it is read out by laser excitation in a raster pattern (scanning successive lines, in the same manner as a TV camera or screen), and the resulting data are transferred into a computer file. A great advantage of imaging plates over film is that their response is a linear function of dose over more than five orders of magnitude in intensity, and most important, they are linear down to the lowest detectable doses. In contrast, film shows a dead zone at very low doses, and it easily saturates at high doses. Imaging plates are reusable, and for the heavy user, the great savings in film that result eventually compensate for the high costs of the imaging plates and the instrument needed to read them out. Although it is possible, in principle, to use several different radioisotopes simultaneously, as is common in liquid scintillation counting, and thus achieve multicolor labeling and detection, in practice, this is rarely done with radioactive DNA sequencing data.

In most contemporary DNA sequencing, radioisotopes have been replaced by fluorescent labels. These can be used on the primers, the terminators, or internally. It may seem surprising that fluorescent detection can be competitive with radioisotopes. However, one can gain enormous amounts of sensitivity in fluorescence by sequential excitation and emission from the same fluorophore until it undergoes some chemical side reaction and becomes bleached. This makes up in large part for the difference in energy between a beta particle and the fluorescent photon. The major determinant of sensitivity in fluorescence detection is, then, not really signal; it is background. Scrupulous care must be taken to avoid the use of reagents, solvents, plastics, glove powder, and detergents that have fluorescent contaminants.

Four different colored fluorescent dyes are used in several of the most common DNA sequencing detection schemes. One dye is used for each base-specific primer extension. The ideal set of dyes would have very similar chemical structures so that their presence would affect the electrophoretic mobility of labeled DNA fragments in identical ways. They would also have emission spectra as distinct as possible, and they would all be excitable by the same wavelength so that a single excitation source would suffice for all four dyes. The dyes would also allow similar very high sensitivity detection so that signal intensities from the four different cleavage reactions would be comparable. Inevitably with currently available dyes there are compromises. For example, a set of nearly identical dye-labeled chain terminators was produced for DNA sequencing that led to very good electrophoretic properties, but the emission spectra of these compounds were too similar for the kind of accuracy needed in reading long sequence ladders. Subsequently a more well-resolved set of fluorescent terminators that are substrates for Sequenase, the most popular enzyme used in Sanger sequencing became commercially available. These have the advantage that all four terminators can be used simultaneously in a single sequencing reaction.

All currently used dyes for four-color DNA sequencing are excited in the UV/visible wavelength range. The limits of this range and the typical widths of emission spectra of high quantum yield dyes make it rather difficult to detect more than four colors simultaneously. The infrared (IR) spectrum is much broader, and work is in progress trying to develop DNA sequencing dyes in this range. If the lower sensitivity of IR detection can be tolerated, such dyes would offer two advantages. The laser sources needed to excite them are inexpensive, and at least eight different colors would be obtainable. This could be used to double the throughput of four-color sequencing, or it could be used to include a known standard in every sequencing lane to improve the accuracy of automatic sequence calling. Recently IR-excited dyes have begun to make an impact on automated DNA sequencing. Multiple IR colors are presumably soon on the horizon.

A significant improvement in fluorescent dyes for automated sequencing is the use of energy transfer methods (Glazer and Mathies, 1997). Primers contain a pair of fluorescent dyes (Fig. 10.6). One dye is common to all four primers. This is optimized to absorb the exciting laser dyes. The second dye is different in each primer, and it is close enough in each case that fluorescence resonance energy transfer is 100% efficient. Thus all the excitation energy migrates to the second dye where it is subsequently emitted. The second dyes are chosen so that they have as different emission spectra as possible to maximize the ability to accurately discriminate the four different colors.

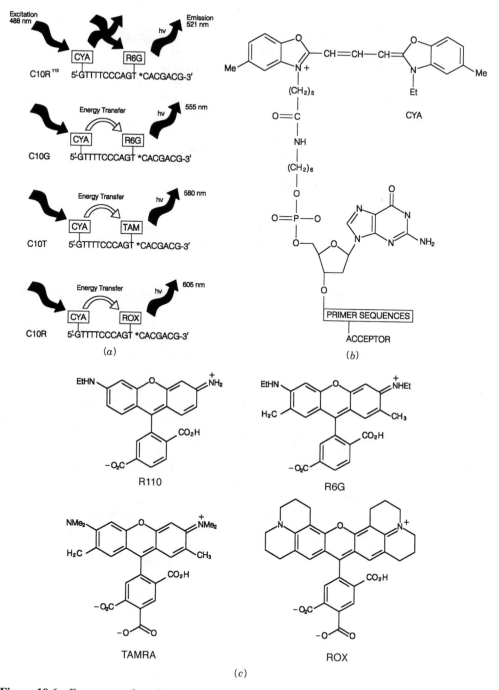

**Figure 10.6** Energy transfer primers (provided by Richard Mathies). (*a*) Schematic design of a set of four primers. (*b*) Structure of the donor dye. (*c*) Structure of four different acceptor fluorescent dyes that can be detected simultaneously in DNA sequencing.

An alternative to fluorescent labels is chemiluminescence. This has the great advantage that no exciting light is needed. Thus the sensitivity can be extremely high, since there is no contamination from scattering of the exciting light used in fluorescence, or the effects of fluorescent impurities. Today, chemiluminescent detection schemes exist that can readily be used in DNA sequencing. They have a few disadvantages. Only one color is currently available, and once the chemiluminescence has been read, it is difficult to use the gel or filter again. While this is not often a problem in most forms of DNA sequencing, it is a problem in most mapping applications where the same filter replica of a gel is frequently probed many times in succession. Nevertheless, the sensitivity of chemiluminescence makes it attractive for some mapping applications. The advantages of four-color fluorescence are also beginning to be felt in some aspects of genome mapping. An example was given in Chapter 8.

## CURRENT FLUORESCENT DNA SEQUENCING

There are two basically distinct implementations of fluorescent detected DNA sequence determination. These are the current commonly available state-of-the-art tools used today in most large-scale DNA sequencing projects. They each can produce more than $10^4$ to $10^5$ bp of raw DNA sequence per laboratory worker per day. Most allow 400 to 800 bases of data to be read per lane; most of the lanes give readable data when proper DNA preparation methods are used. The detection schemes used in the two approaches are illustrated in Figure 10.7. Both are on-line gel readers. These two schemes have a number of serious trade-offs. In the Applied Biosystems (ABI) instrument, based on original developments by Leroy Hood and Lloyd Smith, four different colored dyes are used to analyze a mixture of four different samples in a single gel lane (Fig. 10.7a). This allows four times more samples to be loaded per gel, if the width of the lanes is kept constant. The use of four colors in a single lane avoids the problem of compensating for any differences in the mobility of fragments in adjacent lanes—that is, there is no lane registration problem. In order to do the four-color analysis, a laser perpendicular to the gel is used to excite one lane at a time, and the signal is detected through a rotating four-color wheel to separate the emission from the four different dyes. Thus the effective power of the laser is the time shared among the lanes and the colors. With 20 lanes, the actual time-averaged illumination available is, at most, 1/80 the laser intensity.

In the alternative implementation, embodied in the Pharmacia automated laser fluorescence (ALF) instrument, only a single fluorescent dye is used (Fig. 10.7b). The dye originally selected was fluorescein because it is the most sensitive available for the particular laser exciting wavelength used. In a newer version of the instrument, a different laser and an infrared emitting dye, Cy5, are used. The key feature of the ALF is that the laser excitation is in the plane of the gel, through all the sample lanes simultaneously. This design, which is based on an instrument originally developed by Wilhelm Ansorge, is possible because at the concentrations of label used for DNA sequencing the samples are optically thin. This means that the amount of light absorbed at each lane is an insignificant fraction of the original laser intensity, so all lanes receive, effectively, equal excitation. The emission from all the lanes is recorded simultaneously by an array of detectors, one for each lane. While these could be made four-color detectors, in principle, the cost and complexity is not warranted. Instead the ALF reads data from four closely spaced lanes, one for each base-specific fragmentation. Thus the number of lanes needed for one sample in the

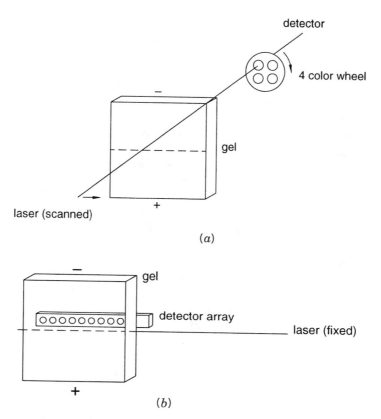

**Figure 10.7**  Schematic operation of current commercially available automated DNA sequencing gel readers. (*a*) ABI four-color instrument. (*b*) Pharmacia one-color instrument.

ALF is four times the number used in the ABI instrument. The use of a single color simplifies the construction of labeled primers, since only a single primer is needed for the four sequencing reactions, whereas with the ABI approach, if fluorescent primers are used, four different ones must be made—one for each color. The basic advantage of the ALF is the higher signal to noise for a given laser strength, since the full intensity of the exciting beam can be used to illuminate each sample continuously. This higher signal to noise allows faster running and, in principle, smaller lanes.

At present the advantages of the two different fluorescent approaches really depend on the application for which they are intended. If massive amounts of sample throughput, on relatively small DNA fragments, is most important, the ABI has an edge because of the larger number of samples that can be run per gel. If longer fragments are important, or if sample amounts are limited, or if the raw data must be scrutinized as in mutation detection (see Chapter 13), the ALF has the edge because of its greater sensitivity.

The remaining status of most state-of-the-art DNA sequencing is easy to summarize. The gels are still made manually. Attempts to manufacture and distribute precast gels have been a dismal failure. Samples can be loaded manually; however, semiautomatic methods are widely available, like multiple-headed microsyringes. Fully automated gel-loading robots will be commonplace soon. The sequencing chemistry done can be done in a fully automated form for almost all choices of templates and tactics (described later in this chapter). Several different robotic systems that perform the sequencing reactions in microtitreplate wells are now available (see Chapter 9). Three basic choices are available for the template. The sample to be sequenced can be cloned into the single-stranded DNA bacteriophage M13, and then a single M13 sequencing primer can be used for all templates. Alternatively, PCR can be used to prepare the template by a process called cycle sequencing in which one strand is differentially labeled or synthesized during cycles of linear amplification and terminators are introduced to allow the sequence to be read. The third general approach is to sequence directly from genomic DNA. Here the primer is used directly on double-stranded plasmid, bacteriophage, cosmid, or even bacterial DNA after that sample has been melted.

## VARIATIONS IN CONTEMPORARY DNA SEQUENCING TACTICS

A tremendous amount of energy and cleverness has gone into attempts to improve and optimize current DNA sequencing technology. Here we cover some recent developments near the cutting edge of conventional DNA sequencing. The first improved the throughput by increasing the running speed. Major improvements in DNA sequencing rates are achievable by the use of thin gel samples. These can be either gel-filled capillaries or thin gel slabs. The advantage of a thin gel is that heat dissipation is more effective. The sample has a more even temperature distribution, and probably more important, higher-field strengths can be used that can increase the running speed by up to an order of magnitude. This increases sample throughput, and it diminishes any residual effects of diffusion. However, greater detection sensitivity is needed to process the fluorescence as the samples whip by the detector. Special gel materials are also available that allow faster running. For example, the Long Ranger™ gel speeds up the electrophoresis; it also appears to have better resolution than standard polyacrylamide when longer than conventional running gels are used. Perhaps most significant, these gels can be reloaded and rerun several times before their performance starts to deteriorate.

The major improvement in recent DNA sequencing chemistry has been the use of engineered polymerases like the modified form of bacteriophage T7 DNA polymerase, called Sequenase (version 2). This genetically modified form of the enzyme has improved processivity, which means it makes more even sequencing ladders. The behavior of the enzyme is further enhanced by using $Mn^{2+}$ ions instead of $Mg^{2+}$ (Fig. 10.8). With Sequenase, the limiting factor in long DNA sequencing reads is at the level of the electrophoresis and not the sequencing chemistry. Different genetically engineered polymerases have been optimized for cycle sequencing. These include Amplitaq FS and Thermosequenase.

In most DNA sequencing the position of a band is used as the sole source of information. The intensity of the band is ignored. With the very even sequencing ladders provided by the use of Sequenase, and the high signal-to-noise ratio of the ALF-type systems, it is possible to use intensity as a base specific label. A typical result is shown in Figure 10.9. Here different amounts of fluorescent and nonlabeled primers were used for the four different base-specific terminations. All samples were combined into a single lane. The result is equivalent to four-color sequencing in terms of throughput and eliminating the need to register adjacent lanes. However, it is not clear how resistant the intensity-labeling process is to various potential errors. Thus this method has not seen widespread use.

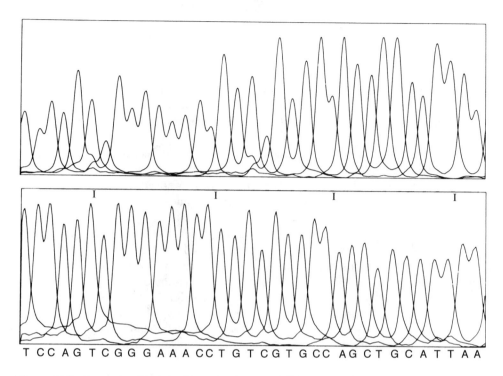

T CC AG TCGG GAAACCTG TCGTGCC AGCT GCATTAA

**Figure 10.8** Example of how engineered polymerases (*bottom*) provide more even DNA sequencing ladders than natural polymerases (*top*). Provided by Wilhelm Ansorge.

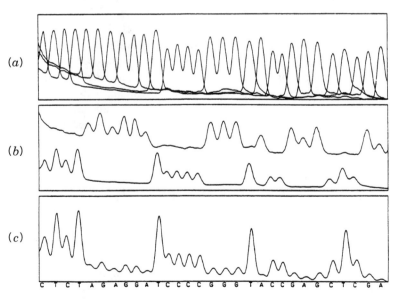

**Figure 10.9** Example of the use of bases labeled with different amounts of a single dye to get information about all four bases in a single lane with only one color. (*a*) Normal sequencing (four separate lanes). (*b*) Two bases at a time (two lanes). (*c*) All four bases in one lane. Taken from Ansorge et al. (1990).

Three other DNA sequencing tricks are worth describing briefly. One solution to the problems of fast on-line analysis is off-line analysis. A very clever way to do this is the bottom wiper. As shown in Figure 10.10, this consists of a short sequencing gel atop a moving membrane. As the electrophoresis proceeds, samples are automatically eluted and transferred to the membrane. The resulting blot can then be analyzed off line in any way one wishes. The unique aspect of this approach is that the spacing between adjacent bands is not only a function of the electrophoresis, it can also be manipulated by how fast the membrane moves. Thus samples that are too close to each other to be well resolved by a given detector, even though the bands are resolved in the electrophoresis, can be separated and analyzed individually.

The second method, developed by Barbara Shaw, is a novel variation on Sanger sequencing chemistry. Here the use of dideoxynucleoside triphosphate terminators is avoided. Instead, trace amounts of boron derivatives of the normal deoxynucleoside triphosphates are added to the cocktail of substrates used for DNA polymerase. Compounds like $5'$-$\alpha$-[P-borano]-triphosphates are well tolerated by most DNA polymerases. These compounds have a $BH_3$ group replacing an oxygen on the alpha phosphate (nearest the sugar). Thus they lead to incorporation of boron-substituted phosphates into the DNA chain. The polymerization process is efficient enough that the derivatives can be introduced as part of PCR amplification. These derivatives are resistant to exonuclease III cleavage. Thus, when the PCR product is digested with exonuclease III, a ladder of fragments is produced that terminates at the first location at which a boron derivative is encountered.

A final trick that appears to be extremely promising is internal labeling. Here a primer is used adjacent to some $3'$ known flanking sequence. Label is introduced into the primer by selective extension in the absence of one of the four ordinary dpppN's (essentially the original Ray Wu sequencing strategy) selected from the known sequence. Either fluorescein-labeled or IR-dye-labeled dU or dA can be used (Fig. 10.11). The latter appears to be superior. There are several advantages in this approach. The fluorophores introduced are internal, which protects them from any exonuclease degradation; there is no background from a great excess of labeled primer, and there is no need to synthesize fluorescent primers. An example of the success in using internal labeling is shown in Figure 10.12. Admittedly this is an extraordinarily good sequencing result. However, recent experience at the EMBL where this technique was developed indicates that even in a teaching setting, the use of internal labeling routinely proved highly accurate raw sequencing

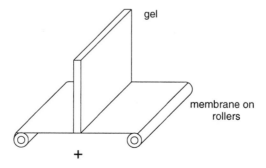

**Figure 10.10**   Schematic illustration of the bottom wiper used to transfer DNA sequencing ladders to a membrane.

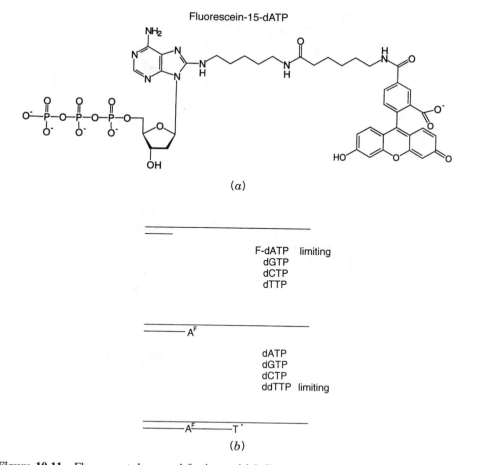

**Figure 10.11**  Fluorescent dyes used for internal labeling. (*a*) Dye structures. (*b*) Typical procedure for internal labeling.

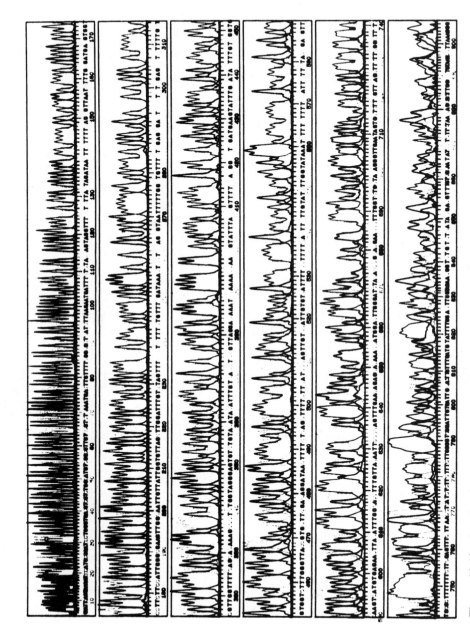

**Figure 10.12** Results of a very good but not exceptional DNA sequencing ladder obtained with internal labeling. Taken from Grothues et al. (1993).

data with typical reads of 400 to 500 bases. Since the required fluorescent materials are readily available, there is to reason why they should not be incorporated into all one-color fluorescent DNA sequencing methods.

## ERRORS IN DNA SEQUENCING

A major factor in automated DNA sequencing is continued improvements in the software used to analyze the data. From manual sequence calling, the state of the art has progressed to automated sequence reading, regardless of whether radioactive, chemiluminescent, or fluorescent sequencing data are recorded. Most software allows manual editing and overriding. Ideal software indicates where bases are known with great accuracy and where there are ambiguities. One recent report cited a 1% automatic calling accuracy in 350-base fluorescent sequencing runs, with significant deterioration beyond this point to 17% error in 500 base runs (Koop et al., 1993). A more recent study (Table 10.1) is more optimistic and demonstrates that manual editing can improve, sometimes substantially, on the accuracy of automated calling software (Naeve et al., 1995). The best current automatic software is claimed to read 500 base runs with less than 1% error. It is very important to realize that most of the issues that have been addressed thus far at the software level are how to deal with cases where adjacent fragments are only partially resolved. The resolution of successive bands in DNA sequencing gel electrophoresis gradually deteriorates as the size of the bands increases. Larger fragments spend more time in the gel and have correspondingly more time to disperse. They also are increasingly subject to electrical orientation (Chapter 5) which leads to a loss in size-dependent electrophoretic mobility. For these reasons one usually tries to sequence both strands of a target, and this provides the most accurate sequence possible at both ends of the target.

Many of the errors and ambiguities that occur in DNA sequencing data are systematic. This encourages the use of clever computer algorithms or artificial intelligence approaches to refine, even more, automated sequence calling. Since the average separation between adjacent fragment lengths varies very gradually in DNA electrophoresis, from the width of a band, one usually knows how many bases it represents even if these are not well resolved, as in the case of a run of the same base at very large fragment sizes. However, when the results are examined more closely, it is apparent that the band spacing is not perfect; it varies slightly depending on the identity of the last base in the chain, and perhaps the one before (Fig. 10.13*a*). Band intensities are also a function of the local sequence, and they are markedly affected by the particular DNA polymerase used. Secondary structure in the single-stranded DNA is not totally eliminated by the denaturing conditions used (Fig. 10.13*b*). This can be partially compensated for by using base analogs like 7-deazaG instead of G, since these form weaker secondary structures. When DNA strands form hairpins under the conditions of sequencing gel electrophoresis, they migrate faster than expected. The result is called a compression; part of the sequence may be missed or may just be impossible to read (Fig. 10.14). Frequently compressions occur on one strand but not on the complementary strand; this is one of the major reasons to sequence both strands. At first, the strand dependence of compressions may seem puzzling. After all, whatever intrastrand base pairs that can be formed by one strand can also be formed by its complement. However, two additional complications arise. First is that any strand with a high local density of G residues can form unusual helical structures with G–G pairing. These are very stable; fortunately the complementary strand will be C-rich

**TABLE 10.1  Accuracy of Automated DNA Sequencing as a Function of the Distance from the Primer**

| Method | Maximum Correct (%) | | | | | | Median Correct (%) | | | | | | Minimum Correct (%) | | | | | |
|---|---|---|---|---|---|---|---|---|---|---|---|---|---|---|---|---|---|---|
| | 1–100 | 101–200 | 201–300 | 301–400 | 401–500 | 501–600 | 1–100 | 101–200 | 201–300 | 301–400 | 401–500 | 501–600 | 1–100 | 101–200 | 201–300 | 301–400 | 401–500 | 501–600 |
| Dye-primer, unedited | 98 | 100 | 100 | 100 | 99 | 96 | 93 | 100 | 100 | 99 | 92 | 44 | 38 | 99 | 96 | 71 | 14 | 9 |
| Dye-primer, edited | 96 | 100 | 100 | 100 | 99 | 65 | 93 | 100 | 199 | 99 | 98 | 61 | 38 | 100 | 100 | 98 | 69 | 0 |
| Dye-terminator, unedited | 100 | 100 | 100 | 100 | 100 | 75 | 97 | 100 | 99 | 98 | 83 | 27 | 51 | 92 | 88 | 36 | 0 | 0 |
| Dye-terminator, edited | 100 | 100 | 100 | 100 | 100 | 84 | 97 | 100 | 100 | 100 | 84 | 15 | 21 | 94 | 95 | 39 | 0 | 0 |

*Source*:  Adapted from Naeve et al. (1995).

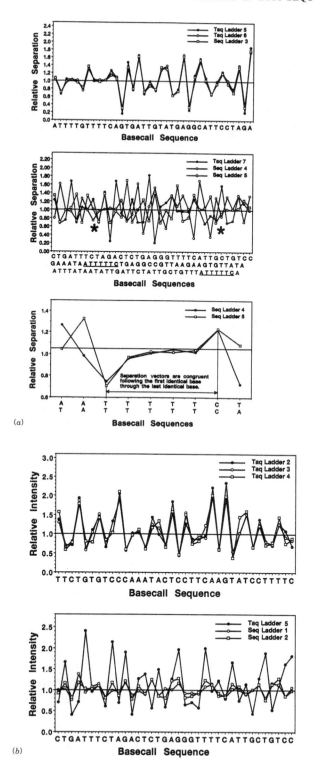

**Figure 10.13**  Effect of local DNA sequences on the band spacing and intensities seen in sequencing data. Taken from Tibetts et al. (1996). (*a*) Relative band spacings are intrinsic to the sequence. (*b*) Relative band intensities depend on the polymerase.

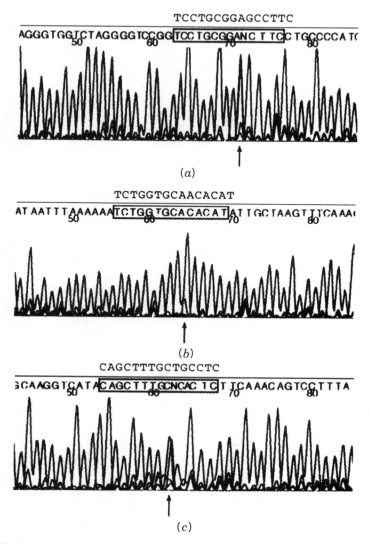

**Figure 10.14** Example of compression artifacts in DNA sequencing caused by stable secondary structures in the single-stranded sample. Shown with the sequence data are the automatic software call, and the true sequence is shown above. Taken from Yamakawa et al., (1996)

instead of G-rich. The use of 7-deazaG presumably eliminates such structures. The second, and more serious complication is the sequence of the loop of the hairpin. It turns out that certain loop sequences promote particularly stable hairpin formation. An example is GCGAAAGC, which forms a hairpin with only two base pairs, but its melting temperature is 76°C in 0.15 M salt (Hirao et al., 1990). A recent study surveyed a large number of natural and synthetic sequences that led to compressions (Yamakama et al., 1996). Remarkably all but 2% of these were formed by only two types of sequence motifs. About a third, which showed up on both strands were hairpins with a G+C-rich stem (≥ 3 bp) connected by a 3–4 base loop. The remaining two-thirds occurred on only 1

**Figure 10.15** Example of background in DNA sequencing ladders generated by mispriming.

strand which carried the consensus sequence YGN1–2AR, where Y and R are complementary. Note that this motif is contained in the extraordinarily stable hairpin just described above. Now that the prevalence of this motif in compressions is understood, it can be used to correct the misread sequence as shown by the example in Figure 10.14.

Other sources of error in DNA sequencing are caused by mispriming. Not uncommonly, there will be a secondary site on the template where the primer can bind and be extended by the polymerase (Fig. 10.15). This adds a bit of low-level, specific noise to the primary sequencing data. Another source of ambiguity arises when the sample is heterozygous or a mixture. The basic point is that most of these errors can be partially or even totally corrected if the software is clever enough to search out these possibilities. As more raw sequence data are obtained, and ultimately corrected into finished sequence, it should be possible to go back to the raw data and refine the algorithms used to process it. In short, the automated analysis of DNA sequencing data ought to be able to improve itself continually with time. The ideal software, which does not yet exist, would actually give use the probability of each of the four bases occurring at a given position. At a given site the result might be

$$A = 0.01$$

$$G = 0.98$$

$$C = 0.00$$

$$T = 0.01$$

This would be the best data to feed back into artificial intelligence approaches to refine the software further. A nice step in this direction is Phil Green's phred algorithm which automatically calls sequences and assigns a quality score, $q$, to each base,

$$q = -10 \log p$$

where $p$ is the estimated error probability for that base. Hence phred scores of 30 or better indicate sequences that are likely to be perfect.

## AUTOMATED DNA SEQUENCING CHEMISTRY

The one remaining area we need to describe, where great progress has been made is the automated preparation of DNA for sequencing. The most success appears to be seen with solid state DNA preparations. These were developed by Mathias Uhlen, and a recent in-

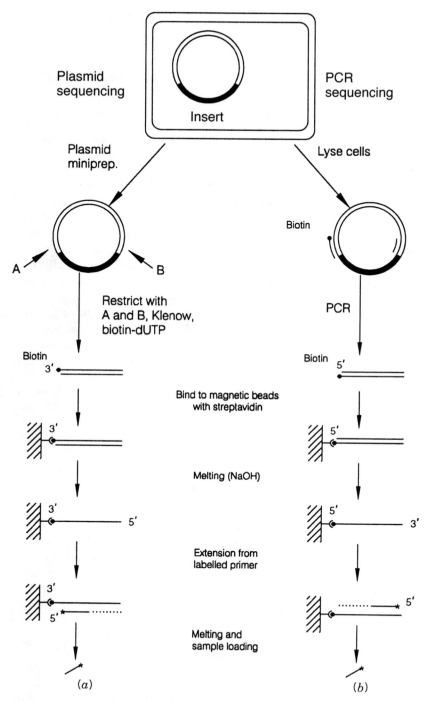

(a)　　　　　　　　　　　　　(b)

**Figure 10.16** Two solid state DNA sequencing schemes. (*a*) From a plasmid DNA miniprep. (*b*) From DNA prepared by PCR. Provided by Mathias Uhlen. See Holtman et al. (1989).

teresting modification has been accomplished by Ulf Landegren. In both methods the idea is to capture DNA onto a solid surface via streptavidin-biotin technology, and then do at least one strand of the DNA sequencing on that surface. Two schemes developed by Uhlen are illustrated in Figure 10.16. They are pretty much self-explanatory. The Uhlen implementation of these solid state preparations uses magnetic microbeads containing immobilized streptavidin. The DNA is biotinylated either by filling in a restriction site with a biotinylated base analog or by using a biotinylated PCR primer. Once the duplex DNA is captured, the nontethered strand is removed by alkali. An essential aspect of the procedure is that the streptavidin-biotin link is resistant to the harsh alkali treatment needed to melt the DNA. Sequencing chemistry is then carried out on the immobilized DNA strand that remains. If desired, the strand released into solution can also be subsequently captured in a different way and sequenced. The great advantage of this approach is the ease with which it can be automated and the very clean DNA preparations that are provided because of the efficient sample washing possible in this format.

Multiple samples can be manipulated with a permanent magnet in a microtitre plate format as shown in Figure 10.17. The alternative implementation, also using immobilized

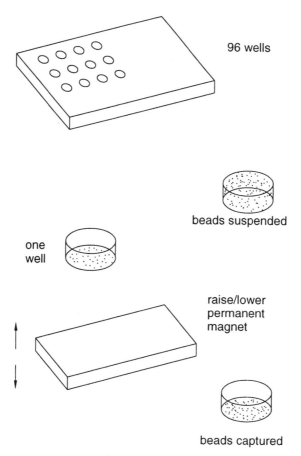

**Figure 10.17** Microtitre plate magnetic separator used by Uhlen for automated solid state DNA sequencing.

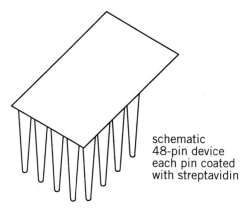

schematic
48-pin device
each pin coated
with streptavidin

**Figure 10.18**   Microtitre plate multipin device used by Landegren for automated DNA sequencing and related automated DNA manipulations.

streptavidin, employs a 48-pin device instead of magnets (Fig. 10.18). Here the immobilized DNA is captured on the ends of plastic pins which have been loaded with streptavidin-conjugated microbeads. A very high density of strepavidin can be generated in this way. It seems clear that by combining magnetic beads and plastic pins, one may be able to automate even more elaborate protocols easily. Recently Landegren reported a very clever variation of this scheme in which the DNA sequencing chemistry is carried out on a plastic comb of the type used to cast the sample slots in a sequencing gel. The teeth of the comb contained immobilized streptavidin beads. Once the chemistry was completed, the contents of the entire comb were loaded onto a DNA sequencing gel by inserting the comb into a gel with wells containing formamide. This solvent disrupts the binding between streptavidin and biotin, denatures the DNA, and releases the DNA samples into the gel. Apparently the formamide has no serious deleterious consequences on the subsequent electrophoresis. Thus, in a very simple way, the problem of automated gel loading has effectively been solved.

## FUTURE IMPROVEMENTS IN LADDER SEQUENCING

Using all the power of current technology, the very best sequencing laboratories can generate more than $10^5$ bp of raw DNA sequence per day per worker. Lanes read to 600 and 700 bases are common. The entire process is fully automated from colony picking to DNA sample preparation to gel loading and running, to the raw sequence analysis and entry into a database. Only gel casting and sequence editing are still manual.

A number of different approaches are being tested to see if the throughput of ladder sequencing can be further improved. Here we will describe some or the more promising or more novel attempts. The basic issues are how to extend a ladder to longer sizes, how to perform the fractionation more rapidly, how to increase the number of different samples that can be handled simultaneously, and how to read the data more rapidly. A number of the approaches share the feature that they use very thin samples. The advantages of such gels for increasing speed were described earlier. A disadvantage is that thin gels mean lower amounts of sample, and this requires greater detection sensitivity.

As detectors are improved, it is to be expected that larger numbers of samples will be loaded on each gel by using closer spaced and narrower lanes. One limitation with the current ALF system is its single color detection; yet the high sensitivity afforded by having the laser in the plane of the gel is a clear advantage. In principle, one could use multiple lasers in the gel, at different positions, and each could be accompanied by a suitable detector array. Ansorge has developed such an instrument, which clearly will have higher throughput since each lane will then be available for multicolor sequencing.

One method of diminishing sample size while retaining sensitive detection is to use a fluorescent microscope as the detector. In Chapter 7 the power of confocal scanning laser microscopy for FISH was described. This microscope also makes an excellent detector for direct scanning of fluorescent-labeled DNA samples in gels. The advantage of the confocal microscope is that it gathers emission very efficiently from a very narrow vertical slice through the sample. Light that emanates from above or below this plane is not imaged. Thus the confocal microscope can detect fluorescence from inside a capillary or thin slab without background due to scattering from the interface between the capillary and the gel, or the interface between the capillary and the external surroundings. This is a major improvement. One consequence is that the capillaries are scanned off line, in order to take full advantage of the scanning speed of the microscope. Dense bundles of capillaries can be made, loaded in parallel with multiple headed syringes, run in parallel, and scanned together (Fig. 10.19). The increase throughput ultimately achievable with this approach may be considerable. A potential limitation is the difficulty in making gel-filled capillaries. This will be alleviated somewhat as it becomes possible to use liquid (noncrosslinked) gels instead of solid (crosslinked) gels. This is because solid gels must be polymerized within the capillary, while liquid gels can just be poured into the capillary. The alternative to capillaries is to use large thin gel slabs. This simplifies the optics needed for on-line detection of the DNA. In anticipation of the considerable demands placed on detector systems by fast running, thin gels, a number of alternative new detectors are being developed as possible readout devices for fluorescence-based DNA sequencing.

## APPROACHES TO DNA SEQUENCING BY MASS SPECTROMETRY

A separate approach to improving ladder sequencing is to change the way in which the labeled DNA fragments are detected. Here considerable attention has been given to mass spectrometry. There are actually three ways in which mass spectrometry might be used, in principle, to assist DNA sequencing. In the simplest case the mass spectrometer is used as a detector for a mass label attached to the DNA strand in lieu of a fluorescent label. Alternatively, the mass of the DNA molecule itself can be measured. In this case the mass spectrometer replaces the need for gel electrophoresis; it separates the DNA molecules and detects their sizes. The most ambitious and difficult potential use of mass spectrometry would involve a fragmentation analysis of the DNA and the determination of all of the resulting species. In this way the mass spectrometer would replace all of the chemistry and electrophoresis steps in conventional ladder sequencing. We are a long way from accomplishing this. In this section we will discuss each of these potential applications of mass spectrometry to enhance DNA sequencing.

Mass spectroscopy is almost as sensitive a detector as fluorescence, with some instruments having sensitivities of the order of thousands of atoms or molecules, and special

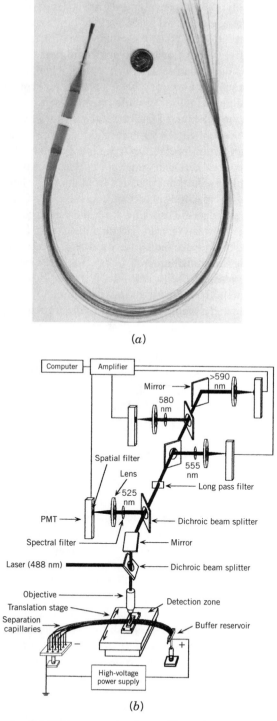

(a)

(b)

**Figure 10.19** Apparatus for DNA sequencing by capillary electrophoresis. (a) An array of gel-filled capillaries used for DNA sequencing. (b) On-line detection by confocal scanning fluorescence microscopy. Figure provided by Richard Mathies. Figure also appears in color insert.

techniques such as ion cyclotron resonance mass spectrometry having even greater sensitivity. However, the principal potential advantages of mass spectra over fluorescence is that isotopic labeling leads to much less of a perturbation of electrophoretic properties than fluorescent labeling, and the number of easily used isotopic labels far exceeds the number of fluorophores that could be used simultaneously. Mass spectrometers actually measure the ratio of mass to charge; the best instruments have a mass to charge resolution of better than 1 part in a million. Thus asking a mass spectrometer to distinguish between, say, two isotopes like $^{34}$S and $^{36}$S is not very demanding if these isotopes reside in small molecules.

One basic strategy in using mass spectrometry as a DNA sequencing detector simply replaces the fluorophore with a stable isotope. Two approaches have been explored. In one case four different stable isotopes of sulfur would be used as a 5' label incorporated, for example, as thiophosphate. In the other case a metal chelate is attached at the 5'-end of the primer, and different stable metal isotopes are used. Some of the possibilities are shown in Table 10.1. Since many of the divalent ions in the table have very similar chemistry, chelates can be built that, in principle, would bind many different elements. Thus, when all the isotopes are considered, there is the possibility of doing analyses with more than 30 different colors. Whether sulfur or metal isotopes are used, the sample must be vaporized and the DNA destroyed so that the only mass detected is that of a small molecule or single atom containing the isotope. With sulfur labeling, one possible role for mass spectrometry is as an on-line detector for capillary electrophoresis. DNA fragments are eluted from the capillary into a chamber where the sample is burned, and the resulting $SO_2$ is ionized and detected.

With metal labeling, a much more complex process is used to analyze the sample by mass spectrometry. This is a technique called resonance ionization spectroscopy (RIS), and it is illustrated in Figure 10.20. Here mass spectrometry would serve to analyze a filter blot, or a thin gel, directly, off line. In RIS just the top few microns of a sample are examined. Either a strong laser beam or an ion beam is used to vaporize the surface of the sample, creating a mixture of atoms and ions. The beam scans the surface in a raster pattern. Any ions produced are pulled away by a strong electric field. Then a set of lasers is used to ionize a particular element of interest; in our case this is the metal atom used as the label. Because ionization energies are higher than the energy in any single laser photon, two or more lasers must be used in tandem to pump the atom up to its ionization state. Then it is detected by mass spectrometry. The same set of lasers can be used to

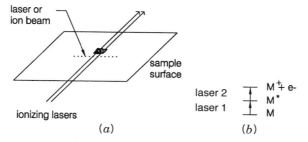

**Figure 10.20**  Resonance ionization mass spectrometry (RIS). (*a*) Schematic design of the instrument used to scan a surface. (*b*) Three electronic states used for the resonance ionization of metal atoms.

excite all of the different stable isotopes of a particular element; however different lasers may be required when different elements are to be analyzed.

An example of RIS as applied to the reading of a DNA sequencing gel is shown in Figure 10.21. The method clearly works; however, it would be helpful to have higher signal to noise. Actually RIS is an extremely sensitive method, with detection limits of the order of a few hundreds to a few thousands of atoms. Very little background should be expected from most of the particular isotopes listed in Table 10.2, since many of these are not common in nature, and in fact most of the elements involved, with the notable exception of iron and zinc, are not common in biological materials. The problem is that gel electrophoresis is a bulk fractionation; very few of the DNA molecules separated actually lie in the thin surface layer that can be scanned. Similarly typical blotting membranes are also not really surfaces; DNA molecules penetrate into them for quite a considerable distance.

To assist mass spectrometric analysis of DNA, it would clearly be helpful to have simple, reproducible ways of introducing large numbers of metal atoms into a DNA molecule and firmly anchoring them there. One approach to this is to synthesize base analogs that have metal chelates attached to them, in a way that does not interfere with their ability to

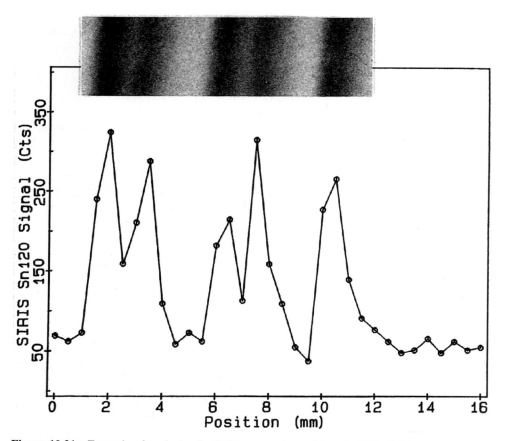

**Figure 10.21**  Example of analysis of a DNA sequencing gel lane by RIS. Gel image appears at top; RIS signal below. Provided by Bruce Jacobson. See Jacobson et al. (1990).

**TABLE 10.2   Stable Metal Isotopes Bound by Metallothionein**

| $_{26}$Fe | $^{54}$Fe | $_{50}$Sn | $^{112}$Sn |
|---|---|---|---|
| | $^{56}$Fe | | $^{114}$Sn |
| | $^{57}$Fe | | $^{115}$Sn |
| | $^{58}$Fe | | $^{116}$Sn |
| | | | $^{117}$Sn |
| $_{27}$Co | $^{56}$Co | | $^{118}$Sn |
| | | | $^{119}$Sn |
| $_{28}$Ni | $^{58}$Ni | | $^{120}$Sn |
| | $^{60}$Ni | | $^{122}$Sn |
| | $^{61}$Ni | | $^{124}$Sn |
| | $^{62}$Ni | | |
| | $^{64}$Ni | $_{79}$Au | $^{197}$Au |
| $_{29}$Cu | $^{63}$Cu | | |
| | $^{65}$Cu | $_{80}$Hg | $^{196}$Hg |
| | | | $^{198}$Hg |
| $_{30}$Zn | $^{64}$Zn | | $^{199}$Hg |
| | $^{66}$Zn | | $^{200}$Hg |
| | $^{67}$Zn | | $^{201}$Hg |
| | $^{68}$Zn | | $^{202}$Hg |
| | $^{70}$Zn | | $^{204}$Hg |
| | | $_{82}$Pb | $^{204}$Pb |
| $_{47}$Ag | $^{107}$Ag | | $^{206}$Pb |
| | $^{109}$Ag | | $^{207}$Pb |
| | | | $^{208}$Pb |
| $_{48}$Cd | $^{106}$Cd | | |
| | $^{108}$Cd | $_{83}$Bi | $^{209}$Bi |
| | $^{110}$Cd | | |
| | $^{111}$Cd | | |
| | $^{112}$Cd | | |
| | $^{113}$Cd | | |
| | $^{114}$Cd | | |
| | $^{116}$Cd | | |
| Total 50 species | | | |

hybridize to complementary sequences. An example is shown in Figure 10.22. An alternative approach is to adapt the streptavidin-biotin system to place metals wherever in a DNA one places a biotin. This can be done by using the chimeric fusion protein shown in Figure 10.23. This fusion combines the streptavidin moiety with metallothionein, a small cysteine-rich protein that can bind 8 divalent metals or up to 12 heavy univalent metals. The list of different elements known to bind tightly to metallothionein is quite extensive. All of the isotopes in Table 10.2 are from elements that bind to metallothionein. The fusion protein is a tetramer because its quaternary structure is dominated by the extremely stable streptavidin tetramer. Thus there are four metallothioneins in the complex, and each retains its full metal binding ability. As a result, when this fusion protein is used to label biotinylated DNA, one can place 28 to 48 metals at the site of each biotin. The use of this fusion protein should provide a very substantial increase in the sensitivity of RIS for DNA detection.

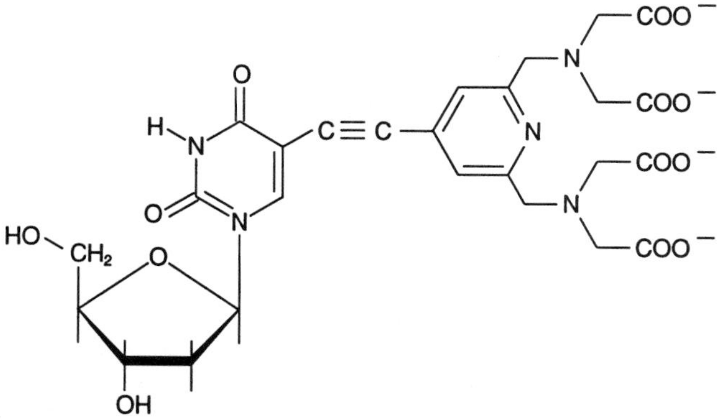

**Figure 10.22**   A metal chelate derivative of a DNA base suitable as an RIS label.

While mass spectrometry has great potential to detect metal labels in biological systems, a drawback of the method is that current RIS instrumentation is quite costly. Another limitation is that RIS destroys the surface of the sample, so it may be difficult to read each gel or blot more than once. Alternative schemes for the use of metals as labels in DNA sequencing exist. One is described in Box 10.1

The second way to use mass spectrometry to analyze DNA sequences ladders is to attempt to place the DNA molecules that constitute the sequencing ladder into the vapor phase and detect their masses. In essence this is DNA electrophoresis in vacuum. A key requirement of the approach is to minimize fragmentation once the molecules have been placed in vacuum, since all of the desired fragmentation needed to read the sequence has already been carried out through prior Sanger chemistry; any additional fragmentation is confusing and leads to a loss in experimental sensitivity. Two methods show great promise for placing macromolecules into the gas phase. In one, called electrospray (ES), a fine mist of macromolecular solution is sprayed into a vacuum chamber; the solvent evaporates and is pumped away. The macromolecule retains an intrinsic charge and can be accelerated by electrical fields, and its mass subsequently measured. In the second approach, matrix-assisted laser desorption and ionization (MALDI), the macromolecule is suspended in a matrix that can be disintegrated by light absorption. After excitation with a pulsed laser, the macromolecule finds itself suspended in the vapor phase; it can be accelerated if it has a charge. These two procedures appear to work very well for proteins. They also work very well on oligonucleotides, and a few examples of high-resolution mass spectra have been reported for compounds with as many as 100 nucleotides.

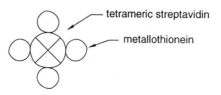

**Figure 10.23**   Structure of a streptavidin-metallothionein chimeric protein capable of bringing 28 to 48 metal atoms to a single biotin-labeled site in a DNA.

**BOX 10.1**
**MULTIPHOTON DETECTION (MPD)**

In contrast to RIS, a much simpler, nondestructive detector has been developed by Andy Druckier that has the capacity to analyze many different metal isotopes simultaneously, but these must be short lived radioisotopes. The principle used in this exquisitely sensitive detector is shown in Figure 10. 24. It is based on certain special radioisotopes that emit three particles in close succession. First a positron and an electron are emitted simultaneously, at 180-degree angles. Shortly thereafter a gamma ray is emitted. The gamma energy depends on the particular isotope in question. Hence many different isotopes can be discriminated by the same detector. The electron and positron events are used to gate the detector. In this way the background produced by gamma rays from the environment is extremely low. The potential sensitivity of this device is just a few atoms. It is a little early to tell just how well suited it is for routine applications in DNA analysis. Prototype many-element detector arrays have been built that examine not just what kind of decay event occurred but also where it occurred. Such position sensitive detectors are in routine use in high-energy physics. The advantage of detector arrays is that an entire gel sample can be analyzed in parallel rather than having to be scanned in a raster fashion. This leads to an enormous increase in sensitivity.

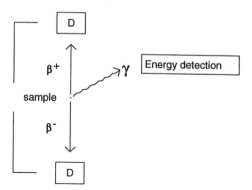

**Figure 10.24**   MPD: Ultra sensitive detection of many different radioisotopes by triple coincidence counting.

However, there appear to be a number of technical difficulties that must be resolved if ES or MALDI mass spectrometry are to become a generally used tool for DNA sequencing. One complication is that a number of different charged species are typically generated. This is a particularly severe problem in ES, but it is also seen in MALDI. The multiplicity of species is an advantage in some respects, since one knows that charges must be integral, and thus the species can serve as an internal calibration. However, each species leads to a different charge/mass peak in the spectrum which makes the overall spectrum complex. The most commonly used mass analyzer for DNA work has been time of flight (TOF). This is a particularly simple instrument well adapted for MALDI work—one simply measures the time lag between the initial laser excitation pulse and the time DNA samples reach the detector after a linear flight. TOF is an enormously rapid measurement, but it does not have as high resolution or sensitivity as some other methods. Today, a typical good TOF mass resolution on an oligonucleotide would be 1/1000. This is sufficient to sequence short targets as shown by the example in Figure 10.25, but DNAs longer than 60 nucleotides typically show far worse resolution that does not allow DNA sequencing. A more serious problem with current MALDI-TOF mass spectrometry is that starting samples of about 100 fmol are required. This is larger than needed with capillary electrophoretic DNA sequencing apparatuses which should become widely available within the next few years.

While current MALDI-TOF mass spectral sequence reads are short, they are extraordinarily accurate. An example is shown in Figure 10.26, where a compression makes a section of sequence impossible to call when the analysis is performed electrophoretically. In

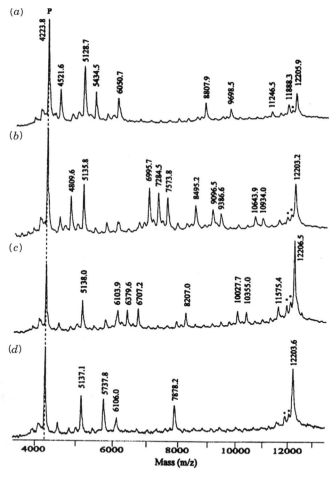

**Figure 10.25** MALDI-TOF mass spectra of sequencing ladders generated from an immobilized 39-base template strand d(TCT GGC CTG GTG CAG GGC CTA TTG TAG TGA CGT ACA). P indicates the primer d(TGT ACG TCA CAA CT). The peaks resulting from depurination are labeled by an asterisk. (*a*) A-reaction, (*b*) C-reaction, (*c*) G-reaction, and (*d*) T-reaction. MALDI-TOF MS measurements were taken on a reflectron TOF MS. From Köster et al. (1996).

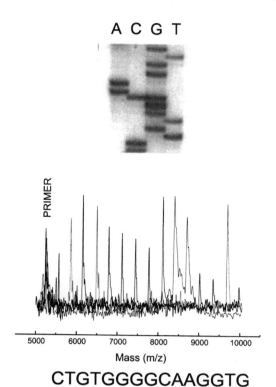

**Figure 10.26** Sequencing a compression region in the beta globin gene by gel electrophoresis (*top*) and MALDI-TOF mass spectrometry (*bottom*). Note in the gel data a poorly resolved set of G's topped by what appears to be a C, G heterozygote. The true sequence, GGGGC, is obvious in the mass spectrum. Provided by Andi Braun. Figure also appears in color insert.

MS, however, all that is measured is the ratio of mass to charge, so any secondary structure effects are invisible. Thus the resulting sequence ladder is unambiguous.

A second mass analyzer that has been used for analysis of DNA is Fourier transform ion cyclotron resonance (FT-ICR) mass spectrometry. Here ions are placed in a stable circular orbit, confined by a magnetic field. Since they continuously circulate, one can repeatedly detect the DNA, and thus very high experimental sensitivity is potentially achievable. The other great advantage of FT-ICR instruments is that they have extraordinary resolution, approaching $1/10^6$ in some cases. This produces extremely complex spectra because of the effects of stable isotopes like $^{13}C$. However, since the mass differences caused by such isotopes are known in advance, the bands they create can be used to assist spectral assignment and calibration. FT-ICR can be used for direct examination of Sanger ladders, or it can be used for more complex strategies as described below. The major disadvantages of FT-ICR are the cost and complexity of current instrumentation and the greater complexity of the spectra. An example of a FT-ICR spectrum of an oligonucleotide is shown in Figure 10.27.

The third strategy in mass spectrometric sequencing is the traditional one in this field. A sample is placed into the vapor phase and then fragmentation is stimulated either by collision with other atoms or molecules or by laser irradiation. Most fragmentation under these conditions appears to occur by successive release of mononucleotides from either

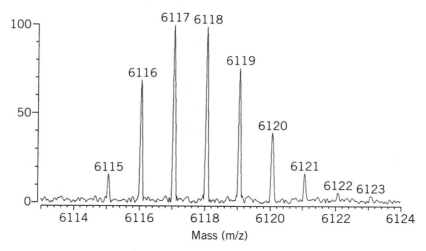

**Figure 10.27**   An example of FT-ICR mass spectrometry of an oligonucleotide. Here the complex set of bands seen for the parent ion of a single compound are molecules with different numbers of $^{13}C$ atoms. Provided by Kai Tang.

end of the DNA chain. While this approach is feasible and has yielded the sequence of one 50-base DNA fragment, the resulting spectra are so complex that at least at present this approach does not appear likely to become a routine tool for DNA analysis.

One great potential power of mass spectroscopy for direct analysis of DNA ladders is speed. In principle, at most a few seconds of analysis would be needed to collect sufficient data to analyze one ladder. This is much faster than any current extrapolation of electrophoresis rates. The major ultimate challenge may be finding a way to introduce samples into a mass spectrometer fast enough to keep up with its sequencing speed.

## RATE-LIMITING STEPS IN CURRENT DNA SEQUENCING

Today, laboratories that can produce $10^4$ to $10^5$ raw base pairs of DNA sequence per person per day are virtually unanimous in their conclusion that at this rate, analysis of the data is the rate-limiting step. So long as one has a good supply of samples worth sequencing, the rate of data acquisition with current automated equipment is not a barrier. A few laboratories have equipment that operates at ten times this rate. This has not yet been used in a steady production mode for sequence determination, but once it enters this mode, there is every reason to believe that data analysis will continue to be rate limiting. Some of the potential enhancements we have described for ladder sequencing promise almost certainly to yield an additional factor of 10 improvement in throughput over the next few years, and a factor of 100, eventually, is not inconceivable. This will surely exacerbate the current problems of data analysis.

The rapid rate of acquisition of DNA sequence data makes it critical that we improve our methods of designing large-scale DNA sequencing projects and develop improved abilities for on-line analysis of the data. This analysis includes error correction, assembly of raw sequence into finished sequence, and interpreting the significance of the sequencing data. In the next chapter we will deal with the first two issues as part of our consideration of the strategies for large-scale DNA sequencing. The issue of interpreting DNA sequence is deferred until Chapter 15.

## SOURCES AND ADDITIONAL READINGS

Ansorge, W., Zimmermann, J., Schwager, C., Stegemann, J., Erfle, H., and Voss, H. 1990. One label, one tube, Sanger DNA sequencing in one and two lanes on a gel. *Nucleic Acids Research* 18: 3419–3420.

Carrilho, E., Ruiz-Martinez, M. C., Berka, J., Smirov, I., Goetzinger, W., Miller, A. W., Brady, D., and Karger, B. L. 1996. Rapid DNA sequencing of more than 1000 bases per run by capillary electrophoresis using replaceable linear polyacrylamide solutions. *Analytical Chemistry* 68: 3305–3313.

Ewing, B., Hillier, L., Wendl, M. C., and Green, P. 1998. Base-calling of automated sequencer traces using Phred. I. Accuracy assessment. *Genome Research* 8: 175–185.

Ewing, B., and Green, P. 1998. Base-calling of automated sequencer traces using Phred. II. Error possibilities. *Genome Research* 8: 186–194.

Glazer, A. N., and Mathies, R. A. 1997. Energy-transfer fluorescent reagents for DNA analyses. *Current Opinion in Biotechnology* 8: 94–102.

Grothues, D., Voss, H., Stegemann, J., Wiemann, S., Sensen, C., Zimmerman, J., Schwager, C., Erfle, H., Rupp, T., and Ansorge, W. 1993. Separation of up to 1000 bases on a modified A.L.F. DNA sequencer. *Nucleic Acids Research* 21: 6042–6044.

H. S. Rye, S. Yue, D. E. Wemmer, M. A. Quesada, R. P. Haugland, R. A. Mathies, and A. N. Glazer. Stable Fluorescent Complexes of Double-Stranded DNA with Bis-Intercalating Asymmetric Cyanine Dyes: Properties and Applications. *Nucleic Acids Research* 20, 3803-3812 (1992).

Hultman, T., Stahl, S., Hornes, E., and Uhlen, M. 1989. Direct solid phase sequencing of genomic and plasmid DNA using magnetic beads as solid support. *Nucleic Acids Research* 19: 4937–4936.

Jacobson, K. B., Arlinghaus, H. F., Schmitt, H. W., Sacherleben, R. A., et al. 1990. An approach to the use of stable isotopes for DNA sequencing. *Genomics* 8: 1–9.

Kalman, L. V., Abramson, R. D., and Gelfand, D. H. 1995. Thermostable DNA polymerases with altered discrimination properties. *Genome Science and Technology* 1: 42.

Koster, H., Tang, K., Fu, D.-J., Braun, A., van den Boom, D., Smith, C. L., Cotter, R. J., and Cantor, C. R. 1996. A strategy for rapid and efficient DNA sequencing by mass spectrometry. *Nature Biotechnology* 14: 1123–1128.

Kustichka, A. J., Marchbanks, M., Brumley, R. L., Drossman, H., and Smith, L. M. 1992. High speed automated DNA sequencing in ultrathin slab gels. *Bio/Technology* 10: 78–81.

Kwiatkowski, M., Samiotaki, M., Lamminmaki, U., Mukkala, V.-M., and Landegren U. 1994. Solid-phase synthesis of chelate-labelled oligonucleotides: Application in triple-color ligase-mediated gene analysis. *Nucleic Acids Research* 22: 2604–2611.

Luckey, J. A., Norris, T. A., and Smith, L. M. 1993. Analysis of resolution in DNA sequencing by capillary gel electrophoresis. *Journal of Physical Chemistry* 97: 3067–3075.

Luckey, J. A., and Smith, L. M. 1993. Optimization of electric field strength for DNA sequencing in capillary gel electrophoresis. *Analytical Chemistry* 65: 2841–2850.

Naeve, C. W., Buck, G. A., Niece, R. L., Pon, R. T., Robertson, M., and Smith, A. J. 1995. Accuracy of automated DNA sequencing: A multi-laboratory comparison of sequencing results. *BioTechniques* 19: 448–453.

Parker, L. T., Deng, Q., Zakeri, H., Carlson, C., Nickerson, D. A., and Kwok, P. Y. 1995. Peak height variations in automated sequencing of PCR products using Taq dye-terminator chemistry. *BioTechniques* 19: 116–121.

Porter, K. W., Briley, J. D., and Shaw, B. R. 1997. Direct PCR sequencing with boronated nucleotides. *Nucleic Acids Research* 25: 1611–1617.

Stegemann, J., Schwager, C., Erfle, H., Hewitt, N., Voss, H., Zimmermann, J., and Ansorge, W. 1991. High speed on-line DNA sequencing on ultathin slab gels. *Nucleic Acids Research* 19: 675–676.

Tabor, S., and Richardson, C. C. 1995. A single residue in DNA polymerases of the *Escherichia coli* DNA polymerase I family is critical for distinguishing between deoxy- and dideoxyribonucleotides. *Proceedings of the National Academy of Sciences USA* 92: 6339–6343.

Yamakawa, H., Nakajima, D., and Ohara, O. 1996. Identification of sequence motifs causing band compressions on human cDNA sequencing. *DNA Research* 3: 81–86.

Yarmola, E., Sokoloff, H., and Chrambach, A. 1996. The relative contributions of dispersion and diffusion to band spreading (resolution) in gel electrophoresis. *Electrophoresis* 17: 1416–1419.

# 11 Strategies for Large-Scale DNA Sequencing

## WHY STRATEGIES ARE NEEDED

Consider the task of determining the sequence of a 40-kb cosmid insert. Suppose that each sequence read generates 400 base pairs of raw DNA sequence. At the minimum two-fold redundancy needed to assemble a completed sequence, 200 distinct DNA samples will have to be prepared, sequenced, and analyzed. At a more typical redundancy of ten, 1000 samples will have to be processed. The chain of events that takes the cosmid, converts it into these samples, and provides whatever sequencing primers are needed is the upstream part of a DNA sequencing strategy. The collation of the data, their analysis, and their insertion into databases is the downstream part of the strategy. It is obvious that this number of samples and the amount of data involved cannot be handled manually, and the project cannot be managed in an unsystematic fashion. A key part of the strategy will be developing the protocols needed to name samples generated along the way, retaining an audit of their history, and developing procedures that minimize lost samples, unnecessary duplication of samples or sequence data, and prevent samples from becoming mixed up.

There are several well-established, basic strategies for sequencing DNA targets on the scale of cosmids. It is still not yet totally clear how these can best be adopted for Mb scale sequencing, but it is daunting to realize that this is an increase in scale by a factor of 50, which is not something to be undertaken lightly. The basic strategies range from shotgun, which is a purely random approach, to highly directed walking methods. In this chapter the principles behind these strategies will be illustrated, and some of their relative advantages and disadvantages will be described. In addition we will discuss the relative merits of sequencing genomes and sequencing just the genes, by the use of cDNA libraries. Current approaches for both of these objectives will be described.

## SHOTGUN DNA SEQUENCING

This is the method that has been used, up to now, for virtually all large-scale DNA sequencing projects. Typically the genomic (or cosmid) DNA is broken by random shear. The DNA fragments are trimmed to make sure they have blunt ends, and as such they are subcloned into an M13 sequencing vector. By using shear breakage, one can generate fragments of reasonably uniform size, and all regions of the target should be equally represented. At first, randomly selected clones are used for sequencing. When this becomes inefficient, because the same clones start to appear often, the set of clones already sequenced can be pooled and hybridized back to the library to exclude clones that come from regions that have already been over sampled (a similar procedure is described later in the chapter for cDNA sequencing; see Fig. 11.17). The use of a random set of small targets ensures that overlapping sequence data will be obtained, across the entire target, and that both strands will be sampled. An obvious complication that this causes is that one does not know a priori to which strand a given segment of sequence corresponds. This has to be sorted out by software as part of problem of assembling the shotgun clones into finished sequence.

Shotgun sequencing has a number of distinct advantages. It works, it uses only technology that is fully developed and tested, and this technology is all fairly easy to automate. Shotgun sequencing is very easy at the start. As redundancy builds, it catches some of the errors. The number of primers needed is very small. If one-color sequencing or dye terminator sequencing is used, two primers suffice for the entire project if both sides of each clone are examined; only one primer is needed if a single side of each clone is examined. With four-color dye primer sequencing the number of primers needed are eight and four, respectively.

The disadvantages of shotgun sequencing fall into three categories. The redundancy is very high, definitely higher than more orderly methods, and so, in principle, optimized shotgun sequencing will be less efficient than optimized directed methods. The assembly of shotgun sequence data is computationally very intensive. Algorithms exist that do the job effectively, such as Phil Green's phrap program, but these use much computer time. The reason is that one must try, in assembly, to match each new sequence obtained with all of the other sequences and their complements in order to compensate for the lack of prior knowledge of where in the target and which strand the sequence derives from. Repeats can confuse the assembly process. A simple example is shown in Figure 11.1. Suppose that two sequence fragments have been identified with a nearly identical repeat. One possibility is that the two form a contig; they derive from two adjacent clones that overlap at the repeat, and the nonidentity is due to sequencing errors. The other possibility is that the two sequences are not a true contig, and they should not be assembled. The software must keep track of all these ambiguities and try to resolve them, by appropriate statistical tests, as the amount of data accumulates.

Closure is a considerable problem in most shotgun sequencing. Because of the statistics of random picking and the occurrence of uncloneable sequences, even at high redundancy, and even if hybridization is used to help select new clones, there will be gaps. These have to be filled by an alternate strategy; since the gaps are usually small, PCR-based sequencing across them is usually quite effective. A typical shotgun approach to se-

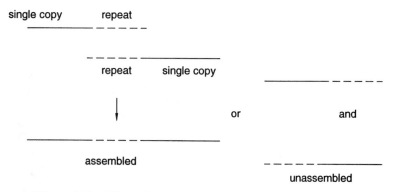

**Figure 11.1**   Effect of a repeated sequence on sequence assembly.

quencing an entire cosmid will require a total of 400 kb of sequence data (redundancy of 10) and 1000 four-color sequencing gel lanes. If 32 samples are analyzed at once, and three runs can be done per day, this will require 10 working days with a single automated instrument. In practice, around 20% sample failure can be anticipated, and most laboratories can only manage two runs a day. Therefore a more conservative time estimate is that a typical laboratory, with one automated instrument at efficient production speed, will require about 18 working days to sequence a cosmid. This is a dramatic advance from the rates of data acquisition in the very first DNA sequencing projects. However, it is not yet a speed at which the entire human genome can be done efficiently. At $8 \times 10^4$ cosmid equivalents, the human genome would require $4 \times 10^3$ laboratory years of effort at this rate, even assuming seven-day work weeks. It is not an inconceivable task, but the prospects will be much more attractive once an additional factor of 10 in throughput can be achieved. This is basically achievable by the optimization of existing gel-based methods described in Chapter 10.

## DIRECTED SEQUENCING WITH WALKING PRIMERS

This strategy is the opposite of shotgun sequencing, in that nothing is left to chance. The basic idea is illustrated in Figure 11.2. A primer is selected in the vector arm, adjacent to the genomic insert. This is used to read as far along the sequence as one can in a single pass. Then the end of the newly determined sequence is used to synthesize another primer, and this in turn serves to generate the next patch of sequence data. One can walk in from both ends of the clone, and in practice, this will allow both strands to be examined.

The immediate, obvious advantage of primer walking is that only about a twofold redundancy is required. There are also no assembly or closure problems. With good sensitivity in the sequence reading, it is possible to perform walking primer sequencing directly on bacteriophage or cosmid DNA. This eliminates almost all of the steps involved in subcloning and DNA sample preparation. These advantages are all significant, and would make directed priming the method of choice for large-scale DNA projects if a number of the current disadvantages in this approach could be overcome.

The major, obvious, disadvantage in primer walking is that a great deal of DNA synthesis is required. If 20-base-long primers are used, to guarantee that the sequence selected is unique in the genome, one must synthesize 5% of the total sequence for each

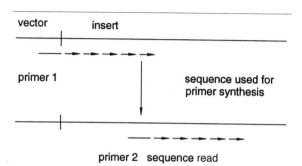

**Figure 11.2**   Basic design of a walking primer sequencing strategy.

strand, or 10% of the overall sequence. By accepting a small failure rate, the size of the primer can be reduced. For example, in 45 kb of DNA sequence, one can estimate by Poisson statistics that the probability of a primer sequence not occurring elsewhere, and only at the site of interest is

| Primer length | 10 | 11 | 12 |
|---|---|---|---|
| Unique | 0.918 | 0.979 | 0.998 |

Thus primers less than 12 bases long are not going to be reliable. With 12 mers, one still has to synthesize 6% of the total sequence to examine both strands. This is a great deal of effort, and considerable expense is involved with the current technology for automated oligonucleotide synthesis.

The problem is that existing automated oligonucleotide synthesizers make a minimum of a thousand times the amount of primer actually needed for a sequence determination. The major cost is in the chemicals used for the synthesis. Since the primers are effectively unique, they can be used only once, and therefore all of the cost involved in their synthesis is essentially wasted. Methods are badly needed to scale down existing automated DNA synthesis. The first multiplex synthesizers built could only make 4 to 10 samples at a time; this is far less than would be needed for large-scale DNA sequencing by primer walking. Recently instruments have been described that can make 100 primers at a time in amounts 10% or less of what is now done conventionally. However, such instruments are not yet widely available. In the near future primers will be made thousands at a time by in situ synthesis in an array format (see Chapter 12).

A second major problem with directed sequencing is that each sample must be done serially. This is illustrated in Figure 11.3. Until the first patch of sequence is determined, one does not know which primer to synthesize for the next patch of sequence. What this means is that it is terribly inefficient to sequence only a single sample at once by primer walking. Instead, one must work with at least as many samples as there are DNA sequencing lanes available, and ideally one would be processing two to three times this amount so that synthesis is never a bottleneck. This means that the whole project has to be orchestrated carefully so that an efficient flow of samples occurs. Directed sequencing is not appropriate for a single laboratory interested in only one target. Its attractiveness is for a laboratory that is interested in many different targets simultaneously, or that works on a system that can be divided into many targets that can be handled in parallel.

At present, a sensible compromise is to start a sequencing project with the shotgun strategy. This will generate a fairly large number of contigs. Then at some point one switches to a directed strategy to close all of the gaps in the sequence. It ought to be possible to make this switch at an earlier stage than has been done in the past as walking becomes more efficient and can cover larger regions. In such a hybrid scheme the initial

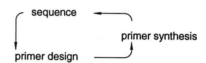

**Figure 11.3** Successive cycles of primer synthesis and sequence determination lead to a potential bottleneck in directed sequencing strategies.

shotgunning serves to accumulate large numbers of samples, which are then extended by parallel directed sequencing.

Three practical considerations enter the selection of particular primer sequences. These can be dealt with by existing software packages that scan the sequence and suggest potentially good primers. The primer must be as specific as possible for the template. Ideally a sequence is chosen where slight mismatches elsewhere on the template are not possible. The primer sequence must occur once and only once in the template. Finally the primer should not be capable of forming stable secondary structure with itself—intramolecularly, as a hairpin, or intermolecularly—as a dimer. The former category is particularly troublesome, since, as illustrated in Chapter 10, there are occasional DNA sequences that form unexpectedly stable hairpins, and we do not yet understand these well enough to generalize and be able to identify and reject them a priori.

## PRIMING WITH MIXTURES OF SHORT OLIGONUCLEOTIDES

Several groups have recently developed schemes to circumvent the difficulties required by custom synthesis of large numbers of unique DNA sequencing primers. The relative merits of these different schemes are still not fully evaluated, but on the surface, they seem quite powerful. The first notion, put forth by William Studier, was to select, arbitrarily, a subset of the $4^9$ possible 9-mers and make all of these compounds in advance. The chances that a particular 9-mer would be present in a 45-kb cosmid are given below, calculated from the Poisson distribution P($n$), where $n$ is the expected number of occurrences:

$$P(O) = 0.709 \quad P(1) = 0.244 \quad P(>1) = 0.047$$

Therefore, with no prior knowledge about the sequence, a randomly selected 9-mer primer will give a readable sequence about 20% of the time. If it does, as shown in Figure 11.4, its complement will also give readable sequence. Thus the original subset of $4^9$ compounds should be constructed of 9-mers and their complements.

One advantage of this approach is that no sequence information about the sample is needed initially. The same set of primers will be used over and over again. As sequence data accumulate, some will contain sites that allow priming from members of the premade 9-mer subset. Actual simulations indicate that the whole process could be rather efficient. It is also parallelizable in an unlimited fashion because one can carry as many different targets along as one has sequencing gel lanes available. An obvious and serious

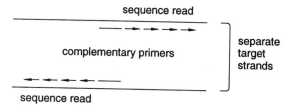

**Figure 11.4**  Use of two complementary 9-base primers allows sequence reading in two directions.

disadvantage of this method is the high fraction of wasted reactions. There are possible ways to improve this, although they have not yet been tried in practice. For example one could pre-screen target-oligonucleotide pairs by hybridization, or by DNA polymerase-catalyzed total incorporation of dpppN's. This would allow those combinations in which the primer is not present in the target to be eliminated before the time-consuming steps of running and analyzing a sequencing reaction are undertaken.

At about the same time that Studier proposed and began to test his scheme, an alternative solution to this problem was put forth by Wacslaw Szybalski. His original scheme was to make all possible 6-mers using current automated DNA synthesis techniques. With current methods this is quite a manageable task. Keeping track of 4096 similar compounds is feasible with microtitre plate handling robots, although this is at the high end of the current capacity of commercial equipment. To prime DNA polymerase, three adjacent 6-mers are used in tandem, as illustrated in Figure 11.5. They are bound to the target, covalently linked together with DNA ligase, and then DNA polymerase and its substrates are added. The principle behind this approach is that end stacking (see Chapter 12) will favor adjacent binding of the 6-mers, even though there will be competing sites elsewhere on the template. Ligation will help ensure a preference for sequence-specific, adjacent 6-mers because ligase requires this for its own function. The beauty of this approach is that the same 6-mer set will be used ad infinitum. Otherwise, the basic scheme is ordinary directed sequencing by primer walking. After each step the new sequence must be scanned to determine which combination of 6-mers should be selected as the optimal choice for the next primer construction.

The early attempts to carry out tandem 6-mer priming seemed discouraging. The method appeared to be noisy and unreliable. Too many of the lanes gave unreadable sequence. This prompted several variants to be developed, as described below. More recently, however, conditions have been found which make the ligation-mediated 6-mer priming very robust and dependable. This approach, or one of the variants below, is worth considering seriously for some of the next generation large-scale DNA sequencing projects. The key issues that need to be resolved are its generality and the ultimate cost savings it affords against any possible loss in throughput. Smaller-scale, highly multiplexed synthesis of longer primers will surely be developed. These larger primers are always

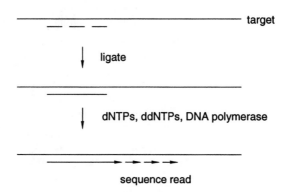

**Figure 11.5** Directed sequencing by ligation-mediated synthesis of a primer, directed by the very template that constitutes the sequencing target.

likely to perform slightly better than ligated mixtures of smaller compounds because the possibility of using the 6-mers in unintended orders, or using only two of the three, is avoided. There is enough experience with longer primers to know how to design them to minimize chances of failure. Similar experience must be gained with 6-mers before an accurate evaluation of their relative effectiveness can be made.

Two more recent attempts to use 6-base oligonucleotide mixtures as primers avoid the ligation step. This saves time, and it also saves any potential complexities arising from ligase side reactions. Studier's approach is to add the 6-mers to DNA in the presence of a nearly saturating amount of single-stranded DNA binding protein (ssb). This protein binds to DNA single strands selectively and cooperatively. Its effect on the binding of sets of small complementary short oligonucleotides is shown schematically in Figure 11.6. The key point is the high cooperativity of ssb to single-stranded DNA. What this does is to make configurations in which 6-mers are bound at separate sites less favorable, energetically, than configurations in which they are stacked together. In essence, ssb cooperativity is translated into 6-mer DNA binding cooperativity. This cuts down on false starts from individual or mispaired 6-mers. An unresolved issue is how much mispairing still can go on under the conditions used. We have faced before the problem in evaluating this; there are many possible 6-mers, and many more possible sets of 6-mer mixtures, so only a minute fraction of these possibilities has yet been tried experimentally. One simplification is the use of primers with 5-mC and 2-aminoA, instead of C and A, respectively. This allows five base primers to be substituted for six base primers, reducing the complexity of the overall set of compounds needed to only 1024.

An alternative approach to the problem has been developed by Levy Ulanovsky. He has shown that conditions can be found where it is possible to eliminate both the ligase and the ssb and still see highly reliable priming. Depending on the particular sequence, some five base primers can be substituted for six base primers. The effectiveness of the approach appears to derive from the intrinsic preference of DNA polymerase for long primers, so a stacked set is preferentially utilized even though individual members of the set may bind to the DNA template elsewhere alone or in pairs. It remains to be seen, as each of these approaches is developed further, which will prove to be the most efficient and accurate. Note that in any directed priming scheme, for a 40-kb cosmid, one must sequence 80 kb of DNA. If a net of 400 new bases of high-quality DNA sequence can be completed per run, it will take 200 steps to complete the walk. As a worst-case analysis, one may have to search the 3'-terminal 100 bases of a new sequence to find a target suitable for priming. Around 500 base pairs of sequence must actually be accumulated per run to achieve this efficiency. Thus a directed priming cosmid sequencing project is still a considerable amount of work.

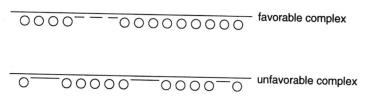

**Figure 11.6**  Use of single-stranded binding protein (ssb) to improve the effectiveness of tandem, short sequencing primers.

## ORDERED SUBDIVISION OF DNA TARGETS

There are a number of strategies for DNA sequencing that lie between the extreme randomness of the shotgun approach and the totally planned order of primer walking. These schemes all have in common a set of cloning or physical purification steps that divides the DNA target into samples that can be sequenced and that also provides information about their location. In this way a large number of samples can be prepared for sequencing in parallel, which circumvents some of the potential log jams that are undesirable features of primer walking. At the same time from the known locations one can pick clones to sequence in an intelligent, directed way and thus cut down substantially on the high redundancy of shotgun sequencing. One must keep in mind, however, that the cloning or separation steps needed for ordered subdivision of the target are themselves quite labor intensive and not yet easily automated. Thus there is no current consensus on which is the best approach. One strategy used recently to sequence a number of small bacterial genomes is the creation of a dense (highly redundant) set of clones in a sequence-ready vector—that is, a vector like a cosmid which allows direct sequencing without additional subcloning. The sequence at both ends of each of the clones is determined by using primers from known vector sequence. As more primary DNA sequence data become known and available in databases, the chances that one or both ends of each clone will match previously determined sequence increase considerably. Such database matches, as well as any matches among the set of clones themselves, accomplish two purposes at once. They order the clones, thus obviating much or all of the need for prior mapping efforts. In addition they serve as staging points for the construction of additional primers for directed sequencing. Extensions of the use of sequence-ready clones are discussed later in the chapter.

Today, there is no proven efficient strategy to subdivide a mega-YAC for DNA sequencing. Some strategies that have proved effective for subdividing smaller clones are described in the next few sections. A megabase YAC is, at an absolute minimum, going to correspond to 25 cosmids. However, there is no reason to think that this is the optimal way to subdivide it. Even if direct genomic sequencing from YACs proves feasible, one will want to have a way of knowing, at least approximately, where in the YAC a given primer or clone is located, to avoid becoming hopelessly confused by interspersed repeating sequences. However, for large YACs there is an additional problem. As described elsewhere, these clones are plagued by rearrangements and unwanted insertions of yeast genomic DNA. For sequencing, the former problem may be tolerable, but not the latter. To use mega-YACs for direct sequencing, we would first need to develop methods to remove or ignore contaminating yeast genomic DNA.

## TRANSPOSON-MEDIATED DNA SEQUENCING

The structure of a typical transposon is shown in Figure 11.7. It has terminal repeated sequences, a selectable marker, and codes for a transposase that allows the transposon to hop from one DNA molecule (or location) to another. The ideal transposon for DNA sequencing will have recognition sites for a rare restriction enzyme, like Not I, located at both ends, and it will have different, known, unique DNA sequences close to each end. A scheme for using transposons to assist the DNA sequencing of clones in *E. coli* is shown in Figure 11.8. The clone to be sequenced is transformed into a transposon-carrying *E.*

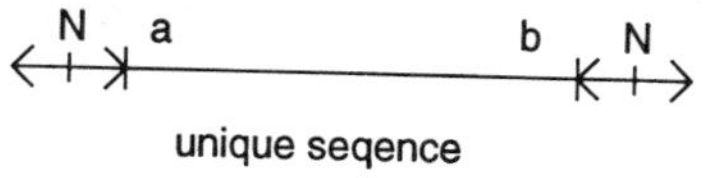

N a b N
unique seqence

**Figure 11.7**   Structure of a typical transposon useful for DNA sequencing. $N$ is a Not I site; $a$ and $b$ are unique transposon sequences. Arrow heads show the terminal inverted repeats of the transposon.

*coli.* It is present in an original vector containing a single infrequent cutting site (which can be the same as the ones on the transposon) and two unique known DNA sequences near to the vector-insert boundary. The transposon may jump into the cloned DNA. If it does so, this appears to occur approximately at random. The nonselectivity of transposon hopping is a key assumption in the potential effectiveness of this strategy. From results that have been obtained thus far, the randomness of transposon insertions seems quite sufficient to make this approach viable.

DNA from cells in which transposon jumping has potentially occurred can be transferred to a new host cell that does not have a resident transposon, and recovered by selection. The transfer can be done by mating or by transformation. Since the transposon carries a selectable marker, and the vector which originally contained the clone carries a second selectable marker, there will be no difficulty in capturing the desired transposon insertions. The goal now is to find a large set of different insertions and map them within the original cloned insert as efficiently as possible. The infrequent cutting sites placed into the transposon and the vector greatly facilitate this analysis. DNA is prepared from all of the putative transposon insertions, cut with the infrequent cutter, and analyzed by ordinary agarose gel electrophoresis, or PFG, depending on the size of the clones. Each insert should give two fragments. Their sizes give the position of the insert. There is an ambiguity about which fragment is on the left and which is on the right. This can be resolved if the separated DNAs are blotted and hybridized with one of the two unique sequences originally placed in the vector (labeled 1 and 2 in Fig. 11.8).

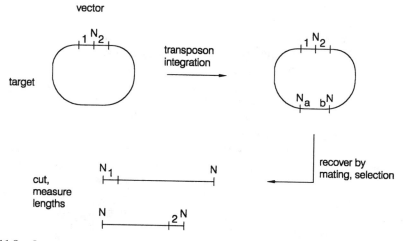

**Figure 11.8**   One strategy for DNA sequencing by transposon hopping. *a, b,* and $N$ are as defined in Figure 11.7; 1 and 2 are vector unique sequences.

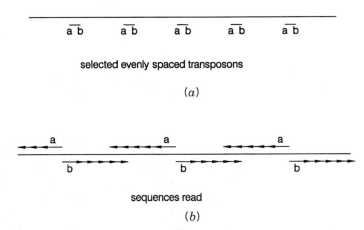

**Figure 11.9**   DNA sequencing from a set of transposon insertions. (*a*) Chosing a set of properly spaced transposons. (*b*) Use of transposon sequences as primers.

Now the position of the transposon insertion is known. One can select samples that extend through the target at intervals than can be spanned by sequencing, as shown in Figure 11.9. To actually determine the sequence, one can use the two unique sequences within the transposon as primers. These will always allow sequence to be obtained in both directions extending out from the site of the transposon insertion (Fig. 11.9). One does not a priori know the orientation of the two sequences primed from the transposon relative to the initial clone. However, this can be determined, if necessary, by reprobing the infrequent-cutter digest fingerprint of the clone with one of the transposon unique sequences. In practice, though, if a sufficiently dense set of transposon inserts is selected, there will be sufficient sequence overlap to assemble the sequence and confirm the original map of the transposon insertion sites.

## DELTA RESTRICTION CLONING

This is a relatively straightforward procedure that allows DNA sequences from vector arms to be used multiple times as sequencing primers. It is a particularly efficient strategy for sampling the DNA sequence of a target in numerous places and for subsequent, parallel, directed primer sequencing. Delta restriction cloning requires that the target of interest be contained near a multicloning site in the vector, as shown in Figure 11.10. The various enzyme recognition sequences in that site are used, singly, to digest the clone, and all of the resulting digests are analyzed by gel electrophoresis. One looks for enzymes that cut the clone into two or more fragments. These enzymes will have cutting sites within the insert. From the size of the fragments generated, one can determine approximately where in the insert the internal cleavage site is located. Those cuts that occur near regions of the insert where sequence is desired can be converted to targets for sequencing by diluting the cut clones, religating them, and transforming them back into *E. coli*. The effect of the cleavage of religation is to drop out from the clone all of the DNA between the multicloning site and the internal restriction site. This brings the vector arm containing

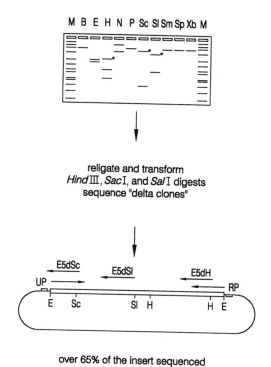

M B E H N P Sc Sl Sm Sp Xb M

religate and transform
*Hind* III, *Sac* I, and *Sal* I digests
sequence "delta clones"

E5dSc

E5dSl

E5dH

UP

RP

E   Sc         Sl  H          H  E

over 65% of the insert sequenced

**Figure 11.10**   Delta restriction subcloning. Top panel shows restriction digests of the target plasmid. Bottom panel shows sequences read from delta subclones. Adapted from Ansorge et al. (1996).

the multicloning site adjacent to what was originally an internal segment of the insert and allows vector sequence to be used as a primer to obtain this internal sequence. In test cases about two-thirds of a 2- to 3-kb insert can be sequenced by testing 10 enzymes that cut within the polylinker. Only a few of these will need to be used for actual subcloning. The problem with this approach is that one is at the mercy of an unknown distribution of restriction sites, and at present, it is not clear how the whole process could be automated to the point where human intervention becomes unnecessary.

## NESTED DELETIONS

This is a more systematic variant of the type of delta restriction cloning approach just described. Here a clone is systematically truncated from one or both ends by the use of exonucleases. The original procedure, developed by Stephen Henikoff, is illustrated in Figure 11.11. A DNA target is cut with two different restriction nucleases. One yields a 3'-overhang; the other yields a 5'-overhang. The enzyme *E. coli* exonuclease III degrades a 3'-overhang very inefficiently, while it degrades the 3'-strand in a 5'-overhang very efficiently. The result is degradation from only a single end of the DNA. After exonuclease treatment, the ends of the shortened insert must be trimmed to produce cloneable blunt ends. The DNA target is then taken up in a new vector and sequenced using primers from

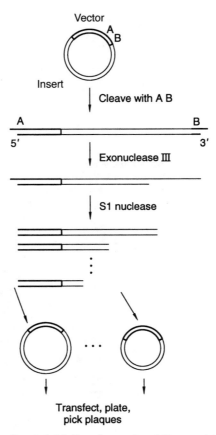

**Figure 11.11**   Preparation of nested deletion clones. *A* and *B* are restriction enzyme cleavage sites that give the overhanging ends indicated. Adapted from Henikoff (1984).

that new vector. In principle, this process ought to be quite efficient. In practice, while this proposed strategy has been known for many years, it does not seem to have found many adherents.

A variation on the original exonuclease III procedure for generating nested deletions has been described by Chuan Li and Philip Tucker. In this method, termed exoquence DNA sequencing, a DNA fragment with different overhangs at its ends is produced and one end is selectively degraded with exonuclease III. At various time points the reaction is stopped, and the resulting template-primer complexes are treated separately with one of several different restriction enzymes and then subjected to Sanger sequencing reactions, as shown in Figure 11.12. The final DNA sequencing ladders are examined directly by gel electrophoresis. Thus no cloning is required. If the restriction enzymes are chosen well, and the times at which the reactions are stopped are spaced sufficiently closely, sufficient sequence data will be revealed to generate overlaps that will allow the reconstruction of contiguous sequence. This is an attractive method in principle; it remains to be seen whether it will prove more generally appealing than the original nested deletion cloning approach.

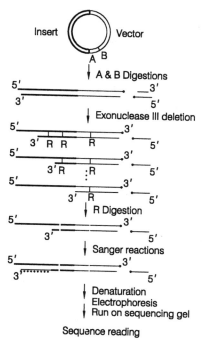

**Figure 11.12**  Strategy for exoquence DNA sequencing. Shown is a relatively simple case; there are more elaborate cases if the restriction enzymes used cut more frequently. *A* and *B* are restriction enzyme sites as in Figure 11.11; *R* is an additional restriction enyzme cleavage site. Taken from Li and Tucker (1993).

## PRIMER JUMPING

This strategy has been discussed quite a bit. However, there are yet no reported examples of its implementation. The basic notion is outlined in Figure 11.13. It is similar in many ways to delta subcloning, but it differs in a number of significant features. PCR is used, rather than subcloning. A very specific set of restriction enzymes is used: one rare cutter which can have any cleavage pattern and an additional pair of restriction enzymes consisting of an eight base cutter and a four or six base cutter; but they have to produce the same set of complementary single-stranded ends. Examples are *Not* I (GC/GGCCGC) and *Sse*8387 I (CCTGCA/GG) for the eight cutters, and *Ene* I (Y/GGCCR) and *Nsi* I (ATGCA/T) or *Pst* I (CTGCA/G), respectively, as more frequent cutters. In principle, the approach shown in Figure 11.13 ought to be applicable to much larger DNA than delta subcloning, based on the past success at making reasonably large jumping libraries (Chapter 8).

For primer jumping the insert is cloned next to a vector fragment containing any desired infrequent cleavage site between two known unique DNA sequences, shown as *a* and *b* in Figure 11.13. The vector fragment is constructed so that it contains no cleavage sites for the 4-base or 6-base cutting enzyme between the unique sequences and the insert, but it does contain a second infrequent cleavage site, *N* in Figure 11.13, as close to the upstream unique sequence as possible. Ideally the vector will be one arm of a

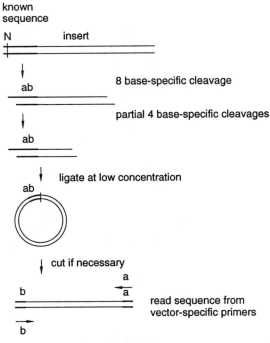

**Figure 11.13** Primer jumping, an untested but potentially attractive method for directed DNA sequencing. Restriction enzyme site *N* must produce an end that is complementary to the end generated by the four-base cutter. Sites *a* and *b* can be anything so long as they are not present in the target, but there must a site for infrequent cleavage between them.

YAC, and the other arm could be treated in a similar fashion. The clone is cut at the distal rare cleavage to completion and then partially digested with the frequently cutting enzyme. The resulting fragments are separated by length, and the separation gel is sliced into pieces. The resulting DNA fragments are diluted to very low concentration and ligated. This will produce DNA circles in which the vector sequence, including segments *a* and *b* in Figure 11.13, is now located next to each site in the partial digest which was cleaved by the frequently cleaving enzyme. Thus the known sequence can now be used for starting a primer walk. The approximate position of the walk within the large clone will be known from the size of the fragment. With the 800-bp to 1-kb sequence reads now being achieved under good circumstances, it is conceivable that one would be able to sequence from the cleaved site up to the next equivalent restriction site without the need to make additional primers, in most cases. If this were the case, one could do a directed walking strategy on a large DNA target using only two primers—one for each vector arm.

In a similar vein to primer jumping, if single-sided PCR ever works well enough (Chapter 4), these methods could be used for directed cycle sequencing by the approaches just described.

## PRIMER MULTIPLEXING

This is a potentially very powerful strategy for large-scale DNA sequencing. It was developed by George Church and has been elaborated, independently, by Raymond Gesteland. There are a number of features that set multiplexing aside from many other approaches. A major peculiarity of the method is that it does not scale down efficiently, so it is best suited for fairly massive projects, typically several hundred kb of DNA sequence or more. The basic scheme for primer multiplexing is shown in Figure 11.14. In the particular case shown, a multiplexing of 40 is used. Forty different vectors are constructed; each has a unique 20-base sequence on each side of the cloning site. The DNA target of interest is shotgun cloned, separately, into all 40 vectors. This produces 40 different libraries. Pools are constructed by selecting one clone from each of the libraries and mixing them. These 40-clone pools are the samples on which DNA sequencing is performed. The pools are subjected to standard DNA sequencing chemistry to generate a mixture of 40 different ladders, but no radioactivity or other label is introduced into the DNA at this stage. The mixture is fractionated by polyacrylamide electrophoresis and blotted onto a membrane. A particularly convenient way to do this is by the bottom wiper described in Chapter 10. The blotted DNA is crosslinked onto the filter by UV irradiation to attach it very stably. This is a key step, since the filters will be reused many times.

To read the DNA sequence from each pool of clones, the filter is hybridized with a probe corresponding to one of the 40 unique 20-base sequences. By this indirect end-labeling method (introduced in Chapter 8), only one of the 40 clones in the sequence ladder is visible. The probe is removed from the filter by washing, and then the hybridization and washing are repeated successively for each of the other unique sequence primers. By this multiplexing approach, most stages of the project are streamlined by a factor of 40.

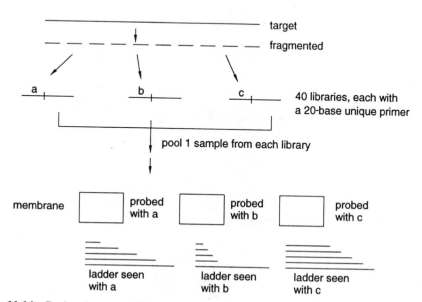

**Figure 11.14**  Basic scheme used for primer multiplexing: *a, b, c,* and so on, represent unique vector primer sequences.

The exceptions are the hybridization, autoradiography, or other color detection, and washing. Thus great care must be taken to automate these steps in the most efficient way. Recently a fairly successful demonstration of the efficiency that can be achieved with primer multiplexing, combined with transposon jumping, was reported by Robert Weiss and Raymond Gesteland.

## MULTIPLEX GENOMIC WALKING

A different approach to multiplex sequencing has been suggested by Walter Gilbert. This is designed to be used for the sequencing of entire small genomes like *E. coli* where direct genomic DNA sequencing is feasible. The method is illustrated in Figure 11.15*a*. The great appeal of this method is that absolutely no cloning is required. The total genome is digested separately with a set of different restriction enzymes. The products of this digestion are loaded onto polyacrylamide gels in adjacent lanes and fractionated. A highly labeled probe with an arbitrary sequence is selected (with a length chosen to occur on aver-

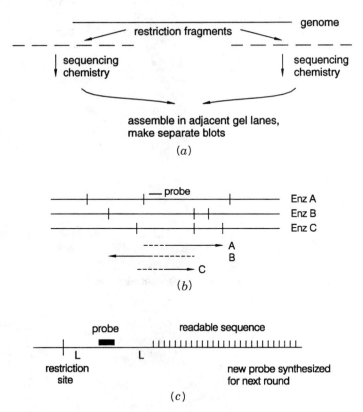

**Figure 11.15**   Multiplex genomic walking. (*a*) Basic outline of the experiment. (*b*) Restriction map in a typical region, and resulting segments of sequence, *A, B, C* revealed by hybridization with one specific probe. (*c*) Sections of readable and unreadable sequence on a particular restriction fragment. The probe is located *L* bases from the end of the fragment.

age once per genome) and used to hybridize with a blot of the separated fragments (Fig. 11.15*b*). In most of the lanes this probe will give a readable sequence. Suppose that the probe lies 60 bp upstream from a given restriction site. The first 60 bases of sequence will be unreadable because data will extend in both directions (Fig. 11.15*c*). However, longer regions of the ladder will be interpretable, since they must lie in the direction away from the nearby restriction site. In general, one will expect to get a number of usable reads in both directions from the probe, just by the fortuitous occurrence of useful restriction sites. These reads are assembled into a segment of DNA sequence. Next probes are designed from the most distal regions of the segment, and these are used to continue the genomic walk.

In principle, multiplex genomic walking is a very elegant and spartan approach to DNA sequencing. One has a choice at any time whether to use additional arbitrary probes, and so increase the number of parallel sequencing thrusts, or whether to focus on directed walking. Thus one has a method with some of the advantages of both random and directed strategies. A potential weakness is the relatively high fraction of failed lanes that will occur unless the probe has a single binding site in the genome. Another problem is the technical demands that genomic sequencing makes. It is also not obvious how easy this strategy will be to automate. It does work, but the overall efficiency needs to be established before the method can be compared quantitatively with others.

## GLOBAL STRATEGIES

A basic issue that has confronted the human genome project since its conception is not how to sequence but what to sequence. From a purely biological standpoint, the most interesting sequencing targets are genes. The choice of genes depends on the sorts of biological questions one is interested in. An evolutionary biologist may want to sequence one homologous gene in a wide variety of organisms. Cell biologists or physiologists may want to focus on a set of functionally related genes or gene families within just a few organisms. However, from the point of view of whole genome studies, the purpose of sequencing is really to find genes and make them available for subsequent biological studies. This puts a very different tilt on the issues that affect the choice of sequencing targets.

For simple gene-rich organisms like bacteria and yeasts, there is little doubt that complete genomic sequencing is desired and worth doing even with existing DNA sequencing technology. Indeed sequencing projects have been completed on many bacteria including *H. influenzae, Mycoplasma genitalium, Mycoplasma pneumoniae, Methanococcus jannaschii, Synechocystis* strain pcc6803, and *Escherichia coli,* and one yeast, *S. cerevisiae* (see Chapter 15). Additional projects are well underway with a number of other microorganisms, including the bacterium *Mycobacterium tuberculosis* and the yeast, *S. pombe. E. coli* is an obvious choice as the focus of much of our fundamental studies in prokaryotic molecular biology. Mycoplasmas represent the smallest known free-living genomes. *Mycobacterium tuberculans* is important because of the current medical crisis with drug-resistant tuberculosis. The two yeasts account for most of our current knowledge and technical power in fungal genetics. They are also very different from each other, so much will be learned from comparisons between them. The real issue that will have to be faced in the future is at what stage in DNA sequencing technology is it desirable and affordable to sequence the genomes of many other simple organisms?

There are a number of more advanced organisms that appear to have relatively high coding percentages of DNA. These include a simple plant, *Arabidopsis thaliana,* a much more economically important plant, rice, the fruitfly, *Drosophila melanogaster,* and the nematode, *Caenorhabditis elegans.* There are strong arguments in favor of obtaining complete DNA sequences on these organisms rapidly. They all are systems where a great deal of past genetics has been done, and a great deal of ongoing interest in biological studies remains. Certain primitive fishes may also have small genomes as does the puffer fish. Here the argument in favor of sequencing is that it will be relatively easy to find most of the genes. However, these organisms are currently pretty much in a biological vacuum.

For more complex, gene-dilute organisms, the selection of sequencing targets is, not surprisingly, also more complex. Here there is little debate that *Homo sapiens* and the mouse, *Mus musculus,* are the obvious first choice. It is much less clear what should come after this. Do we target other primates because they will be most useful in understanding the very large fraction of human genes that are believed to be central nervous system specific? Do we examine genomes of organisms that have long been the focus of physiological studies like rats, dogs, and cats. Or do we aim for a much broader representation of evolutionary diversity? Alternatively, how important should the commercial value of potential genome targets be? Cows, horses, pine trees, maize, and salmon have a much more important economic role than *Arabidopsis* or *C. elegans.* These questions are interesting to ponder, but they really do not require answers at the present time. If sufficiently inexpensive DNA sequencing methods are developed in the future, we will want to sequence every genome of biological interest. For the present, technology pretty much limits us to a few choices.

With most complex organisms, only a few percent of the genome is known to be coding sequence. The function of the rest, which we earlier termed junk, is unknown, today. With limited resources, and relatively slow sequencing technology, most involved groups are choosing to focus on selectively sequencing genes from human or other sources. There are two ways to go about this. One approach is to find a gene-rich region in a genome and sequence it completely. Regions that have been selected include the T-cell receptor loci, immunoglobulin gene families, and the major histocompatibility complex. All of these regions are of intense interest in understanding the function of the immune system. Another region of interest is the Huntington's disease region because it is very gene rich, and in the process of finding the particular gene responsible for the disease a large set of cloned DNA samples from this region has become available.

An alternative to genomic sequencing in a gene-rich region is to sequence cDNAs, DNA copies of expressed mRNAs. These are relatively easy to produce, and many cDNA libraries are available. Each represents the pattern of gene expression of the particular tissue or sample from which the original mRNA was obtained. In sequencing a cDNA, one knows one is dealing with an expressed gene, therefore a functional gene. This is a considerable advantage over genomic sequencing where one has no knowledge a priori that a particular gene found at the DNA level is actually ever used by the organism. With cDNA sequencing, one is always examining genes or nearby flanking sequences. This is another great advantage over genomic sequencing where, even in the best of cases, most of the sequence will not be coding. However, there are some potential difficulties with projects to examine massive numbers of cDNA sequences, as we will demonstrate.

## SEQUENCE-READY LIBRARIES

Today, the notion of sequencing an entire human chromosome from left to right telomere is being considered seriously at a number of Genome Centers. In some cases the plans are based on a preexisting minimum tiling set of clones. Here, as long as the set is complete and exists in a vector like a cosmid or a BAC that allows direct sequencing, the strategy is predetermined. The clones are selected and sequenced one by one by whatever method is deemed optimal at the time for 50- to 150-kb clones.

Suppose, however, that, with sequencing as the eventual goal, one wishes to create an optimal library to facilitate subsequent sequencing of any particular region deemed interesting. There are two basically similar strategies for achieving this objective. If a dense ordered library already exists in an appropriate vector, one can sequence the ends of all of the clones in a relatively easy and cost-effective manner. Since vector priming can be used, the goal is to read into the cloned insert as far as possible in a single pass of raw DNA sequencing. If this is done for all the clones, the result is a sampling of the genomic sequence (Smith et al., 1994). For example, suppose that the initial library is 20-fold redundant 50-kb cosmids. A cosmid end on average would occur every 1.25 kb. A 700-base sequence read at each end would generate a total of 28 kb of sequence. When realistic failure rates and some inevitable overlaps are considered, the result would still be roughly half the total sequence. This is sufficiently dense that almost any cDNA sequence from the region would be represented in some of the available genomic DNA sequence. Thus all sequenced cDNAs could be mapped by software sequence comparisons without the need for any additional experiments. The average spacing between sequenced genomic regions would be short enough so that PCR primers could be designed to close any of the gaps by cycle sequencing.

For many targets, however, there is no existing clone map. The effort to create one de novo is considerable, even by the enhanced methods described in Chapter 9. For this reason, as automated DNA sequencing becomes more and more efficient, strategies that avoid the construction of a map altogether become attractive. One recent proposal for such a scheme also relies on the sequencing of the ends of the clones (Venter et al., 1996). Consider, for example, an ordered tenfold redundant BAC library of the human genome. With 150-kb inserts, 200,000 BACs are required. If each of these is sequenced for 500 bp from both ends, the resulting data set will contain 400,000 sequence reads encompassing 200 Mb of DNA. On average, the density of DNA sequence is a 500-bp block every 7.5 kb. Once created, such a resource would serve two functions. Many cDNAs would still match up with a segment of BAC sequence, and they could serve to correlate the BAC library with other existing genome resources and information. The utility of the BACs in this regard could be improved if, for example, they were created so that their ends had a bias to occur in coding sequence. However, even in the absence of cDNA information, the BACs will serve as a starting point for the genomic sequencing of any region of interest. One could choose any BAC that corresponds to the region of interest and sequence it completely. Then, by inspection, the BACs in the library that overlapped least with the first sequenced BAC could be picked out and used for the next round of sequencing. The process would continue until the region of interest were completed. In this way the sequencing project itself would create the minimum tiling set of BACs needed for the region.

## SEQUENCING cDNA LIBRARIES

Usually cDNA libraries are made by a scheme like that shown in Figure 11.16. To prepare high-quality cDNAs, it is important to start with a population of intact mRNAs. This is not always easy; mRNAs are very susceptible to cleavage by endogenous cellular ribonucleases, and some tissues or samples are very rich in these enzymes. Most eukaryotic mRNAs have several hundred bases of A at their 3'-end. This poly A tail can be used to capture these mRNAs and remove contaminating rRNA, tRNA, and other small cytoplasmic and nuclear RNAs. Unfortunately, one also loses that fraction of mRNAs that lack a poly A tail. An oligo-T primer can then be used with reverse transcriptase to make a DNA copy of the mRNA strand. Alternatively, random primers can be used to copy the mRNAs, or specific primers can be used if one is searching for a particular mRNA or class of mRNAs. There are two general methods to convert the resulting RNA-DNA duplexes into cDNAs. Left to their own devices, some reverse transcriptases will, once the RNA strand is displaced or degraded, continue synthesis, after making a hairpin, until they have copied the entire DNA strand of the duplex. As shown in Figure 11.16a, S1 nuclease can then be used to cleave the hairpin and generate a cloneable end. Unfortunately, the S1 nuclease treatment can also destroy some of the ends of the cDNA. An alternative procedure is to use RNase H to nick the RNA strand of the duplex. The resulting nicks can serve as primer for DNA polymerases like *E. coli* DNA polymerase I. This eventually leads to a complete DNA copy except for a few nicks which can be sealed by DNA lig-

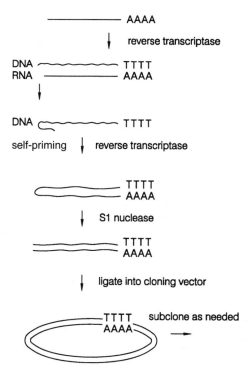

**Figure 11.16**   Approaches to the construction of cDNA libraries: Use of S1 nuclease to generate clonable inserts.

ase. This procedure is generally preferred over the S1 nuclease method because it tends to produce longer, more intact cDNAs.

An unfortunate fact about many cDNA clones is that they are biased toward the 3′-end of the original message because of the poly A capture and the oligo-T priming used to prepare them. The true 5′-end is often missing and needs to be found in other clones or in the genome. Some attempts have been made to take advantage of the specialized cap structure at the 5′-end of eukaryotic mRNAs to purify intact molecules. One possibility is to try to produce high-affinity monoclonal antibodies specific for this cap structure. More effective has been the use of an enzyme called tobacco pyrophosphatase. This cleaves off the cap to leave an ordinary 5′-phosphate-terminated DNA strand that then can serve as a substrate in a ligation reaction, which can be used to add a known sequence. This known sequence will serve as the staging site for subsequent PCR amplification. Several different Japanese groups have recently perfected such strategies to the point where 5′-end-containing cDNA libraries can now be made quite reliably.

## DEALING WITH UNEVEN cDNA DISTRIBUTION

With relatively rare exceptions like rDNAs, genes in the genome are in approximately a 1:1 ratio. In contrast, the relative amount of mRNAs present in a typical cell extends over a range of more than $10^5$. This leads to very serious biases in most cDNA libraries. These will tend to be overrepresented with a relatively small numbers of different high-frequency clones. In addition existing cloning methods will tend to bias the libraries toward short mRNAs. If one attempts to sequence cDNAs at random from a library, in most cases the high copy number clones will be re-sequenced over and over again, while most rare mRNAs will never be sampled. It is important to stress that the problems of random selection and library biases seriously interfere with genomic DNA sequencing projects, even though one is starting with an almost uniform sample of the genome. With cDNAs these problems are much more serious and must be dealt with directly and forcefully.

One simple approach to systematic sequencing of cDNA libraries is shown in Figure 11.17. One starts with an arrayed library. A small number of clones, say 100, are selected and sequenced. All of the sequenced clones are pooled, labeled, and hybridized back to

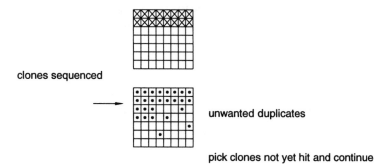

clones sequenced

unwanted duplicates

pick clones not yet hit and continue

**Figure 11.17** Basic scheme for sequencing an arrayed cDNA library, and periodically screening the library to detect repeats of clones that have already been sequenced. The schematic array shown has only 56 clones; a real array would have tens to hundreds of times more.

the library. Any clones that are detected by this hybridization are not included in the next set of 100 to be sequenced. By continuing in this way, most duplication can be avoided. Unfortunately, there will also be a tendency to discard cDNAs from gene families, so many of the members of these families will be underrepresented or missed. As an alternative to handling the clones as arrays, one can carry out a solution hybridization of the entire cDNA library with an excess of sequenced clones, discard all the samples that hybridize, and regrow the remainder. This effectively replaces a screen by a physical selection process.

A more complex approach to compensating for uneven cDNA distribution is to try to equalize or normalize the library. The distribution of mRNAs in a typical cell is shown in Table 11.1. Roughly speaking there are three classes of messages: a few very common species, then approximately equal total amounts of species 20 times more rare, and species another factor of 20 rarer still. The goal is to try to even out these differences. The approach used is to anneal the library to itself and remove all the double-stranded species that are formed. We will do this by allowing the reannealing to occur at a very high $C_0t$: Typically $C_0t = 250$ is used. From Chapter 3, we can write for the fraction of single strand remaining in a hybridization:

$$f_s = \frac{1}{1 + n\, C_0 t\, k_2/2N} = \frac{1}{1 + C_0 t/C_0 t^{1/2}} = \frac{C_0 t^{1/2}}{C_0 t^{1/2} + C_0 t}$$

where all the quantities in this equation have been defined in Chapter 3. When $C_0 t \gg C_0 t^{1/2}$, we can approximate this result as

$$f_s = \frac{C_0 t^{1/2}}{C_0 t}$$

Note that the $C_0 t^{1/2}$ value for a given sequence depends on the ratio of genome size, $N$, and the number of times the sequence is represented, $n$. For a cDNA library, $N$ is the total complexity of the DNA sequences represented in the library, and $n$ is the number of times a given sequence is represented. Thus, for highly frequent cDNAs, $N/n$ will be small so that the $C_0 t^{1/2}$ will be small, and these species will renature relatively more rapidly. Note that the amount of a particular cDNA remaining after extensive annealing will be proportional to its original abundance $n$ and to its hybridization rate, which will scale as $N/n$. Thus, at very long times in the reaction, a relatively even distribution of cDNAs should be produced. We can evaluate the expected results for an attempt to normalize the typical cell mRNAs shown in Table 11.1. This is given in Table 11.2.

**TABLE 11.1    Distribution of mRNA in a Typical Cell**

| Species | Percent | Number of Species | Relative Frequency | $C_0 t^{1/2}$ |
|---|---|---|---|---|
| Common | 10 | 10 | 330 | 0.08 |
| Medium | 45 | 1000 | 15 | 1.7 |
| Rare | 45 | 15,000 | 1 | 25. |

**TABLE 11.2 Effect of Self-Annealing a cDNA Library**

| Class | $f_s$ at $C_0t = 250$ | Initial Frequency | Equalized Frequency |
|---|---|---|---|
| Common | $3 \times 10^{-4}$ | 330 | $9.9 \times 10^{-2}$ |
| Medium | $7 \times 10^{-3}$ | 15 | $1.0 \times 10^{-1}$ |
| Rare | $9 \times 10^{-2}$ | 1 | $9.0 \times 10^{-2}$ |

The predictions in Table 11.2 look very encouraging. However, a serious potential problem is that one has to discard most of the library in order to achieve this result. PCR or efficient recloning must be used to recover the cDNA clones which have not self-annealed.

An actual scheme for efficient cDNA normalization is shown in Figure 11.18. This has been developed by Bento Soares, Argiris Efstratiadis, and their collaborators. It is designed to avoid the preferential loss of long cDNA clones during the self-annealing, and also to avoid the loss of cDNAs from closely related gene families. Long clones would be

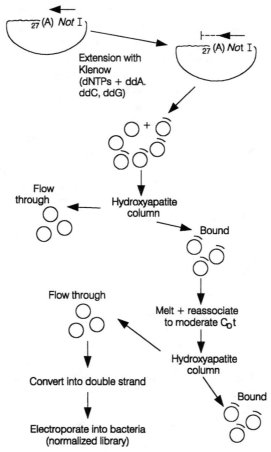

**Figure 11.18** A relatively elaborate scheme for cDNA normalization that attempts to prevent a bias against shorter cDNAs and the loss of cDNAs from gene families. Adapted from Soares et al. (1994).

**TABLE 11.3   cDNA Library Normalization**

| Probes | Original Library | Normalized (HAP-FT) | HAP-Bound |
|---|---|---|---|
| Human $C_0t$ 1 DNA | 10% | 2% | 2.6% |
| Elongation factor 1a | 4.6% | 0.04% | 3.7% |
| a-Tubulin | 3.7%–4.4% | 0.045% | 6% |
| b-Tubulin | 2.9% | 0.4% | 0.85% |
| Mitochondrial 16S rRNA | 1.3% | 1% | |
| Myelin basic protein | 1% | 0.09% | |
| g-Actin | 0.35% | 0.1% | 1.3% |
| Hsp 89 | 0.4% | 0.05% | <0.14% |
| CH13-cDNA#8 | 0.009% | 0.035% | |

*Source:* Data adapted from Soares et al. (1994).

lost if entire cDNA sequences were used for hybridization because, as described in Chapter 3, their rate constants for duplex formation ($k_2$'s) are larger and because there are more places to nucleate potential duplex. The trick used in the scheme of Figure 11.18 is to start with single-stranded cDNA clones and produce a short duplex at the 3'-end of each cDNA clone by primer extension in the presence of chain terminators. By focusing on this region, one will ensure that the new strands synthesized preferentially come from 3' noncoding flanking regions where even closely related genes have significant divergence, since the sequences are not translated and presumably have little function. Any cDNAs that have not successfully templated the synthesis of a short duplex are discarded by chromatography on hydroxyapatite, which specifically binds only duplexes, under the conditions used. These duplex-containing clones are then eluted, melted, and allowed to self-anneal to high $C_0t$. Now any clones with duplexes are removed, and the clones that have remained as single strands represent the normalized library. These are then amplified and sequenced.

Some actual results using the scheme of Figure 11.18 are given in Table 11.3. It is apparent that the equalization is far from perfect. However, it represents a major improvement over nonnormalized libraries, and materials made in this way are currently being used extensively for cDNA sequencing. Two additional schemes for cDNA normalization are described in Box 11.1. It is not clear at the present time just which schemes will ultimately be widely adopted.

## LARGE-SCALE cDNA SEQUENCING

In the past three years at least five separate efforts have been made to collect massive amounts of cDNA sequence. One of these is a collaboration between the Institute for Genome Research (TIGR) and Human Genome Sciences, Inc. At least initially, this effort took an anatomical approach. Libraries of cDNAs from as many different major tissues as possible were collected, and large numbers of clones from each of these were sequenced. The second approach was orchestrated by Incyte Pharmaceuticals, Inc. Here the emphasis was on cell physiology. Sets of cDNA libraries were collected from pairs of cells in known, related physiological states, such as activated or unactivated macrophages. A fixed number of cDNAs, 5000 in the earliest studies, was randomly selected for each of the cell pairs and sequenced. In this way information was obtained about the frequencies of common cDNAs in addition to the sequence information from all classes of cDNAs.

**BOX 11.1**
**ALTERNATE SCHEMES FOR NORMALIZATION OF cDNA LIBRARIES**

Two different schemes for the production of normalized cDNA libraries have been described. The first, proposed by Sherman Weissman and coworkers, is shown schematically in Figure 11.19. First PCR is used to amplify cloned cDNAs. Then, as in the Soares and Efstratiadis scheme described in the text, hydroxylapatite fractionation is used to deplete a reaction mixture of double-stranded products. Next a nested set of PCR primers is used to amplify the single-stranded material that survives hydroxyapatite. Finally this material is cloned to make the normalized library. A survey of typical results is given in Table 11.4.

The scheme developed by Michio Oishi is quite different (Fig. 11.20). Here cDNA immobilized on microbeads is annealed to a vast excess of mRNA from the same source. Under these conditions the kinetics of hybridization become pseudo–first order as described in Chapter 3. The highly overrepresented components in the mRNA will actually deplete the corresponding cDNAs below the level of normalization. The resulting cDNA library will be enriched for rare cDNA sequences. A survey of typical results is given in Table 11.5.

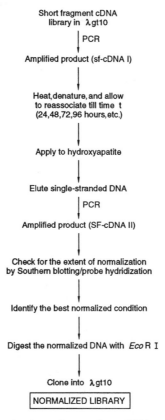

Short fragment cDNA
library in λgt10
│ PCR
↓
Amplified product (sf-cDNA I)
│
↓
Heat, denature, and allow
to reassociate till time t
(24,48,72,96 hours,etc.)
│
↓
Apply to hydroxyapatite
│
↓
Elute single-stranded DNA
│ PCR
↓
Amplified product (SF-cDNA II)
│
↓
Check for the extent of normalization
by Southern blotting/probe hydridization
│
↓
Identify the best normalized condition
│
↓
Digest the normalized DNA with *Eco* R I
│
↓
Clone into λgt10
NORMALIZED LIBRARY

**Figure 11.19** A relatively simple scheme for cDNA normalization. Adapted from Sankhavaram et al. (1991).

*(continued)*

**BOX 11.1** *(Continued)*

**TABLE 11.4   Effect of Normalization on a cDNA Library**

| Probe | Number of Clones Identified per 100,000 Plaques | | |
|---|---|---|---|
| | STH | NSTH I | NSTH II |
| R-DNA | 30,000 | 94 | 12 |
| Blur-8 | 800 | 450 | 360 |
| $\gamma$-actin | 110 | 37 | NT |
| HLA-H | 104 | 80 | 10 |
| CD4 | 28 | 37 | 12 |
| CD8 | 15 | 55 | 12 |
| Oct-1 | 9 | NT | 8 |
| $\beta$-globin | 7 | NT | 10 |
| c-myc | 5 | NT | 11 |
| TCR | 5 | NT | 8 |
| TNF-$\alpha$ | 5 | NT | 6 |
| $\alpha$-fodrin | 3 | NT | 9 |

*Source:* From Patanjali et al. (1991).

Note: cDNAs present at various levels of abundance in STH library become almost identically abundant in the normalized (NSTH) libraries. Increased reassociation times, as indicated by the increased $C_0t$ value, render better normalized libraries. NT, not tested. For NSTH I the $C_0t$ value was 41.7 mol-s/liter, and for NSTH II the $C_0t$ value was 59.0 mol-s/liter.

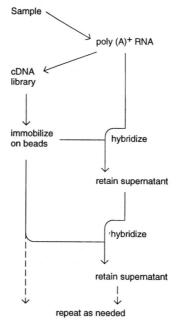

**Figure 11.20**   A scheme for the preferential enrichment of rare cDNAs. Adapted from Sasaki et al. (1994a).

*(continued)*

**BOX 11.1** (*Continued*)

**TABLE 11.5  Change of the Proportion of cDNA Clones Before and After Self-Hybridization**

| Probe | Input[a] | Before | Percentage | After | Percentage | After/Before |
|---|---|---|---|---|---|---|
| Rabbit $\beta$-globin | 1 | 111/10,500 | 1.067 | 5/55,000 | 0.009 | 0.086 |
| $\phi$X174*Hae* III 0.6 kb | 0.01 | 2/30,000 | 0.0067 | 8/35,000 | 0.023 | 3.43 |
| $\phi$X174*Hae* III 0.9 kb | 0.01 | 1/30,000 | 0.0033 | 6/30,000 | 0.02 | 6 |
| neo[r] | 0.0001 | 0/250,000 | <0.0004 | 2/250,000[b] | 0.0008 | >2 |
| $\beta$-actin | | 54/10,000 | 0.54 | 2/10,000 | 0.02 | 0.037 |
| IL-4 | | 0/320,000 | <0.0003 | 3/320,000[b] | 0.0009 | >3 |
| IL-2 | | 0/320,000 | <0.0003 | 0/320,000 | <0.0003 | — |

*Source:* Adapted from Sasaki et al. (1994).

[a]Percent of total RNA (w/w).

[b]The positive clones were confirmed by sequencing approximately 300 bp of the inserts.

The frequency information, when pairs of cells are compared, is often quite interesting, and it suggests potential functional roles for a number of newly discovered genes in the libraries. Incyte has termed such comparisons transcript imaging. A third large-scale cDNA sequencing effort is the Image consortium involving several academic or government laboratories and Merck, Inc. Here normalized libraries are serving as the source of clones for sequencing, and the goal is to collect at least one representative cDNA sequence from all human genes.

At present, the sequencing of human cDNAs is being carried out in large laboratory efforts like the three just described as well as many smaller, more focused efforts. Within each laboratory the amount of duplication seen thus far has been relatively small. Thus the early course of the cDNA sequencing strategy appears to be very effective. At what point it will peter out into a morass of duplicate clones is unknown. It is also really unclear what fraction of the total amount of genes will actually be found first through their cDNAs. The tissues used for the majority of these studies are those where large numbers of different genes are expected to be active. These include early embryos, hytidaform moles, which are differentiated but disordered tumors with many different tissue types, liver, and a number of parts of the brain. Whether many specialized tissues will have to be looked at to get genes expressed only in these tissues, or whether there is a broad enough low-level synthesis of almost any mRNA in one or more of the common tissues to let all genes be found there, is an issue that has not yet been answered. One way to try to extend the cDNA approach to find all of the human genes is described in Box 11.2

---

**BOX 11.2**
**PREPARATION AND USE OF hncDNAs**

A major purpose of making ordered libraries is to assist the finding and mapping of genes. Eugene Sverdlov and his colleagues in Moscow have developed an efficient procedure for preparing chromosome-specific hncDNA libraries. Their method is an elaboration of the scheme originally described by Corbo et al. (1990). The procedure is outlined in Figure 11.21. It uses an initial Alu-primed PCR reaction to make an hncDNA copy of the hnRNA produced in a hybrid cell containing just the chromosome of interest. (See Chapter 14 for details about the Alu repeat and Alu-specific PCR primers.) The resulting DNA is equipped with an oligo-G tail, and then a first round PCR is carried out using an oligo-C containing primer and an Alu primer. Then a second round of PCR is done with a nested Alu primer. The PCR primers are also designed so that the first round introduces one restriction site and the second round another. The resulting products are then directionally cloned into a vector requiring both restriction enzyme cleavage sites. In studies to date, Sverdlov and his coworkers have found that this scheme produces a diverse set of highly enriched human cDNAs. Since these come from Alu's in hnRNA, they will contain introns, but they can be used to locate genes on the chromosome equally well if not better than conventional cDNAs.

Note that the hncDNA clones as produced by the Sverdlov method actually contain substantial amounts of intronic regions. This means that they will be more effective than the ordinary cDNAs in dealing with gene families and in avoiding cross-hybridization with conserved exonic sequences in rodent-human cell hybrids.

*(continued)*

**BOX 11.2** *(Continued)*

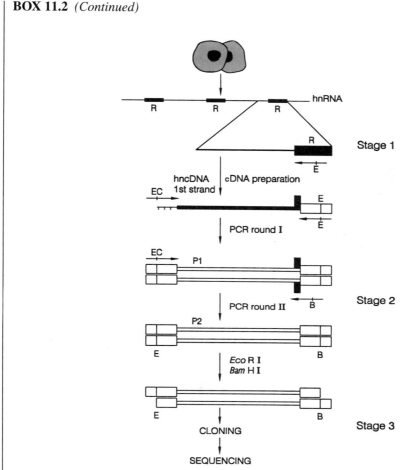

**Figure 11.21** Steps involved in making an hncDNA library. Interspersed repeat elements in hnRNA are represented in the upper line by solid boxes (*R*). Arrows indicate primers. Vertical lines crossing the arrows symbolise the primers with sites for *Eco*R I and *Bam*H I restriction endonucleases (*E*) or (*B*). EC is 5′GAGAATTC(C)203′. The open boxes with similar symbols represent sequences corresponding to primers that are included in PCR products P-1 and P-2. Provided by Eugene Sverdlov.

## WHAT IS MEANT BY A COMPLETE GENOME SEQUENCE?

If the strictest definition is used for a complete genome sequence, namely every base on every chromosome in a cell has been identified, then it is probably safe to say that we will never accomplish this. This is not to say that the task couldn't be accomplished in principle; it could be, but for several reasons it is a foolish task, at least for the human genome. The human genome is quite variable. This will be discussed in more detail later. Suffice it to say here that there are millions of differences in DNA sequence between the set of two homologous chromosomes in a diploid cell. Unless one could separate these into separate, cloned libraries, inevitable confusion will develop as to which homolog one is on. Hybrid cells make these separations for us, and libraries made from chromosomes of hybrid cells are major candidates for eventual large-scale sequencing. However, because of the history behind the construction of such hybrids, we rarely have separate clones of two homologous chromosomes from a single individual. Even more important, most different single chromosome hybrid cell lines have been made from different individuals. So the real answer to the often asked question "Who will be sequenced in the human genome project" is that the sequence will inevitably represent a mosaic of many individuals. This is probably quite appropriate, given the global nature and implications of the project.

There are other bars to total genome sequencing, or even total sequencing of a given chromosome. We have indicated many times already that closure in mapping is a difficult task; closure in large-scale sequencing projects will also be extremely difficult. For whatever reason, there are bound to be a few regions in any chromosome that cannot be cloned by any of our existing methods, or that may not be approachable even by PCR or genomic sequencing. Sequences with very peculiar secondary structures or sequences toxic to the enzymes or cells we must rely on could lead to this kind of problem. The issue of how much effort should be devoted to a few missing stretches has not yet been forced upon us, but placed in any kind of reasonable perspective, it cannot have high priority relative to more productive use of large-scale sequencing facilities.

Finally, some regions of chromosomes are either very variable or dull, at least at the level of fine details. Examples are long simple sequence or tandemly repeating sequence repeats. Human centromeres appear to have millions of base pairs of such repeats. Other heterochromatic regions are occasionally seen on certain chromosomes. Some of these regions show quite significant size variation within the human population. For example, a case is known of an apparently healthy individual with one copy of chromosome 21 that is 50% longer than average. Surely we will not select these extra long variants for initial mapping or sequencing projects. However, the key point is that extensive, large-scale sequencing of simple repeats does not seem to be justified at the present time by any hints that this large amount of sequencing data will be interesting or interpretable. Furthermore our current methods are actually incapable of dealing with such regions of the genome. Thus we will almost certainly have to claim completeness, missing the centromeres and certain other unmanageable genome regions.

## SEQUENCING THE FIFTH BASE

When we have sequenced all of the cloned DNAs from each human single chromosome library, we will not have the complete DNA sequence of an individual, for the reasons cited above. Some of the troublesome regions are almost certainly not in our libraries or, if they are present, they probably represent badly rearranged remnants of what was actually in the genome. If we ever want to look at such sequences in their native state, we may have to sequence them directly from the genome. For the reasons cited above, this may not be a terribly interesting or useful thing to do. However, there is a tremendously important additional reason to perfect methods for direct genomic mammalian sequencing. This is to look at the fifth base, $^mC$, which is lost in all common current cloning systems. It is also lost in PCR amplification. Therefore, to find the location of $^mC$ in mammalian or other higher eukaryotic DNA sequences, it is necessary to immortalize the positions of these residues before any amplification.

PCR can be used to determine DNA sequence directly from genomic samples with mammalian complexity by a ligation technique that is shown in Figure 11.22a. The ge-

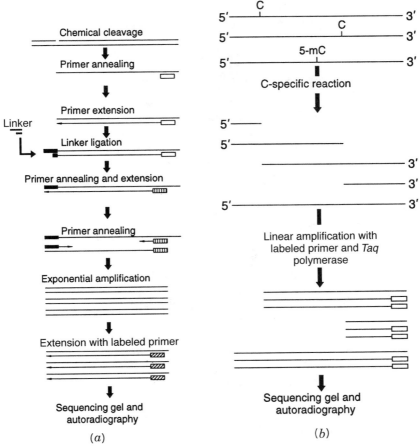

**Figure 11.22**  Genomic DNA sequencing can be used to find the location of methylated bases. (*a*) Basic scheme for PCR-mediated genomic Maxam-Gilbert DNA sequencing. (*b*) A methylated base prevents normal Maxam-Gilbert cleavage and thus alters the appearance of the DNA sequencing ladder. Adapted from Riggs et al. (1991).

nomic DNA is fragmented chemically as in the Maxam-Gilbert procedure. A sequence-specific primer is annealed to the DNA after this cleavage and extended up to the site of the cleavage. This creates a ligatable end. A splint is ligated onto all the genomic DNA pieces. Then primers are used to extend the DNA past the unique region of the splint, as we described earlier for single-sided PCR reactions in Chapter 4. Finally two primers are used for exponential amplification. The sizes of the amplified products reveal where the original cleavages were in the DNA. If all bases in the genome are accessible to the specific cleavage reactions used, then a perfectly normal sequencing ladder should result.

To find the location of $^{m}$C in genomic sequence, one takes advantage of the fact that this base renders DNA resistant to cleavage by the normal C-specific reaction used in the Maxam-Gilbert method. The result, is that the locations of the methylated C's drop out of the sequencing ladder (Fig. 11.22*b*). If the target sequence is already completely known except for the sites of methylation, this is all the information one needs. If not, one can always repeat the sequencing on a cloned sample to confirm the location of the $^{m}$C's by their new appearance as C's. This method is quite powerful; it is providing interesting insights into the patterns of DNA methylation in cell differentiation and in X-chromosome inactivation.

## SOURCES AND ADDITIONAL READINGS

Ansorge, W., Voos, H., and Zimmermann, J., eds. 1995. *DNA Sequencing: Automated and Advanced Approaches.* New York: Wiley.

Azhikina, T., Veselovskaya, S., Myasnikov, V., Potapov, V., Ermolayeva, O., and Sverdlov, E. 1993. Strings of contiguous modified pentanucleotides with increased DNA-binded affinity can be used for DNA sequencing by primer walking. *Proceedings of the National Academy of Sciences USA* 90: 11460–11462.

Beskin, A. D., Sonkin-Zevin, D., Sobolev, I. A., and Ulanovsky, L. E. 1995. On the mechanism of the modular primer effect. *Nucleic Acids Research* 23: 2881–2885.

Borodin, A., Kopatnzev, E., Wagner, L., Volik, S., Ermolaeva, O., Lebedev, Y., Monastyrskaya, G., Kunz, J., Grzeschik, K.-H., and Sverdlov, E. 1995. An arrayed library enriched in hncDNA corresponding to transcribed sequences of human chromosome 19: Preparation and analysis. *Genetic Analysis* 12: 23–31.

Church, G. M., and Kieffer-Higgins, S. 1988. Multiplex DNA sequencing. *Science* 240: 185–188.

Gordon, D., Abajian, C., and Green, P. 1998. Consed: A graphical tool for sequence finishing. *Genome Research* 8: 195–202.

Henikoff, S. 1984. Unidirectional digestion with exonuclease III creates targeted breakpoints for DNA sequencing. *Gene* 28: 351–359.

Hillier, L., et al. 1996. Generation and analysis of 280,000 human expressed sequence tags. *Genome Research* 6: 807–828.

Hunkapiller, T., Kaiser, R. J., Koop, B. F., and Hood, L. 1991. Large-scale and automated DNA sequence determination. *Science* 254: 59–67.

Kieleczawa, J., Dunn, J. J., and Studier, W. F. 1992. DNA sequencing by primer walking with strings of contiguous hexamers. *Science* 258: 1787–1791.

Kotler, L., Sobolev, I., and Ulanovsky, L. 1994. DNA sequencing: Modular primers for automated walking. *BioTechniques* 17: 554–558.

Kozak, M. 1996. Interpreting cDNA sequences: Some insights from studies on translation. *Mammalian Genome* 7: 563–574.

Li, C., and Tucker, P.W. 1993. Exoquence DNA sequencing. *Nucleic Acids Research* 21: 1239–1244.

Makatowski, W., Zhang, J., and Boguski, M. S. 1996. Comparative analysis of 1196 Orthologous mouse and human full-length mRNA and protein sequences. *Genome Research* 6: 846–857.

McCombie, W. R., and Kieleczawa, J. 1994. Automated DNA sequencing using 4-color fluorescent detection of reactions primed with hexamer strings. *BioTechniques* 17: 574–579.

Ohara, O., Dorff, R. L., and Gilbert, W. 1989. Direct genomic sequencing of bacterial DNA: The pyruvate kinase I gene of *Escherichia coli. Proceedings of the National Academy of Sciences USA* 86: 6883–6887.

Patanjali, S. R., Parimod, S., and Weissman, S. M. 1991. Construction of a uniform-abundance (normalized) cDNA library. *Proceedings of the National Academy of Sciences* 88: 1943–1947.

Raja, M. C., Zevin-Sonkin, D., Shwartburd, J., Rozovskaya, T. A., Sobolev, I. A., Chertkov, O., Ramanathan, V., Lvovsky, L., and Ulanovksy, U. 1997. DNA sequencing using differential expression with nucleotide subsets (DENS). *Nucleic Acids Research* 25: 800–805.

Riggs, A., Saluz, H. P., Wiebauer, K., and Wallace, A. 1991. Studying DNA modifications and DNA-protein interactions in vivo. *Trends in Genetics* 7: 207–211.

Saluz, H. P., Wiebauer, K., and Wallace, A. 1991. Studying DNA modifications and DNA-protein interactions in vivo. *Trends in Genetics* 7: 207–211.

Sasaki, Y. F., Ayusawa, D., and Oishi, M. 1994a. Construction of a normalized cDNA library by introduction of a semi-solid mRNA-cDNA hybridization system. *Nucleic Acids Research* 22: 987–992.

Sasaki, Y. F., Iwasaki, T., Kobayashi, H., Tsuji, S., Ayusawa, D., and Oishi, M. 1994b. Construction of an equalized cDNA library from human brain by semi-solid self-hybridization system. *DNA Research* 1: 91–96.

Schuler, G.D., et al. 1996. A gene map of the human genome. *Science* 274: 540–546.

Smith, M. W., Holmsen, A. L., Wei, Y. H., Peterson, M., and Evans, G. A. 1994. Genomic sequence sampling: A strategy for high resolution sequence-based physical mapping of complex genomes. *Nature Genetics* 7: 40–47.

Soares, M. B., Bonaldo, M. D. F., Jelene, P., Su, L., Lawton, L., and Efstratiadis, A. 1994. Construction and characterization of a normalized cDNA library. *Proceedings of the National Academy of Sciences USA* 91: 9228–9232.

Strausbaugh, L. D., Bourke, M. T., Sommer, M. T., Coon, M. E., and Berg, C. M. 1990. Probe mapping to facilitate transposon-based DNA sequencing. *Proceedings of the National Academy of Sciences USA* 87: 5213–6217.

Studier, F. W. 1989. A strategy for high-volume sequencing of cosmid DNAs: Random and directed priming with a library of oligonucleotides. *Proceedings of the National Academy of Sciences USA* 86: 6917–6921.

Szybalski, W. 1990. Proposal for sequencing DNA using ligation of hexamers to generate sequential elongation primers (SPEL-6). *Gene* 90: 177–178.

Venter, J. C., Smith, H. O., and Hood L. 1996. A new strategy for genome sequencing. *Nature* 381: 364–366.

# 12 Future DNA Sequencing without Length Fractionation

## WHY TRY TO AVOID LENGTH FRACTIONATIONS?

All but one of the methods we have described for DNA sequencing in Chapter 10 involved a gel electrophoretic fractionation of DNA. The exception used mass spectroscopy instead of electrophoresis, but a length fractionation was still needed because all of the information about base location had been translated into fragment sizes prior to analysis. There are several motivations to try to get away from this basic paradigm and develop DNA sequencing methods that do not depend on size fractionation. First, size fractionation, except by mass spectroscopy, is really quite a slow and indirect method of reading sequence data. Second, fractionations are intrinsically hard to parallelize. Third, it is not obvious how fractionation methods could ever be used to look efficiently just at sequence differences, and most of the long-term future of DNA sequencing applications probably lies in this key area of differential sequencing. This is true not only in the potential use of DNA sequencing for human diagnostics, but also for evolutionary applications, for population genetic applications, and for ecological screening.

For all of these reasons there is considerable current interest in trying to develop entirely new approaches to DNA sequencing. Many of the techniques that will be mentioned in this chapter are going to be discussed only briefly. While they are probably capable of maturing into methods that can read DNA sequences, they are unlikely to do this soon enough, or ultimately efficiently enough, to be of much use for large-scale DNA sequence processing. However at least one of the second-generation methods treated in this chapter, sequencing by hybridization (SBH), does appear to offer a significant chance of making an impact on the current human genome project, and an even better chance of making a major impact on future DNA sequencing in clinical diagnostics.

## SINGLE-MOLECULE SEQUENCING

A number of different potential DNA sequencing methods require that data be obtained from one single molecule at a time. They include handling DNAs or their reaction products in flow systems, or observing DNAs by microscopy. One approach that has been investigated by Richard Keller, would use an exonuclease to degrade a tethered DNA continually, and detect individual nucleotides as they are cleaved by the enzyme and liberated into a flowing stream. This method exploits the power of flow cytometry for very sensitive detection of a fluorescent target whose location is known rather precisely. A schematic illustration of this approach to single-molecule sequencing is given in Figure 12.1.

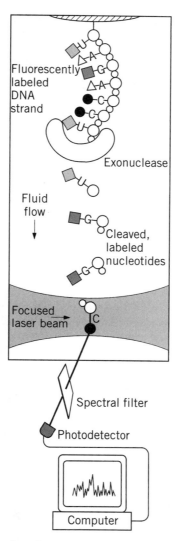

**Figure 12.1**   Schematic illustration of one approach to sequencing single DNA molecules in solution. Provided by Richard Keller.

To see all four bases in a single molecule, each would have to be labeled with a different fluorophore. There have been suggestions to use the intrinsic fluorescence of each base directly, but this fluorescence is weak, so it would require very sophisticated and expensive detection methods to be practical at the single-molecule limit. The challenge is to make fluorescent base analogs that are acceptable substrates for DNA polymerase. Substitution at every single nucleotide position must be accomplished. This requirement has been met with one and two bases, which is an impressive accomplishment. It remains to be seen if it can be met with all four bases. The ideal exonuclease would liberate bases in a kinetically smooth and rapid rate. It would be processive so that a single enzyme would suffice to degrade the DNA molecule. Otherwise, with the arrangement shown in Figure 12.1, there would be pauses during which no product would be appearing, and one would constantly have to replenish the supply of enzyme as molecules fall off and are lost to the flowing stream. The properties of actual exonucleases, for the most part, are not this ideal, but they are rapid, and some are reasonably processive.

It is possible in a flowing stream to detect single fluorophores of the sort used to label nucleic acid bases, which is like that which would be used with a tethered DNA. Some typical results are illustrated in Figure 12.2. The issue that remains unresolved is the chances of missing a base (a false negative) and the chances of seeing a noise fluctuation, by scattering from a microscopic dust particle, or whatever, that imitates a base when none is present (a false positive). It is interesting to examine what the consequences would be if, instead of one molecule, many were tethered together in the flowing stream, and exonucleases were allowed to process them all. If the digestion could be kept in synchrony, the signal-to-noise problem in detection would be alleviated considerably. However, there is no way to keep the reactions in synchrony. The best one can do is to find a way to start the exonucleases synchronously and use the most processive enzymes available. Even in this case, however, there is an inevitable dephasing of the reaction, as it proceeds, because of the basic stochastic (noisy) nature of chemical reactions. Given

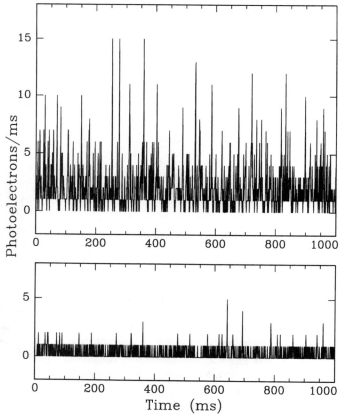

**Figure 12.2**   Typical data obtained in pilot experiments to test the scheme shown in Figure 12.1. Top panel: A dilute concentration of a labeled nucleotide is allowed to flow past the detector. Bottom panel: A control with no labeled nucleotides.

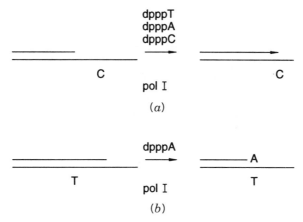

**Figure 12.3**   Plus-minus DNA sequencing. (*a*) Extension until a template coding for a missing dpppN is encountered. (*b*) Degradation of a chain until a specific nucleotide is reached.

some microscopic rate constant for a reaction, at the level of individual molecules the actual reaction times vary in such a way that a distribution of rates actually contributes the average value seen macroscopically. The reaction zone broadens like a random walk as the reaction proceeds down the chain. Its width soon becomes more than several bases, even under ideal kinetic circumstances.

One way to circumvent the statistical problems with sequencing by exonucleases would be to find a method to stop the reaction at fixed points and then allow it to restart. This, in essence, is what is done with DNA polymerase in plus-minus sequencing, the very earliest method used by Ray Wu. If one dpppN is left out, the reaction proceeds up to the place where this base is demanded by the template, and it stalls there (Fig. 12.3*a*). Adding the missing dpppN then allows the reaction to continue. If a DNA polymerase with a 3'-editing exonuclease activity is used, a similar result can be achieved by having only one dpppN present. In this case the enzyme degrades the 3'-end of a DNA chain, until it reaches a place where the dpppN present is called for by the template (Fig. 12.3*b*). As long as a sufficient supply of dpppN remains, the enzyme will stall at this position. These are useful tricks; they work well for sequencing, and there is no reason why they could not be incorporated into strategies for sequencing a single molecule or small numbers of molecules. However, the major potential advantage of the original scheme proposed by Keller is speed, and steps that require changing substrates numerous times are likely to slow down the whole process considerably.

## SEQUENCING BY HIGH-RESOLUTION MICROSCOPY

One of the earliest attempts at the development of alternate DNA sequencing methods was Michael Beer's strategy for determining nucleic acid sequence by electron microscopy. Beer's plan was to label individual bases with particular electron-dense heavy metal clusters and then image these. Two problems made this approach unworkable. First, the nucleic acids were labeled by covalent modification after DNA synthesis. This leads to less than perfect stoichiometry of the desired product, and it undoubtedly also leads to some unwanted side reactions with other bases. The second problem is that sample damage in the conventional electron microscope is considerable; this makes it very difficult to achieve accurate enough images to read the DNA sequence as the spacing between metal-tagged bases, since the structure moves around in response to the electron beam of the molecule. This problem of molecular perturbation by microscopes remains with us today as the greatest obstacle to using high-resolution microscopy for DNA sequencing.

Currently a new generation of ultramicroscopes has reopened the issue of whether DNA could be sequenced, directly or indirectly, by looking at it. The new instruments are scanning tip microscopes; the best studied of these are the scanning tunneling microscope (STM) and the atomic force microscope (AFM). Both of these instruments read the surfaces of samples in much the same way that a blind person reads braille. The surface is scanned with a sharp tip in a raster pattern, as shown schematically in Figure 12.4. In AFM what leads to the image is the force between the tip and the sample. Van der Waals forces will attract the tip to the surface at long distances and repel the tip at short distances. What is usually done is to have a feedback loop via a piezoelectric device. This can be used to place a force on the tip to keep its vertical position constant, and the voltage needed to accomplish this is measured. Alternatively, one can apply a constant force, and measure the vertical displacement of the tip, for example, by bouncing a laser beam off the tip and seeing where it is reflected to. In STM an electrical potential is maintained between the tip and the surface. This leads to a current flow from the tip to the surface at short distances. The current is dependent on the distance between the tip and the surface, and the electrical properties of the surface. In practice, one can adjust the position of the tip to maintain a constant current, and measure the tip position, or keep the vertical height of the tip constant and measure the current.

For AFM or STM to be successful, very flat surfaces are required. With hard samples on such surfaces, atomic resolution is routinely observed, and even subatomic resolution has been reported, where information about the distribution of electron density within the sample is uncovered. DNA is not a hard sample, and it does not easily adhere to most very flat surfaces. These difficulties have produced many frustrations in early attempts to image DNA by AFM or STM. In retrospect, most or all of the spectacular early pictures of DNA have been artifacts, perhaps caused by helixlike imperfections in the underlying surfaces. The best that can be said is that images that looked like DNA were rare and far between, and not generally reproducible. One problem that soon became quite apparent is that the forces used in these early attempts were sufficient in most, if not all, cases to knock the DNA molecules off the surface being imaged.

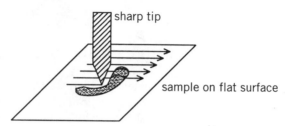

**Figure 12.4** Operating principle of a typical scanning tip microscope. In STM the electrical current between the tip and the surface is measured. In AFM the repulse force between the tip and the surface is measured.

More recent attempts to image DNA with scanning tip microscopes, particular with AFM, have been more successful, at least in the sense that dense arrays of molecules can be seen reproducibly. This is accomplished by using surfaces to which DNA adheres better, like freshly cleaved mica, instead of the graphite or silicon surfaces used earlier. Sharper tips give high enough resolution to be able to measure the lengths of the molecules reliably. The current images are, however, a long way from the resolution needed to read the sequence directly by looking at the bases. A number of severe obstacles will have to be overcome if this is ever to be done. First, the current images are mostly of double-stranded DNA. This is understandable since it is a much more regular structure, much more amenable to detection and quantitation in the microscope. However, in the double strand, only short bits of sequence are readable from the outside, as one is forced to do in AFM. This will lead to a difficult, but apparently not insurmountable, sequence reconstruction problem where data from many molecules will have to be combined to synthesize the final sequence. A second problem is that the DNA molecules could still be distorted quite a bit as the tip of the microscope moves over them. This may or may not be alleviated by newer microscope designs that would allow lower forces to be used.

A third problem with AFMs is that the image seen is a convolution of the shape of the tip and the shape of the molecule, as shown in Figure 12.5. Thus, unless very sharp tips can be made, or tips of known shape, it can be difficult with a soft, deformable molecule to deconvolute the image and see the true shape of the DNA. One approach to circumvent many of these difficulties would be to label the DNA with base-specific reagents that are more distinctive either in AFM, where larger, specific shapes could be used, or in STM, where labels with different electrical properties might serve. As a test of this, and to make sure DNA imaging was now reliable in the AFM, proteins were attached to the ends of DNAs before AFM imaging. Two examples of the sorts of images seen are shown in Figure 12.6. The protein used, purely because it was available and of a size that made it easy to distinguish from DNA, was a chimera between streptavidin and a fragment of staphylococcal protein A, which was already introduced in Chapter 4, where it was used for immunoPCR. The DNA was biotinylated, either on one or both ends. Two different lengths of DNAs were used, and since streptavidin is tetrameric, the resulting images show a progression of structures from DNA monomers up to trimers. Because of the nature of these structures, the proper measured lengths of the DNAs within them, and the expected height difference between the protein label at the ends or vertices of the DNA and the DNA itself, one can be very confident that these are true images of DNA and protein. However, the resolution is still far too low to allow sequencing.

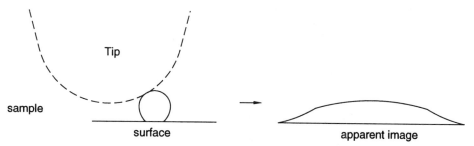

**Figure 12.5**   In scanning tip microscopy, what is actually measured is a convolution of the shape of the object and the shape of the tip.

Biotinylated DNA + Streptavidin

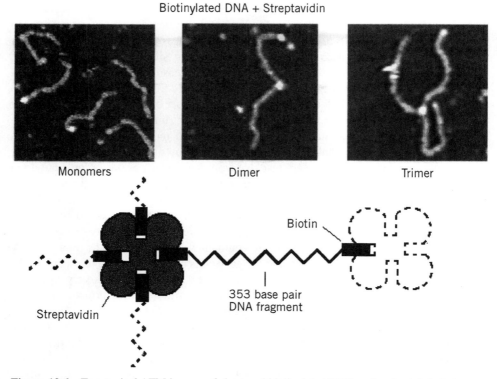

**Figure 12.6**   Two typical AFM images of short-end biotinylated DNA molecules labeled at one or both ends by a chimeric protein fusion of streptavidin and staphylococcal protein A. Since the streptavidin is tetrameric, one can see figures representing more than one DNA molecule bound to the same streptavidin. Reproduced from Murray et al. (1993).

It should be possible to use progressively smaller labels and to increase the resulting resolution. Whether this will lead to direct AFM DNA sequencing soon is anyone's guess. If it does, a real advantage is that one could sequence a wide variety of different molecules in a single experiment without the need to clone or fractionate. The labeling would almost certainly be introduced by PCR using analogs of the four bases. This will be much more accurate than the original chemical modification methods used for electron microscopy. However, the resulting images, as elegant as they may look some day, might have to be analyzed as images to extract the DNA sequence data. By current methods this could become a serious bottleneck. What is still needed is a way to direct the tip of the microscope so that it tracks just over the DNA molecule of interest, rather than scanning a grid that is mostly background. If this can somehow be achieved, the problem of image analysis ought to become much simpler, and the rate of image acquisition also ought to be increased considerably.

## STEPWISE ENZYMATIC SEQUENCING

A major success story in the history of protein sequencing was the development of stepwise chemical degradation. Amino acid residues are removed one at a time from one end of the polypeptide chain and their identity is determined successively. Automated

Edmond degradation currently provides our main source of direct protein sequence data. The yield in each step is the critical variable, since it determines how far from the original end the sequence can be read. Comparable chemical approaches for DNA or RNA sequencing have not been terribly successful. Recently, however, several stepwise enzymatic sequencing approaches have been suggested. As individual processes, they do not at first glance seem all that attractive. However, they have the potential to be implemented in massively parallel configurations, which, if successful, could ultimately provide very high throughput. These schemes are distinct from the single molecule methods described earlier in that any desired number of target molecules of each type can be employed. Thus detection sensitivity is not an issue here.

One strategy, developed by Mathias Uhlen, is to divide the sample into four separate wells, each containing a DNA polymerase without a 3′-proofreading activity. To each well one of the four dpppN's is added. Chain extension will occur only in one well with the concomitant release of pyrophosphate (Fig. 12.7). This product can be detected with great sensitivity; for example, it can be enzymatically converted to ATP, and that can be measured using luciferase, to generate a chemiluminescent signal. The amount of light emitted is proportional to the amount of ATP made. Thus one can quantitate the amount of dpppN incorporated and determine, within limits, how many units the chain was extended by. Sample from the well that was successfully extended is then divided into four new wells, and the process is repeated. Actually three wells would suffice, since one knows that the next base is not the same as the one or ones just added, but it is probably good to have the fourth as an internal control. One obvious complication with the scheme is that the sample keeps getting divided, so one has to either start with a large amount of it or have a sensitive enough assay that only a small aliquot can be removed and assayed. A variation on this basic approach adds dpppNs in a cyclical order. This avoids the problem of sample subdivision. It appears to have considerable promise. The method is called pyrosequencing.

A second strategy is similar in spirit but uses dideoxy pppN's. This is shown in Figure 12.8. In four separate wells containing target DNA and DNA polymerase is added one of the ddpppN's carrying a label. Alternatively, one could use a single well and a mixture of four different fluorescently labeled ddpppN's. Only one of the ddpppN's becomes incorporated. From the location of the well, or the color, the identity of the base just added is known. The base just added is now removed by treating with the 3′-editing exonuclease activity of DNA polymerase I in the presence of all of the dpppN's except the one just

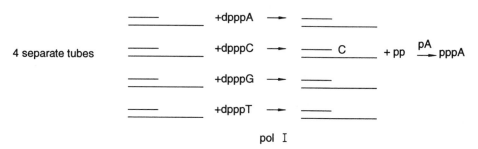

**Figure 12.7**  One scheme for stepwise enzymatic DNA sequencing. Here, when a particular base is added, pyrophosphatase is used to synthesize ATP from the pyrophosphate (pp) released, and the ATP in turn is used to generate a chemiluminescent signal by serving as a substrate for the enzyme, luciferase.

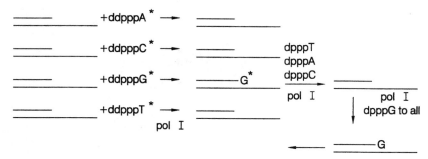

**Figure 12.8**   A second scheme for stepwise enzymatic DNA sequencing. This is similar in spirit to plus-minus DNA sequencing illustrated in Figure 12.3. It uses fluorescent terminators, ddpppN*, like those employed in conventional automated Sanger sequencing.

added. Next the labeled ddpN just removed is replaced by an unlabeled dpppN by using DNA polymerase with only this particular dpppN present. Then the process is repeated. This scheme avoids the sample division or aliquoting problem of the previous strategy.

To detect a run of the same base, one will have to be able to vary the scheme. What will happen in this case is that the exonuclease treatment will degrade the chain back to the location of the first base in the run. To determine the length of the run, one possible approach is to add a labeled analog of the particular dpppN involved in the run, in the presence of DNA polymerase, and detect the amount of synthesis by quantitating the incorporation of label. Then the entire block of labeled dpN's has to be removed, replaced by unlabeled dpN's, and next base after the block can now be determined. This is an unfortunate complication. However, in principle, the entire scheme could be set up in a microtitre plate format and run in a very parallel way. As in the first scheme the whole process would be best carried out in a solid state sequencing format so that the DNA could be purified away from small molecules and enzymes easily and efficiently after each step.

A third strategy, has not been tested to our knowledge, because it depends on finding a dpppNx derivative that has two special properties. Like a ddpppN the derivative must not be extendable by DNA polymerase. However, there must be a way to change the dpNx after incorporation into the DNA chain so that now it is extendable. The scheme then, as shown in Figure 12.9, is to add dpppNx to the target in four separate tubes. Whichever

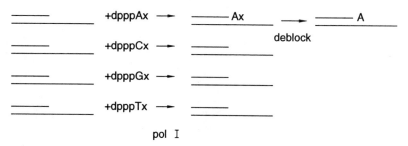

**Figure 12.9**   A third, as yet unrealized scheme for stepwise enzymatic DNA sequencing. Here the key ingredient would be a 3'-blocked pppNx that could be deblocked after each single base elongation step. Incorporation could be detected by pyrophosphate release as in Figure 12.7 or by the use of a fluorescent blocking agent, x.

one extends could be determined by a pyrophosphatase assay as in the first scheme. Then the incorporated dpNx is converted to dpN, and the process continued. One candidate, in principle, for the desired compound is dpppNp, which, after incorporation of the dpNp into DNA could be converted to dpN by alkaline phosphatase. This latter step certainly works; however, it is uncertain how well DNA polymerase will utilize compounds like dpppNp's. Apparently quite a few other reversible dpppN derivatives have been tried as DNA polymerase substrates without much success thus far. This is a pity because the scheme has real appeal.

In deciding how, eventually, to implement any of the above schemes in an automated fashion, one needs to consider an interesting trade-off between the time to sequence one sample and the number of samples that can be handled simultaneously. Instead of trying all four single base additions simultaneously, they could be tried one at a time, in a cyclical pattern, say A, then G, then C, then T, as in pyrosequencing. With an immobilized DNA sample, the target is simply moved from one set of reagents, after washing, to another, and the point where positive incorporation occurs is recorded. The advantage of this is that the logistics and design of the system become much simpler, particularly for the cases where pyrophosphate release is measured. It will take four times as long to complete a given length of DNA sequence, but one could handle precisely four times as many sequences simultaneously.

## DNA SEQUENCING BY HYBRIDIZATION (SBH)

We will devote the rest of this chapter to a number of approaches for determining the sequence of DNA by hybridization. In all of these approaches one uses relatively short oligonucleotides as probes to determine whether the target contains the precise complementary sequences. Essentially SBH reads a word at a time rather than a letter at a time. Intuitively this is quite efficient; it is after all the way written language is usually read. For reasons that will become readily apparent, attempts to perform SBH have usually focused on oligonucleotides with 6 to 10 bases. The conception of SBH appears to have had at least four independent origins. Many groups are now working to develop an efficient practical scheme to implement SBH.

There seems to be a consensus that SBH will eventually work well for some high-throughput DNA sequencing applications, like sequence comparison, sequence checking, and clinical diagnostic sequencing. In all of these cases one is not trying to determine large tracts of sequence de novo; instead, the targets of interest are mostly small differences between a known or expected sequence, and what has actually been found. There is also agreement that SBH will work for determining partial sequences, for fingerprinting, and for mapping. However, SBH may not work for direct complete de novo sequencing unless some of the enhancements or variations that have been proposed to circumvent a number of problems turn out to work in practice.

The two critical features of SBH are illustrated in Figures 12.10 and 12.11. As we demonstrated in Chapter 3, the stability of a perfectly matched duplex is greater than an end mismatched duplex, and much greater than a duplex with an internal mismatch (Fig. 12.10). Thus a key step in SBH is finding conditions where there is excellent discrimination between perfect matches and mismatches. An immediate problem is that for a sequence of length $n$, there is only one perfect match, there are six possible end mismatches (each base on each end of the target can be any one of the three noncomplementary bases), and there are $3(n - 2)$ possible internal mismatches. Unless the discrimination is

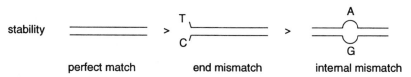

**Figure 12.10**    Effect of mismatches on the stability of short DNA duplexes.

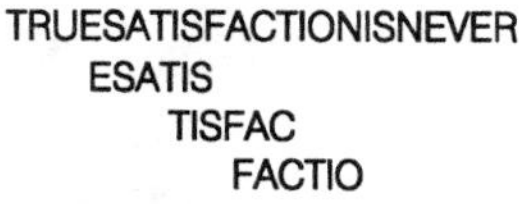

TRUESATISFACTIONISNEVER
ESATIS
TISFAC
FACTIO

**Figure 12.11**    Reading a text by reconstruction from overlapping *n*-tuple words.

very strong, there will be an inevitable background problem where a specific signal is diluted by a large number of weak mismatches.

The second key aspect of SBH is that the sequence can be read by overlapping words, as shown in Figure 12.11. In principle, with perfect data one would not need to try all the words to reconstruct the sequence. The problem of reconstructing a sequence from all *n*-tuple subsegments is highly overdetermined, except for the complications that we will discuss below. Simulations show that reconstructing DNA sequences from oligonucleotide hybridization fingerprints is very robust and very error resistant. Even significant levels of insertions and deletions can be tolerated without badly degrading the final sequence. A key element of SBH that is easily forgotten is that negatives as well as positives are extremely informative. Knowing that a specific oligonucleotide like AACTG-GAC does not exist anywhere in the target provides a constraint that can sometimes be quite useful in assembling the data from words that are found to be present.

## BRANCH POINT AMBIGUITIES

The major theoretical limitation with simple direct implementations of SBH is the inability to determine sequences uniquely if repeating sequences are present. There are two kinds of repeats: tandemly repeated sequences and interspersed repeats. The presence of a tandemly repeated sequence can be detected, but it is difficult to determine the number of repeated sequences present. When the length of the monomer repeat sequence length is longer than the SBH words length, then the number of copies of a tandemly repeated sequence can only be determined when a hybridization signal is quantitated and not simply scored positively. These problems are relatively easily dealt with by conventional sequencing or PCR assays, since the unique sequence flanking the simple sequence or tandem repeat will generally be known.

The more serious problem caused by repeats is called a branch point ambiguity (Fig. 12.12). If the SBH word length used is *n,* these ambiguities arise whenever there is an exact recurrence of any sequence with length *n* − 1 in the target. What happens, as shown in Figure 12.12, is that the data produced by the complete pattern of *n*-tuple words can be assembled in two different ways. In general, there is no way to distinguish between the alternative assemblies. In principle, if one could read the sequence out to the very end of the SBH target, the particular ambiguity shown in Figure 12.12 could be resolved.

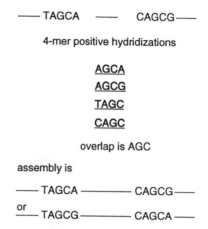

—— TAGCA ——   CAGCG——

4-mer positive hydridizations

AGCA
AGCG
TAGC
CAGC

overlap is AGC

assembly is

—— TAGCA ————— CAGCG——

or
—— TAGCG ————— CAGCA ——

**Figure 12.12**   A tandem repeating sequence results in a branch point ambiguity: Two different reconstructions are possible from the pattern on *n*-tuple words detected.

However, there is no guarantee that this will be possible. The moment that there are more than one recurrences, the ambiguities become almost intractable. For example, the case shown in Figure 12.13, in which a sequence recurs three times, cannot be resolved even if one could read all the way to the ends of the SBH target.

The probability of any particular sequence 8 long recurring in a target of 200 bp is very low: A rough estimate gives $192/4^8 = 3 \times 10^{-3}$. However the probability that the target will have one or more recurrence once all possible recurrences are considered is actually quite high. An analogy is asking what is the probability that two people in a room have the same birthday. If you specify a particular date, or pick a particular person, the odds of a match are very low. However, if you allow all possible pairings to be considered, the odds are quite high that in a room with 30 people, two will share the same birthday. So sequence recurrences are a serious problem. They limit the length of sequence that can be directly and unambiguously read. In general, the chances of recurrences diminish as the length of the word used increases.

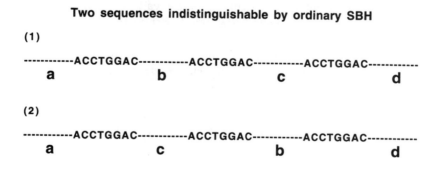

**Two sequences indistinguishable by ordinary SBH**

(1)

------------ACCTGGAC------------ACCTGGAC------------ACCTGGAC------------
**a**                **b**                **c**                **d**

(2)

------------ACCTGGAC------------ACCTGGAC------------ACCTGGAC------------
**a**                **c**                **b**                **d**

**a, b, c, and d are blocks of single copy sequence**

**Figure 12.13**   A more serious branch point ambiguity that leads to uncertainty in the arrangement of two blocks of single-copy DNA sequence.

There are $4^n$ possible words of length $n$, for a four-letter alphabet. Therefore to sequence by hybridization could require examining as many as 65,536 possible 8-mers or 262,144 possible 9-mers. Making complete sets of compounds larger than this and controlling their quality is likely to be challenging with present or currently extrapolated technology. It turns out that an estimate of the average sequence length that can be read before a branch point ambiguity arises is given approximately by the square root of the number of words used. When the words are DNA sequences, this is $4^{n/2}$. For 8-mers, the average length of sequence determined between branch points will be 256. This is quite an acceptable size scale. It seems like a losing proposition to increase the word size much beyond 8, unless technical considerations in the hybridizations demand this. Reducing the word length below 8 will lead to an unacceptably high frequency of branch points, unless some specific additional strategy is introduced to resolve these ambiguities.

Branch point ambiguities have one additional implication that must be dealt with in all attempts to implement a successful SBH strategy. The number of branch points present in a target sequence will grow rapidly as the total length of the target increases. Thus one must subdivide the target into relatively short DNA fragments in order to have a reasonable chance of sequencing each fragment unambiguously. This is a relatively undesirable feature of SBH. However, even if branch points could be resolved some other way, short targets are probably still mandated in order to diminish complications that may arise from intramolecular secondary or tertiary structure in the target. More will be said about this later.

## SBH USING OLIGONUCLEOTIDE CHIPS

It is obvious that SBH cannot be practical if one is forced to look at hybridizations between a single oligonucleotide and a single target one at a time. If 8-mers were used, 65,536 different experiments would have to be done to determine a sequence that on average would be a DNA fragment less than 256 bases in length. The major appeal of SBH is that it seems readily adaptable to highly parallel implementations. There are two very different approaches that are being explored for this. The first is to hybridize a single-labeled sample to an array of all of the possible oligonucleotides it may contain. This is sometimes called format I SBH. The ideal array would be very small to minimize the amount of sample that was needed. Hence it is conventional to call the array a chip, by analogy with a semiconductor chip. A schematic illustration of an SBH experiment using such a chip is shown in Figure 12.14. A real chip would probably contain all 65,536 possible 8-mers, probably each present several times to allow signal averaging and control for reproducibility. The location of each particular oligonucleotide would be known. The actual patterns of oligonucleotides would probably be rather particular, a consequence of whatever systematic method is used to produce them. The chip surface itself could be silicon, or glass, or plastic. The key aspect is that the oligonucleotides must be covalently attached to it, and the surface must not interfere with the hybridization. The surface must not show significant amounts of nonspecific adsorption of the target, and it must not hinder, sterically or electrostatically, the approach of the target to the bound oligonucleotides. The ideal surface will also assist, or at least not interfere with, whatever detection system is ultimately used to quantitate the amount of hybridization that has occurred.

Several approaches are being tested to see how to fabricate efficiently a usable chip containing 65,536 8-mers. One basic strategy is to premake all of the compounds in the

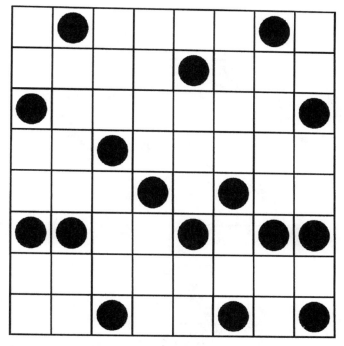

**Figure 12.14**    An example of the expected hybridization pattern of a labeled target exposed to an oligonucleotide chip. The actual chip might contain 65,000 or more probes.

array separately, and then develop a parallelized automated device to spot or spray the compounds onto the right locations on the chip. The disadvantage of this approach is that the rate of manufacture of each chip could be fairly slow. The major advantage of this approach is that the oligonucleotides only have to be made once, and their individual sequence and purity can be checked. There is no consensus at the present time what the optimal way would be to manufacture chips given samples of all the 65,536 8-mers. A key variable is how they will be attached to the chip surface. A long enough spacer must be used to keep the 8-mers well above the surface. Otherwise, the surface is likely to pose a steric restriction for the much bulkier target DNA.

The alternate strategies involve synthesizing the array on the chip. One potential general way to do this is photolithography, a technique that has been very powerful in the construction of semiconductor chips. It has been used quite successfully by Steven Fodor and others at Affymetrix, Inc. to make dense arrays of peptides, and more recently to make dense arrays of oligonucleotides. The basic requirement is that nucleotide derivatives are needed that are blocked from extending, say because the 3' OH is esterified. The block used, however, can be removed by photolysis. A mask is used to allow selective illumination of only those chains that require extension by a particular base in this position (Fig. 12.15). Thus the light activates just a subset of the oligomers on the chip. The chain extension reaction is carried out in the dark. Then, in turn, three other masks are used to complete one cycle of synthesis. The key requirement in this approach is that the photoreaction must proceed at virtually 100% yield. Otherwise, the desired sequences will not be made in sufficient purity. This is a very difficult demand to satisfy with photochemical reactions. Instead, one can still use the principles of masks but just do more standard solid

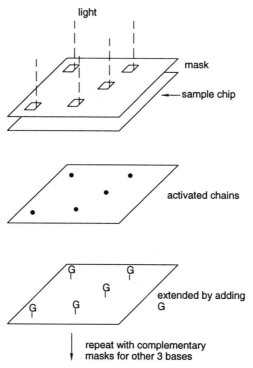

**Figure 12.15**    Construction of an oligonucleotide chip by in situ synthesis using photolithography techniques.

state oligonucleotide synthesis by spraying liquid reagents through the masks. The great power of the lithographic approach is that one can make any array desired—that is, any compound or compounds can be put in any positions on the array.

A different synthetic approach, with more limited versatility, is shown in Figure 12.16. This is the conception of Edwin Southern. It is actually a very simple lithographic approach that makes one particular array configuration efficiently. Figure 12.16 shows the steps in synthesis by stripes that would be needed to generate all possible tetranucleotides. The configuration that results is similar to the way the genetic code is ordinarily written down. Southern actually uses a glass plate as his chip. The reagents needed for the synthesis are pumped through channels between two glass plates as shown in Figure 12.17. It may be hard to miniaturize this design sufficiently to make a really small chip, but the plates made by Southern in this manner were the first dense oligonucleotide arrays actually being tested in real sequencing experiments.

An alternative approach being used to make arrays of 8-mers involves the use of a thin gel rather than a surface. This has the advantage that the sample thickness potentially allows larger amounts of oligonucleotide to be localized. In this approach, developed by Andrei Mirzabekov and coworkers, a glass plate was covered with a 50-micron thin gel. Pre-made oligonucleotides were deposited on the gel in 1 mm spots. The major effect of using a gel rather than a surface is that one has to be concerned about the local concentration of sample during washing steps. On a surface, solvent exchange is quite rapid, so the concentration of free sample can be reduced to zero quickly, and no back reactions of re-

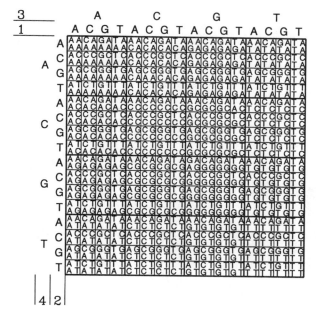

**Figure 12.16**  Pattern of stepwise DNA synthesis used in Southern's procedure for in situ synthesis of an oligonucleotide chip. Four successive synthetic steps are indicated. Within each square of the array, the sequence of the tetranucleotide synthesized is read left to right from the 3' to 5' direction starting from the upper row and continuing with the lower row.

leased material with the chip need be considered. With a gel, if target is released, it will take quite a while to leave the gel, and during this period there is a significant chance of back reaction with the chip if conditions permit duplex formation. This has both advantages and disadvantages, as we will illustrate later.

Regardless of the method of synthesis, the key technical issue that must be overcome is how the oligonucleotides are anchored at their position in the array. Mirzabekov uses direct chemical coupling to an oxidized ribonucleotide placed at the 3'-end of the 8-mer (Fig. 12.18). This is time-honored nucleic acid chemistry, but it does offer some risk of changing the stability of the resulting duplex because of the altered chemical structure at the sugar. The approach used by Southern is to attach a long hydrophilic linker arm to the glass surface (Fig. 12.19). This arm has a free primary hydroxyl group that can be used to

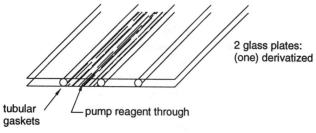

**Figure 12.17**  Glass plates separated by rubber dams are used to direct the reagents in each step of the procedure illustrated in Figure 12.16.

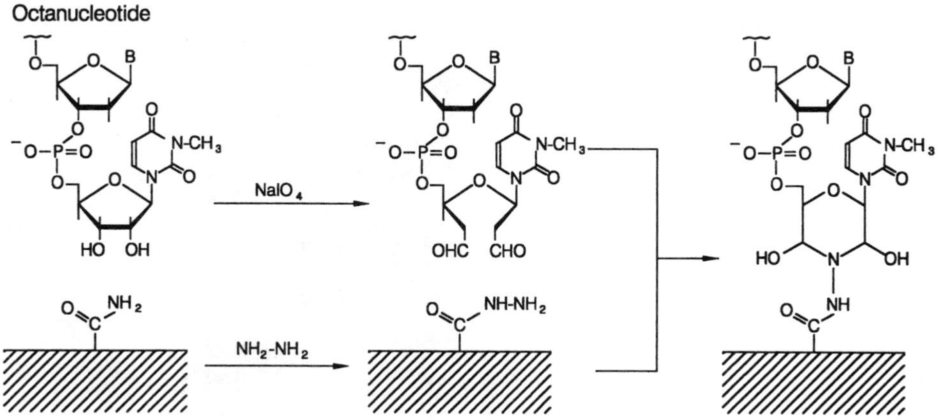

Octanucleotide

**Figure 12.18** Method of attachment of oligonucleotides to polyacrylamide gels used by Mirzabekov and coworkers. From Khrapko et al. (1991).

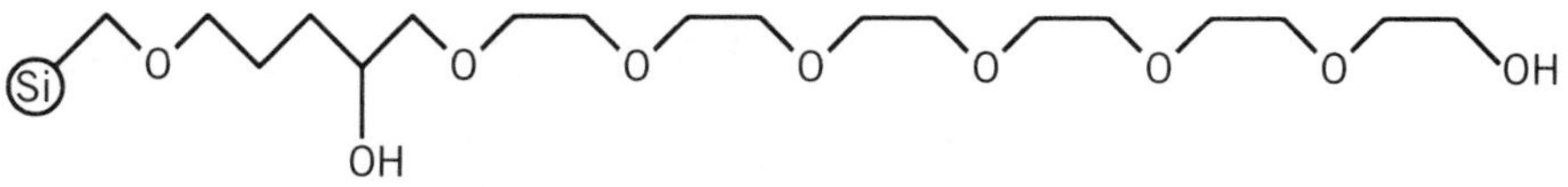

**Figure 12.19** Method of attachment of oligonucleotides to glass plates used by Southern et al. (1992).

initiate the synthesis of the first nucleotide of the 8-mer in standard DNA synthesis protocols. It acts chemically exactly like the 3′ OH of a nucleoside in coupling to an activated phosphate of the next nucleotide.

## SEQUENCING BY HYBRIDIZATION TO SAMPLE CHIPS

The second general SBH approach is to make a large, dense array of samples and probe it by hybridization with one labeled oligonucleotide at a time. This is sometimes called format II SBH. In this format, while it takes a long time to complete the sequence, one is actually sequencing a large number of samples simultaneously. A schematic illustration of this approach is shown in Figure 12.20. It looks deceptively similar to the use of oligonucleotide chips, but everything is reversed. The array might, for example, correspond to an entire cDNA library, perhaps $2 \times 10^4$ clones in all. Because SBH can only use relatively short samples, each cDNA might have to be broken down into fragments. It is not immediately obvious how to do this with large numbers of clones at once. One possibility is to subclone two different restriction enzyme digests of the cDNA inserts. This scrambles up connectivity information in the original clones; however, in most cases that information would be easily restored by the sequencing process itself, or by rehybridization of any ambiguous fragments back to the original, intact clones. If each 1.5-kb average cDNA clone yielded six fragments in each of the two digests, one would want to array $3.6 \times 10^5$ subclones in order to maintain the redundancy of coverage of the original library. This would constitute the sample chip.

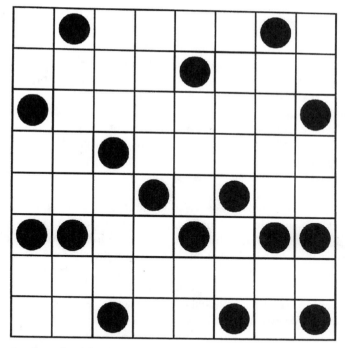

**Figure 12.20**    An example of the expected hybridization pattern of a labeled oligonucleotide to a sample chip. The actual chip might contain 20,000 or more samples.

Since the individual components are available in any desired quantity, one could, in principle, make as many copies as the array as could conveniently be handled simultaneously. In practice, it does not seem at all unreasonable to suppose that 100 copies of the array could be processed in parallel. It is envisaged that the sample chips be made by using the robotic $x-y$ tables common in the semiconductor industry (Fig. 12.21). These are very accurate and fast. It has been estimated that a sample density of $2 \times 10^4$ per 10 to 20

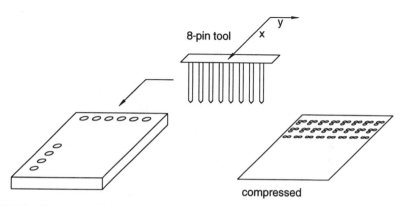

**Figure 12.21**    How a robotic $x-y$ table can be used in offset printing to construct of dense arrays of samples or probes (at right) starting from more expanded arrays (at left).

$cm^2$ is quite practical. Thus the entire array of subclones could be contained in 180 to 360 $cm^2$, which is about 30% to 60% the size of a typical 8.5 × 11 inch piece of paper.

If octanucleotides are used as hybridization probes to this large array, only a small fraction of the samples will show a positive signal. For one 250-bp subclone, the odds of containing any particular 8-mer are $243/4^8 = 3.8 \times 10^{-3}$; the possibility of a positive hybridization is less than 0.4%. When this is multiplied by the number of subclones in the array, there should be an average of $1.4 \times 10^3$ positive subclones per hybridization. Since each yields eight bases of DNA sequence data, the rate of sequence acquisition per single hybridization is $1.1 \times 10^4$ bp. If 100 chips can really be managed simultaneously, and if three hybridizations can be done per day, the overall throughput is $3.3 \times 10^6$. This is quite an impressive rate, and many of the variables used to estimate it are probably conservative.

A major feature of the use of sample chips in sequencing projects is that the approach does not scale down conveniently. Sample chips are only useful if entire libraries are to be sequenced as a unit. Such a method makes good sense for cDNAs and the genomes of model organisms. If all 65,536 8-mers must be used, at the rates we estimated above of 300 hybridizations per day, it will take more than half a year to complete the sequencing. Scaling down would not reduce the time of the effort at all; it would just reduce the amount of sequence data ultimately obtained. In practice, one does not have to use all 65,536 compounds to determine the sequence. Because of the considerable redundancy in the method, one ought to be able to use just a fraction of all 8-mers. The exact fraction will depend on error rates in the hybridization, how branch points will be resolved, and what kind of sequencing accuracy one desires. In some of the enhanced SBH schemes that will be described later, it has been estimated that one might be able to operate close to a redundancy of one rather than eight. However, this remains to be demonstrated in practice.

## EARLY EXPERIENCES WITH SBH

A major difficulty in testing the potential of SBH and evaluating the merits of different SBH strategies or particular variations on conditions, sample attachment, and so on, is that the method does not scale down. A particular problem occurs in the use of oligonucleotide chips. It is difficult to vary parameters using the set of all 65,536 8-mers. Indeed, no one has yet actually made this set of compounds. Instead, several more limited tests of SBH have been carried out.

Southern has used the scheme shown in Figure 12.16 to make a chip containing four copies of all possible octapurine sequences (A and G only). An example of some data obtained with this chip is shown in Figure 12.22. In the actual example used, the labeled target DNA was a specific sequence of 24 pyrimidines (C and T). This contains 17 different 8-tuples, and so 17 positive hybridization spots would be expected. The actual results in Figure 12.22 are much more complex than this. Two problems need to be dealt with, which illustrate some of the basic issues in trying to implement SBH on a large scale. First the amount of oligonucleotide at each position in the array differs. More important, the strength of hybridization to different sequences varies quite a bit. Duplexes rich in G+C will be more stable, under most ordinary hybridization conditions than duplexes rich in A+T. There are ways to compensate for this, as we will illustrate later, and one of these was actually used with the samples in Figure 12.22. But the compensation is not

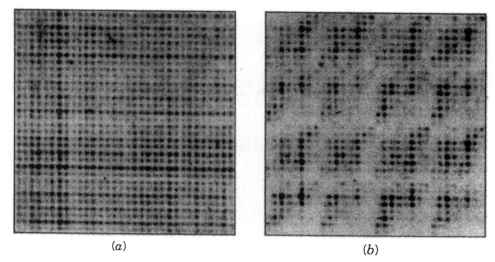

(a)                                        (b)

**Figure 12.22** Properties of an octapurine chip which contains four replicas of all 256 octapurines. (a) Hybridization pattern with an equimolar mixture of all octapyrimidines to show variation in the amount of attached purine. (b) Pattern of hybridization seen with a labeled 24-base target. From Southern et al. (1992).

perfect, and so there is a variability in the signal intensity that needs to be evaluated before a hybridization is scored as positive.

The second basic problem is cross-hybridization with single mismatches. This is a serious problem under the conditions used in Figure 12.22. From the results of these and other experiments, it has been estimated that most of the sample is not hybridized to the correct matches but instead forms a background halo of hybridization with numerous mismatches. Methods are being developed to correct for all these problems and make the best estimates of the right sequence in cases like Figure 12.22. It is too early to judge the effectiveness of these methods. A final potential problem with the test case used by Southern is that homopurine sequences can form triplexes with two antiparallel pyrimidine complements. These triplexes are quite stable, as we will illustrate in Chapter 14. It is conceivable that triplexes could have formed under the conditions used to test the oligopurine arrays, and since they would lead to systematic errors, one could go back and look for them.

Hans Lehrach has shown that oligonucleotide hybridization works well in fingerprinting samples for mapping (Chapter 9). These experiments provide some insight into the potential use of sample arrays for sequencing. Lehrach hybridizes single oligomers, or small pools of compounds, with large arrays of clones. This has successfully led to finished maps, so the sequence specificity under the conditions used must be reasonably good. However, since the mapping systems can tolerate considerable error, this is not a robust test of whether this approach will actually give usable sequence. What greatly expedites these experiments is that for fingerprinting, any oligonucleotide is as good as any other, so a large set of synthetic compounds is not needed to test the basic strategy.

Using the same approach, with a few immobilized samples, Radoje Drmanac and Radomir Crkvenjakov successfully completed two short pilot sequencing projects by SBH. In the first case, the 100-base sequence was known in advance, as was

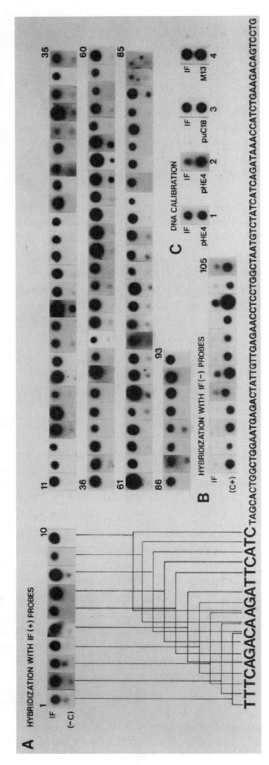

**Figure 12.23** Patterns of hybridization seen when an immobilized target sequence and a control sequence are probed successively with adjacent radiolabeled octanucleotides chosen as complements to the target sequence. Taken from Strezoska et al. (1991).

a comparable negative control sequence. Oligonucleotides were selected on the basis of the known sequence; others were added to serve as negative controls. The results are fairly convincing. As shown in Figure 12.23, the discrimination between true positives and negatives is quite good in most of the individual hybridizations. Of course the obvious criticism of this experiment is that with a sequence known in advance, the test is not a truly objective one.

To address these concerns, Drmanac and Crkvenjakov performed a second pilot test of SBH on three closely related unknown sequences containing a total of 343 bases. The design of the test was based on an uninvolved third party who analyzed these sequences and designed a set of oligonucleotides in which only about half corresponded to the sequences in the target samples. In addition the challenge was to determine all three unknown samples and not generate erroneous composites of them by errors in reconstruction. The test was a total success—all three unknown sequences were correctly determined. However, one caveat needs to be considered. Because all 65,536 8-mers were not provided, this automatically supplies enormous amounts of information about the true sequence. Any compound omitted from the set provided is automatically a true negative. Just this information alone restricts the possible sequences tremendously, even before a single experiment has been done. Thus, while the experimental results that have been achieved are impressive, it cannot yet be said that a definitive test of SBH for de novo DNA sequencing has been done. Indeed, in defense of all who work in this field, it will probably not be possible to test the methods definitively until the gamble is taken to make, directly on chips or in bulk for distribution, all of the 65,536 8-mers.

## DATA ACQUISITION AND ANALYSIS

Three different methods have been used thus far to detect hybridization in pilot SBH experiments. In each case quantitative data are needed so that positive signals can be discriminated as clearly as possible from background. Southern used image plate analyzers to examine radioisotope decay for the results shown in Figure 12.22. Others have used autoradiograms quantitated with a CCD camera. These approaches were discussed in Chapter 9. Fluorescent probes have been used by Fodor and by Mirzabekov. Here a CCD camera can be used in conjunction with a fluorescence microscope to record quantitative signals. Alternatively, a confocal scanning fluorescence microscope can be used. Other approaches such as mass spectrometry (see Chapter 11) are under development. The very notion of an oligonucleotide or sample chip raises the expectation that it should be possible to find a way to read out the amount of hybridization by a direct electronic method. Kenneth Beattie and Mitchell Eggers have developed one approach to this by detecting the mass of bound sample as it changes the local impedance on a silicon surface. In principle, one ought to be able to enhance such detection by providing the DNA probes or targets with attachments that generate more dramatic effects through altered conductivity, as a source of electrons or holes, or through magnetic properties. Perhaps the ultimate notion, as shown by the purely hypothetical example in Figure 12.24, would be to use the stability of the duplex formed in hybridization to directly manipulate elements of a nanoscale chip and thus lead to a detectable electrical signal.

However the data are obtained, current methods for analyzing data are already quite advanced. While it is difficult to convince people to synthesize 65,536 compounds before a method has proved itself, it is much easier to ask people to simulate the results of these

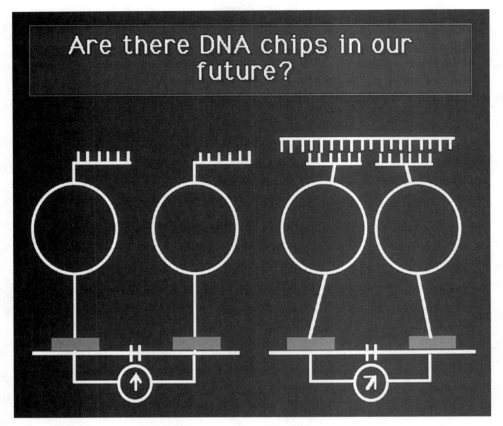

**Figure 12.24**  Possible future direct reading oligonucleotide hybridization chip. Figure also appears in color insert.

experiments and design software to reconstruct sequences from imperfect *n*-tuple word content. We have already indicated that these simulations are very encouraging, and they suggest that SBH will be a very powerful method, especially if the branch point ambiguities can somehow be dealt with. Two different proposals to handling branch points have been discussed. In the first, shown in Figure 12.25, one takes advantage of the fact that it should be possible to make a sample that consists of a dense set of small overlapping

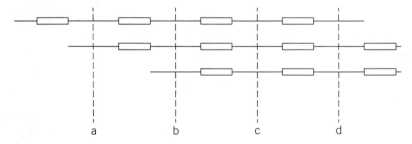

**Figure 12.25**  Overcoming branch point ambiguities by the simultaneous analysis of clones from a dense overlapping library. Recurrent sequences are shown as hollow bars. Unique hybridization probes are indicated by *a, b, c*. Known clone order implies that *b*, and not *c*, follows *a*.

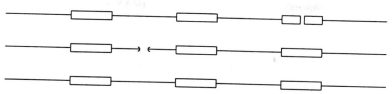

**Figure 12.26**  Overcoming branch point ambiguities by the use of several homologous but not identical DNA sequence targets.

clones. This is what is done for ordinary shotgun ladder sequencing, except that for SBH the clones would probably have to be even smaller. In these clones unique sequences will lie outside and between the repeats that cause branching ambiguities. Matching up these unique sequences not only places the clones in the proper order, it also resolves the ambiguous internal arrangement of sequences on a clone with three repeats, since the order is determined by the identity of these sequences on the flanking clones. This looks like a powerful approach, but it requires a great deal of experimental redundancy with little overall gain.

A second strategy for resolving branch point ambiguities is shown in Figure 12.26. Here the notion is to determine the DNA sequence of several similar but not identical samples. Because of sequence variations among the samples, exact recurrences in one sample will not necessarily be exact in all the others. Any imperfections in the repeats will break the branch point ambiguities in all of the samples because they can be aligned by homology. In principle, one could use different individuals of the same species and take advantage of natural sequence polymorphism. However, simulations show that the most effective application of this approach would use samples that have about 10% divergence on average. In practice, this may mean that it would be more useful to compare three to five similar species, like human and chimp, rather than compare individuals within a species. Here, as in the previous method, the cost of resolving branch point ambiguities is a considerable increase in the number of samples that have to be examined. However, the additional information that will be obtained will be highly interesting if the species are well chosen.

## OBSTACLES TO SUCCESSFUL SBH

The base composition dependence of the melting temperature, $T_m$ poses a very serious challenge to simple and effective implementations of SBH. If a temperature is chosen that allows effective discrimination between perfect matches and mismatches in G+C-rich compounds, many A+T-rich sequences may not form enough duplex to be detected. Alternatively, if one chooses a low enough temperature to stabilize the weakest A+T-rich duplexes, there will not be enough discrimination against mispairing in G+C-rich compounds, and many false positives will result. There are many possible ways to circumvent this problem; quite a few of them are being tested, but no generally acceptable solution has yet been demonstrated in practice.

Ed Southern has been experimenting with the use of high concentrations of tetramethylammonium salts (TMA) instead of more usual low to moderate ionic strength NaCl solutions. These salts have the undesirable feature of slowing down the kinetics of hybridization, but this can be compensated for, if necessary, by adding other agents that speed up hybridizations, such as dextrans which increase the effective concentration of nucleic acids. It has been known for a long time that TMA at the proper concentration can almost equalize the $T_m$ of polynucleotides that are pure A+T and those that are pure G+C. However, when Southern tried TMA in oligonucleotide hybridization, he found that while the $T_m$'s of compounds with extreme base compositions were equalized, a very large effect of DNA sequence on $T_m$ of compounds with intermediate base compositions emerged. Unless this turns out to be an idiosyncracy caused by the use of pure homopurine sequences, it probably means that TMA will have to be abandoned.

An alternative way to even out base composition effects is to use base analogs (Fig. 12.27). One can substitute 2,6-diamino purine for A (an analog that makes three hydrogen bonds with T) and 5-bromoU for T (an analog that has increased vertical stacking energy). This will raise the relative stability of A+T-rich sequences considerably. The base analog 7-deaza G can be used instead of G to lower the stability of G+C-rich sequences. Many more analogs exist that could be tested. The problem is that one really wants to test their effect across the full spectrum of 65,536 8-mers, and there is simply no way to do this efficiently until we have developed much more effective ways to make oligonucleotide chips. Such devices not only provide a way to do SBH, they provide a source of samples that allow the accumulation of massive amounts of duplex $T_m$ data. In model experiments Southern was able to characterize the $T_m$'s of all of the 256 possible homopurine-homopyrimidine 8-mer duplexes under a wide set of experimental conditions. This single set of experiments undoubtedly provided more $T_m$ data than a decade of previous work by several different laboratories.

An alternative approach for compensating for $T_m$ differences has been demonstrated by Mirzabekov. This takes advantage of the fact that chips made of thin gels can rebind significant amounts of released sample at low temperatures. The rate of this rebinding will depend on the concentration of oligonucleotide, since renaturation shows second-order kinetics or pseudo–first-order kinetics (Chapter 3). To reveal these kinetic effects, one first hybridizes a sample to the immobilized probe and then allows a fraction of the duplexes to dissociate with a washing step. By adjusting the relative concentrations of different compounds, one can bring their $T_m$'s very close to the same value. An example is shown in Figure 12.28. These results are very impressive. However, the two samples involved had to be used at a 300-fold concentration difference to achieve them. It is not immediately obvious that this can be done, in general, without leading to serious complications in the detection system used to monitor the hybridization. One will need a system with a very wide dynamic range. It will also be a major effort to try to equalize the melting properties of not just two compounds but 65,536.

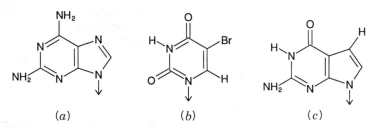

(a)          (b)          (c)

**Figure 12.27**   Base analogs useful in decreasing the differences in stability between A–T-rich and G–C-rich sequences: (a) 2-Aminoadenine. (b) 5-Bromouracil. (c) 7-Deazaguanine.

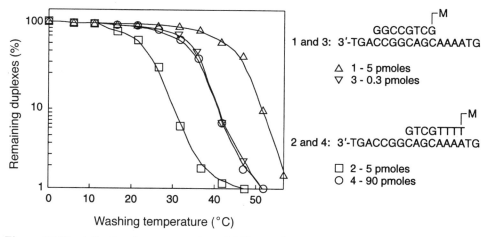

**Figure 12.28**  Adjusting the concentration of different oligonucleotides can compensate for difference in their melting temperatures. Adapted from Mirzabekov et al.

Instead of attempting to compensate for effects of sequence on the stability of duplexes, one can just measure the hybridization across a range of temperatures. This does not increase the number of samples needed. Instead, one would effectively be recording a melting profile for each sample with the entire set of oligonucleotides. This would increase the experimental time by a factor of ten or more, which is tolerable. In the long run, once extensive data on the thermal stability of each of the 8-mer complexes are known, it may be possible to use a much simpler approach. The set of compounds could be split into groups, each studied at a different optimal temperature. In principle, this could still involve a single chip, except that different regions would be kept at different temperatures. The manufacture of such a split chip would require custom placement of each compound, so simple masking strategies like that employed by Southern are unlikely to suffice. However, this is really not a serious additional manufacturing problem. Ultimately a combination of split chips, base analogs, and special solvents may all be needed for the most effective SBH throughput.

Secondary structure in the target is another potential complication in SBH. This is probably easily circumvented in the sample chip strategy. Here the target could be attached at random but frequent places to the surface under denaturing conditions. This would not be expected to interfere with oligonucleotide hybridization very much. It should effectively remove all but the most stable short sample hairpins (Fig. 12.29). The problem of secondary structure is likely to be more serious when oligonucleotide chips are used. The effect of such structures will be to cause a gap in the readable sequence. This is a serious problem, but since the gaps will be small, they can be filled rather easily by PCR-based cycle sequencing, using the sequence flanking the gaps to design appropriate primers. Thus the real issue is how frequent will such gaps be. If one occurs on each target sample, it will be best to forget SBH and just do the entire project by standard cycle sequencing. Presumably conditions will be found where the problem of secondary structure can be reduced to a much lower level. One way to do this would be to place base analogs in the sample that destabilize intramolecular base pairing more than intermolecular base pairing. These might, for example, be bulky groups where one could be tolerated in the groove of a duplex when the target binds to the probe, but two cannot be tolerated,

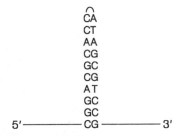

**Figure 12.29**    An example of a hairpin that is too stable to be detected in SBH.

if the target tries to pair with itself. There is undoubtedly room for much development here and much clever chemistry. A second approach would be to use probes with uncharged backbones. Then low ionic strength conditions can be used to suppress target secondary structure without affecting target-probe interactions. One example of such compounds is polypeptide nucleic acids (PNAs; see Chapter 14). Another example is phosphotriesters in which the oxygen that is normally charged in natural nucleic acids is esterified with an alkyl group. However, this creates an addition optically active center at each phosphorous, which leads to severe stereochemical complexities unless optically pure phosphotriesters are available.

The effects of secondary structure or unusual DNA structures are significant but not yet known in any great depth. In Chapter 2 we discussed the peculiar features of a centromere-associated repeat where the single strands may have a more stable secondary structure than the duplex. In Chapter 10 we illustrated the abnormally stable hairpin formed by a particular short DNA sequence. Whether these cases are representive of 1% of all the DNA sequences, or more or less, is simply unknown at the present time. About the only way we will be able to uncover such idiosyncratic behavior, understand it, and learn to deal with it, is to make large oligonucleotide arrays and start to study them. Unfortunately, this appears to be one of those cases in science where a timid approach is likely to be misleading. At some point we will have to dive in.

## SBH IN COMPARATIVE DNA SEQUENCING

Some of the difficulties just described with full de novo SBH approaches have led some experts to doubt that SBH will ever mature into a widespread user-friendly method. For this reason much effort has been concentrated on developing SBH for comparative (or differential) DNA sequencing where one assumes that a reference sequence is known and the objective is to compare it with another sample and look for any potential differences. Comparative sequencing is needed in checking existing sequence data for errors. It is the type of sequencing required for horizontal studies in which many members of a population are examined. This is needed in genetic map construction, genetic diagnostics, the search for disease genes, in mutation detection, and for more biological objectives including ecology, evolution, and profiling gene expression. Some of these applications are discussed in Chapters 13 and 14.

When SBH is considered in the context of sequence comparisons, two problems of the method for de novo sequencing are immediately resolved. It is not necessary to have a probe array consisting of all possible $4^n$ oligonucleotodes of length $n$. Instead the array can be customized to look for the desired target and simple sequence variations of that target. Second, since a reference sequence is known, issues of branch point ambiguities are virtually always resolvable by use of the information in that sequence. A particular powerful version of SBH for comparative sequence has been developed by Affymetrix, Inc. Here a probe array is made that corresponds to all possible strings of length $n$ contained in the original sequence (for a target with $L$ base pairs, $L - n + 1$ substrings are required). For each substring four variants are made corresponding to the expected sequence at the middle position of the substring and all three possible single-base variants there. Thus the array of probes will have $4(L - n + 1)$ elements. This is quite manageable with current photolithographic syntheses for targets in the range of 10 kb.

In actual practice this approach was tested on 16.6-kb human mitochondrial DNA using arrays containing up to 130,000 elements, each of which is a 15- to 25-base probe. (Chee et al., 1996). For convenience these nested targets are arranged, serially, horizontally in the array as shown schematically in Figure 12.30a, with the four possible variants for each central eighth base located vertically. The target is randomly sheared into short fragments (but longer than the length of the probes). A perfectly matched target will hybridize strongly to one member of each vertical set of four probes. A target with a single mismatch will show strong hybridization only to one particular probe in which the central base variant matches the sequence perfectly. For all possible flanking probes, there will be one or two internal mismatches between that target and the probe; hence hybridization will be weak or undetectable. A sample of the actual data seen using this approach is shown in Figure 12.30b. It is impressive. In practice, in most cases a two-color competitive hybridization is used. This allows a sample of the normal sequence (in one color) to be compared with a potential variant (in another color) with most differences in sequence-dependent hybridization efficiency nulled out.

## OLIGONUCLEOTIDE STACKING HYBRIDIZATION

There are a number of ways that could potentially increase the length of sequence that can be read with a fixed length oligonucleotide. This is one major way to improve the efficiency of SBH, since the longer the effective word length, the higher the sequencing throughput and also the smaller the number of branch point ambiguities. One approach, specifically designed by Mirzabekov to help resolve branch point ambiguities, is shown in Figure 12.31. It is based on the fact that once a duplex has been formed by hybridization of the target with an 8-mer, it becomes thermodynamically quite favorable to bind a second oligomer immediately adjacent to the 8-mer. The extra thermodynamic stabilization comes from the stacking between the two adjacent duplexes. This same principle was discussed earlier in schemes for directed primer walking (Chapter 11). In practice, Mirzabekov uses pools of ninety 5-mers, chosen specifically to try to resolve known branch points. A test of this approach, with a single perfectly matched 5-mer or various mismatches, is shown in Figure 12.31. It is apparent that the discrimination power of oligonucleotide stacking hybridization is considerable.

Some calculated $T_m$'s for perfect and mismatched duplexes are given in Table 12.1. These are based on average base compositions. The calculations were performed using the equations given in Chapter 3. In the case of oligonucleotide stacking, it is assumed that the first duplex is fully formed under the conditions where the second oligomer is being tested; in practice, this may not always be the case. It is, however, approximately true for the conditions used for the experiments shown in Figure 12.32. The calculations reveal a number of interesting features about stacking hybridization. Note that the binding

```
5'  ..TGAACTGTATCCGACAT..
3'       tgacatAggctgtag
         tgacatCggctgtag
         tgacatGggctgtag
         tgacatTggctgtag
3'        gacataAgctgtaga
          gacataCgctgtaga
          gacataGgctgtaga
          gacataTgctgtaga


5'  ..TGAACTGTACCCGACAT..
3'       tgacatAggctgtag
         tgacatCggctgtag
         tgacatGggctgtag
         tgacatTggctgtag
3'        gacataAgctgtaga
          gacataCgctgtaga
          gacataGgctgtaga
          gacataTgctgtaga
```

(a)

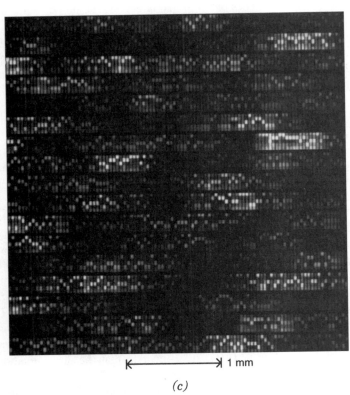

**5' TGAACTGTATCCGACAT**

A
C
G
T

**5' TGAACTGTACCCGACAT**

A
C
G
T

**16,493**

(b)

(c)

**Figure 12.30**  Use of SBH for comparative hybridization. (*a*) Schematic layout of 15 base probes (*b*) Example of actual data probing for differences in human mitochondrial DNA. Top panel shows hybridization with the same sequence as used to design the array. Bottom panel shows hybridization with a sequence with a single T to C transition in position 16,493. (*c*) Example of hybridization to a full array. Panels (*b*) and (*c*) from Chee et al. (1996).

**Figure 12.31**    Basic strategy in oligonucleotide stacking hybridization

**TABLE 12.1    Calculated Thermodynamic Stabilities of Some Ordinary Oligonucleotide Complexes and Other Complexes Involved in Stacking Hybridization**

| Structure[a] | Energetics of Stacking Hybridization | | | |
|---|---|---|---|---|
| | $n =$  8 | 7 | 6 | 5 |
| ▬▬▬ IIIIIIIIIIIII ▬▬▬ | 38 | 33 | 25 | 15 |
| ▬▬▬ ⸌IIIIIIIIIIII ▬▬▬ | 33 | 25 | 15 | 3 |
| ▬▬▬ IIIII⸌⸍IIIII ▬▬▬ | 25 | 15 | 3 | −14 |
| ▬▬ IIIIIIIIIII\|IIIIIIIIIII ▬▬ | 51 | 46 | 40 | 31 |
| ▬▬ IIIIIIIIII\| IIIIIIIIII⸌ ▬▬ | 46 | 40 | 31 | 21 |
| ▬▬ IIIIIIIIII\|⸌IIIIIIIIII ▬▬ | 40 | 31 | 21 | 11 |

Note: Calculated $T_m$ (°C, average base composition).

[a] Structures consist of a long target and a probe of length $n$. The top three samples are ordinary hybridization; the bottom three are stacking hybridization.

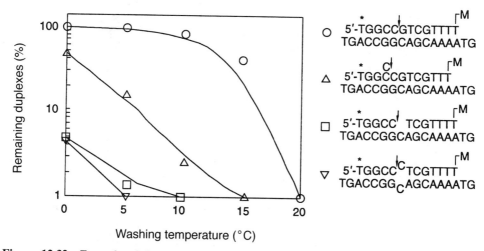

**Figure 12.32**    Example of the ability of oligonucleotide stacking hybridization to discriminate against mismatches. Taken from Mirzabekov et al.

of a second oligomer next to a preformed duplex provides an extra stability equal to about two base pairs. More interesting still is the fact that mispairing seems to have a larger consequence on stacking hybridization than it does on ordinary hybridization. This is consistent with the very large effects seen in Figure 12.32 for certain types of mispairing. Other types of mispairing are less destabilizing, but there may be a way to eliminate these, as we will discuss, momentarily. In standard hybridization sequencing, a terminal mismatch is the least destabilizing event, and thus it leads to the greatest source of ambiguity or background. For an octanucleotide complex, an average terminal mismatch leads to a 6 °C lowering in $T_m$. For stacking hybridization, a terminal mismatch on the side away from the preexisting duplex is the least destabilizing event. For a pentamer, this leads to a drop in $T_m$ of 10 °C. These considerations predict that the discrimination power of stacking hybridization in favor of perfect duplexes might be greater than ordinary SBH. They encourage attempts to modify the notion of stacking hybridization so that it becomes a general, stand-alone method for DNA sequencing.

## OTHER APPROACHES FOR ENHANCING SBH

Once an oligonucleotide has formed a duplex with the target, it ought to be possible to use enzymatic steps to proofread the accuracy of the hybridization and to read further DNA sequence information from the target. For example, the 3'-end of the oligonucleotide could serve as a primer for DNA polymerase to extend. What is needed is a sufficiently stable primer-template complex to allow the polymerase to function at a suitable temperature. An issue that needs to be explored is whether 8-mers are sufficient for this purpose. There are also potential background problems that will need to be addressed. This general approach has been used quite successfully with longer primers and DNA polymerase extension to detect specific alleles adjacent to the primer in a method called genetic bit analysis (Nikiforov et al., 1994). An alternative method for proofreading and extending a sequence read could use DNA ligase to attach a stacked oligonucleotide next to an initial duplex. This would have the potential advantage that ligase requires proper base pairing and might increase the discrimination of the stacking hybridization. In both cases, and in other schemes that can be contemplated, the label is introduced as a result of the enzymatic reaction. This eliminates much of the current background in SBH that arises from imperfect hybridization products. Some specific examples of how these procedures can be implemented in practice will be described in the next section.

A second, general way to enhance the power of SBH is to use gapped oligonucleotides. Two examples of this are shown below:

$$AGCN_4GAC \quad AGCI_4GAC$$

The first case uses a mixture of 256 possible 10-mers that share the same six external bases. The second uses a single 10-mer, but its four central bases are inosine (I) which can base pair with A, C, or T. In both of these cases the stability of the duplex is increased because it has more base pairs: one can read six bases of sequence but with the stability of a decanucleotide duplex. However, of even greater significance is the fact that the effective reach of the oligonucleotide is increased. Branch point ambiguities are less serious with these gapped molecules than with ungapped oligonucleotides with the same number of well-defined bases. William Baines has simulated SBH experiments with gapped

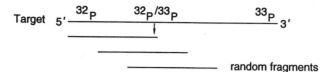

**Figure 12.33**  Resolving branch point ambiguities by using positional information derived from a gradient of two labels.

probes of various types, and these simulations indicate that this approach really improves the efficiency of SBH.

A third general way to enhance the power of SBH is to use oligonucleotide probes with degenerate ends like

$$N_2AGCTTAAGN_2$$

The advantage of this approach is that any mismatches at the ends of the internal 8-base probe sequence are converted to internal mismatches in the actual 12-base probe used.

Another way to enhance the power of SBH is to use the same kinds of pooling and multicolor detection schemes that we discussed in Chapters 9 and 11 for fast physical mapping and enhanced ladder DNA sequencing. There is every reason to use groups of oligomers simultaneously in hybridization to sample arrays, or groups of samples simultaneously, to oligonucleotide chips. Simulations are needed to help design the most effective strategies to do this. However, very simple arguments show that a considerable increase in throughput ought to be achievable. Earlier we calculated that less than 0.4% of the probes or targets score positive in a single hybridization. Performing 16 hybridizations in binary pools will therefore not entail much risk of ambiguities. Doing this in a single color would result in up to a fourfold increase in throughput. Multiple colors could be used to increase the throughput much more.

Alternatively, multiple colors might be used to help resolve branch point ambiguities. Suppose that one had a way of labeling a target with two colors, such that the ratio of these colors depended on the location of the target within a much larger clone. One way to think about doing this is placing the label in the target by a single cycle of primer extension, varying the relative concentrations of two different labeled dpppN's during the extension. When fragments of this target are hybridized to a oligonucleotide chip, the ratio of the labels will tell, roughly, where in the sequence the particular oligonucleotide is located (Fig. 12.33).

## POSITIONAL SEQUENCING BY HYBRIDIZATION (PSBH)

Here we describe a scheme that was developed and refined in our own laboratories as an alternate form of SBH. It is called positional sequencing by hybridization (PSBH). It has a number of potential advantages over conventional SBH but also presents its own set of different obstacles that must be overcome to make the total scheme a practical reality. PSBH relies totally on stacking hybridization. It uses an array of probes constructed as follows, where $X_n$ refers to a single specific DNA sequence of length $n$, and $Y_n$ is the complement of that sequence:

$$\begin{array}{llll} 5' & X_nN_m & 3' & \text{or } 5' & X_n & 3' \\ 3' & Y_n & 5' & 3' & Y_nN_m & 5' \end{array}$$

These probes all share a common duplex stem next to a single-stranded overhang. The details of the duplex sequence are unimportant here. Each element of the probe array will have a different specific overhang. Thus there are $4^m$ possible probes of each type. These probes, which actually resemble PCR splints, are designed to read a segment of target sequence by stacking hybridization. As shown in Figure 12.34, the 5'-overhang probe allows the 5'-end of a target DNA sequence to be read; the 3'-overhang probe will read the 3'-end of a target.

The basic scheme shown in Figure 12.34, can be improved and elaborated by adding to it most of the enhancements described in the previous section. It seems particularly well suited for incorporating many of these enhancements because the duplex stem of the probe can be made long enough to be totally stable under any of the conditions needed for enzymology. For example, it is possible to use DNA ligase to attach the target to the probe covalently, after hybridization (Fig. 12.35). This has several advantages. Any mispaired probe-target complexes are unlikely to be ligated. Any probes that have hybridized to some internal position in the target (like two of the cases shown in Fig. 12.32) will certainly be unable to ligate. All of the nonligated products can be washed away under conditions where the ligated duplex is completely stable. Thus excellent discrimination between perfectly matched targets and single-base mismatches can be achieved (Table 12.2).

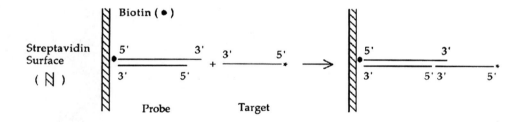

**Basic scheme for positional SBH**

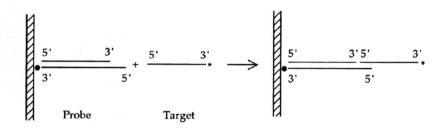

**Alternate scheme**

**Figure 12.34**  Basic scheme for positional SBH to read the sequence at the end of a DNA target.

## Ligation of target DNA with probe

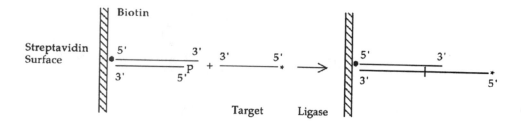

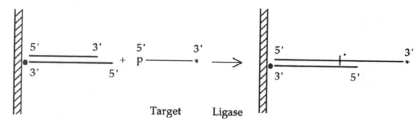

**Figure 12.35**  Use of DNA ligase to enhance the specificity of positional SBH. Note that since the target is ligated to the constant portion of the DNA probe, the ligation product can be melted off and replaced with a fresh constant portion. Thus a sample chip designed with this type of probes is reusable.

**TABLE 12.2   Single-Stranded Target (3′-TCGAGAACCTTGGCT-5′) Annealed and Ligated to Duplexes With 5-base Overhangs with Different Mismatches**

| Probe[a] | Ligation Efficiency (%) | Discrimination Factor |
|---|---|---|
| 3′-CTACTAGGCTGCGTAGTC-5′ | | |
| 5′-b-GATGATCCGACGATCAGCTC-3′ | 17 | |
| 5′-b-GATGATCCGACGCATCAGC**TT**-3′ | 1 | 17 |
| 5′-b-GATGATCCGACGCATCAGC**TA**-3′ | 0.5 | 34 |
| 5′-b-GATGATCCGACGCATCAGC**CC**-3′ | 0.2 | 85 |
| 5′-b-GATGATCCGACGCATCAG**T**TC-3′ | 0.4 | 42 |
| 5′-b-GATGATCCGACGCATCA**A**CTC-3′ | 0.1 | 170 |

*Source:*  Adapted from Broude et al. (1994).

[a]Each probe contained a constant 18-base duplex region formed by annealing the sequences shown with 3′-CTACTAGGCTGCGTAGTC-5′. Mismatches are shown in boldface.

Once the target has been ligated to the probe, it can serve as a substrate for the acquisition of additional DNA sequence data. For example, as shown in Figure 12.36, the 3'-end of the probe can be used as a primer to read the next base of the target by extension with a single, labeled terminator. Alternatively, any of the single nucleotide addition methods described at the beginning of this chapter can now be used on each immobilized target molecule as in Genetic Bit Analysis (Nikoforov et al., 1994). It would also be possible to do plus/minus sequencing on each immobilized target if one had sufficient quantitation with four colors to tell the amounts of each base incorporated. The basic idea is that the probe array can serve to localize a large number of different target molecules, simultaneously, and determine a bit of their sequence. Most probes will capture only a single target, and each of these complexes can then be sequenced in parallel. This should combine some of the best features of ladder and hybridization sequencing. It should produce sequence reads on each target molecule that are long enough to resolve all the common branch point ambiguities, except for those caused by true interspersed repeating sequences.

A major limitation in the PSBH approach we have described thus far is that it only reads the sequence at one end of the target. This would seem to limit its application to relatively short targets. However, one can circumvent this problem, in principle, by making a nested set of targets, as shown in Figure 12.37. One has to be careful in choosing the strategy for constructing these samples, since the ends of the DNAs must still be able to be ligated. Thus dideoxy terminators could be used, but they would have to be replaced by ordinary nucleotides with a single step of plus/minus sequencing, as we described for single-base addition early in the chapter. Alternatively, chemical cleavage could be used, as described when genomic DNA sequencing was used to locate $^m$C's (Chapter 11). The third approach is to use exonuclease digestion to make the nested set. With these nested samples it should be possible to use PSBH to read the entire sequence of a target, limited only by the ability to resolve branch point ambiguities.

A major potential advantage of PSBH over SBH is that stacking hybridization would allow the use of 5-mer or 6-mer overlaps instead of the 8-mer or 9-mer probes required in

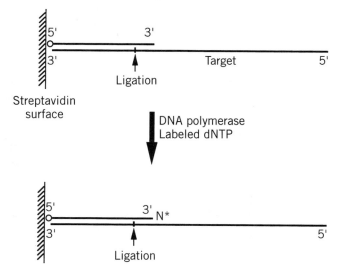

**Figure 12.36**  Extension of the sequence read by a chip by using DNA polymerase. Note that more sequence would be read but the chip would not be reusable.

## Preparation of a nested set of DNAs

**Figure 12.37**   One way to prepare a nested set of DNA samples so that the entire sequence of a target could be read by positional SBH.

ordinary hybridization. This would decrease the size of the sample array needed by a factor of 64. Thus, for 5-mers, an array of only 1024 elements would be needed for unidirectional reading; twice this number is needed for bidirectional reading. However, this advantage will be offset by the increased frequency of branch point ambiguities unless there is some way to resolve them. A potential solution is afforded by the positional labeling scheme discussed in the previous section. A particularly simple way to mark the location of a branch point ambiguity is to combine a fixed end label and an internal label, as shown in Figure 12.38. The amount of end label would be the same on every target. The amount of internal label would vary depending on the length of the target, and thus on the position of the variable end of the target. The ratio between the internal label and the end label would provide the approximate length of the target. This strategy has not yet been tested in practice, but it seems fairly attractive because the reagents needed for two-color end and internal labeling are readily available (Chapter 10).

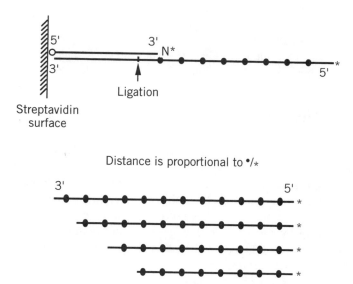

**Figure 12.38**   Determination of the approximate position of a target sequence by combining an end label with an internal label to provide an estimate of the length of the target.

TABLE 12.3    Single-Stranded Targets Ligated to Duplex Probes With the Indicated 5-Base Overhangs With Different A+T Contents

| Probe Overhang[a] (5′ → 3′) | A + T Content | Ligation Efficiency (%) | Discrimination Factor |
|---|---|---|---|
| Match GGCCC | 0 | 30 | |
| Mismatch GGCCT | | 3 | 10 |
| Match AGCCC | 1 | 36 | |
| Mismatch AGCTC | | 2 | 18 |
| Match AGCTC | 2 | 17 | |
| Mismatch AGCTT | | 1 | 17 |
| Match AGATC | 3 | 24 | |
| Mismatch AGATT | | 1 | 24 |
| Match ATATC | 4 | 17 | |
| Mismatch ATATT | | 1 | 17 |
| Match ATATT | 5 | 31 | |
| Mismatch ATATC | | 2 | 16 |

*Source:* Adapted from Broude et al. (1994).

[a]Only the variable overhang portion of the probe sequence is shown. Mismatches are shown in boldface.

The major challenge for PSBH, as for SBH, is to build real arrays of probes and use them to test the fraction of sequences that actually perform according to expectations. Base composition and base sequence dependence on the effectiveness of hybridization is probably the greatest obstacle to successful implementation of these methods. The use of enzymatic steps, where feasible, may simplify these problems, since the enzymes do, after all, manage to work with a wide variety of DNA sequences in vivo. In fact initial results with the ligation scheme shown in Figure 12.35 indicate that the relative amount and specificity of the ligation are remarkably insensitive to base composition (Table 12.3). If further PSBH experiments reveal more significant base composition effects, one potential trick to compensate for this would be to allow the adjacent duplex to vary. Thus for an A+T rich overhang, one could use a G+C rich stacking duplex, and vice versa. This will surely not solve all potential problems, but it may be a good place to begin.

## COMBINATION OF SBH WITH OTHER SEQUENCING METHODS

The PSBH scheme described in the previous section was initially conceived as a de novo sequencing method. However, it may serve better as a sample preparation method for other forms of rapid DNA sequencing. In essence, PSBH is a sequence-specific DNA capture method. A set of all 1024 PSBH probes can serve as a generic capture device to sort out a set of DNA samples on the basis of their 3′-terminal sequences. Such samples can be prepared either by a set of PCR reactions (with individually selected 5-base tags if necessary) or by digestion of a target with restriction enzymes like *Mwo* I that cut outside their recognition sequence as shown below:

GCNNNNN/NNGC

CGNN/NNNNNCG

The resulting captured set of samples is spatially resolved and can now be subjected to Sanger extension reactions to generate a set of sequence ladders. The appeal of this method is that a set of samples can be processed all at once in a single tube without any need for prior fractionation. Capture has been shown to be efficient with mixtures of up to 25 samples (Broude et al., 1994), and high-quality sequencing ladders have been prepared from mixtures of up to eight samples (Fu et al., 1995). The real promise of this approach probably lies in the preparation of samples for MALDI MS DNA sequencing (Chapter 10) where very large numbers of relatively short samples will need to be processed.

## SOURCES AND ADDITIONAL READINGS

Broude, N. E., Sano, T., Smith, C. L., and Cantor, C. R. 1994. Enhanced DNA sequencing by hybridization. *Proceedings of the National Academy of Sciences USA* 91: 3072–3076.

Chee, M., Yang, R., Hubbell, E., Berno, A., Huang, X. C., Stern, D., Winkler, J., Lockhart, D. J., Morris, M. S., and Fodor, S. P. A. 1996. Accessing genetic information with high-density DNA arrays. *Science* 274: 610–614.

Chetverin, A., and Kramer, F. R. 1994. Oligonucleotide array: New concepts and possibilities. *Bio/Technology* 12: 1093–1099.

Dubiley, S. et al. 1997. Fractionation, phosphorylation and ligation on oligonucleotide microchips to enhance sequencing by hybridization. *Nucleic Acids Research* 25: 2259–2265.

Fu, D-J., Broude, N. E., Koster, H., Smith, C. L., and Cantor, C. R. 1995. Efficient preparation of short DNA sequence ladders potentially suitable for MALDI-TOF DNA sequencing. *Genetic Analysis* 12: 137–142.

Fu, D.-J., Broude, N. E., Koster, H., Smith, C. L., and Cantor, C. R. 1995. A DNA sequencing strategy that requires only five bases of known terminal sequence for priming. *Proceedings of the National Academy of Sciences USA* 92: 10162–10166.

Guo, Z., Liu, Q., and Smith, L. M. 1997. Enhanced discrimination of single nucleotide polymorphisms by artificial mismatch hybridization. *Nature Biotechnology* 15: 331–335.

Jurinke, C., van den Boom, D., Jacob, A., Tang, K., Worl, R., and Koster, H. 1996. Analysis of ligase chain reaction products via matrix-assisted laser desorption/ionization time-of-flight mass spectrometry. *Analytical Biochemistry* 237: 174–181.

Khrapko, K. R., Lysov. Y. P., Khorlin, A. A., Ivanov, I. B., Yershov, G. M., Vasilenko, S. K., Florentiev, V. L., and Mirzabekov, A. D. 1991. A method for DNA sequencing by hybridization with oligonucleotide matrix. *Journal of DNA Sequencing and Mapping* 1: 375–368.

Lane, M. J., Paner, T., Kashin, I., Faldasz, B. D., Li, B., Gallo, F. J., and Benight, A. S. 1997. The thermodynamic advantage of DNA oligonucleotide "stacking hybridization" reactions: Energetics of a DNA nick. *Nucleic Acids Research* 25: 611–616.

Li, Y., Tang, K., Little, D.P., Koster, H., Hunter, R. L., and McIver, R. T. Jr. 1996. High-resolution MALDI fourier transform mass spectrometry of oligonucleotides. *Analytical Chemistry* 68: 2090–2096.

Livshits, M. A., Florentiev, M. L., and Mirzabekov, A. D. 1994. Dissociation of duplexes formed by hybridization of DNA with gel-immobilized oligonucleotides. *Journal of Biomolecular Structure Dynamics* 11: 783–795.

Lysov, Y. P. et al. 1994. DNA sequencing by hybridization to oligonucleotide matrix. Calculation of continuous stacking hybridization efficiency. *Journal of Biomolecular Structure Dynamics* 11: 797–812.

Maskos, U., and Southern, E. M. 1992. Parallel analysis of oligodeoxyribonucleotide (oligonucleotide) interactions. I. Analysis of factors influencing oligonucleotide duplex formation. *Nucleic Acids Research* 20: 1675–1678.

Maskos, U., and Southern, E. M. 1992. Oligonucleotide hybridisations on glass supports: A novel linker for oligonucleotide synthesis and hybridisation properties of oligonucleotides synthesised in situ. *Nucleic Acids Research* 20: 1679–1684.

Milosavljevic, A. et al. 1996. DNA sequence recognition by hybridization to short oligomers: Experimental verification of the method on the *E. coli* genome. *Genomics* 37: 77–86.

Murray, M. N., Hansma, H. G., Bezanilla, M., Sano, T., Ogletree, D. F., Kolbe, W., Smith, C. L., Cantor, C. R., Spengler, S., Hansma, P. K., and Salmeron, M. 1993. Atomic force microscopy of biochemically tagged DNA. *Proceedings of the National Academy of Sciences USA* 90: 3811–3814.

Nikoforov, T. T., Rendle, R. B., Goelet, P., Rogers, Y.-H., Kotewicz, M. L., Anderson, S., Trainor, G. L., and Knapp, M. R. 1994. Genetic bit analysis: A solid phase method for typing single nucleotide polymorphisms. *Nucleic Acids Research* 22: 4167–4175.

Pease, A. C., Solas, D., Sullivan, E., Cronin, M. T., Holmes, C. P., and Fodor, S. P. A. 1994. Light-generated oligonucleotide arrays for rapid DNA sequence analysis. *Proceedings of the National Academy of Sciences USA* 91: 5022–5026.

Ronaghi, M., Uhlen, M., and Nyren, P. (1998). Real-time pyrophosphate detection for DNA sequencing. *Science* 281: 363–365.

Shalon, D., Smith, S. J., and Brown, P. O. 1996. A DNA microarray system for analyzing complex DNA samples using two-color fluorescent probe hybridization. *Genome Methods* 6: 639–645.

Southern, E. M., Maskos, U., and Elder, J. K. 1992. Analyzing and comparing nucleic acid sequences by hybridization to arrays of oligonucleotides: Evaluation using experimental models. *Genomics* 13: 1008–1017.

Southern, E. M., Green-Case, S. C., Elder, J. K., Johnson, M., Mir, K. U., Wang, L., and Williams, J. C. 1994. Arrays of complementary oligonucleotides for analysing the hybridisation behavior of nucleic acids. *Nucleic Acids Research* 22: 1368–1373.

Strezoska, Z., Paunesku, T., Radosavljevic, D., Labat, I., Drmanac, R., and Crkvenjakov, R. 1991. DNA sequencing by hybridization: 100 bases read by a non-gel method. *Proceedings of the National Academy of Sciences USA* 88: 10089–10093.

# 13 Finding Genes and Mutations

## DETECTION OF ALTERED DNA SEQUENCES

Genomic DNA maps and sequences are a means to an end. The end is to use this information to understand biological phenomena. At the heart of most applications of mapping and sequencing is the search for altered DNA sequences. These may be sequences involved in an interesting phenotypic trait, an inherited disease, or a noninherited genetic disease due to a DNA change in somatic (nongermline) cells. The way in which maps and sequences can be used to identify altered DNA sequences very much depends on the context of that alteration. Here we will briefly survey the range of applications of maps and sequences, and then we will cover a few examples in considerable depth. However, the emphasis of much of this chapter will be the development of more efficient methods to find any sequence differences between two DNA samples.

Some DNA differences are inherited. There are three levels at which we characterize inherited DNA differences. DNA maps and sequences greatly assist the finding of genes responsible for inherited diseases or other inherited traits. Once a disease gene has been identified, we attempt to develop DNA-based tests for the clinical diagnosis of disease risk. The success of these tests will depend on the complexity of the disease and normal alleles. Even before a disease gene has been identified, DNA-based analyses of linked markers can sometimes offer considerably enhanced presymptomatic or prenatal diagnosis, or carrier screening. Finally DNA tests, in principle, provide a way for us to look for new germline mutations, either at the level of sperm (and ova in principle, but not very easily in practice) or anytime after the creation of an embryo. These mutations are referred to, respectively, as gametic mutations and genetic mutations. The distinction is a subtle one. Any mutations that destroy the ability of a gamete to function will not be inheritable because this gamete will produce no progeny.

Some DNA differences are important at the level of organism function, but they do not affect the germ cells, so they are not passed to the offspring. Examples in normal development occur frequently in the immune system. Both the immunoglobulin genes and the T-cell receptor genes rearrange in lymphocytes, and they also have a high degree of point mutagenesis in certain critical regions. These processes are used to generate the enormous repertoire of immune diversity needed to allow the immune system to detect and combat a wide variety of foreign substances. It has been speculated that DNA rearrangements might also occur in other normal somatic tissues, like the brain, but thus far, evidence for any such functionally significant rearrangements is not convincing. DNA changes in abnormal development appear to be commonplace. Most cancer cells contain DNA rearrangements that somehow interfere with the normal control of cell division. As the resulting cells multiply and spread, they frequently accumulate many additional DNA alterations. Other somatic DNA differences occur when chromosomes segregate incorrectly during mitosis.

A final example where DNA sequence information plays an important role in clinical diagnosis is in infectious disease. For example, strain variations of viruses and bacteria can be of critical importance in predicting their pathogenicity. Examples include virulent versus nonvirulent forms of bacteria like *Mycobacterium tuberculosis,* and various drug-resistant strains of HIV, the virus that causes AIDs. Other examples are quite common in parasitic protozoa, since these organisms, like HIV, use rapid DNA sequence variation as a way of escaping the full surveillance of the immune system of the host. Thus DNA sequence analysis is important in understanding the biology of *Plasmodium falciparum,* the organism that causes malaria, *Trypanosoma brucii* and *Trypanosoma cruzii,* which cause sleeping sickness, and many other organisms that pose significant public health hazards.

In this chapter we will describe the sorts of DNA analyses that can be done to detect genomic changes with present technology, and we will try to extrapolate to see what improvements will be likely in the future.

## FINDING GENES

The approach used to find genes based on their location on the genetic map has been called reverse genetics, but a more accurate term is positional cloning. The basic strategy is to use the genetic map to approximate the position of the gene (Fig. 13.1). Then a physical map of the region is constructed if it is not already available. The physical map should provide a number of potential sequence candidates for the gene of interest. It also helps to find additional useful polymorphic markers that narrow the location of the desired gene further. Ultimately one is reduced to a search for a particular set of DNA sequence differences that correlates with a phenotype known to be directed by an allele of the gene. In contrast to positional cloning, genes can sometimes be found by functional cloning. Here an altered biochemical function is traced to an altered protein. This is sequenced, and the resulting string of amino acids is scanned to find regions that allow relatively nondegenerate potential DNA coding sequences to be synthesized and used as hybridization probes to screen genomic or cDNA libraries.

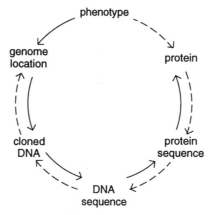

**Figure 13.1**  Contrasting stages in strategies to find genes by positional cloning (solid line) and by functional cloning (dashed line). Adapted from a slide displayed by Hilger Ropers.

In the past few years there have been many dramatic successes in human positional cloning. Among these are the genes responsible for Duchenne muscular dystrophy, cystic fibrosis, some forms of familial Alzheimers disease, myotonic dystrophy, familial colon cancer, two forms of familial breast cancer, HD, one form of neurofibromatosis, and several genes involved in fragile X-linked mental retardation. Some of the ways in which the genetic map has helped locate and clone these disease genes were discussed in Chapter 6. Here we review, briefly, some of the aspects of this process, with the particular goal of showing where DNA sequencing plays a useful or necessary role. In most gene searches, thus far there have been unexpected benefits in that interesting biological or genetic mechanisms became apparent as correlations became possible between genotype and phenotype. These serendipitous findings may have occurred because so few human disease genes were known previously. However, it is still possible that many additional basic biological surprises remain to be uncovered as much larger numbers of human disease genes are identified. In Box 13.2 we will illustrate one of the most novel disease mechanisms seen in several of these diseases which is caused by unstable repeating trinucleotide sequences.

A successful genetic linkage study within a limited set of families is just the first step in the search for a gene. It reveals that there is a specific single gene involved in a particular disease or phenotype, and it provides the approximate location of that gene in the genome. However, genetic studies in the human can rarely locate a gene to better than 1 to 2 cM. In typical regions of the human genome, this corresponds to 1 to 2 Mb; the problem is that such regions will usually contain 30 to 60 genes. To narrow the search, it is usually necessary to isolate the DNA of the region (Fig. 13.2). Until the advent of YACs and other large insert cloning systems, this was a very time-consuming and costly process. It frequently consisted of parallel attempts at chromosome microdissection and microcloning and attempts at cosmid walking or jumping from the nearest markers flanking the region of interest. Now these steps can usually be carried out much more efficiently by using the larger size YACs and mega-YACs that span most of the human genome. While these have some limitations, discussed in Chapter 9, the DNA of most regions is available just by a telephone call to the nearest YAC repository.

With the DNA of a particular region in hand, one can search for additional polymorphic markers fairly efficiently. For example, simple tandem repeating sequences can be selected by hybridization screening or sequence-specific purification methods (see Chapter 14). These new markers can be used to refine the genetic map in the region. However, a more effective use of nearby markers, as we discussed in Chapter 6, is to pinpoint the location of any recombinants in the region. This is illustrated in Figure 13.3.

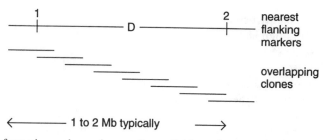

**Figure 13.2**  Information and samples usually available at the start of the end game in the search for a disease gene.

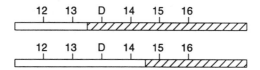

**Figure 13.3** The nearest recombination breaking points flank the true location of the gene. Hatched and hollow bars indicate chromosome segments inherited from different parental homologs. D must lie to the right of marker 13 and to the left of marker 15.

Before the gene of interest was successfully linked to markers, any recombinants were damaging, since they subtracted from the statistical power of the linkage tests. Now, however, once the locale of the gene is established beyond doubt, the recombinants are a very valuable resource, and it is often very profitable to search for additional recombinants. As shown in Figure 13.3, the gene location can be narrowed down to a position between the nearest set of available recombinants.

In an ideal case the gene of interest is large, and it occupies a considerable portion of the region. Then frequently a disease allele can be found that contains a large enough size polymorphism to be spotted by PFG analysis of DNA hybridized with available probes in the region. The polymorphism may arise from an insertion, a deletion, or a translocation. Such an association of a disease phenotype with one or more large-scale rearrangements almost always rapidly pinpoints the location of the gene because finer and finer physical mapping can rapidly be employed to position the actual disrupted gene relative to the precise sites of DNA rearrangements. An example of this approach was the search for the gene for Duchenne muscular dystrophy where roughly half of the disease alleles are large deletions in the DNA of the region.

In typical cases one is not lucky enough to spot the gene of interest by using low-resolution mapping approaches. Then it is usually safest to take a number of different approaches simultaneously. This is especially true if, as in many cases of interest, the search for the gene is a competitive one. The genetic approach useful at this point is linkage disequilibrium. This was described in detail in Chapter 6. To reiterate, briefly, if there is a founder effect, that is, if most individuals carrying a disease allele have descended from a common progenitor, they will tend to have similar alleles at other polymorphic sites in the region. This is true even though the individuals have no apparent familial relationships. The closer one is to the gene, the greater the tendency of all individuals with the disease to share a common haplotype. This gradient of genetic similarity allows one to narrow down the location of the gene, but there are many potential pitfalls, as described in Chapter 6.

A second useful approach is to search for individuals who display multiply genetic disorders including the particular disease of interest. Such individuals can frequently be found, and they will often be carriers of microscopic DNA deletions. As shown in Figure 13.4, one can use these individuals to narrow the location of the gene. Low-resolution physical maps of each individual can often reveal the size and position of the deletions. Pooling data from several individuals with different deletions will indicate the boundaries on the possible location of the gene of interest. The process is easiest in cases like X-linked disease, since here, in males, there is only one copy of the region of interest. In somatic disease, there will be two copies, and the altered chromosome will have to be distinguished and analyzed in the presence of the normal one. This general approach can be very productive because after one gene is found, the genes for the additional inherited disorders must lie nearby, and it will be much easier to find them.

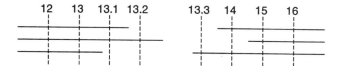

**Figure 13.4** The nearest available chromosome breaking points, frequently seen in patients with multiple inherited disorders, flank the true location of the gene. Horizontal lines show markers present in three individuals with a common genetic disorder. The disease gene must lie between markers 13.2 and 13.3.

A third parallel approach is to map and characterize the transcripts coded for by the region. This can be done by using available DNA probes in hybridization against Northern blots (electrophoretically fractionated mRNAs) or against cDNA libraries. If the disease is believed to be predominantly localized in particular tissues, this approach can be very effective, since one can compare mRNAs or cDNAs from tissues believed to be significant sites of expression of the gene of interest with other samples where this gene is not likely to be expressed. With cystic fibrosis, for example, hypotheses about gene expression in sweat glands and in the pancreas were very helpful in narrowing the location of the gene in this way. Alternatively, genes in the target region may already be known as a large number of expressed sequence tags (ESTs) from known tissues are being added to the EST database at GenBank daily and are being mapped to chromosomal regions. Note, however, that considerable pitfalls exist with this approach, since hypotheses about the sites of expression can easily be wrong, and, even if they are correct, the gene of interest may be expressed at too low a level to be seen as mRNA or represented in a typical cDNA library.

DNA of the region can be used in a number of different ways to help find the location of the genes in the region, even where no prior hypotheses about sites of likely expression exist. YACs have been used as hybridization probes to directly isolate corresponding mRNAs or cDNAs, a technique sometimes referred to as fishing (Lovett, 1994). Techniques, such as exon trapping, have been developed to allow specific subcloning of potentially coding DNA sequences from a region (see Box 13.1). Another frequently effective strategy is to look for regions of DNA that are conserved in different mammals or even more distant species. Genes are far more likely to be conserved than noncoding regions. However, this approach is not guaranteed because there is no reason to expect that every gene will be conserved or even exist among a set of species tested. Even genome scanning by direct sequencing has revealed the location of genes.

In some types of disease, other strategies become useful. For example, in dominant lethal disease, most if not all affected individuals are new mutations. These will most likely occur in regions of DNA with high intrinsic mutations rates. While we still have much to learn about how to identify such regions, at least one class of unstable DNA sequence has emerged in recent years that appears to play a major role in human disease. Tandemly repeated DNA sequences have intrinsically high mutation rates because of the possibilities for polymerase stuttering or unequal sister chromatid exchange, as described in Chapter 7. Repeats like $(GAG)_n$ occur in coding regions; $(GAA)_n$ and $(GCC)_n$ occur outside of coding regions. These can shrink or grow rapidly in successive generations and lead to disease phenotypes. Examples of this were first seen in myotonic dystrophy, fragile X-linked mental retardation, and Kennedy's disease (see Box 13.2). A systematic search is now underway to map the locations of these and other trinucleotide repeats, since they may well underlie the cause of additional human diseases. The repeats appear to be fairly widespread as shown by the examples already found (see Table 13.1).

**BOX 13.1**
**EXON TRAPPING METHODS**

Exon trapping methods are schemes for selective cloning and screening of coding DNA sequences. Several different approaches have been described (Duyk et al., 1990; Buckler et al., 1991; Hamaguchi et al., 1992). Here we will illustrate only the last of these because it seems to be relatively simple and efficient. The vector used for this exon trapping scheme is shown in Figure 13.5a. It contains intron 10 of the p53 gene, which includes a long pyrimidine tract (which appears to prevent exon skipping), and consensus sequences for the 5'- and 3'-splicing sites (AG/GTGAGT and AG, respectively), and the branch site (TACTCAC) used in an intermediate step in RNA splicing. The intron contains a *Bgl* II cleavage site used for cloning genomic DNA. Surrounding the intron are two short p53 exons, flanked by SV40 promoters known to be transcriptionally active in COS-7 cells. Reverse transcriptase is used to make a cDNA copy of any transcripts, and then PCR with two nested sets of primers is used to detect any transcripts containing the two p53 exons. When the vector alone is transfected into COS-7 cells, only a 72-bp transcript is seen. Cloned inserts containing other complete exons will produce longer transcripts after transfection. In practice, fragments from 90 to 900 bp are screened for because most exons are shorter than 500 bp. These new fragments will arise by two splicing events as shown in Figure 13.5b. For an example of recent results using exon trapping, see Chen et al. (1996).

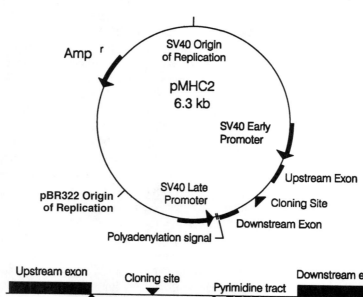

(a)

*(continued)*

**BOX 13.1** (*Continued*)

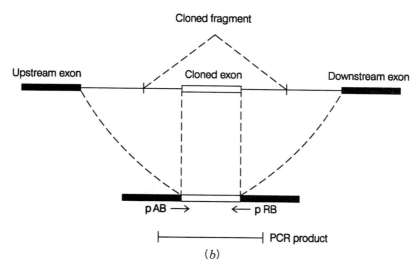

(*b*)

**Figure 13.5** Exon trapping to clone expressed DNA sequences. (*a*) Vector and procedures used. Adapted from Hamaguchi et al. (1992). pA, pAB, pRB, and pR are primers used for nested PCR. (*b*) Schematic of the PCR product expected from a cloned exon.

**TABLE 13.1    Trinucleotide Repeats in Human Genes**

| Gene or Encoded Protein | Copy Number[a] | Location |
|---|---|---|
| *znf6* (zinc finger transcription factor) | 8, 3, 3 | 5′ Untranslated region |
| CENP-B (centromere autoantigen) | 5 | 5′ Untranslated region |
| c-*cbl* (proto-oncogene) | 11 | 5′ Untranslated region |
| Small subunit of calcium-activated neutral protease | 10, 6 | Coding region (N-terminal) |
| *CAMIII* (calmodulin) | 6 | 5′ Untranslated region |
| *BCR* (breaking point cluster region) | 7 | 5′ Untranslated region |
| Ferritin H chain | 5 | 5′ Untranslated region |
| Transcription elongation factor SII | 7 | 5′ Untranslated region |
| Early growth response 2 protein | 5 | Coding region (central) |
| Androgen receptor | 17 | Coding region (central) |
| FMR-1 (fragile X disease) | 6–60 | Not certain yet |
| $(AGC)_n$ androgen receptor (Kennedy's disease) | 13–30 | Coding region (central) |
| DM-1 myotonic dystrophy | 5–27 | 3′ Untranslated region |
| IT 15 Huntington's disease | 11–34 | Coding region (N-terminal) |

*Source:* Updated, from Sutherland and Richards (1995).

[a]In normal individuals

## BOX 13.2
## DISEASES CAUSED BY ALTERED TRINUCLEOTIDE REPEATS

Fragile sites on chromosomes have been recognized, cytogenetically, for a long time. When cells are growth under metabolically impaired conditions, some chromosomes, in metaphase, show defects. These give the superficial appearance that the chromosome is broken at a specific locus, as shown in Figure 13.6. Actually it is most unlikely that a real break has occurred; instead, the chromatin has failed to condense normally. A particular fragile spot on the long arm of the human X chromosome, called fraXq27, shows a genetic association with mental retardation. About 60% of the chromosomes in individuals with this syndrome show fragile sites; the incidence in apparently normal individuals is only 1%. Fragile X-linked mental retardation is actually the second most common cause of inherited mental retardation. It occurs in 1 in 2000 males and 4 in 10,000 females. Earlier genetic studies of fragile X syndrome showed a number of very peculiar features that were inexplicable by any simple classical genetic mechanisms.

Now that the molecular genetics of the fragile X has been revealed, and similar events have been seen in many other diseases, including Kennedy's disease, myotonic dystrophy, and Huntington's disease, we can rationalize many of the unusual genetic features of these diseases. A number of fundamental issues, however, remain unresolved. The basic molecular genetic mechanism common to all four diseases and many others is illustrated in Figure 13.7 (Sutherland and Richards, 1995). In each case, near or in the gene, a repeating trinucleotide sequence occurs. Like other variable number tandem repeats (VNTRs), this sequence is polymorphic in the population. Normal individuals are observed to have relatively short repeats: 6 to 60 copies in fragile X syndrome, 13 to 30 in Kennedy's disease, 5 to 27 copies in myotonic dystrophy, and 11 to 34 in Huntington's. Individuals affected with the disease have much larger repeats: more than 200 copies in fragile X, more than 39 in Kennedy's, more than 100 in myotonic dystrophy, and more than 42 in Huntington's.

The case studied in most detail thus far is the fragile X syndrome, and this will be the focus of our attention here. Individuals who are carriers for fragile X, that is, individuals whose offspring or subsequent descendants display the fragile X phenotype, have repeats larger than the 60 copies, which represents the maximum in the normal population, but smaller than 200, the lower bound of individuals with discernable disease phenotypes. This progressive growth in the size of the repeat, from normal to car-

**Figure 13.6**  Appearance of a fragile X chromosome in the light microscope.

*(continued)*

BOX 13.1 *(Continued)*

FINDING GENES    **441**

**BOX 13.2** *(Continued)*

HD - Huntington's Disease

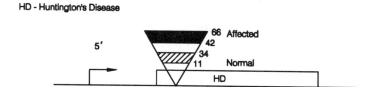

FMR-1 - Fragile X Syndrome

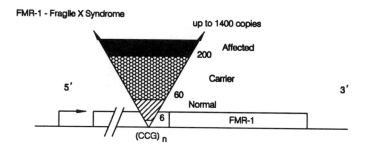

AR - Kennedy's Disease

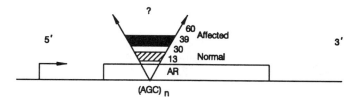

DM-1 - Myotonic Dystrophy

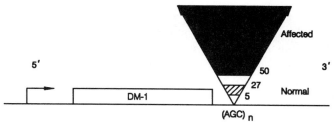

**Figure 13.7** Summary of the VNTR expansions seen in four inherited diseases. Shown are repeat sizes to normal alleles and disease-causing alleles. Adapted from Richards and Sutherland (1992).

*(continued)*

**BOX 13.2** *(Continued)*

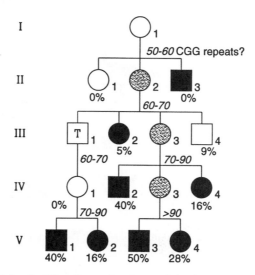

**Figure 13.8**  Typical fragile X pedigree showing anticipation and a nontransmitting carrier male (T). Black symbols denote mentally retarded individuals; gray symbols are carrier females. Arabic number %'s are the risk of mental retardation based on the general statistics for pedigrees of this type; italic numbers are the copies of CGG repeats present in particular individuals. Adapted from Fu et al. (1991).

rier to affected, explains some of the unusual genetic features of the disease that so puzzled early investigators. Figure 13.8 shows a typical fragile X pedigree. It reveals two nonclassical genetic effects. The individual labeled T in the figure is called a non-transmitting male. He is unaffected, and yet he is an obligate carrier because two of his grandchildren developed the disease. One of these is a male who must have received the disease-carrying chromosome from his mother. The second, more general feature of the pedigree in Figure 13.8 is called anticipation. As the generations proceed, a higher percentage of all the offspring develop the disease phenotype. This is because the number of copies of the repeated sequence in the carriers keeps increasing, until the repeat explodes into the full-blown disease allele. Note that in this pedigree, as is usual, the affected males do not have any offspring. This is because the disease is un-treatable and severely disabling. It is effectively a genetic lethal, and thus can only ex-ist at the high frequency observed because the rate of acquisition of new mutations must be high.

Something about the gradual increase in the size of the repeat must eventually trig-ger a molecular mechanism that leads to a much greater further expansion. Thus above some critical size the sequence is genetically unstable. Figure 13.9 shows two alternate mechanisms that have been proposed to account for this instability. In the first of these, it is postulated that somewhere else in the genome, there is a sequence that normally has no effect on the trinucleotide repeat. However, in a founder chromosome (one that will lead to the carrier state and eventually produce the disease) a mutation occurs in this sequence. This acts, either in cis or in trans, to destabilize the repeat, which then

*(continued)*

**BOX 13.2** *(Continued)*

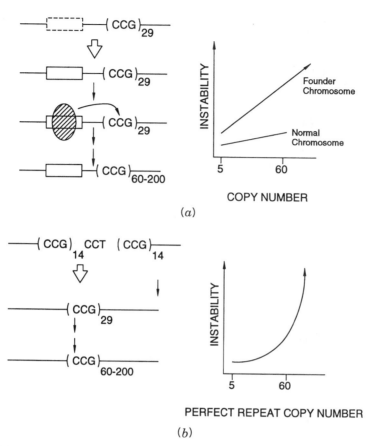

(a)

(b)

**Figure 13.9** Two possible mechanisms for the generation of a chromosome with an unstable trinucleotide repeat. (*a*) A mutation affects a site outside the repeat that then acts in cis (directly) or trans (by attracting some other component such as a protein) to destablize the repeat. (*b*) A mutation changes an imperfect repeat into a perfect one, which is assumed then to be intrinsically unstable. Adapted from Richards and Sutherland (1992).

grows larger until, above some critical size, a new mechanism (some abnormality in chromatin packaging, recombination, or replication) leads to the explosive increase in repeat size that produces the disease. There is no a priori reason in this mechanism why the initial mutation rate might be so high.

The second mechanism shown in Figure 13.9 postulates that in a normal chromosome, the repeat is imperfect. This somehow intereferes with processes that lead to expansion in the size of the repeat. If a mutation occurs that makes the sequence a perfect trinucleotide repeat, a founder chromosome is created that is progressively more and more unstable as the repeat size grows. This second mechanism also explains what is seen experimentally, and it may offer clues as to why the observed mutation rate is so high, since an imperfect repeat could be converted to a perfect repeat by a number

*(continued)*

**BOX 13.2** *(Continued)*

of processes including unequal sister chromatid exchange or recombination, or gene conversion (Chapter 6).

Whatever processes lead to the progressive increase in the size of the repeat, the available data clearly show that the degree of instability is strongly affected by the size of the repeat. Figure 13.10 shows the distribution of repeat sizes in the normal population (where a few specific sizes predominant) and in carriers where a much broader range of larger repeat sizes is seen). When the repeat size of parent and offspring is compared, there is a clear correlation between the size of the parental allele and the increase in size seen in the allele of the offspring. These results are shown in Figure 13.11 for offspring that are still carriers and not affected. However, the effects are much more dramatic when affected are also included. This is illustrated in Table 13.2. (It is impractical to include the affected in Fig. 13.11 because some fragile X disease alleles have more than 1000 copies of the trinucleotide repeat, which makes them difficult to display on the same scale.)

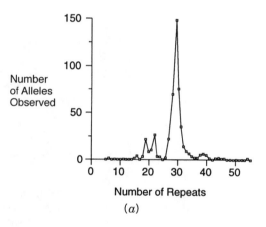

(a)

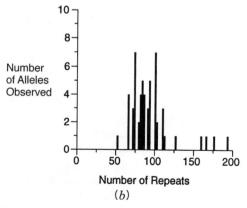

(b)

**Figure 13.10**   Distribution of CGG repeat lengths in normal individuals (*a*) and premutation carriers (*b*). Adapted from Fu et al. (1991).

*(continued)*

**BOX 13.2** *(Continued)*

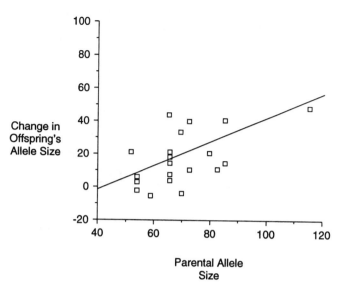

**Figure 13.11**  Effect of the CGG repeat length in the fragile X gene of an individual on the length of the corresponding repeat in an offspring of that individual. Adapted from Fu et al. (1991).

The data in Table 13.2 also demonstrate that as the size of the carrier allele grows, the probability that it will convert during the next female transmission to the full disease state increases steadily until, above carrier alleles of 90 repeats, the offspring are always affected. In contrast, the expansion of the allele in male transmission is slight, explaining the existence of nontransmitting males shown in Figure 13.8. Thus some difference between male and female transmission must underlie the basic mechanism of expansion of this allele. Repeat expansion in myotonic dystropy, like the fragile X syndrome, requires female transmission. In contrast, Huntington's disease is more severe if it is inherited as a result of male transmission. Since we know that meiotic recombination in the male and female are significantly different (Chapter 6), it is tempting to speculate that meiotic recombination hot spots may somehow be involved in the generation of these disease alleles. However, in most fragile X affected males and females, a mosaic pattern of triplet repeat sizes is seen. In adult tissues, and even in a fetus as young as 21 days, this mosaic pattern is mitotically stable. The inescapable conclusion from these findings is that expansion of the repeat occurs after fertilization, at some early stage in embryogenesis. Perhaps some form of imprinting (methylation pattern) in the X chromosome derived from the mother plays a key role in the subsequent expansion of the repeat to form the disease allele. The mosaicism of the fragile X syndrome is also responsible for the failure of male fragile X patients to transmit the disease to their offspring. (Rare as these offspring may be, a few cases are known.) When sperm from these patients are examined, none contain the full expanded fragile X disease allele. Either this allele interferes with some key process in spermatogenesis,

*(continued)*

**BOX 13.2** *(Continued)*

**TABLE 13.2    Comparison of Offspring's Fragile X Alleles with Premutation Alleles in Parents**

|  | Parental Allele | Child's Allele |
|---|---|---|
| Male transmissions | 66 | 70, 83 |
|  | 86 | 100 |
|  | 116[a] | 163 |
|  |  |  |
| Female transmissions | 52 | 73 |
|  | 54[b] | 58, 60, 57, 58, 52 |
|  | 59[b] | 54 |
|  | 66 | 73, 86, f |
|  | 66 | 60, 73, 110 |
|  | 70 | 66, 103 |
|  | 70 | f, f |
|  | 73 | f, f, f |
|  | 73 | f, 170[c] |
|  | 73 | 113, f |
|  | 77 | f, f, f |
|  | 80 | f |
|  | 80 | f, f |
|  | 80 | 100 |
|  | 83 | f, f |
|  | 83 | f, f |
|  | 83 | f |
|  | 83 | 93, f, f, f |
|  | 86 | f, f, f |
|  | 86 | 126/193[a] |
|  | 90 | f, f |
|  | 90 | f, f |
|  | 93 | f, f |
|  | 93 | f |
|  | 93 | f, f |
|  | 93 | f, f |
|  | 93 | f |
|  | 100 | f, f |
|  | 100 | f |
|  | 100 | f |
|  | 110 | f, f |
|  | 113 | f |

*Source:* Adapted from Fu et al. (1991).

Note: Numbers indicate number of CGG repeats found in each individual. Those children marked with f received a full mutation.

[a] These individuals are mosaic, and the most prominent allele(s) are indicated.

[b] These transmissions are in a family (CEPH 1408) with no history of fragile X syndrome; all others are fragile X chromosomes segregating in fragile X disease pedigrees.

[c] This allele was measured by Southern blot analysis, and the number of repeats estimated.

*(continued)*

**BOX 13.2** (*Continued*)

or the germ cells all derive from an early embryonic presursor cell that cannot tolerate the full-blown fragile X mutation.

The actual mechanism by which the expanded fragile X allele leads to altered expression of the fragile X gene is a bit mysterious, since the expanded repeat lies outside the coding region. However, a tantalizing hit about this mechanism is provided by the observation that in chromosomes with highly expanded triplet repeats, a nearby HTF island is hypermethylated. Furthermore severity of the illness is modulated by the degree of methylation. Under ordinary circumstances this will be associated with inactivation of gene expression. What is less obvious is the cause and effect relationship between allele expansion and the hypermethylation. For instance, some T-males have expanded repeats but not hypermethylation of the nearby CpG island. A variety of recent studies indicate that the CTG and CGG repeating sequences adopt unusual structures as duplexes, while the separated strands can form stable hairpins that can interfere with replication and lead to a number of other anamolous properties (Wells, 1996). In fact the characteristics of long repeats are so unusual that it is difficult to carry out conventional PCR amplifications of such sequences. Methods have been developed to exploit these unusual characteristics in an attempt to make the detection of new expanded repeats more efficient (Schalling et al., 1993, Broude et al., 1997).

Suppose that all of the approaches described above fail in the search for a particular disease gene. Such a case, for a long time, was Huntington's disease. This was one of the first inherited diseases identified by linkage to polymorphic markers. However, for almost a decade after the first evidence of linkage, and after extensive characterization of the DNA of the region, no specific gene had emerged as a strong contender for the cause of the disease. Huntington's is a case where the disease phenotype is premature death of certain cells in the brain. Thus genes with expression specific to the brain were plausible candidates, but many more indirect mechanisms could not be ruled out. Eventually the gene for Huntington's was identified because a coding sequence isolated from the region by exon trapping (Box 13.1) was found to contain a polymorphic $(CAG)_n$ sequence. However, in less fortunate cases, unless particular biological hypotheses emerge that can be tested, one is left with the particularly unpalatable alternative of sequencing the DNA of the region in a set of affected and unaffected individuals. The disease allele should be one that is common to all affected and not common to others. The problem is that in a Mb region of DNA, there will be several thousand polymorphisms. Only one of these is the disease allele, itself. The remainder are presumably harmless, normal, silent variations in the population.

A sensible strategy would focus on sequencing coding regions first, in the hope that the disease allele is directly expressed as an altered protein sequence. However, there is plenty of precedent for disease alleles that lie outside the coding region and exert their influence by altering RNA splicing or mRNA expression levels. Thus in the worst case one might have to resort, ultimately, to sequencing the entire genomic DNA of the region of interest in a set of individuals. This would not be impossible, even with existing methods, but it would be extremely costly and inefficient. The prospects of such tasks, strongly encourage the development of methods for finding and then sequencing just the differences between pairs of DNA samples. (See Chapter 14.)

A few guidelines exist that can help identify particular polymorphisms that might be responsible for producing disease alleles. DNA sequence changes that disrupt protein sequence are likely to lead to significant disease phenotypes. Prime suspects are frame shifts that will cause massive changes in protein sequence, and usually premature chain termination when a stop codon in the new coding region is reached. Insertions, deletions, and nonconservative amino acid changes are also likely suspects for causing disease. For example, the protein collagen consists mostly of a extensive pro-pro-gly repeat, and the existence of gly every third residue is essential for proper formation of the tertiary structure of the collagen fiber. Thus any mutations of the glycine are likely to have serious consequences, and it has been observed in many inherited collagen disorders studied thus far that mutations of these glycines are involved.

It is important not to trust any guidelines about disease alleles too strictly. It is too easy to get fooled. For example, many disease alleles in familial Alzheimer's disease turn out to be very conservative amino acid changes like isoleucine to valine. Note that disease alleles that are truly catastrophic will usually not be found except in carriers for a recessive disease. The homozygous affected individuals is unlikely to survive beyond early embryogenesis and will usually not be detected because the only result will be an early, usually unnoticed, spontaneous abortion.

In several years the human genome project should provide the DNA sequence of all human genes, including those responsible for all forms of inherited disease. Once all these sequence data are available, the process of linking specific DNA sequence alterations to particular diseases should become much simpler. In some cases comparison of DNA sequence in the human to sequences that have already been studied functionally, in the human or in other organisms, will provide direct clues to possible function (see Chapter 15). In many cases the DNA sequence will allow the construction of useful hybridization probes or PCR primers to examine the pattern of gene expression at the mRNA level. This will help find disease genes where the result of a DNA alteration is a change in mRNA expression at any level, including transcription efficiency or tissue distribution, splicing, or mRNA stability.

In some cases mRNA levels or other characteristics could remain essentially unaltered, even though there was a disruption in the nature of the protein product produced, or the level or cellular location of this product. In such cases the DNA sequence can be used to design peptide antigens that will elicit the production of high-affinity monoclonal antibodies specific for the protein of interest. Such powerful antigenic peptides are called immunodominant epitopes. Their location can often be predicted from the sequence, and the peptides can be synthesized automatically and used for immunization. The resulting antibodies are excellent reagents for examining the level and location of the particular protein in specific cells or tissues. Thus, all in all, a fairly powerful arsenal of approaches is accessible for finding genes with potential functions, once DNA sequence information is available.

## DIAGNOSTICS AT THE DNA LEVEL

We have discussed the role that DNA physical mapping and sequencing can play in finding genes involved in human diseases. Once such genes are found (or in more limited cases, once the approximate location of a disease gene is known), it is extremely desirable to develop diagnostic tests for the presymptomatic disease state or for the carrier state.

There are a number of distinct advantages to the direct examination of DNA for diagnostics. Any cell of the body can be used to detect an inherited alteration in DNA. Thus there is no need to sample what may be a relatively inaccessible target tissue where the disease effects will be most pronounced. Since all cells will have the same alteration in DNA, whether or not this produces a disease phenotype in these particular cells, any cells can be investigated. In practice, it is most convenient to use easily accessed samples like blood or epithelia sloughed off in saliva.

In principle, diagnosis at the DNA level will be less sensitive than many tests at the RNA, protein, or metabolic level. The DNA is present in only a few copies per cell; most RNAs and proteins or small molecules are much more frequent, especially if a cell in which the gene or its products are metabolically active is selected. However, for genomic analysis, it turns out that the sensitivity of DNA testing is more than sufficient. This is especially true in most cases because PCR amplification can be used, and thus the samples of cells required can be quite modest in size. The problem is quite different when tests are applied to tumors or infectious diseases. Here only a small fraction of the target cells may have altered DNA sequences present. This demands much higher sensitivity, and sometimes DNA tests may not be sensitive enough for the desired analysis, unless potentially affected cells or infectious agents are purified prior to PCR analysis. However, in some cases sensitive PCR methods have been able to detect low levels of circulating tumor cells (Nawroz et al., 1996).

The perfect DNA test will discriminate precisely between all normal alleles in the population and all disease alleles. How close we can come to this ideal is very different for different inherited diseases. Sickle cell anemia represents an example of the clearest case for DNA analysis. Here the physical diagnostic criteria are clear enough that a single disease allele in the population is responsible for the bulk of the phenotype we call sickle cell disease. The actual change in the gene of HbA, the normal beta chain of hemoglobin, to HbS, with a single altered amino acid, occurs in a restriction site for two enzymes. Thus the disease allele is an RFLP, and it can be analyzed this way to distinguish HbA normals, HbA HbS carriers (sickle cell trait), and HbS individuals affected with sickle cell disease.

| | | | |
|---|---|---|---|
| HbA | CCTGAGGAG | Bsu36 I | CCTNAGG |
| HbS | CCTGTGGAG | Dde I | CTNAG |

Alternatively, allele-specific PCR, as illustrated in Chapter 4, can be used to make the analysis. There are other hemoglobin disorders, collectively called hemoglobinopathies, which produce anemia or other impairments in oxygen transport. Many of these can be successfully analyzed prenatally or in the carrier state by DNA diagnostics. A long catalog of specific abnormalities with particular clinical presentations is now known. Most are quite rare.

Cystic fibrosis is an example of a more complex case for DNA diagnostics. This relatively common disease is undoubtedly far more representative of what will be found for the majority of human inherited disorders, and it raises a number of severe complications. In cystic fibrosis, one particular disease allele, a three-base deletion, resulting in the deletion of a single, key amino acid in the protein chain, phenylalanine 508, accounts for most of the observed disease cases. In populations of northern European extraction, this allele occurs in about 70% of all affected individuals. (Note the implicit complication caused by the heterogeneity of the partially outbred human population. The actual statistics of disease occurrence and alleles responsible are highly dependent on the particular subpopulation of origin of the individuals in question.) In this same group another 10 alleles account for about 20% more of the disease state. Hundreds of additional alleles must be considered, however, to account for all cystic fibrosis cases that have been examined thus far at the DNA level. Ironically some of these produce disease that is phenotypically so different from a classical cystic fibrosis presentation that until DNA analysis of this gene defect became prevalent, these cases were not even recognized as being cystic fibrosis at all. We can extrapolate and conclude that for a large gene like cystic fibrosis, the number of potential disease alleles in our population will be so large that it may never be possible to identify all of them a priori. This will certainly be the case if some are the result of new mutations.

The complexity of disease alleles at the cystic fibrosis locus leads to serious problems when prenatal or postnatal diagnosis for this disease is considered. Suppose that two non-affected potential parents are tested for the major allele. If neither is positive, it is fairly likely their offspring will be unaffected by cystic fibrosis. A negative result in a test for the next ten most likely alleles would certainly all but rule out the risk of a child with cystic fibrosis. If, instead, both parents test positively for the major allele, or for other frequent disease alleles, the risk of their producing an affected child is one in four (assuming that the homologous allele in each is normal; otherwise, they would be affected individuals). In this case, prenatal diagnosis of a fetus conceived by this couple would be strongly indicated.

A third, not uncommon, potential outcome of genetic tests of the parents leads to a more complicated situation. Suppose that one parent tests positive for the major cystic fibrosis allele (or other common alleles) while the second parent tests negative. There is still a small but significant chance that this second parent is a carrier for a rare disease allele. Thus a prospective child from this couple has a significant risk for cystic fibrosis. This risk can be partially assessed by performing prenatal diagnosis. If the fetus shows an absence of all the major cystic fibrosis alleles, it is very likely that the child is safe from cystic fibrosis. However, if the fetus has inherited the detectable allele from the first parent, one may really want to scrutinize the second parent for minor cystic fibrosis alleles. We need to develop much more cost-effective and accurate ways to do this.

The above example illustrates a major benefit of genetic testing at the DNA level. In a family known to be segregating a disease like cystic fibrosis, the result of a fairly harmless test will, in most cases, rule out any significant chance of a child affected with the disease. Thus the family is spared a great deal of unnecessary anxiety, and the child is spared the trauma of more invasive diagnostic tests. A particularly compelling case for such analysis which is not yet practical but will probably be doable soon is neurofibromatosis. This is a very unpleasant and disabling disease (the elephant man syndrome). The usual earliest clinical presentation of the disease is the presence, at birth, of occasional cafe au lait (light brown pigmentation) spots. Many less severe syndromes can also result in such spots. Hence the real power of genetic tests, once available, will be to rule out the likelihood of a terrible disease in all but a small percentage of those who display these symptoms.

Genetic analysis at the DNA level becomes much more complex for dominant lethal diseases like neurofibromatosis. The problem, discussed earlier, is that most if not all of the alleles responsible for such diseases have to be new mutations. Unless there is a restricted pattern of mutation hot spots in the gene responsible for the disease, one is faced with the difficult prospect of having to search the entire gene for all possible DNA changes. When the enormous size of some genes is considered, the task in some cases could be truly formidable. The gene responsible for Duchenne muscular dystrophy has more than 2.5 million bases; the coding region is more than 15 kb, and it is divided into more than 50 exons. The coding regions for an early onset BRCA2, an early onset breast cancer gene, cover 11,000 base pairs with 26 exons spread over 100 kb.

DNA analyses are also potentially applicable to somatic mutations such as the DNA alterations seen in most kinds of cancer. Some of the types of alterations seen, which are believed to be primary events in the ultimate progression that leads to tumor formation, are shown in Figure 13.12. Some tumor genes are dominant lethals. Once a single copy of the gene responsible for cancer, an oncogene, is affected by a mutation in a somatic cell, uncontrolled growth, or the accumulation of further mutations which lead, in turn, to uncontrolled growth, is triggered (Fig. 13.12a). In other cases the alleles affecting oncogenes are recessive. A single somatic mutation does not immediately lead to cancer. Instead, a second DNA rearrangement must occur to expose the effect of the first gene (Fig. 13.12b). The requirement for a second hit is clear if one imagines the function of the recessive oncogene to be suppression of a gene or a process that otherwise would lead to uncontrolled growth. Loss of one homolog still leaves the other normal allele intact and functional. However, if the normal allele is lost by a second event, which could be local DNA damage, a deletion, or mitotic nondisjunction, the resulting cell is now triggered for

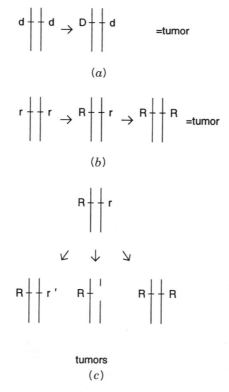

**Figure 13.12** DNA changes seen in various types of cancer. (a) A dominant oncogene. (b) A recessive oncogene. Reduction to homozygosity must occur by a second mutation. (c) In familial cancers only one new mutation is required.

uncontrolled growth. This cell now has a selective advantage, and hence its progeny develop into a large population, a tumor.

Many cancers are familial: They have inherited components. Examples include a subset of the individuals affected with several common cancers like colon and breast cancer, as well as many rarer forms like retinoblastoma, a tumor of the eye, and neurofibromatosis, which we have already discussed. In inherited retinoblastoma, the child receives one defective recessive oncogene from the parents; both eyes are highly susceptible to an additional event that damages this gene. As a result bilateral tumors are common. In contrast, with the purely somatic form of the disease, two independent damage events are needed. It is very unlikely for these to occur in both eyes; hence this form of retinoblastoma is usually unilateral.

Mutations in the DNA from many individuals affected by familial (inherited) and spontaneous (somatic) forms of several kinds of cancer have been investigated. The results for colon cancer somatic mutations seen in the adenomous polyposis coli (ApC) gene are summarized in Table 13.13. It is evident, as expected, for such new mutations, that there are a very large number of different disease alleles. This will make it difficult to use DNA analysis to find all possible disease cases unless the entire gene can be scanned. Mutation sites for somatic and familial colon cancer are summarized in Figure 13.13. It is apparent that no significant hot spots are seen for the somatic form. The familial form does display a hot spot area, but the actual mutations here are still widely distributed. Interestingly, regions of the gene that predominant in the familial form are different from those that are common in the somatic form. Thus the challenges of DNA analysis for this fairly common and deadly cancer are truly formidable.

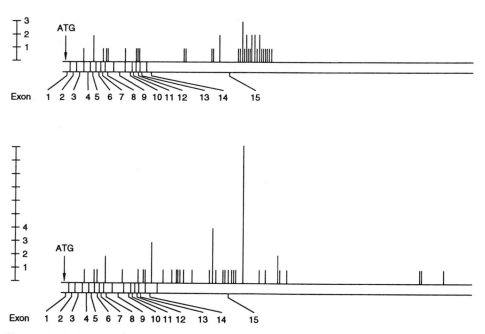

**Figure 13.13** Locations of mutations in the APC gene, responsible for many human colon cancers. Shown are numbers of individuals found with (*a*) somatic mutations and (*b*) inherited mutations. Adapted from Miyoshi et al. (1992).

**TABLE 13.3   Forty-Three Somatic Mutations of Colorectal Tumors in the APC Gene**

| | | | |
|---|---|---|---|
| C113(F) | 142 | a*a*tag/CTG → atag/CTG | Splice acceptor |
| C31, C124(F) | 213 | CGA → *T*GA | Arg → Stop |
| C24 | 279 | aatttttag/GGT → agtttttag/GGT | Splice acceptor |
| C47 | 298 | C*AC*TC → CTC | AC deletion |
| C108(F) | 302 | CGA → *T*GA | Arg → Stop |
| C135 | 438 | CAA/gtaa → CAA/g*c*aa | Splice donor |
| C33 | 516 | AAG/gt → AAG/*t*t | Splice donor |
| C28 | 534 | AAA → *T*AA | Lys → Stop |
| C10 | 540 | TTA → TTA*T* | A insertion |
| C37 | 906 | T*C*TG → TTG | C deletion |
| A128(F) | 911 | GAA → G*G*A | Glu → Gly |
| C23 | 1068 | *TCAA*GGA → GGA | TCAA deletion |
| C11, C15 | 1114 | CGA → *T*GA | Arg → Stop |
| C20 | 1286 | GAA → *T*AA | Glu → Stop |
| A53 | 1287 | ATA → A*A*TA | A insertion |
| C27 | 1293 | *ACACAGGAAGCAGATTCT* | 31 bp deletion |
| | | *GCTAATACCCTGC*AAA → AAA | |
| C7, C21 | 1309 | G*AAAGA*T → GAT | AAAGA deletion |
| C14 | 1309 | GAA → *T*AA | Glu → Stop |
| A41 | 1313 | ACT → *G*CT | Thr → Ala |
| C31, C42 | 1315 | TCA → T*A*A | Ser → Stop |
| A44 | 1338 | CAG → *T*AG | Gln → Stop |
| C22 | 1353 | GAA*TTTTC*TTC → TTC | 8 bp deletion |
| A56 | 1356 | TCA → T*G*A | Ser → Stop |
| C4, C27 | 1367 | CAG → *T*AG | Gln → Stop |
| C10 | 1398 | A*GT*CG → TCG | AG deletion |
| C19 | 1398 | AG*T*C → AGC | T deletion |
| A43 | 1411 | A*G*TG → ATG | G deletion |
| C16 | 1420 | C*C*CA → CCA | C deletion |
| C40, A52(F) | 1429 | GAA → *T*AA | Gln → Stop |
| C29 | 1439 | C*C*TC → CTC | C deletion |
| C37 | 1446 | G*CTCAAACCAA*GC → GGC | 10 bp deletion |
| A50(F) | 1448 | T*T*AT → TAT | T deletion |
| A49(F) | 1465 | A*G*TGG → TGG | AG deletion |
| C23 | 1490 | C*A*TT → CTT | A deletion |
| C12 | 1492 | GC*C*A → GCA | C deletion |
| A41 | 1493 | *ACAGAAAGTAC*TCC → TCC | 11 bp deletion |
| C3 | 1513 | GAG → *T*GAG | T insertion |

*Source:* Adapted from Miyoshi et al. (1992).

If we had the ability to do very large-scale DNA sequencing, genetic diseases could be diagnosed by this technique with great power but still not without difficulties. We would still have to develop effective ways to distinguish, for newly found alleles, whether they were just harmless polymorphisms or true disease-causing alleles. While some guidelines for how this might be done were presented earlier in the chapter, it will be hard to do this in general without considerable information about the function of the protein product of the gene. An example of the complex spectrum of spontaneous mutations seen in the human factor IX gene responsible for hemophilia B is shown in Table 13.4. These results

**TABLE 13.4   Summary of Sequence Change in 260 Consecutive Cases of Hemophilia B**

|  | Number | Percentage |
|---|---|---|
| 1. Number with sequence changes in the eight regions of likely functional significance | 249 | 96 |
| 2. Of those with sequence changes, number of independent mutations[a] | 182 | 73 |
| 3. Of independent mutations, number with a second sequence change | 6 | 3 |
| 4. Type of independent mutation: | | |
|    Transitions at CpG | 48 | 26 |
|    Transitions at non-CpG | 65 | 36 |
|    Transversions at CpG | 8 | 4 |
|    Transversions at non-CpG | 35 | 19 |
|    Small deletions and insertions ($\leq$50 bp) | 15 | 8 |
|    Large deletions ($>$50 bp) | 10 | 6 |
|    Large insertions | 1 | 0.6 |
| 5. Location of independent mutations:[b] | | |
|    Promoter | 1 | 0.5 |
|    Coding sequence[c] | 163 | 86 |
|    Splice junctions | 12 | 6 |
|    Intron sequences away from splice junctions | 1 | 0.5 |
|    Poly A region | 0 | — |
|    Unlocalized (total gene deletions) | 5 | 3 |
|    Unknown[b] | 8 | 4 |
| 6. Functional consequences of observed independent mutations | | |
|    Protein with amino acid substitutions | 114 | 63 |
|    Garbled protein (truncated, frameshifted or partial or full deletion of amino acids) | 54 | 30 |
|    Abnormal splicing | 13 | 7 |
|    Decrease expression | 1 | 0.6 |

*Source:* Adapted from Sommer et al. (1992).

[a]Recurrent mutations were judged independent if the haplotypes differed. In a few cases recurrent mutations with the same haplotype were judged independent because the origin of mutation was determined. In four patients recurrent mutations were judged independent because the races of the individuals were different.

[b]Assumes that 11 patients with unknown mutations have the same frequency of independent mutations as patients in which the mutation could be defined (73%). Thus eight independent mutations should be unknown.

[c]Includes partial gene deletions that affect the coding region.

clearly indicate that no single test at the DNA level, even complete DNA sequencing, would be capable of 100% certain diagnoses. It is difficult to measure heterozygous positions with typical gel-based sequencing, and haplotypes of compound heterozygotes cannot be determined at all.

The difficulty of DNA analysis of cancers includes all of these problems, but it is confounded by the heterogeneity of typical tumor tissue. For very early onset diagnosis of somatic cancer, one would need to distinguish, potentially, any altered nucleotide, in a complex gene, present in only a minute fraction of the cells in a sample. It is not easy to see how this could be accomplished by directly DNA sequencing alone. In all these cases, the task is much simpler if only a finite number of specific sequence alleles are correlated with the disease. Then one can set up specific assays for the alleles, and some such assays seem capable of dealing with the sorts of heterogeneous samples encountered in tumors. The emergence of specific cancer or disease-associated alleles can also be tested for in easily assayable fluids. For instance, the fingerprint of prostate tumor cells has been detected in urine, and those of head and neck tumors have been detected in saliva and blood.

## ANALYSIS OF DNA SEQUENCE DIFFERENCES

The examples described in the previous section clearly indicate the need to be able to analyze a stretch of DNA sequence to look for abnormalities that might be as small as single base pair. If the DNA target is just a few thousand base pairs, this search can be done by direct sequencing. If the target is much larger, direct sequencing with current methods is impractical. In this section we explore the present status of methods that can detect changes in DNA as small as single base pairs, with less effort than would be required for total DNA sequencing. Many of these methods resort to the formation of DNA heteroduplexes to facilitate the screening for differences between a test sequence and a standard.

Figure 13.14*a* shows, schematically the three possible genotypes that must be distinguished in making a genetic diagnosis. In general, an individual tested could be normal (dd), heterozygous for the disease allele (dD), or homozygous for the disease allele (DD). If the disease is dominant, one usually expects the affected to be a heterozygote. If one is testing for a carrier status, the test is really looking to see if the individual in question is a heterozygote. In either of these cases, a normal DNA duplex is present that can serve as an internal control for the possible presence of an altered duplex. DNA from the region of interest can be prepared directly from genomic material by PCR, assuming enough known sequence exists to design suitable primers. If this DNA is melted and the separated strands are allowed to reanneal, four distinct products will be formed (Fig. 13.14*b*). Two of these are the perfectly paired normal and abnormal duplexes. The remaining two are heteroduplexes composed of one normal and one abnormal strand. Any DNA sequence differences between these species will lead to imperfections in the duplex because one or more base pairs will be mismatched.

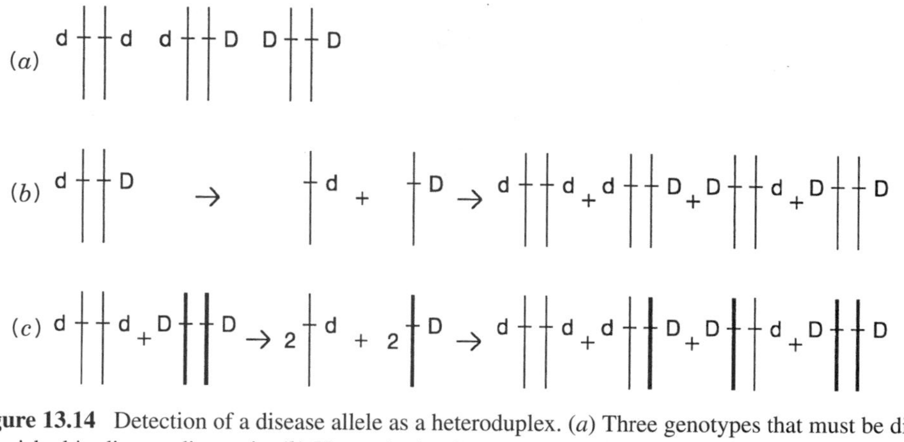

**Figure 13.14** Detection of a disease allele as a heteroduplex. (*a*) Three genotypes that must be distinguished in disease diagnosis. (*b*) Heteroduplex formation by melting and reannealing DNA from a recessive carrier or a dominant heterozygote. (*c*) Heteroduplex formation by mixing DNA from a homozygous recessive with DNA from a normal individual.

If the disease in question is recessive, the usual question asked is whether the particular individual being tested is homozygous for the disease allele or heterozygous. Thus a minimal test would be to proceed exactly as described above, except that now, presence of heteroduplex indicates that the individual is a carrier, unaffected for the disease. However, this test would be ambiguous because absence of heteroduplex would imply that either the individual in question is a homozygous normal or a homozygous affected. An additional test is required to resolve this ambiguity. This second test is designed as shown in Figure 13.14c. DNA from the individual to be tested is mixed with a standard DNA sample from a person previously shown to be normal, not a carrier. If the test individual carries the disease, four DNA products will be formed; two of these will be heteroduplexes. If the test individual is normal, no heteroduplex products will be produced. Thus the tests shown in Figure 13.14 reduce the problem of detecting DNA sequence alterations to the problem of detecting heteroduplex DNA.

We will shortly describe a wide variety of techniques that have varying success in distinguishing between DNA heteroduplexes and homoduplexes. One caveat in using these methods for genetic testing must be noted. Successful heteroduplex detection will find any DNA alterations, whether or not these are disease alleles or harmless polymorphisms. Thus what heteroduplex detection does is indicate the presence of a DNA sequence variant. Once this is found, DNA sequencing will frequently be needed to examine the characteristics of the particular variant discovered. Thus the strategy used is to apply a very simple test that can scan large DNA regions to see if any sequence variations exist. If none are found, and the test is reliable, one need go no further. If differences are discovered, then usually a more robust test will need to be applied, but the screening will have narrowed down the DNA target to a much smaller region. For example, PCR can be used to examine the exons of a complex gene one at a time. If a sequence variation is discovered in a single exon, at worst one would have to sequence the DNA of that exon to complete the diagnosis.

## HETERODUPLEX DETECTION

The difficulty in designing schemes to detect heteroduplexes is that many possibilities can arise, even from a single altered base pair. This is shown in Figure 13.15. Any mixture of two DNAs with a single base pair difference produces two different heteroduplexes with a single base mismatch. In all, there are eight possible single base mismatches: A–C, A–A, A–G, C–C, C–T, T–T, T–G, G–G, and an acceptable test would have to be able to detect them all. The ideal test would not only detect them, but it would also reveal which exact mismatches were present. A potential complication is that each heteroduplex occurs

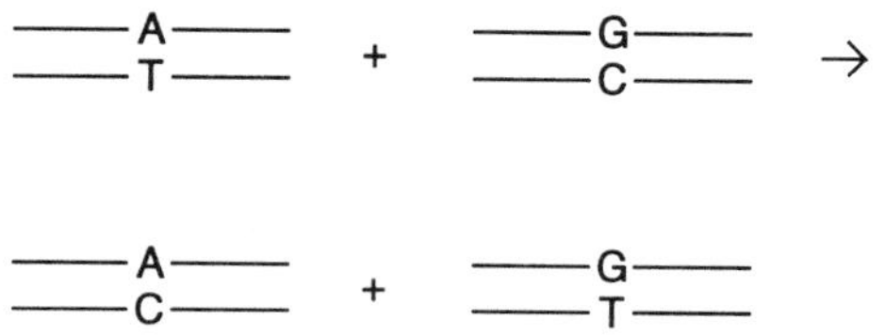

**Figure 13.15** A single site mutation will serve to generate the formation of two different heteroduplexes.

within the context of a specific DNA sequence, and the identity of the neighboring base pairs could easily modulate the properties of the heteroduplex. Not much is known about this at present.

In practice, the formation of heteroduplexes from a diploid sample will produce a pair of mismatches. For single base pair differences there are only four possibilities, and each gives a different and discrete set of heteroduplexes. Thus a test that detected a specific half of the possible heteroduplexes would suffice:

| Mutation | Heteroduplexes |
|---|---|
| A–T to T–A | A–A and T–T |
| G–C to C–G | G–G and C–C |
| A–T to G–C | G–T and A–C |
| A–T to C–G | A–G and C–T |

Thus, for example, a method that could identify a mispaired T or G but not A or C would suffice to spot the presence of a heteroduplex, but it would probably not have enough resolving power to identify the exact heteroduplex present.

The single-base, mismatched heteroduplexes just illustrated are actually the most difficult case to detect by the methods currently available. Larger mismatches or heteroduplexes arising from insertions or deletions lead to much larger perturbations in the DNA double helical structure, and these are easier to reveal by physical and chemical or enzymatic methods. The principle complication is that the number of possible heteroduplexes becomes rather larger. There are four possible single base insertions or deletions, but these are likely to be susceptible to complexities caused by the local sequence context. For example, in the sequence shown below, two altered mismatched structures compete with each other.

$$
\begin{array}{c}
\text{A–T} \\
\text{G–C} \\
\text{G–C} \\
\text{T–A}
\end{array}
\longrightarrow
\begin{array}{c}
\text{A–T} \\
\text{G–C} \\
\text{T–A}
\end{array} \!\text{C}
\longleftrightarrow
\text{C}\!\begin{array}{c}
\text{A–T} \\
\text{G–C} \\
\text{T–A}
\end{array}
\longleftarrow
\begin{array}{c}
\text{A–T} \\
\text{G–C} \\
\text{G–C} \\
\text{T–A}
\end{array}
$$

The genetic consequences of the two different structures shown above are identical; however, the presence of two alternate heteroduplex structures could complicate the analysis. Much more work needs to be done to characterize the properties of such structures in more detail. This will have to be done before the overall accuracy of any proposed method of heteroduplex detection can be validated for clinical use.

At least six different basic methods for detecting heteroduplex DNA have been described. All of these tests work well in some cases. However, none have yet been proved to be generally applicable to all possible heteroduplexes. A key issue in these tests is how large a DNA target can be examined directly. The larger the target, the fewer fragments will be needed to cover a whole gene. However, if targets are too large, they may have such a high probability of containing a phenotypically silent polymorphism that the advantage of the test as a primary screen will be lost, since many fragments will test positive. The first four tests, illustrated in Figure 13.16, all have potentially similar characteristics, and all will work, in principle, on very large DNA targets. A straightforward and direct approach is to use single-strand-specific DNases like S1 nuclease to cleave at the

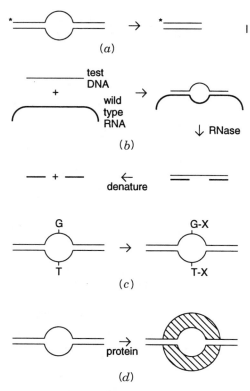

**Figure 13.16** Four different methods for direct detection of heteroduplex DNA. (*a*) Nicking with S1 nuclease. (*b*) Trimming and nicking an RNA-DNA hybrid with ribonuclease. (*c*) Reaction of unpaired bases with organic molecules like carbodiimides. (*d*) Binding of *mut*S protein to the site of the mismatch.

mismatched bases in a heteroduplex (Fig. 13.16*a*). Then the resulting shortened DNA fragment could be detected by a Southern blot or by PCR. The difficulty with this approach is that no known combination of enzyme and reaction conditions allows reliable cleavage at single base mismatches. Larger mispaired targets are needed before cleavage becomes efficient and selective. Other nucleases that look very promising for such studies include bacteriophage T4 endonuclease VII.

RNase nicking can be used as an alternative to DNase nicking. Here a single strand of the DNA to be tested is annealed with RNA made from a DNA sample representing the normal allele (Fig. 13.16*b*). This is easily accomplished, for example, by subcloning that allele downstream from a strong in vitro promoter like that for T7 RNA polymerase. One can produce either internally labeled RNA, and look for two shorter fragments as a sign of fragmentation, or end-labeled RNA, and look for one shorter fragment. In this latter case it is necessary to have the test DNA extend beyond the labeled end of the RNA; otherwise, the RNase will remove the label. Some workers swear by the reliability and sensitivity of the RNase approach, but many others have apparently been unable to use it successfully.

Chemical methods can be used instead of enzymes to mark or to cleave at the site of a mismatch. A particularly effective approach has been the use of water-soluble carbodiimides which can react with mismatched T or G (Fig. 13.16*c*). These compounds are available radiolabeled, so one can detect the presence of a mismatch by the incorporation of radioisotope. Alternatively, there are monoclonal antibodies available that are specific for the carbodiimide reaction product. This is a very powerful analytical tool, since it allows physical fractionation of any heteroduplexes, which are thus purified and concentrated if needed for subsequent analysis. The carbodiimide approach does have two distinct disadvantages. It cannot detect all possible mismatches, and it has been reported to be a difficult test to master in the laboratory.

The fourth test based on protein recognition of heteroduplexes is rather different because it does not involve enzymatic cleavage. In *E. coli,* a protein is made by the *mut*S gene that recognizes mismatches and binds to them (Fig. 13.16*d*). This is an early step in the excision and repair of mismatched bases. The binding of *mut*S protein can be detected in a number of different ways. DNA, once bound by *mut*S, will stick to nitrocellulose filters, while free DNA passes through. Alternatively, *mut*S fusions to other proteins involved in color-generating reactions have been made, and monoclonal antibodies against *mut*S are also available. Thus a variety of different methods to exist to detect *mut*S-DNA complexes. An attractive feature of *mut*S, like carbodiimides, is that it allows the selective isolation of intact heteroduplexes. The disadvantage shared by both systems is that not all mismatches are detected. For example, *mut*S fails to recognize a C–C mismatch, and some others. The full extent of the advantages and limitations of analysis of mismatches with *mut*S has not yet been described. However, this general kind of approach is attractive because it mimics a natural biological mechanism, and proteins with properties analogous to *mut*S, but perhaps with even broader mismatch recognition, may well exist in other organisms.

Two additional methods for detection of heteroduplexes are based on the altered electrophoretic properties of these structures. The simplest of these is direct separation of heteroduplexes from homoduplexes of the same length by using specialized gels. A very effective method uses a modified polyacrylamide called MDE, a term that stands for mutation detection electrophoresis. This gel has a somewhat hydrophobic character which alters the mobility of heteroduplexes selectively. Some move faster; most  move slower. In good cases a single mismatched base in a 900 base pair duplex is sufficient to give an easily detectable mobility shift. An example of the improved ability of MDE compared with ordinary gel media to resolve heteroplexes from homoduplexes is shown in Figure 13.17. A more complex example of the use of MDE is illustrated in Figure 13.18, where it is clear that different heteroduplexes within the same basic DNA fragments show different mobilities. Thus, in principle, the method offers some promise of revealing, not just that a heteroduplex is present but additional information about its characteristics. The appeal of this method is that it is very easy to perform, since aside from the special properties of the gel, the electrophoretic procedures used are quite ordinary. It is also easy to analyze many samples in parallel. However, the full generality of the method has yet to be proved. For example, it would be good to know how the nature of the mismatch and its location within the duplex affect the ability to detect it. Already we know that deletions lead to large mobility shifts, and DNA molecules with more than one heteroduplex region show very complex behavior. In general, a heteroduplex combination of a deletion and a separate, distant single-base mismatch leads to much larger mobility shifts than expected from the effects seen with the two mutations separately. The reason for this synergistic behavior is not currently understood.

A second electrophoretic method was originally developed by Leonard Lerman and his collaborators, and several variations on this general theme now exist. The original method

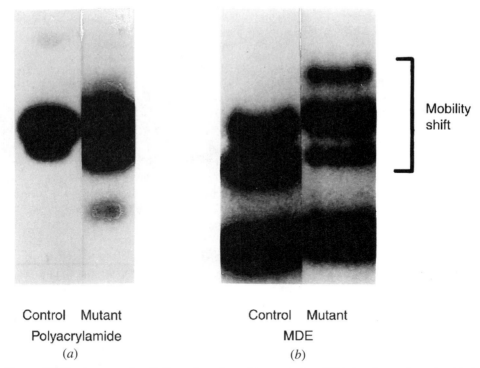

Control   Mutant          Control   Mutant
Polyacrylamide              MDE
(*a*)                        (*b*)

**Figure 13.17** An example of direct detection of heteroduplex DNA by electrophoresis. (*a*) On polyacrylamide. (*b*) On MDE gel. Provided by Avitech, Inc.

was called denaturing gradient gel electrophoresis (DGGE). The basic idea behind the method is illustrated in Figure 13.19. DGGE appears to be a general method capable of detecting any mismatch in a DNA sequence that is not at the very end. An internal mismatch in a duplex leads to a substantial decrease in the thermodynamic stability; this is manifested by a drop in the $T_m$ of the duplex (Fig. 13.19*a*). To test for mismatches, DNA fragments are electrophoresed in a gel through a gradient of increasing denaturant like urea, or a gradient of increasing temperature. At some critical point the section of the duplex containing the destabilizing mismatch reaches conditions above its local $T_m$, and it

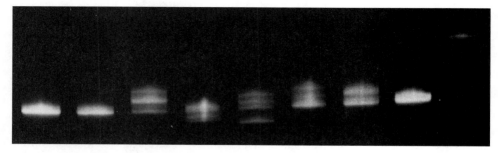

**Figure 13.18** Many different single mismatches can be distinguished by electrophoresis on MDE. Each lane is a different cystic fibrosis disease allele except for the left lane which is a normal allele, and shows no heteroduplex formation. Provided by Avitech, Inc.

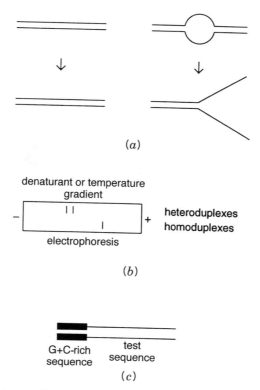

(a)

(b)

(c)

**Figure 13.19**  Denaturing gradient gel electrophoresis (DGGE) detection of heteroduplexes. (a) Destabilization of a duplex by mismatching. (b) Electrophoresis in a temperature or denaturant gradient. (c) Use of a GC clamp to prevent strand separation in DGGE.

melts. The resulting Y-shaped structure (or one with a large internal loop) has little or no electrophoretic mobility, and so the molecule is trapped in the gel near or at the site where it melted (Fig. 13.19b). Thus from a single experiment one can determine not only whether heteroduplexes are present but also what kinds of mismatches were present, since most give characteristic and different shifts in $T_m$.

It is important that the mismatch does not destabilize the duplex so much that the entire structure melts into separate strands. These would still be free to migrate in the gel. To prevent this unwanted effect, a G+C-rich sequence is usually placed at one end of the duplex to be tested. This is easily done by using PCR primers with an extra overhanging G+C-rich sequence. The use of this so-called GC clamp (Fig. 13.19c) prevents complete melting under typical DGGE conditions. A major advantage of DGGE is its generality. Another advantage is that rather complex samples can be analyzed by running a gel composed of a restriction enzyme digest of a clone and then blotting it with particular probes. A major disadvantage of DGGE is that specialized apparatus is needed for this method.

A third electrophoretic method we will describe for detection of altered DNA sequences is also based on electrophoresis, and it takes advantage of the effects of altered DNA sequence on thermodynamic stability. However, in detail, this method is actually quite different from DGGE; it does not involve heteroduplexes. Called single-strand conformational polymorphism (SSCP), the method is based on nondenaturing gel electrophoresis of melted and rapidly cooled samples. Under these conditions individual

DNA strands fold back on themselves to form whatever combination of stems and loops (and other secondary structures, e.g., pseudoknots) that the particular DNA sequence allows. Changes in even a single base can substantially alter the spectrum of secondary structures formed. RNA can be used instead of DNA to enhance the stability of secondary structure formation, and this apparently increases the fraction of heteroduplexes that can be detected. The use of MDE gels also enhances the resolution of SSCP. The results in a typical SSCP analysis are complex. However, SSCP is serving as a very easy and sensitive method to detect a reasonable fraction of all possible heteroduplexes. Two variations of SSCP have been described by Steven Sommer that increase the probability of detecting any mutation in the target. The first of these, called dideoxy fingerprinting (ddf), is a hybrid between Sanger sequencing and SSCP (Sarkar et al., 1992). A Sanger ladder is produced with a single dideoxy pppN, and it is analyzed on a native polyacrylamide sequencing gel. In the second method, restriction endonuclease fingerprinting, the nucleases and the products from these digests are pooled and analyzed together by SSCP (Liu and Sommer, 1995).

A fourth electrophoretic method takes advantage of the power of automated fluorescent DNA sequences. A mutation shows up as an unexpected peak. The sensitivity of the method, called orphan peak analysis (Hattori et al., 1993), is sufficient to allow multiple samples to be probed in each gel lane.

The final method we will mention for detecting altered DNA sequences is based on recent findings by Sergio Pena and coworkers (1994). In using short random DNA primers (RAPD; see Chapter 4) for PCR analysis of human DNA, they noted that the pattern of amplified bands seen was exquisitely sensitive to DNA sequence variations in the neighborhood of the primers such that virtually all alleles tested led to a different, distinct pattern of amplified DNA lengths.

In current practice, faced with a gene to search for mutants, the simplest and most general existing method is probably to do both SSCP and MDE-heteroduplex analysis. The real difficulty that remains is the size of many genes of interest. A typical 3- to 5-kb gene could be scanned in 5 to 10 pieces by selective PCR. In order to do this, however, one has to have available mRNA from the individual to be tested. This may not always be available. A further caveat is that some altered mRNAs may be selectively degraded in a cell if they are not functional. Thus a mutation may make itself invisible at the RNA level. The alternative is do the analysis at the DNA level. Here one can use PCR to look only at the exons. However, the difficulty is that some genes have 30 or more exons, and some exons can be very small. Given the large size of typical introns, each exon will have to be analyzed by a separate PCR reaction; thus the overall test becomes quite complicated.

## DIAGNOSIS OF INFECTIOUS DISEASE

DNA sequence is proving to be quite useful in the diagnosis of the presence of infectious organisms. Usually a small bit of DNA sequence will suffice to indicate the presence of a virus, bacterium, protozoan, or fungus. Different species can be identified definitively, once their characteristic DNA sequences are known. The major problem in using DNA analysis for detection of infectious agents is sensitivity; this is the same problem we encountered earlier in examining the prospects of DNA diagnosis for cancer. There may be only a few copies of the DNA (or RNA, for some viruses) genome per organism. The number of infected cells (or the number of organisms free in the blood stream or other body fluids may be very small. PCR is very helpful in amplifying whatever DNA is present, but the real problem is distinguishing a true positive signal from the frequent, unwanted background caused by PCR side reactions. Several strategies have been used to enhance the DNA analysis of infection. If the type of target cell is known, one can often use immobilized monoclonal antibodies against this particular cell type to purify it away from the rest of the sample. This will substantially reduce subsequent PCR background.

Another approach is to screen for the ribosomal RNA of the infectious agent. This is applicable to all agents except viruses, since they do not have ribosomes. Some regions of rRNA vary sufficiently to allow a wide variety of organisms to be distinguished easily. However, the major advantage of looking at rRNA directly is that there are typically $10^4$ to $10^5$ copies per cell versus only a few copies of most DNA sequences. An alternative, undoubtedly worth exploring for protozoa, will be to use repeating DNA sequences specific to a given species. This is unlikely to work for most simpler organisms because they have very few repeats.

The potential utility of DNA analysis in infectious disease is staggering. We will consider in detail the case of HIV, the virus that causes AIDS. One difficulty in the clinical management of this disease is that the virus has a very high mutation rate. Thus most people have different viral mixtures, and some components of these mixtures are resistant to particular drugs, because of mutations within the HIV reverse transcriptase or protease genes. A brute force approach recently described, which appeared to have some success (although the generality of this success is now disputed), is to treat with a mixture of several different drugs simultaneously. However, each of the drugs has potentially serious side reactions. Furthermore, by treatment with all of the effective agents at once, there is a real possibility of selecting for a viral variant resistant to them all. The complete DNA sequence of the virus is known, and the sequence of many drug-resistant variants has also been determined. Thus we know which sites in the viral reverse transcriptase and protease are likely to mutate and confer resistance to particular drugs.

By direct PCR cycle sequencing of blood samples from AIDs patients undergoing drug therapy, Mathias Uhlen and his coworkers have been able to monitor the course of the disease with a precision not before obtainable (Wahlberg et al., 1992). The single-color, four-lane fluorescent sequencing used by Uhlen is sufficiently quantitative that it not only can distinguish pure viruses, it also can analyze the composition of mixtures of viruses as seen as apparent fractional populations of particular bases at given sequence locations (Fig. 13.20). When the analysis shows that the population of a particular drug-resistant variant is beginning to climb, the physician is alerted to alter the therapy by switching to a different drug. After a while, it is typical to see a relapse in the viral population back to the original major strain, and since this is sensitive to the original drug used, one can switch back to that drug to help control the infection. As this kind of precise diagnostics becomes more affordable and more readily available, it could have a major impact on the practice of medicine.

## DETECTION OF NEW MUTATIONS

By definition, a new mutation can occur at any site within a target DNA sequence. To detect new mutations by current methods is very difficult, and it is much more demanding than most of the problems we have discussed earlier in this chapter. New mutations must be detected in the analysis of autosomal dominant lethal diseases, as we have discussed

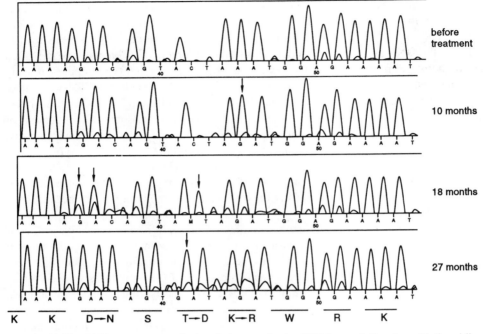

**Figure 13.20** DNA sequence analysis of changes in the HIV-1 population in azidothymidine-treated AIDs patients. Shown is the raw sequence data A's (dotted dashed line) and G's (solid line) for a portion of the reverse transcriptase gene before treatment and at various times after treatment. Corresponding amino acid changes are shown below. Adapted from Wahlberg et al. (1992).

before. It is also necessary to detect new mutations if we are to be able to estimate the intrinsic, basal human mutation rate, and how this may be influenced by exposure to various agents in our environment including radioactivity, sunlight, exposure to various chemicals, diet, and various types of radiation such as emissions from electrical power lines, microwave ovens, and color televisions. These types of environmental damage raise serious issues of liability and responsibility which can only be properly assessed if we can monitor, directly and quantitatively, their effect on our genes. Hence the motivation to be able to monitor human new mutations is very high. The sensitivity of different animal species to many of these environmental agents is known to be quite variable. Thus, unfortunately, here we have a case where humans must be studied directly.

There are actually three different types of mutation rates that have to be considered in judging the relative effects of various environmental agents. The three are illustrated in Figure 13.21. Genetic mutations are the type of events we have been discussing throughout most of this text. A new genetic mutation means a change in the genomic DNA of a child resulting in the presence of a sequence that could not have been inherited from either parent. One obvious way that this can come about is mis-paternity. Clearly this trivial explanation must be ruled out for any putative new mutation. Fortunately the power of current DNA personal identity testing makes such screening quite easy and accurate. The second type of mutation one must consider is a gametic mutation. Here one can look in a sperm cell for a DNA sequence not present in the father's genomic DNA. Alternatively, one can compare single sperm, by methods described in Chapter 7, and look for the oc-

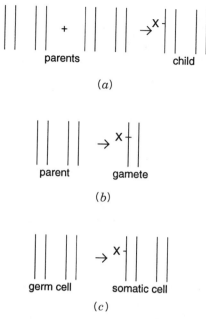

**Figure 13.21** Three different types of new mutations that one would like to be able to detect. (*a*) Genetic mutations are inherited by the offspring. (*b*) Gametic mutations are present in the gametes but are lethal, so no offspring are produced. (*c*) Somatic mutations are present in a subset of cells and are not passed to offspring.

currence of sequences that could not have arisen by simple meiotic recombination or gene conversion. The third form of mutation we must consider is at the somatic level. Here we need to look for DNA sequences that are present in some cells but not others. A key element of the problem is that effects of some environmental agents may be different for each type of mutagenesis. For example, an agent that was lethal to gametes would not show up as a genetic mutagen, and yet it would be a source of considerable damage if a significant decrease in fertility resulted. Specific types of cells may be particularly susceptible to certain agents. A few generalities can be made—it seems that rapidly dividing cells are more sensitive to agents that damage DNA; highly transcribed genes seem to be more easily mutagenized. However, in general, our current knowledge of these effects is very slight.

The major difficulty in studying new mutations is that, unless they are accompanied by an easily scored phenotype, the target is a small needle in a very large haystack. The basal, spontaneous mutation rate in the human is estimated to be $10^{-8}$ per base per meiosis. This means only 30 new mutations per haploid genome. What fraction of these are point mutations or more complex DNA rearrangements is unknown at the present time, although a guess that about half are in each category is probably reasonable. Since the location of the new mutants is unknown a priori, the magnitude of the search required to find them is staggering. Environmental agents will raise the rate of mutations above the basal level. What little evidence we have for typical agents of concern indicates that the increases in mutation caused by typical exposure levels are small. Thus, to quantitate these, we will either need a very good strategy or extraordinary sensitivity.

Two different basic scenarios must be considered that can arise and demand a search for new mutations. In the first of these one may have a small number of individuals exposed to a local, and perhaps very high, dose of toxic agent. An example would be a chemical waste spill. This is a very difficult situation to analyze. One choice is to examine a large percentage of the genomes of the exposed individuals in order to have sufficient sensitivity to see an effect. This is not practical with existing methods. The alternative approach would be to look at potential hot spots for action of the particular agent if enough is known for us to be able to identify such hot spots. Cases where this may eventually be possible are agents that cause very specific kinds of tumors such as acetyl-aminofluorene, an aromatic hydrocarbon that is a selective hepatic carcinogen.

The second scenario is a more favorable case, at least from the limited point of view of DNA analysis. Here one wants to look for mutations in a large number of people at risk (or a large number of sperm at risk) by exposure to a particular agent. An example may be to screen for the genetic effects of depletion of the ozone in the earth's atmosphere. Here a sensible approach is to design assays around regions that are easily tested in large numbers of different samples. Then one can apply these assays to a large population of individuals (or sperm). A risk in this approach is that the region selected may not be representative of the genome as a whole. Some portions of the genome are known to be mutation hot spots, such as VNTRs, and the mitochondrial D loop (origin of replication) because there are no genes there. Many other regions are likely to be identified as we learn more about both the sequence of the human genome and the molecular mechanisms of mutagenesis.

Two types of easy assays can be imagined. In the first strategy, one needs to find a region of DNA that is homozygous in a male, or in both parents. Then a mixture of DNA from parent and child (or sperm) is melted and reannealed. Any heterozygotes are purified away from perfect DNA duplexes. These heterozygotes must represent new mutations. This can now be tested more conclusively by sequence comparisons of the DNAs of interest. The important feature is that a physical purification step is used to examine a very complex mixture of DNA species simultaneously and select just a small fraction of it for subsequent analysis. In this way one begins to approach the ability to handle the amounts of DNA needed to see effects in the $10^{-8}$ range.

An alternative approach which is potentially very powerful but is only applicable to very particular regions of the genome is shown in Figure 13.22. Here PCR is used to assay for mutations in a restriction enzyme recognition site. Amplification will occur only when the site is not intact. By starting with a set of DNA sequences that contain restriction sites in between PCR primers, only DNA that contains mutations will be amplified.

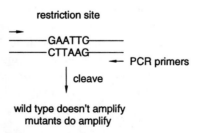

**Figure 13.22**  A potentially very sensitive PCR method for detecting a mutation in the site recognized by a restriction endonuclease.

This is a most effective way to purify the DNA of interest. However, the difficulty with this approach is that in each sample one only looks at the small numbers of bases that make up a single restriction site. For the assay to be effective at very low mutation frequencies, it will probably be necessary to perform the initial PCR from a large mixture of different sequences, and this makes things potentially complex and noisy. Nevertheless, this sort of approach is methodologically quite attractive, and perhaps variants can be conceived that will be even more sensitive.

## SOURCES AND ADDITIONAL READINGS

Birkenkamp, K., and Kemper, B. 1995. In vitro processing of heteroduplex loops and mismatches by endonuclease VII. *DNA Research* 2: 9–14.

Broude, N. E., Chandra, A., and Smith, C. L. 1997. Differential display of genome subsets containing specific interspersed repeats. *Proceedings of the National Academy of Sciences USA* 94: 4548–4553.

Buckler, A. J., Chang, D. D., Graw, S. L., Brook, J. D., Haber, D. A., Sharp, P. A., and Houseman, D. E. 1991. Exon amplification: A strategy to isolate mammalian genes based on RNA splicing. *Proceedings of the National Academy of Sciences USA* 88: 4005–4009.

Chen, X., and Kwok, P.-Y. 1997. Template-directed dye-terminator incorporation (TDI) assay: A homogeneous DNA diagnostic method based on fluorescence resonance energy transfer. *Nucleic Acids Research* 25: 347–353.

Chi, N.-W., and Kolodner, R. D. 1994. Purification and characterization of MSH1, a yeast mitochondrial protein that binds to DNA mismatches. *Journal of Biological Chemistry* 269: 29984–29992.

Cotton, R. G. H. 1997. Slowly but surely towards better scanning for mutations. *Trends in Genetics* 13:43–46.

Duyao, M., et al. 1993. Trinucleotide repeat length instability and age of onset in Huntington's disease. *Nature Genetics* 4: 387–392.

Duyk, G. M., Kim, S., Myers, R. M., and Cox, D. R. 1990. Exon trapping: A genetic screen to identify candidate transcribed sequences in cloned mammalian genomic DNA. *Proceedings of the National Academy of Sciences USA* 87: 8995–8999.

Fu, Y.-H., Kuhl, D. P. A., Pizzuti, A., Pieretti, M., Sutcliffe, J. S., Richards, S., Verkerk, A. J. M. H., Holden, J. J. A., Fenwick Jr., R. G., Warren, S. T., Oostra, B. A. Nelson, D. L., and Caskey, C. T. 1991. Variation of CGG repeat at the fragile X site results in genetic instability: Resolution of the Sherman paradox. *Cell* 67: 1047–1058.

Ganguly, A., and Prockop, D. J. 1990. Detection of single-base mutations by reaction of DNA heterduplexes with a water-soluble carbodimide followed by primer extension: Application to products from the polymerase chain reaction. *Nucleic Acids Research* 18: 3933–3939.

Hacia, J. G., Brody, L. C., Chee, M. S., Fodor, S. P. A., and Collins, F. S. 1996. Detection of heterozygous mutations in BRCA1 using high density oligonucleotide arrays and two-colour fluorescence analysis. *Nature Genetics* 14: 447.

Hamaguchi, M., Sakamoto, H., Tsuruta, H., Sasaki, H., Muto, T., Sugimura, T., and Terada, M. 1992. Establishment of a highly sensitive and specific exon-trapping system. *Proceedings of the National Academy of Sciences USA* 89: 9779–9783.

Hattori, M., Shibata, A., Yoshioka, K., and Sakaki, Y. 1993. Orphan peak analysis: A novel method for detection of point mutations using an automated fluorescence DNA sequencer. *Genomics* 15:415–417.

Ketterling, R. P., Vielhaber, E., and Sommer, S. S. 1994. The rates of G:C → T:A → C:G transversions at CpG dinucleotides in the human factor IX gene. *American Journal of Human Genetics* 54: 831–835.

Khrapko, K., Hanekamp, J. S., Thilly, W. G., Belenkii, A., Foret, F., and Karger, B. L. 1994. Constant denaturant capillary electrophoresis (CDCE): A high resolution approach to mutational analysis. *Nucleic Acids Research* 22: 364–369.

Knight, S. J. L., Flannery, A. V., Hirst, M. C., Campbell, L., Christodoulou, Z., Phelps, S. R., Pointon, J., Middleton-Price, H. R., Barnicoat, A., Pembrey, M. E., Holland, J., Oostra, B. A., Bobrow, M., and Davies, K. E. 1993. Trinucleotide repeat amplification and hypermethylation of a CpG island in *FRAXE* mental retardation. *Cell* 74: 127–134.

Lishanski, A., Ostrander, E. A., and Rine, J. 1994. Mutation detection by mismatch binding protein, MutS, in amplified DNA: Application to the cystic fibrosis gene. *Proceedings of the National Academy of Sciences USA* 91: 2674–2678.

Liu, Q., and Sommer, S. S. 1995. Restriction endonuclease fingerprinting (REF): A sensitive method for screening mutations in long, contigous segments of DNA. *BioTechniques* 18: 470–477.

Lovett, M. 1994. Fishing for complements: Finding genes by direct selection. *Trends in Genetics* 10: 352–357.

Miyoshi, Y., Nagase, H., Ando, H., Horii, A., Ichii, S., Nakatsuru, S., Aoki, T., Miki, Y., Mori, T., and Nakamura, Y. 1992. Somatic mutations of the APC gene in colorectal tumours: Mutation cluster in the APC gene. *Human Molecular Genetics* 1: 229–233.

Nawroz, H., Koch, W., Anker, P., Stroun, M., and Sidransky, D. 1996. Microsatellite alterations in serum DNA of head and neck cancer patients. *Nature Medicine* 2: 1035–1037.

Oliveira, R. P., Broude, N. E., Macedo, A. M., Cantor, C. R., Smith, C. L., and Pena, S. D. 1998. Probing the genetic population structure of *Trypanosoma cruzi* with polymorphic microsatellites. *Proceedings of the National Academy of Sciences USA* 95: 3776–3780.

Pena, S. D. J., Barreto, G., Vago, A. R., De Marco, L., Reinach, F. C., Dias Neto, E., and Simpson, A. J. G. 1994. Sequence-specific "gene signatures" can be obtained by PCR with single specific primers at low stringency. *Proceedings of the National Academy of Sciences USA; Biochemistry* 91: 1946–1949.

Petruska, J., Arnheim, N., and Goodman, M. F. 1996. Stability of intrastrand hairpin structures formed by the CAG/CTG class of DNA triplet repeats associated with neurological diseases. *Nucleic Acids Research* 24: 1992–1998.

Richards, R. I., and Sutherland, G. R. 1992. Dynamic mutations: a new class of mutations causing human disease. *Cell* 70: 709–712.

Sarkar, G., Yoon, H.-S., and Sommer, S. S. 1992. Dideoxy fingerprinting (ddF): A rapid and efficient screen for the presence of mutations. *Genomics* 13: 441–443.

Schalling, M., Hudson, T. J., Buetow, K. H., and Housman, D. E. 1993. Direct detection of novel expanded trinucleotide repeats in the human genome. *Nature Genetics* 4: 135–138.

Smith, J., and Modrich, P. 1996. Mutation detection with MutH, MutL, and MutS mismatch repair proteins. *Proceedings of the National Academy of Sciences USA* 93: 4374–4379.

Snell, R. G., MacMillan, J. C., Cheadle, J. P., Fenton, I., Lazarou, L. P., Davies, P., MacDonald, M. E., Gusella, J. F., Harper, P. S., and Shaw, D. J. 1993. Relationship between trinucleotide repeat expansion and phenotypic variation in Huntington's disease. *Nature Genetics* 4: 393–397.

Sommer, S. S. 1992. Assessing the underlying pattern of human germline mutations: Lessons from the factor IX gene. *FASEB Journal* 6: 2767–2774.

Soto, D., and Sukumar, S. 1992. Improved detection of mutations in the p53 gene in human tumors as single-stranded conformation polymorphs and double-stranded heteroduplex DNA. *PCR Methods and Applications* 2: 96–98.

Sutherland, G., and Richards, R. I. 1995. Simple tandem DNA repeats and human genetic disease. *Proceedings of the National Academy of Sciences USA* 92: 3636–3641.

Wahlberg, J., Albert, J., Lundeberg, J., Cox, S., Wahren, B., and Uhlen, M. 1992. Dynamic changes in HIV-1 quasispecies from azidothymidine (AZT)-treated patients. *FASEB Journal* 6: 2843–2847.

Wells, R. D. 1996. Molecular basis of genetic instability of triplet repeats. *Journal of Biological Chemistry* 271: 2875–2878.

Youil, R., Kemper, B. W., and Cotton, R. G. H. 1995. Screening for mutations by enzyme mismatch cleavage with T4 endonuclease VII. *Proceedings of the National Academy of Sciences USA* 92: 87–91.

Yu, A., Dill, J., Wirth, S. S., Huang, G., Lee, V. H., Haworth, I. S., and Mitas, M. 1995. The trinucleotide repeat sequence d(GTC)$_{15}$ adopts a hairpin conformation. *Nucleic Acids Research* 23: 2706–2714.

# 14 Sequence-Specific Manipulation of DNA

## EXPLOITING THE SPECIFICITY OF BASE-BASE RECOGNITION

In this chapter various methods will be described that take advantage of the specific recognition of DNA sequences to allow analytical or preparative procedures to be carried out on a selected fraction of a complex DNA sample. For example, one can design chemical or enzymatic schemes to cut at extremely specific DNA sites, to purify specific DNA sequences, or to isolate selected classes of DNA sequences. Methods have been developed that allow the presence of repeated DNA sequences in genomes to be used as powerful analytical tools instead of serving as roadblocks for mapping and DNA sequencing. Other methods have been developed that allow the isolation of DNAs that recognize specific ligands. Finally a large number of programs are underway to explore the direct use of DNA or RNA sequences as potential drugs.

In almost all of the objectives just outlined, a fundamental strategic decision must be made at the outset. If the DNA target of interest can be melted without introducing unwanted complications, then the single-stranded DNA sequence can be read, directly, and the full power of PCR can usually be brought to bear to assist in the manipulation of the DNA target. PCR has been well described in Chapter 4, and there is no need to re-introduce the principles here. In those cases where it is not safe or desirable to melt the DNA, alternative methods are needed. Such cases include working with very large DNA, which will break if melted, and working in vivo. Here a very attractive approach is to use DNA triplexes that are capable of recognizing the specific sequence of selected portions of an intact duplex DNA. Triplexes may not have been encountered by some readers before, and so their basic properties will be described before their utility is demonstrated.

## STRUCTURE OF TRIPLE-STRANDED DNA

Unanticipated formation of triple-stranded DNA helices was a scourge of early experiments with model DNA polymers. Most of the first available synthetic DNAs were homopolymers like poly dA and poly dT. Contamination of samples or buffers with magnesium ion was rampant. DNAs love to form triplexes under these conditions, if the sequence permits it. Many homopolymeric or simple repeating sequences can form triple-stranded complexes consisting of two purine-rich and one pyrimidine-rich strand or one purine-rich and two pyrimidine-rich strands, depending on the conditions. This is true for DNAs, RNAs, or DNA-RNA mixtures. Eventually conditions were found where the unwanted formation of these triplexes could be suppressed. The whole issue was forgotten and lay dormant for more than a decade. Triplexes were rediscovered, under much more

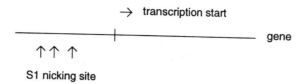

**Figure 14.1**    Appearance of S1 nuclease hypersensitive sites upstream from the start of transcription of some genes.

interesting circumstances when a decade ago investigators began to explore the chromatin structure surrounding active genes.

The key observation that led to a renaissance of interest in triplexes is a phenomenon called S1 hypersensitivity. S1 nuclease is an enzyme that cleaves single-stranded DNA specifically, usually at slightly acidic pH. It will not cleave double strands; it will not even cleave a single-base mismatch efficiently, although it will cut at larger mismatches. Investigators were using various nucleases to examine the accessibility of DNA segments near or in genes as a function of the potential for gene expression in particular tissues. Unexpectedly, many genes showed occasional sites where S1 could nick one of the DNA strands, upstream from the start of transcription, quite efficiently (Fig. 14.1). The phenomenon was termed S1 hypersensitivity. Its implication was that some unusual structure must exist in the region, rendering the normal duplex DNA susceptible to attack. To identify the sequences responsible for S1 hypersensitivity, upstream sequences were cloned and tested for S1 sensitivity. Fortunately they were initially tested within the plasmids used for cloning. These plasmids were highly supercoiled, and S1 hypersensitive sites were found and rapidly localized to complex homopurine stretches like the example shown in Figure 14.2. The S1 nicks were found to lie predominantly on the purine-rich strand. It was soon realized that the S1 hypersensitivity, under the conditions used, required a supercoiled target. The effect was lost when the plasmid was linearized, even by cuts far away from the purine block.

The problem that remained was to identify the nature of the altered DNA structure responsible for S1 hypersensitivity. The dependence of cleavage on a high degree of superhelicity implied that the sites must, overall, be unwound relative to the normal B DNA duplex. Obvious possibilities were melted loops, left-hand helix formation, or cruciform extrusion (formation of an intramolecular junction of four duplexes like the Holliday structure illustrated in Chapter 1). None of these, however, were consistent with the particular DNA sequences that formed the S1 hypersensitive sites, and none could explain why only the purine strand suffered extensive nicking. The key observation that resolved this dilemma was made by Maxim Frank-Kamenetskii, then working in Moscow. He noted that there was a direct correlation between the amount of supercoiling needed to reveal the S1 hypersensitivity and the pH used for the S1 treatment. A quantitative analysis of this effect indicated that both the amount of unwinding that occurred when the S1 hypersensitive site was created, and the number of protons that had to be bound during this

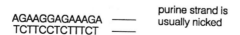

**Figure 14.2**    DNA sequence of a typical S1 hypersensitive site.

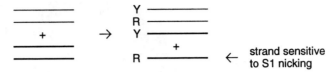

**Figure 14.3**   Formation of a DNA triplex by disproportionation of two homopurine-homopyrimidine duplexes.

process could be explained by a simple model, which involved the formation of a specific intramolecular pyrimidine-purine-pyrimidine (YRY) triple helix.

It is easiest to examine intermolecular triplex formation before considering the ways in which such structures might be formed intramolecularly at the S1 hypersensitive site. Figure 14.3 illustrates a disproportionation reaction between two duplexes that results in a triplex and a free single strand. This is precisely the sort of reaction that occurred so frequently in early studies with DNA homopolymers and led to the presence of unwanted DNA triplexes. If such a reaction is assayed by S1 sensitivity, disproportionation by the appearance of a single-stranded polypurine would be detected. The corresponding possible intramolecular reactions are illustrated in Figure 14.4. Here a block of homopyrimidine sequence folds back on itself (spaced by a short hairpin) to make an intramolecular triplex; the remaining homopurine stretch, not involved in the triplex, is left as a large single-stranded loop. It is this loop that is the target for the S1 nuclease. The net topological effect is an unwinding of roughly half of the homopurine-homopyrimidine duplex stretch. This is consistent with what is seen experimentally. In order to form base triplets between two pyrimidines and one purine, a T:A–T complex can form directly, but a CH$^+$:G–C complex requires protonation of the N$_3$ of one C, as shown in Figure 14.5.

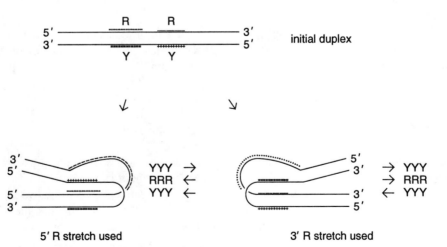

**Figure 14.4**   Two intramolecular routes for formation of triplexes from a long homopurine-homopyrimidine duplex. The structure shown on the right is the one consistant with a large body of available chemical modification data. *Y* and *R* refer, respectively, to homopyrimidine and homopurine tracts.

CH + : G-C                                        T : A-T

(a)

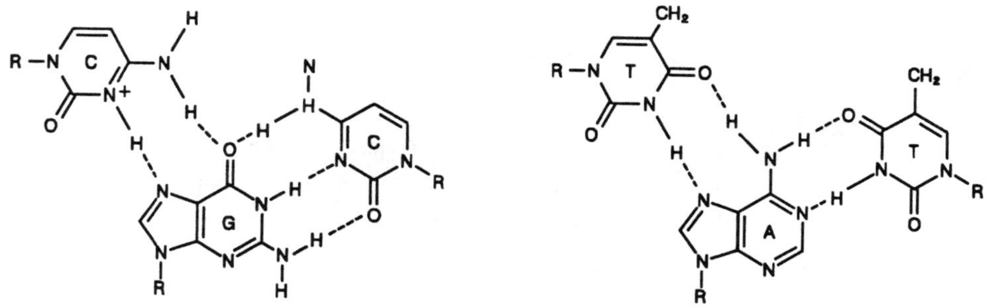

(b)

**Figure 14.5**  Acid-stabilized triplex base pairing schemes. A dash (-) indicates the normal Watson-Crick base pairing scheme while a colon (:) indicates the base pairs which involve the third strand. (a) As written schematically. (b) Actual proposed structures.

There are two possible isomeric models consistent with the unwinding and pH dependence of formation of the YRY triple strand. In both the two pyrimidine strands run antiparallel to each other; the Watson–Crick pyrimidine strand is antiparallel to the purine strand; the triplex pyrimidine strand is parallel to the purine strand. The specific structural models proposed require that the pyrimidine sequences have mirror symmetry. This can be tested by manipulating particular DNA sequences, and it turns out to be valid. In addition the two models in Figure 14.4 can be evaluated by looking at the pattern of accessibility of the S1 hypersensitive structure to various agents that chemically modify DNA in a structure-dependent manner. These studies reveal that the correct model for the S1 hypersensitive structure is the one shown on the right in Figure 14.4, where the 3' segment of the purine stretch is the one incorporated into the triplex. The reason why this structure predominates is not known.

With the principles of triplex formation in S1 hypersensitive sites understood, a number of research groups began to explore the properties of simple linear triplexes more systematically. The need for superhelical density to drive the formation of triplex can be avoided simply by working at a low enough pH. In practice, pH 5 to 6 suffices for most sequences capable of forming triplexes at all. A surprise was the remarkable stability of triplexes. They can survive electrophoresis, even with lengths as small as 12. Proof that the third strand lies in the major groove of the Watson–Crick duplex, and that the third, pyrimidine, strand is antiparallel to the Watson–Crick pyridine strand, was obtained by an elegant series of experiments in which agents were attached to the ends of the third strand that were capable of chemically nicking bases on the duplex. The specific pattern of nicks provided a detailed picture of the structure of the complex (Fig. 14.6).

A second type of DNA triplex, stable at pH 7 was soon rediscovered. This purine-purine-pyrimidine (RRY) was precisely the form known two decades before, stabilized

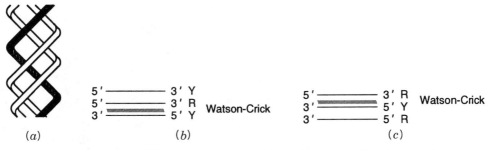

**Figure 14.6** Structure of DNA triplexes determined from chemical modification experiments. *(a)* The third strand lies in the major groove of the duplex helix. *(b)* Strand directions in structures with two pyrimidine strands and one purine strand. *(c)* Strand directions in structures with two purine strands and one pyrimidine strand. Shading in *(b)* and *(c)* indicates the Watson-Crick duplex.

by $Mg^{2+}$ or other polyvalent cations. This structure could also lead to an S1 hypersensitive site in supercoiled plasmids. However, here the pyrimidine-rich strand was nicked by the enzyme instead of the purine-rich strand. Some DNA sequences can actually form both types of triplexes, depending on the conditions. The structure of the S1 hypersensitive sites favored by divalent ions is shown in Figure 14.7. This particular isomer is the one consistent with the observed pattern of modification with various chemical agents that react with DNA covalently. Three types of base triples can be accommodated in this structure: G:G−C, A:A−T, and T:A−T. Their patterns of hydrogen bonding are shown in Figure 14.8. The two non−Watson−Crick base-paired strands in these complexes are antiparallel; this is supported by studies on particular DNA sequences. As in the type of triplex described earlier, the third strand, in this case an additional purine or an additional pyrimidine strand, lies in the major groove of the Watson−Crick duplex (Fig. 14.6). Studies using circular oligonucleotides can help confirm assignments about the direction of strands in triplexes (Kool, 1995).

More complex triple helices can also be made. An example is shown in Figure 14.9. Here all three strands must contain blocks of alternating homopurine and homopyrimidine sequences. The third strand lies down in the major groove of the Watson−Crick duplex, and alternate blocks made triplexes with two pyrimidine and one purine strand and triplexes with one pyrimidine and two purine strands. As our knowledge of triplex structures increases, and as base analogs are tested, it will undoubtedly be possible to design a wealth of triplexes in which a third strand can be used to recognize a wide variety of DNA duplex sequences. Based on experience to date, these triplexes are likely to be quite stable. One strong caveat to using them in various biological applications must be noted. The kinetics of triplex formation and dissociation are very slow, much slower than the rates of corresponding processes in duplexes.

**Figure 14.7** Intramolecular triplex structure formed at neutral pH in the presence of $Mg^{2+}$ ions.

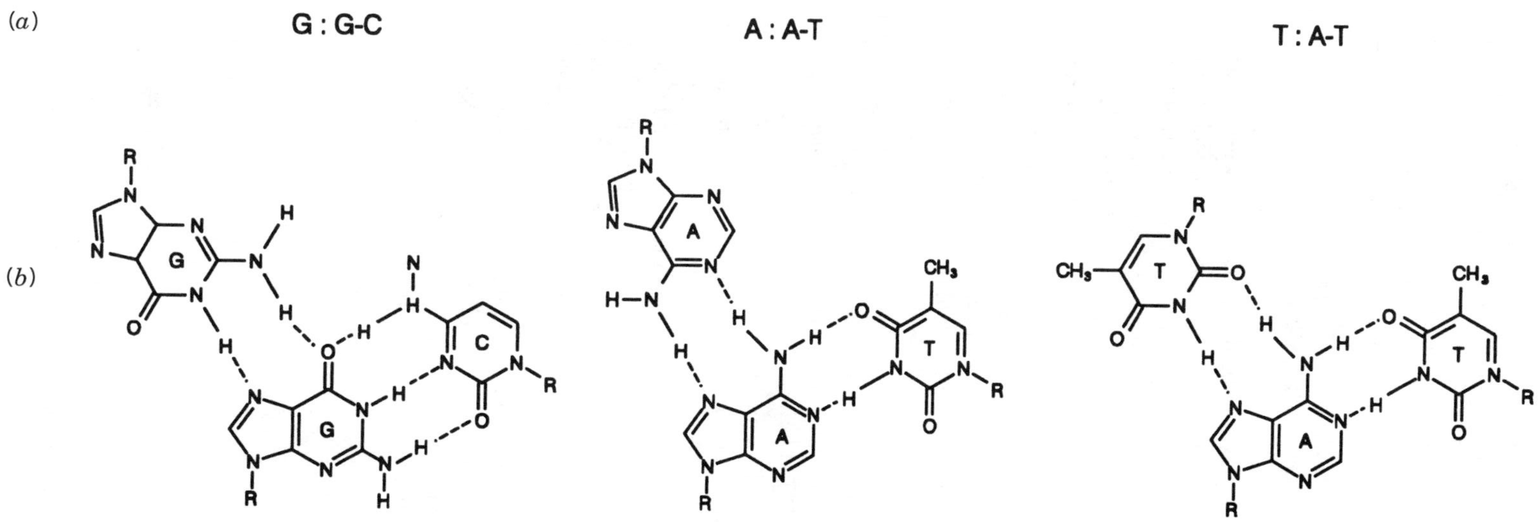

**Figure 14.8** Triplex base pairing schemes favored by $Mg^{2+}$ ions. (*a*) As written schematically. (*b*) Actual proposed structures.

G : G-C

A : A-T

T : A-T

(*a*)

(*b*)

475

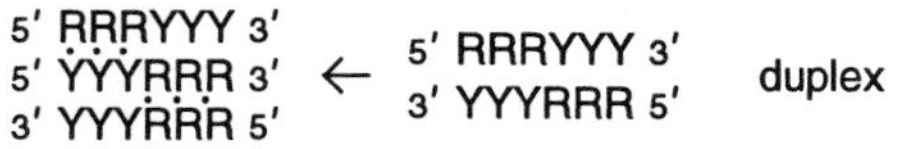

5′ RRRYYY 3′
5′ ẎẎẎRRR 3′   ←   5′ RRRYYY 3′     duplex
3′ YYYṚṚṚ 5′         3′ YYYRRR 5′

**Figure 14.9**   An example of a more complex triplex structure formed by alternating blocks of purines and pyrimidines. The third strand lies in the major groove of the Watson-Crick duplex. Its interactions with the duplex are indicated by dots.

## TRIPLEX-MEDIATED DNA CLEAVAGE

The first application of triplexes to be discussed is their use in recognizing particular duplexes and rendering these susceptible to specific chemical or enzymatic cleavage. This potential was already described briefly in the previous section when chemical derivatives of the third strand were used to help analyze the structure of the triplex. The appeal of this approach is that it will be relatively easy to find or introduce a unique DNA sequence capable of forming triple strands into a target of interest. Subsequent cleavage at this sequence would represent the sort of cut that is extremely useful for executing any of the Smith-Birnstiel-like mapping strategies we described in Chapter 8.

Chemical cleavage agents that have been tried include Cu-phenanthroline complexes, iron-EDTA-ethidium bromide complexes, and others shown elsewhere in the chapter. The types of reactions one would like to be able to carry out with these modified oligonucleotides are shown schematically in Figure 14.10. Rather good yields and specificities have been observed when the chemical cleavage is used to cut the complementary strand of a duplex (Fig. 14.10a). Much less success has been had with direct triplex-mediated cleavage of a duplex (Fig. 14.10b). Generally, nicking of one strand of the duplex proceeds very well, but it is difficult to make the second cut needed to affect a true double-strand cleavage. The reason for this is that many of the chemical agents used are stoichiometric rather than catalytic. They have to be reactivated or replaced by a fresh reagent in order to be able to perform a second strand cleavage. While elegant chemical methods have been proposed to circumvent this problem, to date, specific efficient duplex chemical cleavage has been an elusive goal. However, this has not proved to be a serious roadblock, because alternative methods for using triplexes to promote specific enzymatic cleavage of duplexes have been very successful.

Achilles's heel strategies are based on the general notion of using restriction methylases to protect all except a single or small set of protected recognition sites. This renders most of the potential sites in the sample resistant to the conjugate restriction nuclease. Then the protecting agent is removed, and the nuclease added. Cleavage only occurs at

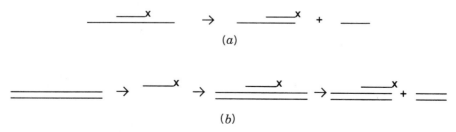

**Figure 14.10**   Triplex-mediated DNA cleavage using a DNA strand containing a chemically reactive group *(x)* that generates radicals. *(a)* Cleavage of a single strand. *(b)* Cleavage of a duplex.

the sites that escaped the initial protection. These strategies were named by one of their developers, Wacslaw Szbalski, at the University of Wisconsin, by analogy to the myth of Achilles. As an infant Achilles was dipped into the river Styx by his mother which rendered him immune to all physical harm except for his heel, which was masked from the effects of the Styx because his mother was holding him there. Two straightforward examples of Achilles's heel specific cleavage of DNA are shown in Figure 14.11. These approaches are applicable to very large DNA or even intact genomic DNA, since they can be carried out in situ in agarose. In one case it is necessary to find a tight binding site for a protein that masks a restriction site (Fig. 14.11a). Examples of suitable protein and binding sites are lac repressor, lambda repressor, E. coli lexA protein, or a host of eukaryotic transcription factors, particularly viral factors like the NFAT protein. To be useful, the site must contain an internal or nearby flanking restriction site. There is no guarantee that such a site will conveniently exist in a target of interest. However, given the large number of potentially useful restriction sites, there are many possibilities. If necessary, for some applications the desired site can always be designed and introduced.

Instead of proteins, triplexes can be used to mask restriction sites (Fig. 14.11b). It turns out that embedding a four-base restriction site in a homopurine-homopyrimidine

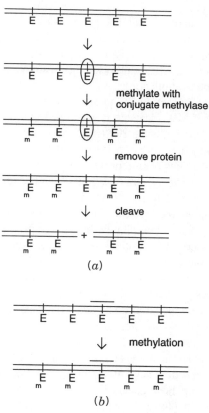

**Figure 14.11** Achilles's heel strategies for specific DNA cleavage. *(a)* Blocking a restriction enzyme cleavage site *E* with a DNA binding protein. *(b)* Blocking a restriction enzyme cleavage site *E* with a triplex. *M* indicates methylation sites.

stretch destabilizes the resulting triplex only slightly. However, triplex formation renders the site totally unaccessible to the restriction methylase. After the remaining restriction sites have been methylated, conditions are altered to dissociate the triplex. Then the restriction enzyme is added, and cleavage is allowed to occur. This Achilles's heel approach works very well, even at the level of single sites in the human genome. However, it still suffers from the limitation that only a small subset of sequences within a target will be potential sites of triplex-mediated specific cleavage.

A generalization of the Achilles's heel approach is possible by the use of the *E. coli* recA protein. Developed by Camerini-Otero and elaborated by Szybalski (Koob et al., 1992), this method has been called recA-assisted restriction endonuclease (RARE) cleavage (Fig. 14.12). The method is applicable to genomic DNA because all of the steps can be carried out in agarose. The recA protein has a number of different activities. One of these is a cooperative binding to single-stranded DNA, leading to a completely coated complex containing about one recA monomer for every five bases (Fig. 14.12*a*). The coated complex will then interact with double-stranded DNA molecules in a search for sequences homologous to the single strand. In *E. coli* this process constitutes one of the early steps that eventually leads to strand invasion and recombination. In the test tube, without accessory nucleases, the reaction stops if a homologous duplex sequence is found, and the third strand remains complexed to this homolog, even if the recA protein is subsequently removed (Fig. 14.12*b*). The mechanism of the sequence search is unknown. Similarly the actual nature of the complexes formed with recA protein present or after recA protein removal are still not completely understood despite intense efforts to study these processes because of their importance in basic *E. coli* biology. Some sort of triple strand is believed to be involved, although this has never been proven. What is key, however, for Achilles's heel applications, is that the recA protein-mediated complex blocks the access of restriction methylases to duplex DNA sequences contained within it.

A schematic outline of RARE cleavage is given in Figure 14.12*c*. The technique has worked well to cut at two selected sites 200 to 500 kb apart in a target to generate a specific internal fragment, and generation of a 1.3-Mb telomeric DNA fragment that requires only a single RARE cleavage has been reported. In practice, it has been more efficient to use a six-base specific restriction system like rather than the four-base systems used with other Achilles's heel methods. The reason is that a common source of background in these approaches is incomplete methylation. This produces a diverse distribution of hemi-methylated sites which are cut, albeit slowly, by the conjugate nuclease. The result is a significant background of nonspecific cleavage. This background can be markedly reduced by going to the six-base enzyme, since its sites are 16 times less frequent, on average. Since recA-mediated cleavage is applicable, in principle, to any selected DNA sequence, the rarity of six-base cleavage sites does not pose a particular obstacle.

The recA protein-coated single strands can be as short as 15 bases for RARE cleavage, although in practice targets two to four times this length are usually employed. One makes a trade-off between the increased efficiency and specificity obtained with longer complexes, and the lowered efficiency of their diffusion into agarose-embedded DNA samples. Yields of the desired duplex of 40% to 60% have been reported in early experiments. It remains to be seen how generally obtainable such high yields will be. The power of RARE cleavage in physical mapping is that given two DNA probes spaced within about 1 Mb, RARE cleavage should provide the DNA between these probes as a unique fragment free from major contamination by the remainder of the genome.

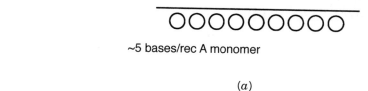

~5 bases/rec A monomer

(*a*)

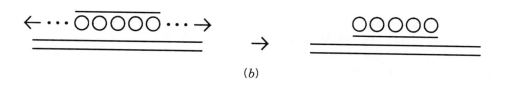

(*b*)

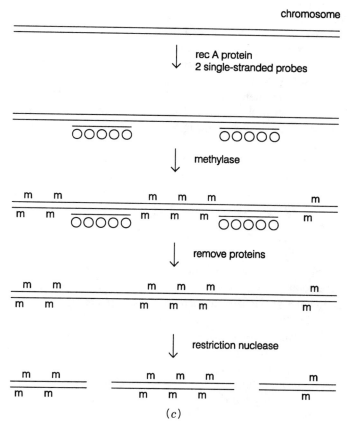

(*c*)

**Figure 14.12** RARE cleavage of DNA. *(a)* Complex formed between *E. coli* recA protein and single-stranded DNA. *(b)* Complex between a recA-protein coated DNA strand and the homologous sequence in duplex DNA. *(c)* Outline of the procedure used to generate a large DNA fragment between two known sequences by RARE cleavage.

## SEQUENCE-SPECIFIC DNA CAPTURE

In the majority of cases, the motivation behind attempts to develop ways for sequence-specific DNA cleavage is to provide access to particular DNA sequences, as just described above for RARE cleavage. A more direct approach would be to develop methods for the specific capture of DNA sequences of particular interest. There are a number of different approaches that have been used to purify DNAs based on sequence rather than size. Biological methods we have already described include PCR or differential cloning. These are extremely effective, but because the steps are usually carried out in solution and involve a number of enzymes, the targets of such purifications are usually limited in size. Large insert cloning systems have been described, but these are not yet applicable to differential cloning. For example, YAC cloning can be used to immortalize a set of large DNA fragments, but then this set has to be screened to identify the clone containing a particular sequence of interest. Such screening can be quite time-consuming. Instead of screening, we are concerned here with methods that select the target of interest. Purification techniques are a natural way to select molecules with desired properties such as specific DNA sequences.

In principle, a DNA sequence could be selected by cloning, by PCR, by binding to a specific protein, by binding to the complementary strand to form a duplex, or by binding to a third strand to form a triplex. In each of these approaches the desire is a simple sequence-specific purification method. In this section we will limit our attention to physical purifications based solely on DNA base interactions. Such methods have been called in the past, perhaps somewhat inelegantly, DNA fishing. Differential PCR and cloning methods will be described later in the chapter. The advantage of the pure, physical methods is that if successful, they are very easy to implement and easy to scale up to large numbers of samples or large quantities. A key obstacle in the use of such methods is that once the DNA target is captured by a probe, there either must be an efficient way to release it, or one has to have an efficient way of working directly with the immobilized DNA (as in solid state DNA sequencing, described earlier in Chapter 9).

## TRIPLEX-MEDIATED DNA CAPTURE

The major advantage in using a third strand to capture a DNA duplex is that there is no need to melt the duplex. This avoids potential damage arising from any preexisting nicks in the DNA strands. It also helps avoid many of the complications caused by the extensive interspersed repeats in the DNA of higher organisms. There are a number of potential applications for triplex capture of DNA. For example, triplex formation could be used, in principle, to isolate from a library all those clones that contained specific homopurine-homopyrimidine sequences such as $(AG)_n$, $(AAG)_n$, or $(AAAG)_n$. These clones are all potentially useful genetic probes, since all are VNTRs. Such clones have been isolated in the past by screening libraries by hybridization with the repeated seqeunce. However, if one is interested in a large set of such clones, each one detected by hybridization has to be picked by hand for future use. It would be far simpler to isolate the entire set of VNTR-containing clones in a single physical purification step. A survey of simple repeated sequence motifs in the GenBank database several years ago is shown in Table 14.1. This indicates that a significant fraction of all simple sequence VNTRs known to date could be amenable to triplex-mediated capture because they involve homopurine repeats.

**TABLE 14.1    Repeating Single Sequence Motifs in the Genbank Database (1991)**

| | | | | | | |
|---|---|---|---|---|---|---|
| Dinucleotide | AG | 24 | | | | |
| | AT | 18 | | | | |
| | AC | 86 | | | | |
| | TG | 1 | | | | |
| Trinucleotide | AAC | 14 | | | | |
| | CCG | 12 | | | | |
| | AGG | 9 | | | | |
| | AAT | 7 | | | | |
| | AGC | 10 | | | | |
| | AAG | 3 | | | | |
| | ATC | 2 | | | | |
| | ACC | 2 | | | | |
| Tetranucleotide | AAAG | 20 | AGCC | 1 | AACT | 1 |
| | AAAT | 22 | AGAT | 9 | ACAT | 1 |
| | ACAG | 3 | ATAG | 3 | ACGC | 1 |
| | ACAT | 2 | AGCG | 1 | ACTG | 2 |
| | AAGG | 8 | ATCC | 8 | AGGG | 1 |
| | AAGG | 13 | AATC | 4 | AGCT | 1 |
| Lohger | AAAAG | 1 | | | | |
| | AAAAAAAAAG | 1 | | | | |
| | TTTTG | 1 | | | | |

*Source:* Adapted from results summarized by Lincoln McBride.

Note: Underlined sequences are triplex selectable.

Two other potential applications for triplex capture concern the purification of a specific DNA sequence. If a genomic fragment contains a known specific homopurine sequence, one should be able to capture this sequence away from the entire genome by triplex formation on a solid support. Alternatively, such a sequence could be introduced into a desired section of a genome to allow subsequent capture of DNA from this region. Such approaches, once they have matured, could potentially expedite the analysis of selected megabase regions of DNA considerably. The second potential use of triplex capture would be to separate clones from host cells. Here the triplex forming sequence would be built into the vector used for cloning. After cell lysis, triplex capture would be used to retain the cloned DNA and discard the DNA of the host. This has many appealing features. Once host DNA is gone, direct analysis of cloned DNA is possible without the need for radiolabeled DNA probes or specific PCR primers. Host DNA leads to considerable background problems when low-copy number vectors like YACs are used. The sensitivity of hybridization or PCR analyses would be increased considerably if host DNA could be removed in advance.

A number of different ways to use triplexes to purify DNA have been explored. Although much more development work needs to be done, the preliminary findings are very promising. In our work, magnetic microbeads were used as the solid support. The power of this general approach was described earlier in applications for solid state DNA sequencing. Magnetic microbeads are available commercially that are coated with streptavidin. Streptavidin will bind specifically to virtually any biotinylated macromolecule. Any desired small DNA sequence can be prepared with a 5'-biotin by the use of biotin phosphoramidite in ordinary automated oligonucleotide synthesis. Longer 5′-biotinylated probes can be made by PCR using 5′-biotinylated primers. Alternatively, internally biotinylated DNA sequences can be made by PCR using biotinylated base analogs as dpppN substrates. Some of these compounds are incorporated quite efficiently by DNA polymerases. Once attached to the streptavidin microbeads, biotinylated DNA probes are still quite accessible for hybridization or for triplex formation. A schematic illustration of the way microbeads have been used to develop a simple triplex capture method of DNA purification is shown in Figure 14.13. The key point is that a permanent magnet can be used to hold the beads in place while supernatant is removed and exchanged. This allows very rapid and efficient exchanges of reagents. At pH 5 or 6 any duplex DNA captured by the complementary sequence on a bead will remain attached. At pH 8 the captured DNA will be readily released and removed with the supernatant.

Three different experiments have been used to test the ease and efficiency of triplex DNA capture with magnetic microbeads. In the first, an artificial mixture was made of two plasmids. One contained just a vector with no known forming capability, which included a lacZ gene. When this plasmid is transformed into *E. coli,* blue colonies are produced on the appropriate indicator plates, which contain a substrate for the beta-galactosidase product of this gene that yields a blue-colored product. The second plasmid, initially

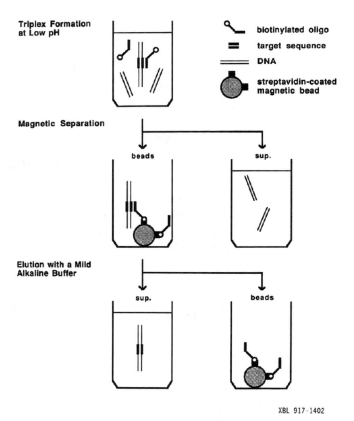

XBL 917-1402

**Figure 14.13**   Use of streptavidin-coated magnetic microbeads for affinity purification of triplex DNA. Adapted from Ito et al. (1992).

present at much lower concentrations, contained an insert of $(AG)_n$, cloned into the lacZ gene in such a way as to disrupt its translation. This results in white colonies after transformation of *E. coli*. Thus the ratio of plasmids in a mixture can be determined just by comparing the number of colonies of each color after transformation (making the reasonable assumption that the transformation efficiencies of the two almost identical plasmids are the same). When a mixture of the two types of DNA was subjected to a single cycle of purification via triplexes formed with magnetic beads containing $(TC)_n$, the result is a purification of the $(AG)_n$-containing plasmid by a factor of 140,000 fold with an overall yield of 80%, as shown in Table 14.2. Such a purification is sufficient for almost any need. However, if necessary the purity could surely be increased much further by a second cycle with the magnetic beads. The entire process took only a few hours. Recently Lloyd Smith and coworkers have demonstrated that a similar process can be carried out in only 15 minutes. G- and T-containing sequences can also be used in an analogous way to capture homopurine-homopyrimidine sequences in the presence of $Mg^{2+}$ or other polyvalent cations (Takabatake et al., 1992).

A second test of triplex-mediated capture with magnetic microbeads is shown schematically in Figure 14.14. Here the challenge was to purify $(AG)_n$-containing DNA clones away from the remainder of a chromosome 21-specific library. The original bacteriophage library was first subcloned into a plasmid vector. Magnetic purification proceeded as described in Figure 14.13. Then the purified DNA was used to transform *E. coli*. Individual colonies were picked and screened for the presence of $(AG)_n$ by PCR between each of the vector arms and the internal $(AG)_n$ sequence. A positive signal should be seen for each $(AG)_n$-containing clone with one of the two vector arms depending on the orientation of the insert (Fig. 14.15). Some of the actual results obtained are shown in Figure 14.16. The effectiveness of the procedure is quite clear. Overall, 17 of the first 18 colonies tested showed vector-insert PCR signals, and each of these was different, implying that different genomic $(AG)_n$-containing clones had been selected.

**TABLE 14.2  Triplex-Mediated Purification of Target Plasmids from a Reconstituted Library**

| Number of Colonies (%) | |
| --- | --- |
| White (pTC45) | Blue (pUC19) |
| *Before Enrichment* | |
| $5.0 \times 10^4$ | $1.1 \times 10^7$ |
| (0.5) | (99.5) |
| *After Enrichment* | |
| $4.0 \times 10^4$ | $0.5 \times 10^2$ |
| (99.9) | (0.1) |

*Source:* Adapted from Ito et al. (1992).

Note: Plasmids prepared from a reconstituted library were used for transformation of *E. coli* with or without enrichment by triplex affinity capture. pTC45 (target) and pUC19 give white and blue colonies, respectively, on indicator plates. In this experiment the enrichment was $1.8 \times 10^5$-fold with a recovery of $\approx 80\%$.

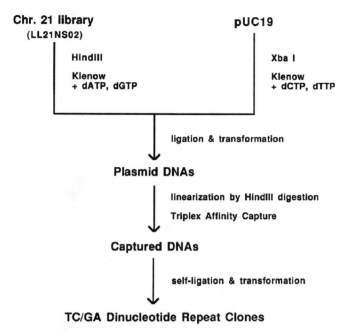

**Figure 14.14** Outlined of the procedure used to purify $(AG)_n$-containing clones from a chromosome 21 library.

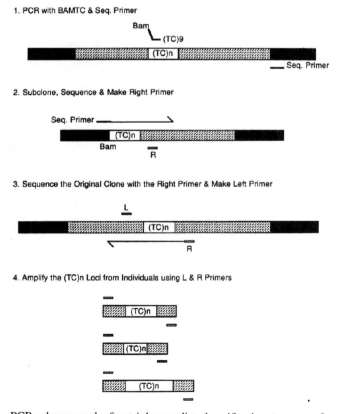

**Figure 14.15** PCR scheme used, after triplex-mediated purification, to screen for clones that contain an $(AG)_n$ insert.

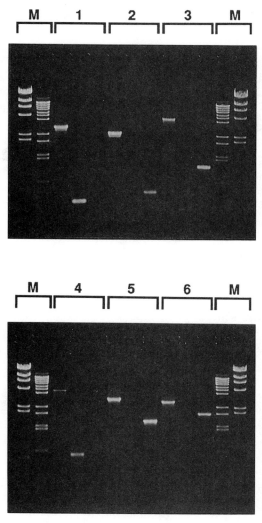

**Figure 14.16**   Results of the analysis of a number of triplex-purified clones by the PCR assay described in Figure 14.15. Adapted from Ito et al. (1992).

A third test of triplex capture took advantage of a strain of yeast that Peter Dervan and his co-workers had constructed, containing an insert of a 40-base homopurine-homopyrimidine sequence in yeast *(S. cerevisiae)* chromosome III. This strain was originally made to test the efficiency of triplex-mediated DNA cleavage. We were able to obtain and use the same strain to check the efficiency of triplex-mediated DNA capture. Instead of working with yeast genomic DNA, which would suffer considerable shear damage under the conditions used to manipulate the magnetic microbeads, we first subcloned this strain of yeast into a plasmid vector. Then plasmid DNAs were selected using magnetic microbeads containing the appropriate homopyrimidine sequence. In practice, 50% of the clones isolated contained the correct DNA insert as determined by DNA sequencing. The other contaminating clones had a similar, but not identical, sequence that was a natural component of the yeast genome. Presumably this contaminant could have been selected against more efficiently by the use of slightly more stringent conditions for triplex formation and washing.

## AFFINITY CAPTURE ELECTROPHORESIS

A major limitation with all of the magnetic microbead methods is that the shear damage generated by liquid phase handling of DNA in this way restricts targets to DNAs less than a few hundred kb in size. To work with larger DNAs, it is necessary that most or all manipulations take place in an anticonvective medium like agarose. This encourages using electrophoresis in agarose to carry DNA past an immobilized triplex capture probe (Ito et al., 1992). Such an approach has been termed affinity capture electrophoresis (ACE). To test this method, we used streptavidin-containing microbeads (no need for magnets here) embedded in agarose. A DNA sample containing a potential target for triplex formation with the sequence on the beads was loaded in a sample well and electrophoresed at pH 5 past the potential capture zone, as shown in Figure 14.17. Then the pH was raised to 8, and any material captured was released and analyzed in a subsequent electrophoresis step.

This simple procedure works, but at present, its efficiency is less than desired. The poor efficiency arises from nonspecific binding between streptavidin and DNA at pH 5. Streptavidin is a protein with an isolectric point of about pH 7. Therefore at pH 5 the protein is positively charged and, as such, binds nonspecifically to DNA. To destabliize these electrostatic interactions, high ionic strength buffers can be used. Such buffers worked very well with the magnetic separations described earlier. However, they lead to serious complications in electrophoretic procedures because the high salt buffers have high conductivity, and as a result there is considerable heating and band broadening. These problems can probably be circumvented by changing the properties of the surface of streptavidin to introduce more negative charges. Since the three-dimensional structure of the protein is known (Chapter 3), the molecular design and engineering needed to accomplish this change in charge should be relatively straightforward.

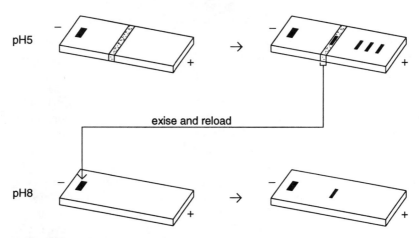

**Figure 14.17**   Schematic illustration of a procedure for ACE: Affinity capture electrophoresis. The lightly shaded portion contains gel-embedded immobilized triplex-forming oligonucleotides corresponding to the target of interest.

To avoid the problems with the currently available form of triplex-mediated ACE, an alternative capture scheme was developed that could be implemented at neutral pH. This scheme lacks the attractive generality of triplex capture because it is applicable only to the ends of DNA fragments, and it requires that information be available about the DNA sequence at one end, or at least that a clone be available that overlaps the ends of the DNA to be captured. This is a major limitation, in principle, but it is less confining in practice. A major potential application for ACE is to try to purify specific large restriction fragments from complex mixtures. If a linking library that corresponds to the ends of these fragments is available, as described in Chapter 8, then the necessary clones or sequences will already be in hand.

The basic scheme of an alternative ACE procedure, termed end-capture electrophoresis, is shown in Figure 14.18. In this scheme, the ends of a long DNA duplex are treated with an exonuclease like *E. coli* Exonuclease III, or DNA polymerase in the absence of triphosphates, to remove a small portion of the 3′-ends of the duplex and expose the complementary 5′ sequence as single strand. The affinity capture medium is made, as before, with the sequence complementary to the target. In this case the beads will contain the authentic 3′-end of the large DNA fragment, synthesized from known sequence or isolated as the appropriate half of a linking clone. Capture consists of ordinary duplex formation. The challenge is to find an efficient and nondisruptive way to release the target after it has been captured. An effective way to do this, illustrated in Figure 14.19, is to prepare the capture probe so that it contains the base dU instead of T. This still allows efficient, sequence-specific strand capture. However, subsequent treatment with Uracil DNA glycosidase, an enzyme that participates in the repair of DNA (Chapter 1), will release the captured target. The same general idea was described in Chapter 4 as a method for minimizing PCR contamination.

Figure 14.20 shows an actual example of end capture electrophoresis. Here, with DNA fragments of the order of 10 kb in size, the method works quite well. It still has not been successful in the capture of much larger targets. This may reflect the much slower rate at which these targets will find a probe during the electrophoresis. Because of excluded volume effects, the end of a large DNA will be accessible for hybridization only a small fraction of the time. If this is the problem, it should be resolvable in principle by using much slower electrophoresis rates.

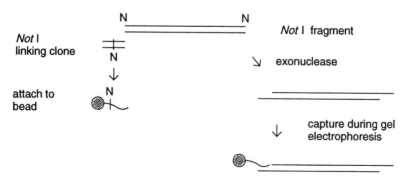

**Figure 14.18** Schematic procedure for end capture electrophoresis. The appropriate single-stranded probes needed will be easily obtained from half-linking clones (see Chapter 8).

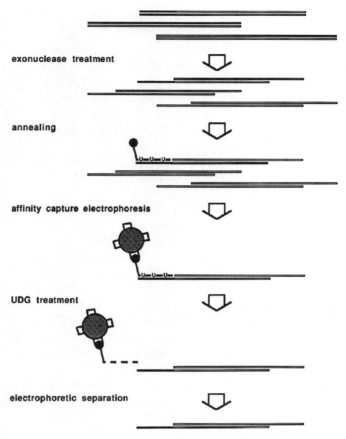

**exonuclease treatment**

**annealing**

**affinity capture electrophoresis**

**UDG treatment**

**electrophoretic separation**

**Figure 14.19** An end capture electrophoresis scheme that allows easy release of the captured DNA.

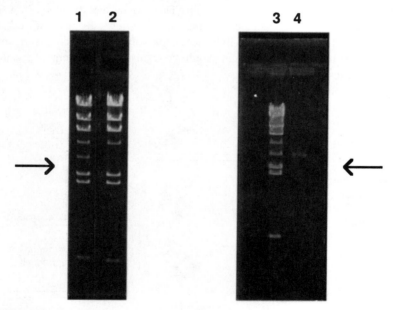

**Figure 14.20** Example of the successful implementation of the scheme shown in Figure 14.19. Adapted from Ito et al. (1992). Lane 2 (arrow) shows a band removed by capture. Lane 4 (arrow) shows elution of the captured band.

## USE OF BACKBONE ANALOGUES IN
## SEQUENCE-SPECIFIC DNA MANIPULATION

A number of backbone analogues of DNA have been described in recent years. The principle force motivating the development of such compounds is their presumed usefulness as DNA or RNA drugs. The expectation is that some of these analgoues will be resistant to intracellular enzymes that interfere with attempts to use natural RNA or DNA oligonucleotides to modulate ordinary cellular processes. We will discuss some of these efforts at the end of the chapter. One feature of a number of the backbone analogues that have been made is replacement of the highly charged phosphodiesters by neutral groups. The simplest approach is the use of phosphotriesters, in which the POO$^-$ group of the normal backbone is replaced by POO-R. The advantage conferred by such a substitution is that, in a duplex or triplex containing such an altered backbone, the natural electrostatic repulsion between the DNA strands will be minimized. Thus the resulting complexes ought to be much more stable, and one might expect that the backbone analog would preferentially form duplex or triplex at the expense of the normal strands present in a cell. This is why such compounds are attractive as potential drugs. Normal triplexes with two homopurine and one homopyrimidine strands, for example, are seriously destabilized at low-salt concentration because they have three intra-strand sets of electrostatic repulsion compared to only one for the duplex and a separated third strand (Fig. 14.21).

Phosphotriesters have a feature that considerably complicates their use. Since each phosphate now has four different substituents, as shown in Figure 14.22, it is optically active. This means that each alkyl group R can occupy one of two positions on each phosphate. The resulting number of different stereoisomers for a chain with $n$ phosphates is $2^n$. This is a very discouraging prospect for in vivo or in vitro studies, and phosphotriesters are likely to see rather limited use until the problem of synthesizing specific isomers can be solved. Until this has been accomplished, attention has focused on other backbone analogues which have the disadvantage of being less like natural nucleic acids but

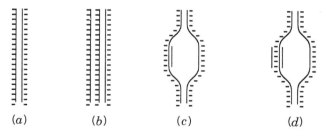

(a)          (b)          (c)          (d)

**Figure 14.21**   Electrostatic repulsion expected in various DNA structures. *(a)* Normal DNA duplex. *(b)* Normal DNA triplex. *(c)* DNA duplex with one strand displaced by an uncharged DNA analog. *(d)* DNA duplex with one strand displaced and a triplex formed at the other strand by binding of two uncharged DNA analogs.

**Figure 14.22**   Stereoisomeric pair of phosphotriesters.

which avoid the problem of multiple stereoisomers. Among the compounds that have been made and tested are polypeptide nucleic acids, in which a peptide backbone is used to replace the phosphodiester backbone, and polyamide nucleic acids (PNAs). The schematic structure of one example of this latter class of compounds is shown in Figure 14.23.

An amusing series of accidents clouded the first attempt to characterize the interactions of PNAs with duplex DNA. The actual compound used was R1-T10-R2. The notation T10 means that 10 thymine bases and backbone units were present. R1 was chosen to contain a positive charge for extra stability of interaction of the short PNA with DNA. R2 contained an intercalating acridine, also present to enhance binding stability, and a p-nitrobenzoylamide group. This latter promotes radical-induced cleavage of the DNA upon irradiation with near-UV (30 nm) light. This PNA was designed with the expectation that it would bind to a dA-dT stretch in a duplex and form a triple strand. What actually happened was more complex. A number of different chemical and enzymatic probes were used to examine the resulting PNA-DNA complex. All were consistent with the idea that the PNA had displaced the dT-containing strand of the natural duplex and formed a more stable duplex with the uncovered dA-stretch. A number of the results that led to this conclusion are shown in Figure 14.24a. The prospects raised by this outcome were extremely exciting because a duplex strand displacement mechanism would be applicable to any DNA sequence, not just to the more limited set capable of forming triplexes.

Further investigations of the properties of the PNA-DNA complex reveal an additional level of complication. It turns out that the complex contains not just one stoichiometric equivalent of PNA, but two instead. The result, as shown schematically in Figure 14.24b, is a complex in which displacement of one of the strands of the original DNA duplex has occurred, but the displaced strand is captured as a triplex with two PNAs. Thus this sort of reaction will be limited to sequences with triple-strand forming capabilities. Perhaps other backbone analogues will be found that work by displacing strands and capturing

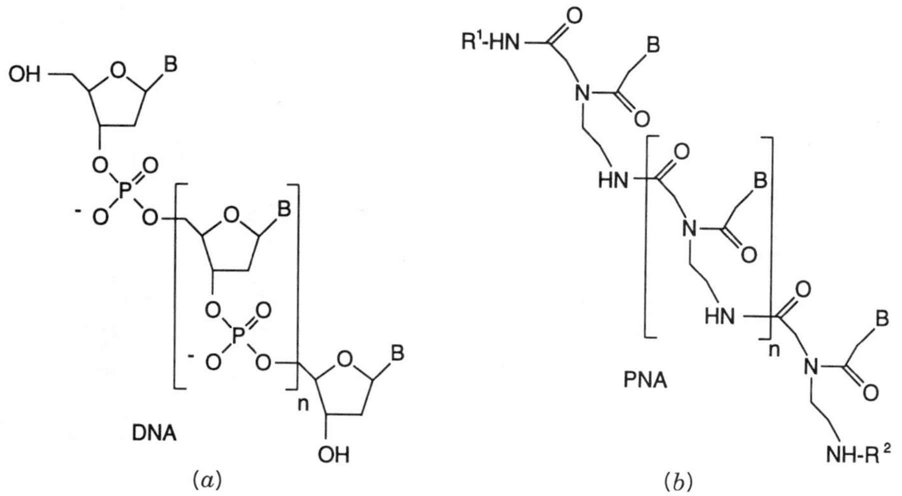

**Figure 14.23**  Chemical structures of the normal DNA backbone and polyamide nucleic acid (PNA) analogs. Adapted from Nielson et al. (1991).

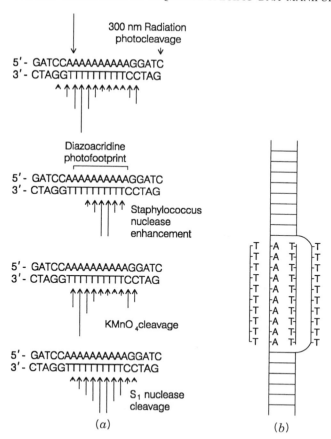

**Figure 14.24**  PNA binding to DNA. *(a)* Chemical evidence for the displacement of a (dT) 10 sequence by a corresponding PNA derivative. *(b)* The structure of the complex actually formed. Adapted from Nielson et al. (1991).

them as duplexes. An alternative scheme for in vitro work would be the use of recA protein-coated single strands. However, this is most unlikely to be useful in vivo.

In the past few years, a considerable amount of work has been done to explore the properties of PNAs and their potential usefulness in DNA analysis or clinical diagnostics. A few of these findings are briefly summarized here. The stability of PNA-DNA or DNA-PNA duplexes is essentially salt-independent (Wittung et al., 1994). Thus low salt can be used in hybridization procedures such as SBH to supress the interference caused by stable secondary structures in the target. PNAs are capable of forming sequence-specific duplexes that mimic the properties of double-stranded DNA except that the complexes are completely uncharged. Because there is no chirality in the PNA backbone, the duplexes are optically inactive; they have no preferred helical sense. However, attachment of a single chiral residue such as an amino acid at the end of the PNA strand leads to the formation of a helical duplex (Wittung et al., 1995). The ability of PNAs to bind tightly to specific homopurine, homopyrimidine duplexes leads to an effective form of Achilles's heel cleavage (Veselkov et al., 1996). Triplets that are located near restriction enzyme cleavage sites block these sites from recognition by the conjugate methylase. After removal of the triplex, the restriction nuclease will now cleave only at the sites that were previously protected, as in Figure 14.11*b*. The PNA-mediated protection appears to be quite effi-cient. A final novel use of PNAs is for hybridization prior to gel electrophoresis (Perry-O'Keefe et al., 1996). Since PNA is uncharged, it can be used to label ssDNA without in-terfering with subsequent high-resolution electrophoretic fractionations.

## SEQUENCE-SPECIFIC CLONING PROCEDURES

Instead of physical isolation of particular DNA sequences, cloning or PCR procedures can be used to purify a desired component from a complex mixture. Direct PCR is very powerful if some aspect of the target DNA is known at the sequence level (Chapter 4). Where this is not the case, less direct methods must be used. Here several procedures will be described for specific cloning based indirect information about the desired DNA se-quences to be purified. Several PCR procedures that take advantage of the possession of only a limited amount of DNA sequence information will be described later in the chapter.

Subtractive cloning is a powerful procedure that has played an important role in the search for genes. It can be carried out at the level of the full genome with much difficulty, or at the level of cDNAs with much greater ease. In subtractive cloning the goal is to iso-late components of a complex DNA sample that are missing in a similar, but not identical, sample. One strategy for doing this, which illustrates the general principles, is shown in Figure 14.25. In this case, which is drawn from the search for the gene for Duchenne muscular dystrophy (DMD), two cell lines were available. One had a small deletion in the region of the X chromosome believed to contain the gene responsible for the disease. This deletion was actually found in a patient who displayed other inherited diseases in addition to DMD (the utility of such samples was discussed in Chapter 13). The objective of the

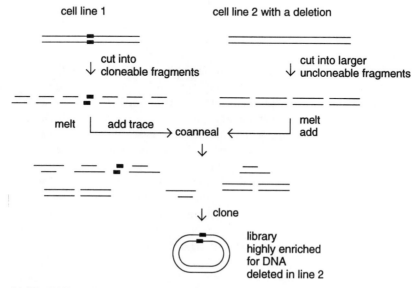

**Figure 14.25** Differential cloning scheme originally used to obtain clones corresponding to the region of the genome deleted in Duchenne muscular dystrophy.

differential cloning was to find DNA probes that derived from the region that was deleted in this patient, since these would be candidate materials for the DMD gene itself.

A small amount of DNA from a normal individual was used as the target. This was cut with a restriction enzyme to give DNA fragments with cloneable ends. A large excess of DNA from the patient with the small deletion was prepared and cut into longer fragments than the target sample. This was done with an enzyme that would not give cloneable ends in the vector ultimately used. The two samples were melted and mixed together to co-anneal. Because the normal DNA was limiting, the DNA from the sample with the deletion acted as a driver. It rapidly formed duplexes with itself, and with corresponding fragments of the normal DNA. In contrast, DNA from the region of the deletion was present at very low concentrations in the mixture, and it renatured very slowly. Once renatured, however, the resulting duplexes had cloneable ends, unlike all of the rest of the DNA fragments in the sample. The mixture was then ligated into a vector and transformed into a suitable *E. coli* host. The resulting clones were, indeed, highly enriched for DNA from the desired deletion region.

The major difficulty inherent in the scheme shown in Figure 14.25, is that the desired DNA fragments are at low concentration and form duplex very slowly and inefficiently. In fact, to achieve an acceptable yield of clones, the renaturation had to be carried out in a phenol-water emulsion, which raises the effective DNA concentration markedly. This is not an easily managed or popular approach. More recent analogs of subtractive genomic cloning have been described that look powerful, and they should be more easy to adopt to a broad variety of problems (see Box 14.1). The potential power of such schemes is shown by the mathematical analysis of the kinetics of differential cloning in Box 14.2.

---

## BOX 14.1
## NEWER SCHEMES FOR DIFFERENTIAL GENOMIC DNA CLONING

Three schemes for cloning just the differences between two DNA samples will be described. The first two were designed to clone DNA corresponding to a region deleted in one available source but not in another. These schemes are similar to that described in Figure 14.25 except that they first use biotinylated driver DNA to facilitate the separation of target molecules from undesired contaminants, and then they use PCR to amplify the small amount of target molecules that remain uncaptured. In one scheme, developed by Straus and Ausubel (1990; Fig. 14.26), an excess of biotinylated driver DNA is used to capture and remove most of the target DNA by repeated cycles of hybridization and affinity purification with streptavidin-coated beads. Then the remaining desired target DNA is amplified and subsequently cloned by ligation of appropriate PCR adapters.

In a related scheme, developed by Eugene Sverdlov and co-workers, it is the target DNA that is biotinylated by filling in the ends of restriction fragments with dpppN derivatives (Fig. 14.27; Wieland et al., 1990). This target is then provided with PCR adapters by ligation. Excess driver DNA is used to deplete most of the target by cycles of hybridization and hydroxylapatite chromatography to remove any DNA duplexes formed. After several such cycles, streptavidin affinity chromatography is used to capture any biotinylated target remaining. The target molecules are then amplified by

*(continued)*

**BOX 14.1** *(Continued)*

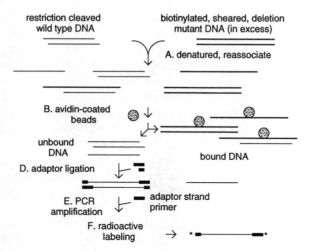

**Figure 14.26**   Differential cloning scheme based on repeated cycles of hybridization and biotin-affinity capture followed by PCR amplification. Adapted from Straus and Ausubel (1990).

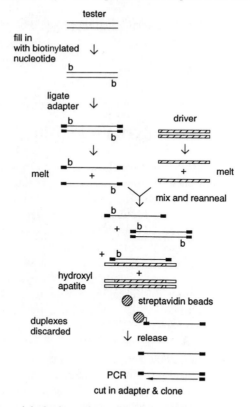

**Figure 14.27**   Differential cloning scheme based on repeated cycles of hybridization and hydroxyl apatite chromatography followed by biotin affinity capture and PCR amplification. Based on a method described by Wieland (1990).

*(continued)*

**BOX 14.1** *(Continued)*

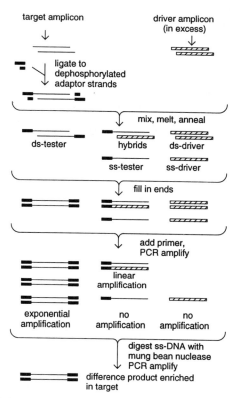

**Figure 14.28** A method for cloning the differences between two genomes (RDA). Adapted from Lisityn et al. (1993).

primers complementary to the adapter sequences and cloned. Both of these methods appear to be quite satisfactory, and both can be enhanced, if necessary, by repeating the steps involved.

Recently a scheme has been described by the Lisityn et al. (1994) for cloning polymorphic restriction fragments. This scheme is illustrated in Figure 14.28. It has been called representation difference analysis (RDA). First, the complexity of both target and driver genomes is reduced by PCR to allow more effective subsequent hybridizations (Chapter 4). This is done by ligating on adapters and removing them after the PCR amplification. Then, as shown in Figure 14.28, the target is provided with new PCR adapters by ligation. Target is mixed with excess driver, melted, and reannealed. The ends of the duplexes formed are filled in with DNA polymerase. PCR is now used to amplify the entire reaction mixture. The key point is that target duplexes will show exponential amplification because they contain two adapters. Heteroduplexes will show only linear amplification, while driver DNA will not be amplified at all. Any single-stranded molecules remaining are destroyed by treatment with mung bean nuclease, a single-strand specific enzyme similar to S1. Then the cycles of hybridization and amplification are repeated.

**BOX 14.2**
**SUBTRACTIVE HYBRIDIZATION**

The purpose of subtractive hybridization is to purify a target DNA strand, symbolized by $T$, from other DNA, called tracer DNA, symbolized by $S$. This is accomplished by the use of driver DNA strands flanked by different primers, symbolized by $D$. The procedure is illustrated schematically below. In general, genomic or cDNA samples would be digested to completion with a restriction nuclease and ligated to splints to prepare sequences for subsequent PCR amplification.

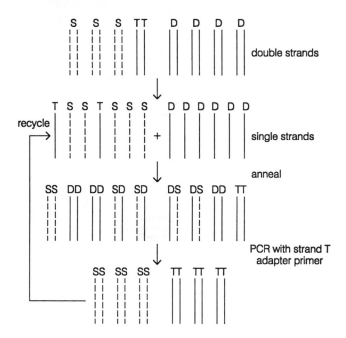

The mathematics behind this procedure is based on an equation developed in Chapter 3 to describe the kinetics of double-stranded DNA formation. If we call the initial concentration of single-stranded DNA segments is $c_0$, then the fraction of DNA that has formed double-stranded segments, $f_{ds}$, is given by the equation

$$f_{ds} = \frac{k_2 c_0 t}{1 + k_2 c_0 t}$$

where $t$ is the time and $k_2$ is a constant for that particular sequence of DNA. Using this equation, we can determine the concentration of double-stranded segments by multiplying the initial concentration:

$$c_{ds} = f_{ds} c_0 = \frac{k_2 c_0^2 t}{1 + k_2 c_0 t}$$

*(continued)*

**BOX 14.2** *(Continued)*

**First Round of Subtraction**

Consider two DNA samples. The first contains $S$- and $T$-type DNA; the second contains only $S$-type DNA. The sample containing only $S$-type DNA, however, is flanked by different primers, and will be designated by $D$. When these two samples are denatured and mixed, they will form double-stranded segments. The concentrations of various species can be determined by the equations above.

Since the $T$-type single strand will bind only with other $T$-type single strands, we can ignore the presence of $S$- and $D$-type strands in calculating the concentration of the double-stranded $T \cdot T$ duplexes formed. If $c_T$ is the initial concentration of $T$ single strands, the concentration of $T \cdot T$ double-stranded segments, is

$$(c_{T_{ds}})_{first\ round} = \frac{k_2 c_T^2 t}{1 + k_2 c_T t}$$

To calculate the concentration of the double-stranded tracer $S \cdot S$, we assume that the concentration of $S$ is insignificant compared to the concentration of $D$. Double-strand formation of $S \cdot S$, $S \cdot D$, and $D \cdot D$ will occur indiscriminantly. Thus we can first compute the kinetics of $D \cdot D$ formation and then extract the amount of $S \cdot S$ by multiplying by the mole fraction of $S \cdot S$, denoted by $X_{S \cdot S}$:

$$X_{S \cdot S} = \frac{c_{S_1}^2}{c_{D_1}^2}$$

Where $c_{S_1}$ is the initial concentration of S-type strands and $c_{D_1}$ is the initial concentration of $D$-type strands. Therefore

$$(c_{S_{ds}})_{first\ round} = (c_{D_{ds}})_{first\ round} \times X_{S \cdot S} = \frac{k_2 c_{D_1}^2 t}{1 + k_2 c_{D_1} t} \times \frac{c_{S_1}^2}{c_{D_1}^2} = \frac{k_2 c_{S_1}^2 t}{1 + k_2 c_{D_1} t}$$

$(c_{D_{ds}})_{first\ round}$ and $(c_{S_{ds}})_{first\ round}$ are the concentrations of double-stranded $D \cdot D$ and $S \cdot S$ structures.

The next step in the subtraction protocol shown above is to amplify the strands by PCR. Since only the $T \cdot T$ and $S \cdot S$ strands have matching primers, only these strands will be amplified exponentially. This results in the effective removal of $S \cdot D$ and $D \cdot D$ strands, reducing the amount of $S$ strand contamination, while increasing the concentration of $T$ strands.

The ratio of concentration of $T \cdot T$ versus the concentration of $S \cdot S$ can now be calculated. This ratio is the enrichment resulting from the first subtraction step:

$$E_{first\ round} = \left(\frac{c_{T_{ds}}}{c_{S_{ds}}}\right)_{first\ round} = \frac{k_2 c_T^2 t}{1 + k_2 c_T t} \times \frac{1 + k_2 c_{D_1} t}{k_2 c_{S_1}^2 t} = \frac{c_T^2}{c_{S_1}^2} \times \frac{1 + k_2 c_{D_1} t}{1 + k_2 c_T t}$$

Since the initial round of subtraction is performed with samples directly from the genome, the concentration of $T$ strands is the same as the concentration of $S$ strands.

*(continued)*

**BOX 14.2** *(Continued)*

This means that the ratio $c_T^2/c_{S_1}^2$ is equal to one for the first round. So, for large $t$, the enrichment ratio is

$$E_{first\ round} = \left(\frac{c_{T_{ds}}}{c_{S_{ds}}}\right)_{first\ round} = \frac{c_{D_1}}{c_T}$$

indicating that by the end of the first round, the ratio between $T$ and $S$ will be as large as the initial ratio of $D$ and $T$ DNAs used. Simply by using a much higher concentration of $D$ strands than $T$ strands, the presence of $S$-type strands can be significantly reduced.

**Second Round of Subtraction**

In following the procedure illustrated above for a second round of subtraction, it is important to note that the initial concentrations of $S$, $T$, and $D$ for the second round are related to their concentrations at the end of the first round. The initial concentration of $T$ strands for the second round can be assumed to be the same as the initial concentration of $T$ strands for the first round. The concentration of $S$ strands, however, will be less than that of the first round and will be denoted by $c_{S_2}$. Since the PCR amplification in the first round should not discriminate between $S$ and $T$, we can assume that

$$\frac{c_{S_2}}{c_T} = \left(\frac{c_{S_{ds}}}{c_{T_{ds}}}\right)_{first\ round} = \frac{c_T}{c_{D_1}}$$

The concentration of $D$ strands for the second round is much greater than either $c_{S_2}$ or $c_T$. This value will be called $c_{D_2}$.

Calculations for the annealing kinetics in the second round proceed exactly the same way as in the first, except that we now use the values $c_T$, $c_{S_2}$, and $c_{D_2}$:

$$(c_{T_{ds}})_{second\ round} = \frac{k_2 c_T^2 t}{1 + k_2 c_{D_2} t}$$

$$(c_{S_{ds}})_{second\ round} = \frac{k_2 c_{S_2}^2 t}{1 + k_2 c_{D_2} t}$$

PCR amplification at this point replenishes the amount of $T$-type strands, while effectively removing some of the $S$-type strands. The ratio between the concentrations of $T$ and $S$ can be calculated, as in the first round, from the initial concentrations for the second round:

$$\left(\frac{c_{T_{ds}}}{c_{S_{ds}}}\right)_{second\ round} = \frac{k_2 c_T^2 t}{1 + k_2 c_T t} \times \frac{1 + k_2 c_{D_2} t}{k_2 c_{S_2}^2} = \frac{c_T^2}{c_{S_2}^2} \times \frac{1 + k_2 c_{D_2} t}{1 + k_2 c_T t}$$

*(continued)*

**BOX 14.2** *(Continued)*

For large *t*, the final ratio of concentrations in the second round can be simplified to

$$E_{second\ round} = \left(\frac{c_{T_{ds}}}{c_{S_{ds}}}\right)_{second\ round} = \frac{c_T^2}{c_{S_2}^2} \times \frac{c_{D_2}}{c_T} = E_{first\ round}^2 \times \frac{c_{D_2}}{c_T}$$

Since $c_{D_2}$ is chosen arbitrarily, if we use the value $c_{D_1}$ again, the ratio of the final concentration of the second round simplifies even further:

$$E_{second\ round} = E_{first\ round}^2 \times \frac{c_{D_1}}{c_T} = E_{first\ round}^3$$

This is a remarkable result which shows that multiple subtraction protocols have a purification power that increases unexpectedly (adapted from notes provided by Eugene Sverdlov, as formulated by Ron Yaar).

If the starting DNA samples are genomic restriction fragments, the resulting amplified products eventually recovered will be those fragments in the subset of originally amplified material that had one restriction site that differed in the driver DNA. Such polymorphisms identified have been called polymorphic amplifiable restriction endonuclease fragments (PARFs). There are estimated to be around 1000 such *Bam*H I fragment differences between any two human genomes. Thus PARFs offer a potentially very powerful way to obtain useful genetic probes near preselected regions if DNA from appropriate individuals is available. For example, suppose that one has a population of individuals heterozygous for a dominant trait of interest. Subtraction of the DNAs of subsets of this population with DNAs from related individuals who lacked the trait should offer a reasonable chance of producing clones that contain polymorphisms linked to the trait or even responsible for the trait. A number of interesting variations on the original differential cloning scheme have been described (Yokata and Oishi, 1996; Rosenberg et al., 1995; Inoue et al., 1996). It remains to be seen how well such potentially very exciting new strategies actually perform in practice.

## IDENTIFICATION OR CLONING OF SEQUENCES BASED ON DIFFERENCES IN EXPRESSION LEVEL

Once DNA sequences of potential interest have been identified, a frequent next step in understanding their function is to determine when and where in the organism they are expressed. For a simple sequence of interest, a suitable analytical method is the Northern blot. Here mRNAs from tissues or other samples of interest are fractionated by length using gel electrophoresis, transferred to a membrane and hybridized with a probe specific to the gene of interest. This is called a Northern blot, and it is a widely used procedure for accessing the expression level of individual genes. An alternative method is quantitative PCR (qRT-PCR). qRT-PCR requires much lower sample amounts but is difficult to standardize because of the intrinsically variable characteristics of PCR. Some protocols add a standard amount of target mimic to the PCR reaction. Since the mimic is usually shorter than the true target, the ratio of true target to mimic products can be determined. These approaches fall far short of the mark when the goal is to analyze many genes of interest. A number of interesting methods have been developed to sample mRNAs or corresponding cDNAs and look for expression differences or patterns in biological states. These are still evolving rapidly.

In differential display a degenerate short PCR primer is used to amplify the cellular population of cDNAs. This can be a poly dT complementary to the 3′ poly A sequence of messages, it can be an anchored $dT_n$ (Liang et al., 1994), a short oligonucleotide, or a short sequence complementary to the end of DNA fragments generated by type II-S restriction enzymecleavage (Kato, 1995). Type II-S enzymes cut outside of their recognition sequences to yield a mixture of different single-stranded overhangs. A subset of this mixture can be captured by ligation to complementary overhangs, a technique that has been called molecular indexing (Unrau and Deugau, 1994). Regardless of the method used, the result is to reduce the complexity of the mRNA population and amplify the resulting cDNA to produce a discrete set of species to analyze quantitatively. The analysis can be performed by conventional or fluorescence-detected gel electrophoresis, or by hybridization (or two-color competitive hybridization) to an array of DNA probes analogous to the arrays used for SBH (Chapter 12).

These differential display methods are extremely powerful and broadly applicable. They suffer from a common limitation that whenever PCR amplification must be applied to a complex sample, the actual population of preexisting mRNAs will be distorted by differences in their ability to sustain multiple rounds of amplification. Two very different approaches for avoiding PCR-induced distortions are actively being explored. Picking cDNA clones at random and identifying them after sequencing is a powerful but expensive method. Alternatively, cDNAs can be sampled prior to amplification by cutting out a specific short fragment. The fragments are co-ligated into concatemers and amplified as a group. Then sequencing of cloned concatemers reveals relative abundances in a clever approach termed serial analyses of gene expression (SAGE).

An alternative to differential display is differential cloning. Differential cDNA cloning can proceed by schemes very similar to the one shown in Figure 14.25, or they can involve more elaborate techniques such as illustrated for the preparation of normalized cDNA libraries in Chapter 11. The objective is to recover cDNAs that represent messages present in one cell type but not another. The more carefully the cell types are selected, to differ just in the desired characteristics, the more efficient will be the search for the genes responsible for those characteristics. Differential cDNA cloning can be used to prepare genes specific for particular tissue types, developmental stages, chromosome origin, or even subchromosomal origin. These procedures are particularly effective when the target differences are very small and precisely defined. Examples are differences between unactivated and activated lymphocytes, to recover specific immune response genes, or differences between regenerating and nonregenerating tissue to isolate specific growth factors. As in genomic subtractive cDNA cloning, PCR can be used very effectively to solve most problems caused by small samples or rare messages. In principle, PCR should allow cDNAs to be made and subtracted from targets as small as single cells.

## COINCIDENCE CLONING

Subtractive cloning allows the selective isolation of DNAs that differ in two samples. A related, but technically somewhat more demanding approach is coincidence amplification, involving either cloning or PCR, which is designed to allow the selective isolation of DNAs that are the same in two samples. In contrast to subtractive hybridization, which tries to purify unique homoduplexes (i.e., unique differences between samples), coincidence amplification targets unique similarities or homoduplexes between samples. Coincidence amplification will be most useful when two samples are available that contain only a small amount of DNA in common. A number of such situations exist. For example, suppose that one has isolated a chromosome or just a fragment of a chromosome in a hybrid cell. A successful coincidence cloning procedure would allow one to clone out just the human component by using DNA from normal human cells as the second sample. Two hybrids that contain only a small overlap region on a single human chromosome could allow the selective cloning of just that overlap region. Large DNA fragments cut out from a PFG fractionation could be used in coincidence cloning experiments with hybrid cell DNA to purify human components that lie, specifically, on a pre-chosen fragment size. Finally large DNA fragments could be used in coincidence with other large fragments to selectively clone just regions of overlap. Alternate PCR procedures exploiting human-specific interspersed repeating sequences may be used to isolate the human specific DNA. We will describe these methods in detail later in this chapter. However, they are not as general or as powerful, in principle, as coincidence cloning.

The basic task that has to be accomplished in any coincidence amplification is shown in Figure 14.29. DNAs from the two samples to be tested are melted, mixed, and allowed to reanneal. Most fragments in the samples will form homoduplexes because most of the DNA in two well-chosen samples will be different. Occasional heteroduplexes will be

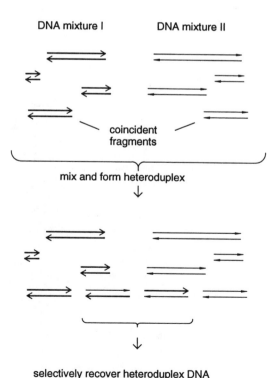

**Figure 14.29**   Basic requirements for coincidence cloning or coincidence PCR.

formed when sequences in the two samples match very accurately. What is needed, then, is a way of specifically cloning or amplifying the heteroduplexes. Note that this is a different problem than the heteroduplex detection we described in Chapter 13. In that case the desired heteroduplexes were those with one or more mismatches, and the mismatches were used as a specific handle to detect or capture the heteroduplexes. Here the desired heteroduplexes will in general be perfect matches. Quite a few different schemes have been tested for coincidence cloning (Box 14.3). None appear to work totally satisfactorily yet.

---

**BOX 14.3**
**PROPOSED SCHEMES FOR COINCIDENCE AMPLIFICATION**

Although these schemes for coincidence cloning are largely unproved, we will describe them in some detail because they illustrate some of the available arsenal of tricks for manipulating DNA sequences (Brooks and Porters, 1992). Similar schemes can be conceived of for coincidence PCR (Barley et al., 1993).

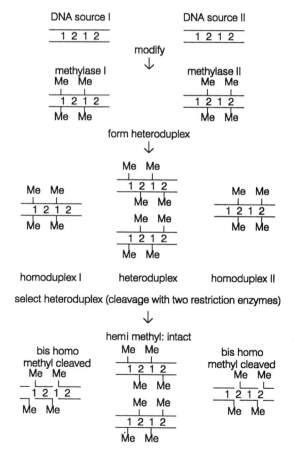

**Figure 14.30**   A scheme for coincidence cloning based on preferential cleavage of methylated homoduplexes. Adapted from Brooks and Porteus (1992).

*(continued)*

**BOX 14.3** (*Continued*)

Figure 14.30 shows a scheme for differential cloning of heteroduplexes based on DNA methylation and the unusual properties of restriction enzymes like *Dpn* I. This enzyme has already been described in Chapter 8, where it was used to generate large DNA fragments by selective cutting. Here it will be used to destroy homoduplexes selectively. The scheme in Figure 14.30 requires two different restriction enzymes with the ability, like *Dpn* I, to cut only fully methylated DNA duplexes. DNA from one sample is methylated with one conjugate methylase; the second conjugate methylase is used to methylate DNA from the second sample. Then the two samples are mixed, denatured, and allowed to renature. The key feature of the resulting mixture is that all homoduplexes will be fully methylated at their respective sites, and they will be cut into small pieces when treated with a mixture of the two corresponding restriction nucleases. Only heteroduplexes will be hemi-methylated at all of the restriction sites in question, and so they should be much less sensitive targets for cleavage. Thus a size fractionation after the digestion should allow the preferential isolation and subsequent cloning or PCR of the heteroduplexes. The flaw in this scheme, as in the *Dpn* I-mediated specific DNA cleavage discussed in Chapter 8, is that it requires nucleases that do not cut hemi-methylated DNA. *Dpn* I at least does not have this necessary property.

A second scheme for coincidence cloning is illustrated in Figure 14.31. This scheme is based on the frequently used method of directional cloning. A vector is employed that requires two different ligatable ends for efficient cloning of a target. DNA from one source is amplified with a PCR primer with an extension that generates one of the necessary ends. DNA from the second source is amplified with the same PCR primer with an extension having different restriction enzyme cleavage sites. The best way to do this would be to start with separate libraries of the two source DNAs in the same vector. In this way tagged single primers corresponding to flanking vector sequence could be used for efficient PCR. Alternatively, tagged random primers or tagged short specific primers could be used (see Chapter 4). This is a more general approach that could be applied directly to genomic DNA, but it is likely to be much less efficient. After amplification the samples are treated with the restriction enzymes needed to cleave within the sites introduced by the primer extensions. Then the two samples are melted, mixed, and reannealed. In principle, only heteroduplexes will have the necessary ends required for efficient directional cloning.

In practice, this approach is likely to have problems of low yield and significant contamination with unwanted homoduplexes. Unless all of the cloned DNA is dephosphorylated, it will be quite common to co-clone homoduplex and heteroduplex fragments. Since the former are present in vast excess, they will contaminate most samples. Even if the restriction fragments are dephosphorylated prior to ligation to the vector arms, homoduplexes will be the major initial ligation product with the vast majority of the vector. This will consume most of the vector, and although the subsequent cloning of the resulting linear products will be relatively inefficient, it is likely to occur often enough to lead to a very serious background of unwanted homoduplex products.

A third strategy for coincidence cloning is presented in Figure 14.32. This appears to be potentially powerful, but it is also fairly complex. DNAs from two sources are separately treated with the same two restriction enzymes to form mixture of double

*(continued)*

**BOX 14.3** *(Continued)*

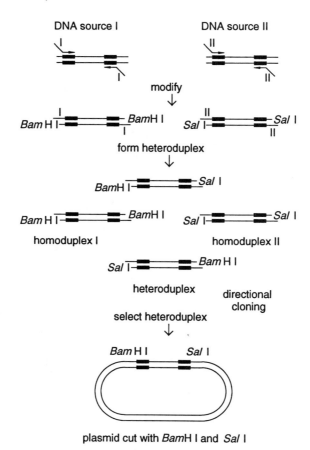

**Figure 14.31**   A scheme for coincidence cloning based on preparation of heteroduplexes to facilitate directed cloning. Adapted from Brooks and Porteus (1992).

digest products. One of the samples is left as restriction fragments. The other is directionally subcloned into the bacteriophage vector M13. Double-stranded flanking vector sequences, containing additional DNA segments, that will subsequently be used for PCR, are annealed to DNA from the M13 clones. Then both samples are melted, mixed, and allowed to reanneal. Heteroduplexes will be formed when restriction fragments from the first sample match clones from the second sample exactly. These are ligated to the tagged vector arms. Then PCR is used, with primers corresponding to the tagged sequences, to amplify the heteroduplexes specifically. The samples will be contaminated at this point by large amounts of M13 clones, but since these are all significantly larger than the PCR products, it should be possible to remove them by size selection.

*(continued)*

**BOX 14.3** *(Continued)*

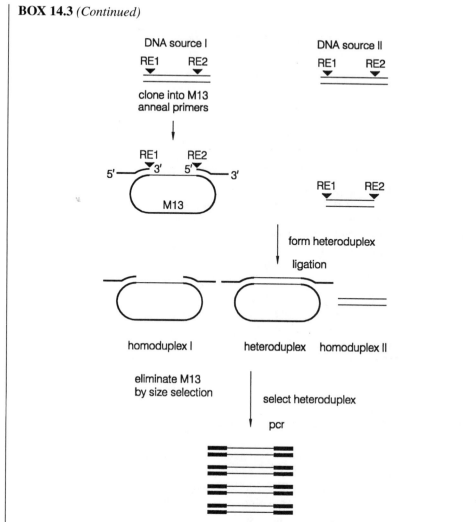

**Figure 14.32**    A scheme for coincidence cloning based on preparation of a complex that allows selective PCR of heteroduplexes. Adapted from Brooks and Porteus (1992).

There would be many interesting applications of robust coincidence cloning amplification procedures. For example, coincidence cloning of a cDNA library and a YAC should allow efficient capture of all of the genes on the YAC in a single step. Coincidence cloning of cDNAs from two very different tissue types should result in a highly enriched population of housekeeping genes that are not tissue specific. Finally coincidence cloning could be a very effective way to isolate very specific human DNAs from hybrid cell lines. An example of this is shown in Figure 14.33. Here probes of interest detect PFG-fractionated large human restriction fragments in a rodent background, and the goal is to obtain clones for the human DNAs contained on these fragments. The key point is that the human DNAs recognized by the same probes in the two different digests must have corresponding DNA sequences, while the rodent DNA that contaminates each human fraction

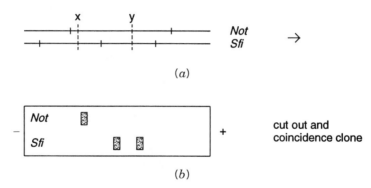

(a)

(b)

**Figure 14.33**    How coincidence cloning might be used to enrich the DNA from particular large restriction fragments seen by hybridization in rodent-human hybrid cells. *(a)* Restriction map of the region; *X* and *Y* indicate two human-specific probes. *(b)* Southern blot of a PFG fractionation hybridized with the two available probes. Intact material from a duplicate PFG run would be used to cut out the regions detected by hybridization, and one digest would be used in coincidence cloning with the other.

is likely to be different. Thus coincidence cloning of DNAs from a pair of gel slices from the two different digests should preferentially yield the human material. Even more complex logical cloning schemes consisting of various separation and amplification steps can be conceived of. Their ultimate utility will depend on how effective the more simple straightforward procedures become.

## HUMAN INTERSPERSED REPEATED DNA SEQUENCES

Most repeated sequence DNA in the human genome and other complex genomes appears to be interspersed with single-copy DNA. Although plenty of tandem repeats exist, such as VNTRs, except for centromeres these actually make up only a small fraction of the class of highly repeated sequences. The original demonstration that the bulk of the human repeats was interspersed was done in a series of classic experiments by Britten and Davidson. Subsequently the type of analysis they used has been refined and elaborated by Robert Moyzis and his colleagues. The basic scheme that underlies these approaches is shown in Figure 14.34*a*. A trace amount of labeled total human DNA is sheared randomly to various average lengths, *L*. This DNA is hybridized in solution with a vast excess of much shorter driver DNA. The driver DNA consists of unlabeled total human DNA isolated as all of the duplex that forms at a $C_0t$ of less than or equal to 50. This sample will contain all significant high-copy number human repeats but will contain very little single-copy DNA or infrequent repeats such as those seen in gene families. The hybridization of the driver with the labeled DNA is carried out at a $C_0t$ of 12.5. The goal of the experiment is to determine how much of the total labeled DNA can be captured by the repeated DNA driver during the hybridization.

The basic idea behind the experiment in Figure 14.34*a* is that all the DNA in clustered repeats will be captured very easily and efficiently, regardless of the length of DNA target used. In contrast, when *L* is small, most interspersed DNA will be captured without flanking single-copy sequences. As *L* increases, more and more single-copy DNA will be cap-

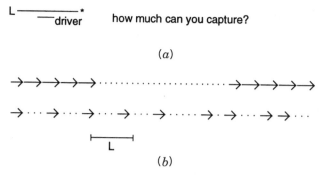

*(a)*

*(b)*

**Figure 14.34**   Analysis of the pattern of repeats in human DNA by renaturation kinetics. *(a)* Design of the basic experiment. Purified repeat is used in excess as the driver to capture indexed genomic DNA fragments of length *L*. *(b)* Two possible extreme patterns of repeating sequence.

tured by virtue of its neighboring repeated DNA (Figure 14.34*b*). A typical experimental result from such a procedure is shown in Figure 14.35. It is evident that only about a quarter of the genome is captured when *L* is small; this provides an estimate for the total amount of highly repeated DNA. Almost all of the genome is captured by the time the average fragment size reaches 8 kb. Thus many single-copy DNAs have a repeated sequence within 1 kb, and almost all single-copy DNAs have a repeat within 8 kb. The simplest fit to the data in Figure 14.35 suggests the occurrence of an interspersed repeat every 3 kb. A more elaborate and more accurate fitting procedure suggests that the distribution of repeats is bimodal. About 58% of the genome has a repeat on average every 1 kb; the remainder of the genome has a repeat on average every 8 kb. Of course this analysis does not reveal any information about the nature of the repeats or the number of different basic kinds of repeats and their particular distribution.

In fact there are just a few known major types of interspersed repeats in the human genome. Their properties are summarized in Table 14.3. By far the most common human repeat is a sequence called Alu because it contains two cutting sites for the restriction enzyme Alu I. Thus when human DNA is cut with Alu I and fractionated by size, the resulting material shows a bright, specific size band standing out from a background broad smear of other fragment sizes. There may be as many as $10^6$ Alu sequences in the human genome. Very similar sequences are seen in other primates, but in more diverged species, like rodents, the Alu-like sequences are sufficiently different that Alu's can be used as

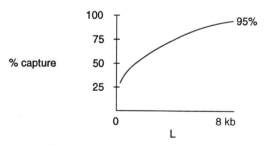

**Figure 14.35**   Fraction of total genomic DNA captured at low $C_0 t$ with excess repeated DNA driver, as a function of the length, *L*, of the genomic DNA used. Adapted from Moyzis et al. (1989).

**TABLE 14.3   Major Known Human Repeats**

| Type | Abundance (Copies) | Hybridization Average Spacing | GenBank Average Spacing | Average Length | Total Mass (Est.) |
|------|--------------------|-----------------------------|-------------------------|----------------|-------------------|
| 3' line (L1) | $5 \times 10^4 - 10^5$ | 30–60 kb | 27 kb | 1.1 kb | 55–110 Mb |
| Intact L1 | $4 \times 10^3 - 10^4$ | 150–480 kb | — | 6.4 kb | Trivial |
| (GT)$_n$ | $5 \times 10^4 - 10^5$ | 30–60 kb | 54 kb | 0.04 kb | Trivial |
| Alu | $5 \times 10^5 - 10^6$ | 3–6 kb | 4 kb | 0.24 kb | 120–240 Mb |

general species-specific DNA probes. A typical Alu sequence is around 0.24 kb and actually consists of an approximate tandem repeat of a shorter sequence. This is illustrated in Figure 14.36, which shows the consensus among known Alu sequences. Alu is an example of a class of repeating DNA sequences called short interspersed repeat sequences (SINES). Alu sequences in the human are far from identical. Known sequences have been classified into at least five different families, and even within a family the average difference between two Alu's is roughly 10%.

```
HS CON    GGCCGGGCGC  GGTGGCTCAC  GCCTGTAATC  CCAGCACTTT  GGGAGGCCGA      50
pPD39     ..........  ..........  ..........  ..........  ..........
BLUR 8    XXXXXXXXXX  XXXXXXXXXX  XXXX......  ..........  ........A.


HS CON    GGCGGGCGGA  TCACGAGGTC  AGGAGATCGA  GACCATCCCC  CCTAAAACGG     100
pPD39     ..........  ..........  ..........  ..........  ..........
BLUR 8    ..A....A..  ....CT.AAGTC .....T.T..  .....G..T.  ..C..C.T..


HS CON    TGAAACCCCG  TCTCTACTAA  AAATACAAAA  AATTAGCCGG  GCGTAGTGGC     150
pPD39     ..........  ..........  ..........  ..........  ..........
BLUR 8    ......T..A  ........G.  ..........  .X......A.  ..A.G...AT


HS CON    CGGCGCCTGT  AGTCCCAGCT  ACTTGGGAGG  CTGAGGCAGG  AGAATGGCGT     200
pPD39     ..........  ..........  ..........  ..........  ..........
BLUR 8    .C.T.....G  .A........  ....A.....  .....A...A  .....CC.T.


HS CON    GAACCCGGGA  GGCGGACCTT  GCAGTGAGCC  GAGATCCCGC  CACTGCACTC     250
pPD39     ..........  ..........  ..........  ..........  ..........
BLUR 8    A....AAX..  ..T....G..  ..........  ......G.A.  GG........


HS CON    CAGCCTGGGC  GACAGAGCGA  GACTCCGTCT  CAAAAAAAAA                 290
pPD39     ..........  ..........  ..........  ..........  A12
BLUR 8    ........TX  ..........  ......A...  ........X
```

**Figure 14.36**   Sequence of a typical human Alu repeat. Shown are the consensus sequence for many known Alu's, and two particular clones often used as Alu probes. Taken from Batzer et al. (1994).

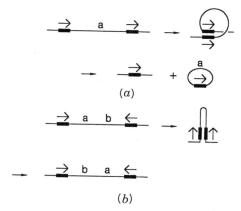

**Figure 14.37** DNA rearrangements produced by recombination between interspersed repeats like the Alu sequence. *(a)* A direct repeat leads to a DNA deletion because the short fragment produced is unlikely to be retained or replicated. *(b)* An inverted repeat leads to an inversion in the DNA sequence.

The next most common human repeat is the L1 sequence. It sometimes called Kpn because it has characteristic restriction sites for the enzyme Kpn I. The Kpn family is quite complex. The total sequence is about 6.4 kb in length. However, only 10% to 20% of the repeats are full length. Most others retain only the 3'-terminal kb of the L1. These 3' L1's occur at about a tenth the frequency of the Alu sequence. L1's are more similar in primates and rodents than Alu's. Thus they must be used cautiously as species-specific probes. L1 sequences are an example of the class of repeats called long interspersed repeating sequences or LINES. Altogether the repeated sequences listed in Table 14.3 add up at most to 350 Mb of DNA. Centromeric satellite is estimated to account for another 150 to 300 Mb of DNA. This material is tandemly repeating (Chapter 2). Thus at least 100 Mb of the total estimated 750 Mb (25%) of repeated human DNA remains unaccounted for, and we may be missing as much half of all the repeated sequences.

Interspersed repeats appear to act as recombination hot spots. The sequence of most examples of Alu repeats is similar enough that recombination between pairs of such repeats is sometimes seen as a cause of disease alleles. As shown in Figure 14.37, depending on whether the repeats are head to head (tandem) or head to tail (inverted), the result of an intrachromosomal recombination event will be an inversion or a deletion. Recombinations between repeats on different chromosomes will produce translocations. The low-density lipoprotein receptor is one of the genes where a deletion caused by inter-Alu recombination has been seen. In this case the result is to produce a disease allele for familial hypercholesterolemia.

## DISTRIBUTION OF REPEATS ALONG CHROMOSOMES

In Chapter 2 we illustrated the fact that light bands and dark bands of human chromosomes had some very different general properties. A commonly accepted mechanism for the spread of interspersed repeats is that at least some of the copies of these sequences are mobile elements and can spread themselves or copies to other sites. This has definitely been shown to be the case for the intact L1 repeat. Given this general picture, it is not surprising that certain types of repeats appear to cluster in certain genome regions. While the exact mechanism for this clustering is not understood, a formal mechanism to explain it would simply be to state that the transposition of the mobile elements favors genomic regions with certain properties.

Alu sequences are preferentially located in Giemsa light, G + C-rich bands. Note that the Alu's, themselves are G + C rich. Their average base composition is 56% G + C, and they have a CpG content that is only 64% of that expected statistically for such a G + C content. This is remarkable given the overall suppression of CpG throughout most of the genome. Thus Alu's behave a bit like HTF islands (Chapter 8). It is not surprising then, that like HTF islands. Alu's seem to preferentially associated with genes. The Giemsa light bands are very gene rich, and within many of these genes there are truly remarkable numbers of Alu repeats (in the introns, of course). These Alu's greatly complicate attempts to sequence genomic, gene-rich regions by shotgun strategies (Chapter 10) because of the difficulty of assembling the sequence.

L1 sequences occur preferentially in dark bands. These bands are A + T rich. The L1s themselves are A + T rich (58%). They are also extremely deficient in CpG sequences with only 13% of what would be expected statistically. Thus we can conclude from both the patterns of Alu's and L1's that like attracts like, but the mechanism behind this remains unclear. Earlier we indicated that the pattern of Alu distribution was really biphasic. Presumably the Alu-rich phase seen in renaturation experiments corresponds to DNA from light bands, but this has not been formally proved.

The final class of well-characterized interspersed repeats are the VNTRs. These are preferentially located in telomeric light Giemsa bands, although they are spread well enough through the genome to be generally useful as genetic markers. The reason why VNTRs cluster near the telomeres is unknown. However, it is worth noting that the frequency of meiotic recombination appears to be very high in the telomeric regions and that one mechanism of VNTR growth and shrinkage is recombination. There is no way to tell at present whether any causal relationships existing among these observations. However, they represent a tantalizing area for future study.

## PCR BASED ON REPEATING SEQUENCES

Sequences in the Alu and L1 families are similar enough so that a single PCR primer can be used to initiate DNA synthesis within a large number of these elements. Some of the common Alu primers are summarized in Figure 14.38. These are chosen to try to focus on the most-conserved regions of the repeats within known human sequences without selecting sequences that are also conserved in rodents. Some Alu primers are tagged with extensions to allow more efficient amplification after the first few rounds of PCR where an inexact or very short match between primer and target template may be occurring (Chapter 4). The general situation in which these primers are applicable is shown in Figure 14.39. Neighboring copies of a repeat can have inverted configurations (head to head or tail to tail) or tandem configurations (head to tail). In the former case, a single PCR primer will serve to amplify the DNA between the repeats. In the latter case, two primers must be used. Inter-Alu PCR is a very powerful tool because so much of the human genome is dense in Alu sequences, and Alu sequences in humans are well diverged

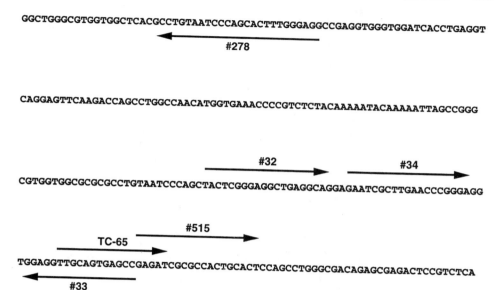

**Figure 14.38**   DNA sequences of some of the primers commonly used to amplify DNA between repeated Alu sequences by PCR.

from those of rodents. Thus a significant fraction of the human genome is potentially amplifiable by inter-Alu PCR. The L1 repeat is less useful in this regard because it is rarer and because its sequence is more conserved in rodents and human. Despite this limitation inter-L1 PCR or PCR between L1 and Alu sequences can still be helpful tools.

It would take most of a chaper to describe the myriad applications of inter-Alu or inter-L1 PCR in detail. We will just list a number of the most prominent applications, and then illustrate a few in more detail. Inter-Alu PCR is helpful whenever one needs to selectively amplify the human component in a nonhuman background. Not all the human DNA will be amplified. In general, one can expect good amplification wherever two Alu sequences with close sequence homology to the primers used lie in the correct orientation within a few kb of each other. This will be quite frequent in Alu-rich regions of the genome, much rarer in other regions. For example, inter-Alu PCR will selectively amplify the human component in a rodent hybrid cell line. It will preferentially amplify the YAC DNA in a background of yeast DNA. Combining YAC vector primers and Alu primers

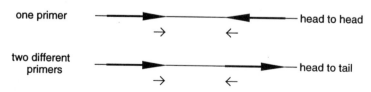

**Figure 14.39**   Some of the arrangements of interspersed repeats (arrows) that can be amplified by PCR using primers selected from the sequence of the repeat.

will preferentially amplify human DNA cloned near the ends of YAC inserts. These are exactly the samples most desirable as probes for YAC walking techniques. Single-sided Alu amplification procedures have also been described for regions where Alu's are too dilute to allow efficient inter-Alu PCR (Quereshi et al., 1994). Single-sided Alu PCR is also a very useful method to treat chromosome-specific hncDNA libraries, as described in Chapter 11.

The pattern of bands amplified by inter-Alu PCR from a whole human chromosome is often too complex to analyze (although inter-L1 PCR is helpful in such applications). Sometimes, with particular primers and conditions, the number of amplified bands can be reduced to of the order to 40. In such cases a significant fraction of these bands appears to be polymorphic in the population, and thus the Alu PCR provides a very convenient and easily used set of multiplexed genetic markers (Fig. 14.40).

In contrast to most attempts to amplify DNA from the whole genome, the pattern of bands from a few Mb of DNA is usually quite clear and diagnostic. Thus inter-Alu PCR is a powerful fingerprinting method. The added incentive is that the PCR-amplified bands seen in an electrophoretic analysis can be cut out and used as single-copy hybridization probes (competing any residual Alu sequences, as needed). Thus inter-Alu PCR finger-printing has been applied to YACs, chromosome fragments, radiation hybrids, and PFG gel slices, and it has been used to isolate single-copy DNA probes from all of these kinds of samples. The products of inter-Alu PCR reactions are very useful for FISH mapping. They are usually complex and concentrated enough to yield good results and allow the rough map position of the sample to be identified. Inter-Alu PCR products are also very useful for cross-connecting libraries.

An example of inter-Alu PCR applied to the analysis of PFG gel slices is shown in Figure 14.41. Here a *Not* I-digested hybrid cell line containing chromosome 21 as its only human component was fractionated by PFG under a number of different conditions to maximize the resolution of indivdiual size regions. Each gel lane was sliced into 40 fragments, and each fragment was subjected to inter-Alu PCR. In most cases lanes showed discrete patterns of several amplified bands, and adjacent slices often showed quite different patterns (Fig. 14.42). Thus the carryover of material from slice to slice during the PFG and subsequent steps is not so serious as to obscure the fractionation. This means that individual PCR products from analytical gels like the one shown in Figure 14.42 can be cut out, reamplified, and used as immortal specific single-copy DNA probes. Even if the PFG gel slice contained more than one genomic human *Not* I fragment, the individual PCR products from that slice are each likely to derive from only a single genomic fragment. Thus they constitute a very convenient source of new single-copy human DNA probes.

The method illustrated in Figures 14.41 and 14.42 is very helpful in the later stages of physical mapping where most fragments of a chromosome are located and the goal is to obtain new probes for unassigned bands as efficiently as possible. One problem with the kind of results shown in Figure 14.42 is that the number of new probes provided by a single experiment is very large, frequently a hundred or more. Before selecting probes for further study, usually one would like to know something about their regional location on the chromosome of interest. The standard way to do this is to take a probe of interest and hybridize it to a mapping panel of chromosome deletions as we described in Chapter 8. However, this is far too inefficient when a hundred or more probes must be mapped at once. An alternative approach, useful for YACs or slices of PFG fractionations, is shown in Figure 14.43. Ideally what one would like to do is take DNA from hybrid cell lines

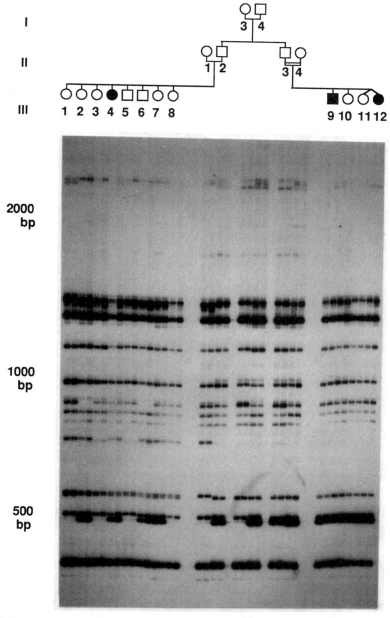

**Figure 14.40**    An example of Alumorphs: Polymorphic genomic DNA sequences amplified by in-ter-Alu PCR. Analysis of two pseudo–vitamin D-deficient rickets (PDDR) families (affected indi-viduals are indicated by filled symbols). The $^{32}$P-labeled products of PCR amplification using an Alu-specific primer were analyzed by electrophoresis in nondenaturing 6% polyacrylamide gel. Each individual from the pedigree shown on the top of the autoradiogram was analyzed in duplicate by two independent PCR amplifications, shown in two adjacent lanes on the gel. Molecular size markers are indicated at left. Taken from Zietkiewicz et al. (1992).

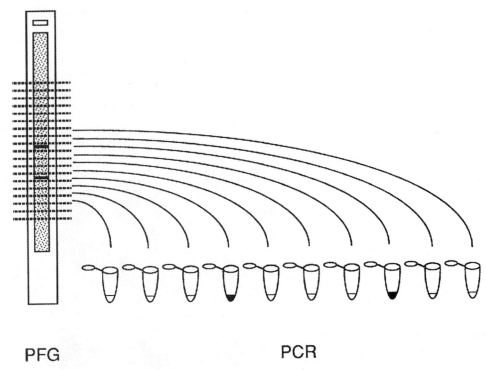

PFG                          PCR

**Figure 14.41** Schematic example of the use of inter-Alu PCR to preferentially amplify human DNA from PFG-fractionated large restriction fragments in a hybrid cell line.

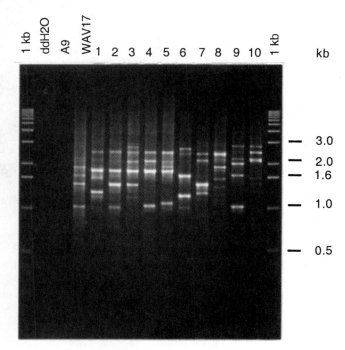

**Figure 14.42** Inter-Alu PCR products from 10 consecutive slices of a PFG-fractionated *Not* I digest of a hybrid cell line containing chromosome 21 as its only human component. From Wang et al. (1995).

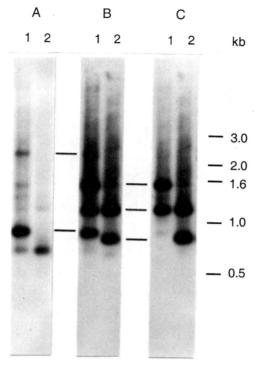

**Figure 14.43**   Example of how inter-Alu PCR products can be assigned, en masse, to chromosome regions. Inter-Alu probes generated from DNA from cell lines 8q⁻ (A), R2-10W (B), or 21q⁺ (C) were used to assign inter-Alu gel slice products regionally. Conventional gel lanes (lanes 1 and 2) containing inter-Alu products generated from template DNA contained in two different PFG slices are shown. Note that each cell line specific probe is hybridized to a different set of inter-Alu PCR products. From Wang et al. (1995).

containing human chromosome fragments and use this as a hybridization probe against a blot of a gel-fractionation of all the PCR products from a set of YACs or Not I fragments. The difficulty is that the complexity of the source DNA is too large, and one needs a way to reduce it and label the human component selectively in order to obtain efficient hybridization. However, inter-Alu PCR products from these cell lines provide the precise DNA subpopulation needed for efficient hybridization to inter-Alu PCR products from YACs or PFG fractions. By selectively amplifying the same segments of the DNA in both the hybridization probe and the hybridization target, one achieves enormously rapid and specific hybridizations. This same principle can be applied whenever inter-Alu PCR products are used for fingerprinting or for cross-connecting libraries.

The pattern of PCR products between repeating sequences can also be used to provide information about the distribution of repeats in the genome. With Alu, the patterns are too complex to analyze on a whole genome level. The situation is much more favorable with the 5′ L1 sequence. To estimate the number of PCR products expected with a single 5′ L1 primer, Yoshiyuki Sakaki assumed that there were about 3000 copies of the L1 repeat or one per Mb. This estimate is on the low side of the range reported by others (Table 14.3). Suppose that the PCR range is 2 kb. Only a quarter of the L1's within this range will be oriented head to head and thus amplified by the primer. So the expected frequency of PCR products per genome can be estimated as

Number $(3 \times 10^3) \times$ Spacing $(10^{-6}) \times$ Range $(2 \times 10^3) \times$ Orientation $(0.25) = 1.5$ per genome

In actuality about 20 genomic PCR products are seen. This presumably indicates that L1's are clustered, which is in accord with observations we have described previously. Note, however, that if we took a higher estimate for the number of 5′ L1 sequences, say $1.5 \times 10^4$ copies, then the expected number of products would be 7.5 per genome, and the evidence for clustering, from this one experimental result alone, would be much less compelling.

PCR amplification schemes can also be based on tandemly repeating dinucleotide or trinucleotide sequences. These are too infrequent to allow amplification between repeats. Instead, single-sided amplification methods are used. Alternatively, repeating-sequence-containing fragments are captured by hybridization with an immobilized single-strand, and then the released repeats are amplified in a number of different ways. These procedures are quite efficient (Broude et al., 1997; Kandpal et al., 1994).

## REPEAT EXPANSION DETECTION

A final example of the use of DNA amplification based on interspersed repeating sequences is shown in Figure 14.44. This illustrates a newly developed technique called repeat expansion detection (RED) which is designed as a way to specifically isolate very large tandemly repeating DNA sequences such as the expanded triplet repeats found in fragile X syndrome and other human disease alleles (Chapter 13). RED uses the ligase chain reaction (Chapter 4) instead of PCR. Oligonucleotide probes consisting of 11 to 17 tandem triplet repeats are annealed to target DNA in the presence of a thermostable DNA ligase. Repeated cycles of denaturation and renaturation are carried out. Long repeated triplet alleles in the target will promote more effective ligation of the probes than short alleles, and a more complex set of ligated products will result. This can be detected by electrophoresis after hybridization with the complementary triplet repeating sequence. This method is a very promising approach to the discovery of new genes, where unstable triplet repeats may be responsible for producing disease alleles. Other methods for repeat expansion detection may be based on the observation that PCR amplification of many long triplet repeats is inefficient at best and often fails completely (Broude et al., 1997).

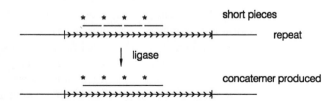

**Figure 14.44** Schematic illustration of the repeat expansion detection (RED) procedure used to identify cells with large, potentially disease-causing repeated triplet alleles. Adapted from Schalling et al. (1993).

## APTAMER SELECTION STRATEGIES

A relatively recently developed set of strategies combines physical purification and PCR to select DNAs (or RNAs or proteins) with desired sequences or binding properties. These methods appear to be powerful enough that in some cases one can start with all $4^n$ possible nucleic acids of length $n$ and find the one or few with the optimal affinity for a given target. Among the potential applications are:

- Purification of DNA sequences with the highest affinity for a given protein, ligand, or drug (such molecules have been termed aptamers)
- Purification of RNA sequences with the highest affinity for a given DNA (via duplex or triplex), RNA, protein, ligand, or drug
- Purification of protein sequences with the highest affinity for a given receptor, ligand, or drug (also called aptamers)

It is relatively easy to make all $4^n$ possible DNAs of length $n$ just by adding all four dpppN's at each step in automated DNA synthesis. More restricted mixtures can be made by an obvious extension of this approach.

The general principle behind select strategies is shown in Figure 14.45. A complex mixture is allowed to bind to an immobilized target of interest, usually at fairly low stringency. Those species that do bind are eluted, and PCR amplification is used to regenerate a population of molecules comparable to the initial total concentration. Now, however, this population should be enriched for molecules that have some affinity for the target. The cycles of affinity purification and amplification can be repeated as often as needed, until the complexity of the mixture becomes small enough to analyze. In the perfect case only a single species would remain, and if necessary, the stringency of the affinity step could be progressively increased during successive cycles. In actual cases a mixture of molecules will be seen, but this will eventually attain a small enough complexity so that individual components can be cloned and examined. Alternatively, a powerful approach is to sequence the mixture of molecules remaining after a large number of cycles of select purification. If certain positions within the DNA (or RNA or protein) are required for affinity, and others are not, the sequence of the mixture will show conserved residues at some positions, which will be unambiguously identified, and mixtures of residues (usually refractory to analysis) at other positions.

For the select approach to work, one needs a fairly good affinity purification with relatively little nonspecific background. Three potential implementations of the select strategy are shown in Figure 14.46. These allow for purification of DNAs, RNAs, or proteins with particular affinity properties. A number of highly successful examples of the application of selection strategies are summarized in Table 14.4. Such strategies are also effec-

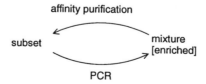

**Figure 14.45**  Basic principle behind a select strategy for purification of sequences with specific affinity properties.

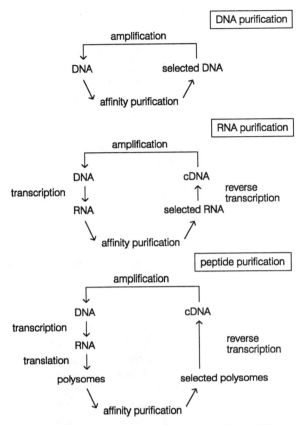

**Figure 14.46** Select strategies that have been proposed for purifying DNAs, and DNAs coding for RNAs, and proteins with particular properties that can be converted to differential affinities. Adapted from Irvine et al. (1991).

### TABLE 14.4   High-Affinity Aptamers

| Target | Library | Nucleotides[a] | Rounds[b] | Motif | $K_d$ (nM) |
|---|---|---|---|---|---|
| *E. coli* rho factor | RNA | 30 | 8 | Hairpin | 1 |
| *E. coli* metJ protein | RNA | 40 | 15 | Unknown | 1 |
| HIV-1 rev protein | RNA | 32 | 10 | Bulge | 1 |
| sPLA2 | RNA | 30 | 12 | Complex | 1 |
| | Modified RNA | 30 | 10 | Pseudoknot | 1 |
| Basic fibroblast growth factor | RNA | 30 | 13 | Hairpin | 0.20 |
| Vascular endothelial growth factor | RNA | 30 | 13 | Hairpin/bulge | 0.20 |
| SLE monoclonal antibody | ssDNA | 40 | 8 | Unknown | 1 |

*Source:* Adapted from Gold et al. (1995).

[a] Length of the random region.

[b] Rounds of selection used.

tive for finding consensus nucleic acid binding sequences to known proteins such as transcription factors (Pollack and Treisman, 1990; Nallur, et al., 1996). For proteins the select strategy shown may well be too cumbersome to use in practice. However, as an alternative to PCR amplification, one can use in vivo amplification for proteins instead. An elegant way to do this is bacteriophage display, shown schematically in Figure 14.47. Here the random coding sequence of interest is subcloned as a fusion with a surface coat protein of the bacteriophage M13. Either a minor coat protein is used with only five copies or the major coat protein with thousands of copies is used. Each bacteriophage plaque will be a clonal population representing one particular variant. Mixtures of plaques can be subjected to cycles of selection by physical affinity to the target of interest, and the successive populations of bacteriophage that remain will begin to be populated more and more with clones with the desired affinity properties.

Note that there is no reason why one must start with totally random sequences in select or bacteriophage display strategies. In many cases there will be existing structures with properties similar to the optimum behavior desired. In this case random mutagensis of just a small portion of an existing macromolecule can be used as a starting point try to select a more desirable variant. Despite the intrinsic attractiveness of bacteriophage display, this approach does have a number of limitations. Only monomeric target proteins can be examined by bacteriophage display. The protein targets must be able to fold properly within *E. coli,* and they must be oriented in the fusion so that the site that generates their affinity for the ligand or target is accessible.

A generalization of select strategies has been proposed by Sydney Brenner and Richard Lerner; it is called encoded combinatorial chemistry. The basic idea involves tagging linear oligomers of any type of residue, with a PCR-amplifiable specific DNA sequence that is a unique identifier of the particular oligomer. The general chemical structure needed is shown in Figure 14.48. Here the variable DNA identifier is placed between two constant PCR primers, and one of these is attached via a hub to the oligomer, which could be a nucleic acid, a peptide, an oligosaccharide, or really any kind of organic species that can be built up in a stepwise fashion. The number of different types of

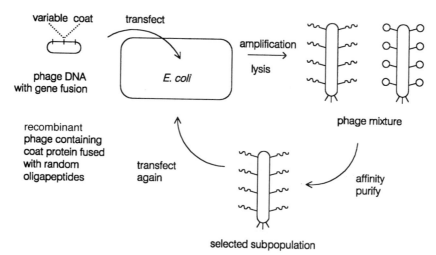

**Figure 14.47**  Bacteriophage display method for purifying DNA coding for proteins with selectable affinity properties.

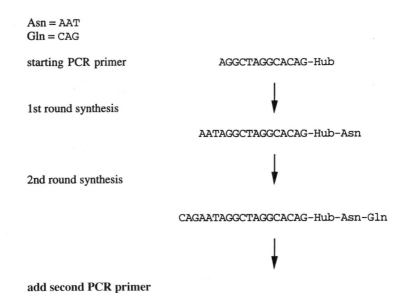

Figure 14.48 General construct needed for encoded combinatorial chemistry. Adapted from Brenner and Lerner (1992).

monomeric units in the oligomer will determine the complexity of the coding scheme needed in the DNA. A particularly nice trick, if small numbers of monomer units are involved, is to use a comma-less code. For example to specify the 20 amino acids, only a subset of 20 of the 64 possible triplet codons is needed. One can choose these, for example, so that if AAT and CAG represent two different amino acids, which can occur in either order: AATCAG and CAGAAT, then ATC, TCA, AGA, and GAA are not assigned to any amino acids. This makes the code resistant to frame shift errors and other ambiguities.

The chimeric DNA-oligomer compounds (Fig. 14.48) are screened for whatever activity is desired in the oligomer; then PCR is used just as in the select strategy in order to identify those components that have the desired affinity for a target. It is relatively easy to synthesize the full set of sequence identifiers and oligomers in a systematic way. This is shown in Figure 14.49. The actual efficiency of such schemes needs to be tested experimentally. Undoubtedly new schemes and variations on existing schemes will proliferate. However, the important feature of all of these approaches is that they illustrate the immense power that DNA analysis can bring to conventional chemistry.

## OLIGONUCLEOTIDES AS DRUGS

A large number of young biotechnology companies are betting their futures on the prospect that nucleic acids or nucleic acid analogs will function effectively as drugs. Most of this effort is not based on conventional ideas about gene therapy, where a underactive or inactive defective gene might be supplemented by an active one, or an overactive or inappropriately active gene might be substituted with a normal one. Such unconventional therapies are attractive, especially for many tissue-specific disorders, and such somatic

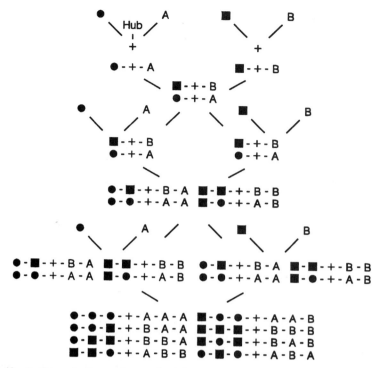

**Figure 14.49**  Split pool scheme for synthesizing a set of oligomeric compounds linked to their specific oligonucleotide identifiers. Adapted from Brenner and Lerner (1992).

gene therapy is already being tested in a few clinical trials. Here we are concerned with the much more limited and conventional approach of attempting to use short pieces of nucleic acids or their analogues as drugs.

The intrinsic attractiveness of nucleic acids as drugs is their sequence specificity. One can imagine that placed in the appropriate cell, an oligonucleotide could interfere with RNA or DNA function either by binding directly to these species or by competing with them for targets such as proteins. One approach is to simply use the sequence-specific binding affinity of the oligonucleotide to physically occlude a site or target. Here DNA is a potentially attractive target because it exists in very low copy numbers. An ideal anti-gene scenario would be to design an oligonucleotide that would form a very stable triplex under physiological conditions with an unwanted promoter and so turn off the transcription of the gene controlled by this promoter. Alternatively, the oligonucleotide could be used as an affinity reagent to carry a photochemical or other covalent modifier to a target of interest. Again this seems potentially most effective with a DNA target. An extreme version of oligonucleotide therapy would be to use catalytic RNAs to find and destroy multiple target molecules. If this can be realized, it will be an extremely effective way to deal with infections by viruses with RNA genomes, or to attack other RNA targets.

A number of obstacles must be overcome before successful oligonucleotide therapy will be achieved. First, the materials must be delivered effectively to the correct target cells and in sufficient quantities to be therapeutically active. Side reactions with other cells must be kept to a minimum. If cells lacked receptors for uptake of oligonucleotides, the problem would be to develop a targeting mechanism for the specific cells of interest.

Unfortunately, at least some cells in the body, T lymphocytes, have a natural pathway for oligonucleotide uptake. How general this phenomenon is remains to be seen. The implication is that unless cells with intrinsic uptake pathways are the desired target, this uptake may have to be suppressed, possibly by competition with a harmless oligonucleotide and possibly by shielding the therapeutic compound in some way.

Once cellular uptake is achieved, the oligonucleotide must then be targeted successfully to the desired intracellular location. This will be the nucleus for a reagent directed against DNA, and it might be the nucleus or endoplasmic reticulum for reagents directed against RNA. Such targeting is not a simple matter. Most extracellular macromolecules taken up by cells are automatically targeted to the lysosome where they are destroyed. This natural pathway must be interfered with to successfully deliver a nucleic acid elsewhere. There is no doubt that one should be able to do this by exploiting the same sorts of processes that various viruses use to enter cells and infect the nucleus or the cytoplasm. However, many of these processes are not yet well understood, and we may have to learn much more about them before successful oligonucleotide delivery mechanisms can be created.

Within the cell, or in intercellular fluids, a plethora of agents exist that can destroy or inactivate foreign nucleic acids. This is not surprising. Such agents must have evolved as antiviral defense mechanisms. To circumvent the action of these agents, oligonucleotide drugs would either have to be introduced in large quantities or rendered resistant to a variety of nucleases and other enzymes of nucleic acid metabolism. For example, antisense messenger RNAs have been proposed as therapeutic agents to interfere with the translation of an unwanted message, arising perhaps from a virus or a tumor cell. These would be expected to act by binding to the normal mRNA and inactivating it for translation by occlusion or by a more active destructive process. The difficulty in vivo is the presence of substantial amounts of RNA helicase activity. This is an enzyme that specifically recognizes double-stranded RNAs and unwinds the double helix. To be an effective drug under most circumstances, the backbone or bases of an antisense RNA will have to be altered so that this molecule is no longer recognized by RNA helicases.

Given the constrants mentioned above, the ideal oligonucleotide drug is probably likely to have an altered backbone to render it immune to normal nucleases and other enzymes and to increase its binding affinity to natural nucleic acids. Compounds with uncharged backbones like PNAs seem particularly attractive in this regard. It may also be desirable to equip potential oligonucleotide-analog drugs with additional chemical func-

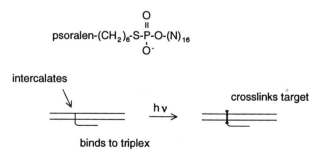

**Figure 14.50**   A potential oligonucleotide drug, designed by Claude Helene, that can bind to a DNA duplex and permanently inactivate it by photocrosslinking. For examples of results with this kind of approach, see Giovannangeli et al. (1992).

tionalities in order to further enhance their binding and effectiveness at inactivating the cellular target. An interesting example of such a potential drug is shown in Figure 14.50. Designed by Claude Helene, this compound consists of a triplex-forming oligonucleotide attached to a psoralen by a flexible chain. The psoralen enhances binding to DNA duplexes because it is an intercalator. More importantly, psoralen is a DNA photocrosslinking, so near-UV irradiation, after formation of the triplex, results in irreversible crosslinking of the target DNA duplex. This procedure has been demonstrated to work effectively in cells. It may be a prototype of the sorts of materials that we will eventually see in actual therapeutic use.

## SOURCES AND ADDITIONAL READINGS

Bailey, D. M. D., Carter, N. P., de Vos, D., Leversha, M. A., Perryman, M. T., and Ferguson-Smith, M. A. 1993. Coincidence painting: A rapid method for cloning region specific DNA sequences. *Nucleic Acids Research* 21: 5117–5123.

Batzer, M. A., Alegria-Hartman, M., and Deininger, P. L. 1994. A consensus Alu repeat probe for physical mapping. *Genetic Analysis: Techniques and Applications* 11: 34–38.

Brenner, S., and Lerner, R. A. 1992. Encoded combinatorial chemistry. *Proceedings of the National Academy of Sciences USA* 89: 5381–5383.

Brookes, A. J., and Porteous, D. J. 1992. Coincident sequence cloning: A new approach to genome analysis. *Trends in Biotechnology* 10: 40–44.

DeRisi, J., Penland, P., Brown, P. O., Bittner, M. L., Meltzer, P. S., Ray, M., Chen, Y., Su, Y. A., and Trent, J. M. 1996. Use of a cDNA microarray to analyse gene expression patterns in human cancer. *Nature Genetics* 14: 457–460.

Eriksson, M., and Nielsen, P. E. 1996. PNA-nucleic acid complexes. Structure stability and dynamics. *Quarterly Review of Biophysics* 29: 369–394.

Feigon, J., Dieckmann, T., and Smith, F. W. 1996. Aptamer structures from A to Z. *Chemistry and Biology* 3: 611–617.

Ferguson, J. A., Bowles, T. C., Adams, C. P., and Walt, D. R. 1996. A fiber-optic DNA biosensor microarray for the analysis of gene expression. *Nature Biotechnology* 14: 1681–1684.

Ferrin, L. J., and Camerini-Otero, R. D. 1991. Selective cleavage of human DNA: RecA-assisted restriction endonuclease (RARE) cleavage. *Science* 254: 1494–1497.

Ferrin, L. J., and Camerini-Otero, R. D. 1994. Long-range mapping of gaps and telomeres with RecA-assisted restriction endonuclease (RARE) cleavage. *Nature Genetics* 6: 379–383.

Giovannangeli, C., Thuong, N. T., and Helene, C. 1992. Oligodeoxynucleotide-directed photo-induced cross-linking of HIV proviral DNA via triple-helix formation. *Nucleic Acids Research* 20: 4275–4281.

Gnirke, A., Iadonato, S. P., Kwok, P.-Y., and Olson, M. V. 1994. Physical calibration of yeast artificial chromosome contig maps by recA-assisted restriction endonuclease (RARE) cleavage. *Genomics* 24: 199–210.

Inque, S., Kiyama, R., and Oishi, M. 1996. Construction of highly extensive polymorphic DNA libraries by in-gel competitive reassociation procedure. *Genomics* 31: 271–276.

Irvine, D., Tuerk, C., and Gold, L. 1991. Selexion: Systematic evolution of ligands by exponential enrichment with integrated optimization by non-linear analysis. *Journal of Molecular Biology* 222: 739–761.

Ito, T., Kito, K., Adati, N., Mitsui, Y., Hagiwara, H., and Sakaki, Y. 1994. Fluorescent differential display: Arbitrarily primed RT-PCR fingerprinting on automated DNA sequencer. *FEBS Letters* 351: 231–236.

Ito, T., Smith, C. L., and Cantor, C. R. 1992. Sequence-specific DNA purification by triplex affinity capture. *Proceedings of the National Academy of Sciences USA* 89: 495–498.

Ito, T., Smith, C. L., and Cantor, C. R. 1992. Affinity capture electrophoresis for sequence-specific DNA purification. *Genetic Analysis: Techniques and Applications* 9: 96–99.

Ji, H., Smith, L. M., and Guilfoyle, R. A. 1994. Rapid isolation of cosmid insert DNA by triple-helix-mediated affinity capture. *Genetic Analysis: Techniques and Applications* 11: 43–47.

Kato, K. 1996. RNA fingerprinting by molecular indexing. *Nucleic Acids Research* 24: 394–395.

Kandpal, R. P., Kandpal, G., and Weissman, S. M. 1994. Construction of libraries enriched for sequence repeats and jumping clones, and hybridization selection for region-specific markers. *Proceedings of the National Academy of Sciences USA* 91: 88–92.

Kool, E. T. 1996. Circular oligonucleotides: New concepts in oligonucleotide design. *Annual Review of Biophysical and Biomolecular Structure* 25: 1–28.

Liang, P., and Pardee, A. B. 1992. Differential display of eukaryotic messenger RNA by means of the polymerase chain reaction. *Science* 257: 967–971.

Liang, P., Zhu, W., Zhang, X., Guo, Z., O'Connell, R. P. O. Averboukh, L., Wang, F., and Pardee, A. B. 1994. Differential display using one-base anchored oligo-dT primers. *Nucleic Acids Research* 22: 5763–5764.

Lisitsyn, N., Lisitsyn, N., and Wigler, M. 1993. Cloning the difference between two complex genomes. *Science* 259: 946–951.

Lockhart, D. J., Dong, H., Byrne, M. C., Follettie, M. T., Gallo, M. V., Chee, M. S., Mittmann, M., Wang, C., Kobayashi, M. Horton, H., and Brown, E. 1996. Expression monitoring by hybridization to high-density oligonucleotide arrays. *Nature Biotechnology* 14: 1675–1680.

Mathieu-Daude, F., Cheng, R., Welsh, J., and McClelland, M. 1996. Screening of differentially amplified cDNA products from RNA arbitrarily primed PCR fingerprints using single strand conformation polymorphism (SSCP) gels. *Nucleic Acids Research* 24: 1504–1507.

Moyzis, R. K., Torney, D. C., Meyne, J., Buckingham, J. M., Wu, J.-R., Burks, C., Sirotkin, K. M., and Goad, W. B. 1989. The distribution of interspersed repetitive DNA sequences in the human genome. *Genomics* 4: 273–289.

Nallur, G. N., Prakash, K., and Weissman, S. M. 1996. Multiplex selection techniques (MuST): An approach to clone transcription factor binding sites. *Proceedings of the National Academy of Sciences USA* 93: 1184–1189.

Nielson, P. E., Egholm, M., Berg, R. H., and Buchardt, O. 1991. Sequence-selective recognition of DNA by strand displacement with a thymine-substituted polyamide. *Science* 254: 1497–1500.

Perry-O'Keefe, H., Yai, X.-W., Coull, J. M., Fuchs, M., and Egholm, M. 1996. Peptide nucleic acid pre-gel hybridization: An alternative to Southern hybridization. *Proceedings of the National Academy of Sciences USA* 93: 14670–14675.

Pollock, R., and Treisman, R. 1990. A sensitive method for the determination of protein-DNA binding specificities. *Nucleic Acids Research* 18: 6197–6204.

Qureshi, S. J., Porteous, D. J., and Brookes, A. J. 1994. Alu-based vectorettes and splinkerettes more efficient and comprehensive polymerase chain reaction amplification of human DNA from complex sources. *Genetic Analysis: Techniques and Applications* 11: 95–101.

Rosenberg, M., Przybylska, M., and Straus, D. 1994. "RFLP subtraction." A method for making libraries of polymorphic markers. *Proceedings of the National Academy of Sciences USA* 91: 6113–6117.

Schena, M. Shalon, D., Davis, R. W., and Brown, P. O. 1995. Quantitative monitoring of gene expression patterns with a complementary DNA microarray. *Science* 270: 467–470.

Sosnowski, R. G., Tu, E., Butler, W. F., O'Connell, J. P., and Heller, M. J. 1997. Rapid determination of single base mismatch mutations in DNA hybrids by direct electric field control. *Proceedings of the National Academy of Sciences USA* 94: 1119–1123.

Straus, R., and Ausubel, F. M. 1990. Genomic subtraction for cloning DNA corresponding to deletion mutations. *Proceedings of the National Academy of Sciences USA* 87: 1889–1893.

Strobel, S. A., Doucette-Stamm, L. A., Riba, L., Housman, D. E., and Dervan, P. B. 1991. Site-specific cleavage of human chromosome 4 mediated by triple-helix formation. *Science* 254: 1639–1642.

Takabatake, T. et al. 1992. The use of purine-rich oligonucleotides in triplex-mediated DNA isolation and generation of unidirectional deletions. *Nucleic Acids Research* 20: 5853–5854.

Unrau, P., and Deugau, K. V. 1994. Non-cloning amplifications of specific DNA fragments from whole genomic DNA digests using DNA "indexers." *Gene* 145: 163–169.

Velculescu, V. E., Zhang, L., Vogelstein, B., and Kinzler, K. W. 1995. Serial analysis of gene expression. *Science* 270: 484–487.

Veselkov, A. G., Demidov, V. V., Nielsen, P. E., and Frank-Kamenetskii, M. D. 1996. A new class of genome rare cutters. *Nucleic Acids Research* 24: 2483–2487.

Wan, J. S., Sharp, S. J., Poirier, G. M.-C., Wagaman, P. C., Chambers, J., Pyati, J., Hom, Y.-L., Galindo, J. E., Huvar, A., Peterson, P. A., Jackson, M. R., and Erlander, M. G. 1996. Cloning differentially expressed mRNAs. *Nature Biotechnology* 14: 1685–1691.

Wittung, P., Kim, S. K., Buchart, O., Nielsen, P., and Norden, B. 1994. Interactions of DNA binding ligands with PNA-DNA hybrids. *Nucleic Acids Research* 22:5371–5377.

Wittung, P., Eriksson, M., Lyng, R., Nielsen, P., and Norden, B. 1995. Induced chirality in PNA-PNA duplexes. *Journal of the American Chemical Society* 117: 10167–10173.

Yokota, H., and Oishi, M. 1990. Differential cloning of genomic DNA: Cloning of DNA with an altered primary structure by in-gel competitive reassociation. *Proceedings of the National Academy of Sciences USA* 87: 6398–6402.

# 15   Results and Implications of Large-Scale DNA Sequencing

The accumulation of completed DNA sequences and the development and utilization of software tools to analyze these sequences are changing almost daily. It is extremely frustrating to attempt an accurate portrait of this area in the kind of static snapshot allowed by the written text. The authors know with certainty that much of this chapter will become obsolete in the time interval it takes to progress from completed manuscript to published textbook. This is truly an area where electronic rather than written communication must predominate. Hence, while a few examples of original projections of human genome project progress will be given, and a few examples of actual progress will be summarized, the emphasis will be on principles that underlie the analysis of DNA sequence. Here it is possible to be relatively brief, since a number of more complete and more advanced treatments of sequence analysis already exist (Ribskaw and Devereaux, 1991; Waterman, 1995). The interested reader is also encouraged to explore the databases and software tools available through the Internet (see the Appendix).

## COSTING THE GENOME PROJECT

When the lectures that formed the basis of this book were given in the Fall 1992, there were more than 37 Mb of finished DNA sequence from the six organisms chosen as the major targets of the U.S. human genome project. The status in February 1997 of each of these efforts is contrasted with the status five years earlier in Table 15.1. The clear impression provided by Table 15.1 is that terrific progress has been made on *E. coli, S. cerevisiae,* and *C. elegans,* but we have a long way to go to complete the DNA sequence of any higher organism. In addition to the six organisms listed in Table 15.1, a few other organisms like the plant model system *Arabidopsis thaliana* and the yeast *S. pombe* will surely also be sequenced in the next decade along with a significant number of additional prokaryotic organisms. Other attractive targets for DNA sequencing would be higher plants with small genomes like rice, and higher animals with small genomes like the puffer fish. If methods are developed that allow efficient differential sequencing (see Chapter 12), methods that look just at sequence differences, some of the higher primates become of considerable interest. Although these genomes are as complex as the human genome, DNA sequences differences between them are only a few percent, and most of these will lie in introns. Thus a comparison between gorilla and human would make it very easy to locate small exons that might be missed by various computerized sequence search methods.

A different cast is provided by the figures in Table 15.2, which illustrate the average rate of DNA sequencing per investigator up until 1990 since the first 24 DNA bases were determined almost 30 years ago. The results in Table 15.2 show a steady rise in the rate of

**TABLE 15.1    Progress Towards Completion of the Human Genome Project**

| Organism | Finished DNA Sequence (Mb) | | | Comment |
|---|---|---|---|---|
| | Complete Genome | June 1992 | February 1997 | |
| E. coli | 4.6 | 3.4 | 4.6 | Complete |
| S. cerevisiae | 12.1 | 4.0 | 12.1 | Complete |
| C. elegans | 100 | 1.1 | 63.0 | Cosmids[a] |
| D. melanogaster | 165 | 3.0 | 4.3 | Large contigs only |
| M. musculus | 3000 | 8.2 | 24.0 | Total assuming 2.5 × redundant |
| H. sapiens | 3000 | 18.0 | 31.0 | In contigs, > 10Kb |
| | | | 116.0 | Total assuming 2.5 × redundant |

[a] Completion is expected at the end of 1998.

DNA sequencing. However, they understate this rise, since the data are derived from all DNA sequencing efforts. In practice, the majority of these sequencing efforts use simple, manual technology, and many are performed by individuals just learning DNA sequencing. The common availability and widespread use of automated DNA sequencing equipment probably had little impact on the results in Table 15.2 because these advances are too recent. Despite this fact, the general impression, based on DNA sequences deposited into databases, is that the total accumulation of DNA sequences is increasing nearly exponentially with time, and this trend certainly has continued to the present.

A stated goal of the human genome project in the United States is to complete the DNA sequence of one haploid-equivalent human genome by the year 2005. Two immediate cautionary notes must be struck. First, the sequence is unlikely to really be completed by then or perhaps by any time in the foreseeable future because it is unlikely anyone would want to, or could, sequence through millions of base pairs of centromeric tandem repeats, looking for occasional variations. Second, it may not be a single genome that is sequenced. A considerable amount of the material for DNA sequencing is likely to come from cell lines containing particular human chromosomes. In general, each line represents chromosomes from a different individual. Thus, to whatever degree of completion the first human DNA sequence is obtained, it will surely be a mosaic of many individual genomes. In view of the nature of the task, and its importance to humanity, this outcome actually seems quite appropriate. It also answers, once and for all, the often-asked question "who will be sequenced in the human genome project?"

Setting the above complications aside, the chance of achieving the DNA sequencing goals of the human genome project depends principally on three variables: the amount of money available for the project as a whole, the percent of it used to support large-scale genomic sequencing, and the efficiency of that sequencing, namely the cost per base pair and how it evolves over the course of the project. In most current genomic DNA sequencing efforts, labor still appears to be the predominant cost, and all other considerations can be scaled to the number of individuals working. This is changing with the incorporation of more and more highly automated methods; supplies and materials are becoming the dominant cost. However, since we cannot yet accurately estimate the impact of automation on sequencing costs, we will assume the continuation of dominant labor costs in order to make some projections.

**TABLE 15.2   Summary of Progress Made in DNA Sequence Analysis Between 1967 and 1990**

| DNA Sequence Determined | Method(s) | Time Period | Nucleotides Determined | Number of Investigators Involved | Percent Time for Sequencing Steps | Nucleotides per Year, per Investigator | Relative Speed |
|---|---|---|---|---|---|---|---|
| Cohesive ends of λ DNA | Partial incorporation; partial digestion | 1967–1970 | 24 | 2 | 80–90 | 4 | 1 |
| Cohesive ends of 186p DNA | Same as above | 1971–1972 | 38 | 2 | 80–90 | 12 | 3 |
| φX174, f1 | Mobility shift | 1972–1973 | 140 | 5 | 40–50 | 28 | 7 |
| φX174 | Plus-and-minus | 1973–1977 | 5000 | 9 | 20–30 | 138 | 35 |
| Over 500 sequences | Dideoxy chain termination | 1977–1982 | 66,000 | 350 | 20–30 | 380 | 94 |
| Over 7000 sequences | Recombinant DNA M13 vectors, etc. | 1982–1986 | 9,000,000 | 3,500 | 15–25 | 640 | 160 |
| Over 40,000 sequences | Same as above, automatic DNA sequencer | 1986–1989 | 27,000,000 | 9,000 | 10–20 | 1000 | 250 |
| Over 23,000 sequences | Same as above | 1989–1990 | 14,000,000 | 12,000 | 10–20 | 1160 | 290 |

*Source:* Adapted from Wu (1993).

In most U.S. academic or pure research settings, an average person working in the laboratory costs about $100,000 a year to support. This includes salary, fringe benefits, chemicals, supplies, and academic overhead such as administrative costs, light, heating, and amortization of laboratory space. Let's make the reasonably hard-nosed estimate that the average genome sequencer will work 250 days per year. Dividing this into the yearly cost results in a cost per working day of $400. A reasonable estimate of state-of-the-art DNA sequencing rates is about $10^4$ raw base pairs per day per individual. Allowing for 20% waste, and dividing by a factor of 8 for the redundancy needed for shotgun strategies, we can estimate that the sequencing rate for finished base pairs per day is $10^3$. When the daily cost is divided by the daily output, the result is a cost per finished base pair of $0.40. This is far lower than the current commercial charges for DNA sequencing services, which average several dollars per base, or the cost of more casual sequencing efforts which, based on the most recent rates shown in Table 15.2, weould be about $90 per base. The cost per finished base of the *H. influenza* project recently completed was $0.48 in direct supplies and labor. This is equivalent to at least $1.00/bp when overhead and instrument depreciation costs are added.

Although the relatively low cost of current automated DNA sequencing is impressive, compared to the cost of less systematic efforts, it falls far short of the economies that will need to be achieved to complete the  human genome project within the allowable budget and time scale. The initial design of the U.S human genome project called for $3000 million to be spent over 15 years. This would translate into a cost of $1 available per human base pair if all one did was sequence. Such a plan would be ridiculous because it ignores mapping, which provides the samples to be sequenced, and it does not allow for any model organism work or technology development. The U.S. human genome budget in 1992 was $160 million per year. If we had ramped up to large-scale DNA sequencing immediately, starting in October 1993, we would have had 12 years to complete the project. A steady state sequencing rate model would have required sequencing at a rate of 250 million base pairs per year. At current best attainable costs this would require $100 million per year to be spent on human genomic DNA sequencing. So this way the project could be completed, but to proceed with such a plan that anticipates no enhancements in technology would be lunacy.

It is more sensible to scale up sequencing more gradually and to build in some assumptions about improved efficiency. The caveat to this approach is that the slower the scale up, the more efficient the final sequencing rates must become. Suppose that we arbitrarily limit the amount of funds committed to human genomic DNA sequencing to the $100 million annual costs required by the steady state model. Table 15.3 shows one set of cost projections, developed in 1992 by Robert Robbins. At first glance this may seem like an extremely optimistic scenario, since it starts with average costs that are $1.50 per base pair and requires only a factor of ten decrease in unit cost over a seven-year time period to reach $0.15 per finished base pair by 2001. However, these are average costs, and at these costs an average work force of 1000 individuals will be needed for the last five years of the project, just for human DNA sequencing. The bottom line is that the scenario in Table 15.3 seems reasonable and achievable, but one hopes that some of the potential improvements in DNA sequencing described in Chapters 11 and 12 will be realized and will result in considerably faster rates of sequence acquisition. From this point of view, the scenario in Table 15.3 is actually quite pessimistic.

One way to view the cost effectiveness of the genome program is to ask how much additional DNA sequencing will be accomplished beyond the stated goal of one human

**TABLE 15.3   One Model for DNA Sequencing Costs in the Human Genome Project**

| Year | Finished, per-Base Direct Cost | Annual Sequencing Budget ($millions) | Genomic Sequence (Mb) | | Percent of Genome Completed |
|------|------|------|------|------|------|
| | | | Year | Cumulative | |
| 1995 | $1.50 | 16 | 11 | 11 | 0.33 |
| 1996 | $1.20 | 25 | 21 | 32 | 0.96 |
| 1997 | $0.90 | 35 | 39 | 71 | 2.15 |
| 1998 | $0.60 | 50 | 84 | 155 | 4.71 |
| 1999 | $0.45 | 75 | 168 | 324 | 9.81 |
| 2000 | $0.30 | 100 | 337 | 660 | 20.01 |
| 2001 | $0.15 | 100 | 673 | 1334 | 40.42 |
| 2002 | $0.15 | 100 | 673 | 2007 | 60.82 |
| 2003 | $0.15 | 100 | 673 | 2681 | 81.23 |
| 2004 | $0.15 | 100 | 673 | 3354 | 101.63 |

genome. If the development of new sequencing methods proceeds very well, it may be possible to complete extensive portions of the mouse genome, and perhaps even other model organisms with large genomes, under the cost umbrella of funding for the human sequence. This is not inappropriate, since the more relevant model organism sequence data that we have available, the more powerful will be our ability to interpret the human DNA sequence.

Whether one adopts an optimistic or a pessimistic scenario, the inevitable conclusion is that by October 2005 or thereabouts, $3 \times 10^9$ base pairs of human genomic DNA encoding for something like 100,000 human genes will be thrust upon the scientific community. The challenge will be to find any genes in the sequence that are not already represented as sequenced cDNAs, translate the DNA sequence into protein sequence, make some preliminary guesses about the function of some of these proteins, and decide which ones to study first in more detail. The remainder of this chapter deals with these challenges. Because of the rapid advances in cDNA sequencing described earlier in Chapter 11, some of these challenges already confront us today.

## FINDING GENES

There are two very different basic approaches to finding the genes encoded for by a sample of genomic DNA. The experimental approach is to use the DNA as a probe (or a source of probes or PCR primers) to screen available cDNA libraries. One can also use available DNA materials to look directly at the mRNA population in different cell types. This is most easily accomplished by Northern blots, as described in Chapter 13. None of these approaches require that the DNA sequence be completed, and except for the synthesis of PCR primers, none require any known DNA sequence. These pure experimental methods will succeed only if the mRNAs or cDNAs in question are present in available sources in sufficient quantities to be detected above the inevitable background caused by nonspecific hybridization or PCR artifacts.

The second basic approach is the one that we will concentrate on here. It requires that the sequence of the DNA be known. Genes have characteristic sequence properties that distinguish them, to a considerable extent, from nongenic DNA or from random sequences. The goal is to optimize the methods for making these discriminations. One trivial procedure should always be tried to an newly revealed piece of genomic DNA. This is to compare its sequence with all known sequences. A near or exact match could immediately reveal genes or other functional elements. For example, large amounts of partial cDNA sequence information are accumulating in publicly accessible databases. Frequently these cDNAs have already been characterized to some extent, such as their location on a human physical map or their pattern of expression in various tissues. Finding a genomic match to such a cDNA fragment clearly indicates the presence of a gene and provides a jump start to further studies. The problem that arises is what if the match between a portion of the genomic DNA sequence and other molecules with known sequence is not exact, or what if there is no significant detectable match at all?

A number of basic features of genomic DNA sequence can be examined to look for the location of genes. The most straightforward of these is to search for open reading frames (ORFs). This is illustrated in Figure 15.1. A given stretch of DNA sequence can potentially be translated into protein sequence in no less than six different ways. The triplet genetic code allows for three possible reading frames, and a priori either DNA strand (or both) could be coding. Computer analysis is used to translate the DNA sequence into protein in all six reading frames, and assemble these in order along the DNA. The issue is then to decide which represent actual segments of coding sequence and which are just noise.

Almost all known patterns of mRNA transcription occur in one direction along a template DNA strand. Segments of the mRNA precursor are then removed by splicing to make the mature coding mRNA. The exceptions to this pattern are fairly rare in general, and the few organisms in which they are more frequent are fairly restricted. Some parasitic protozoa and some worms have extensive trans-splicing where a mRNA is composed from units coded on different chromosomes. Editing of mRNAs, where DNA-templated bases are removed, individually, and bases not templated by DNA are added, individually, also appears to be a rare process concentrated in a few lower organisms. Thus it is almost always safe to assume that a true mRNA will be composed of one or more successive segments of DNA sequence organized, as shown schematically in Figure 15.2. The challenge is to predict where the boundaries of the possible exons are, and where individual genes actually begin and end. This requires the simultaneous examination of aspects of the RNA sequence as well as aspects of the coded protein sequence.

True exons must have arisen by splicing. Thus the sequences near each splice site must resemble consensus splicing signal sequences. While these are not that well defined, they have a few specific characteristics that we can look for. True exon coding segments can have no stop codons, except at the very end of the gene. Thus the presence of UAA, UAG, or UGA can usually eliminate most possible reading frames fairly quickly. A true gene must begin with a start codon. This is usually AUG, although it can also be GUG,

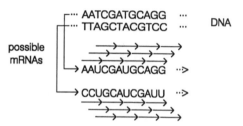

**Figure 15.1** Six possible reading frames from a single continuous stretch of DNA sequence shown earlier in Figure 1.13.

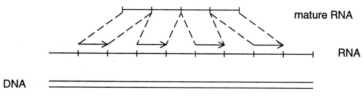

**Figure 15.2**   A message is read discontinuously from a DNA, but all of the examples known to date are read from the same strand, in a strict linear order (except for the very specialized case of transsplicing seen in some simple organisms).

UUG, or AUA. True start codons are context dependent. For example, a true AUG used for starting translation must have nearby sequences that are used by the ribosome or protein synthesis initiation factors (proteins) to bind the message and initiate translation. These residues usually lie just upstream from the AUG, but they can also extend a bit downstream.

The key factor in determining the correct points for the initiation of translation is that the context of starting AUGs is very species dependent. Thus, since the species of origin of a DNA will almost always be known in advance, a great deal of information can be brought to bear to recognize a true start. For example, in *E. coli,* a given mRNA can have multiple starting AUGs within a message because *E. coli* commonly makes and uses polycistronic messages. In eukaryotes, messages are usually monocistronic. Most frequently it is the first AUG in the message that signals the start of translation. This means that information about transcription starting signals or splicing signals must be used to help locate the true starting points for translation of eukaryotic DNA sequences.

A true mRNA must also have a transcription stop. This is relatively easy to find in prokaryotes. However, we still know very little about transcription termination in eukaryotes. Fortunately, most eukaryotic mRNAs are cleaved, and a polyA tail is added. The consensus sequence for this process is AATAAA. This is distinct, but it is very short, and such a sequence can easily occur by random fluctuations in A + T-rich mammalian DNA. This polyA addition signal is an example of a fairly general problem in sequence analysis. Many control sequences are quite short; they often work in concert with other control elements, and we do not yet know how to recognize the overall pattern.

## MORE ROBUST METHODS FOR FINDING GENES BY DNA SEQUENCE ANALYSIS

If the methods just described represented all of the available information, the task of finding previously unknown genes by genomic DNA sequence analysis alone would be all but hopeless. Fortunately a great deal of additional information is available. Here we will describe some of the additional characteristics that distinguish between coding and noncoding DNA sequences. The real challenge is to figure out the optimal way to combine all of this additional knowledge into the most powerful prediction scheme. A simple place to start is to consider the expected frequency of each of the 20 amino acids. Some, like tryptophan, are usually very rare in proteins; others are much more common. A true coding sequence will, on average, contain relatively few rare amino acids. Similarly the overall average amino acid compositions of proteins vary, but they usually lie within certain bounds. A potential coding sequence that led to a very extreme amino acid composition would usually be rejected if an alternative model for the translation of this segment of the DNA led to a much more normal amino acid composition. Note that the argument we are using is one of plausibility, not certainty. Statistical weights have to be attached to such arguments to make them useful. The same sort of considerations will apply to all of the other measures of coding sequence that follow.

The genetic code is degenerate. All amino acids except methionine and tryptophan can be specified by multiple DNA triplet codons. Some amino acids have as many as six possible codons. The relative frequency at which these synonymous codons are used varies widely. In general, codon usage is very uneven and is very species specific. Even different classes of genes within a given species have characteristic usage patterns. For example in highly expressed *E. coli* genes the relative frequencies of arginine codon use per 1000 amino acids are

| | | | |
|------|-----|------|------|
| AGG | 0.2 | CGG | 0.3 |
| AGA | 0.0 | CGA | 42.1 |
| CGA | 0.2 | CGC | 13.9 |

These nonrandom values have a significant effect on the selection of real open reading frames.

Some very interesting biological phenomena underlie the skewed statistics of codon usage. To some extent, the distribution of highly used codons must match the distribution of tRNAs capable of responding to these codons; otherwise, protein synthesis could not proceed efficiently. However, many other factors participate. Codon choice affects the DNA sequence in all reading frames; thus the choice of a particular codon may be mediated, indirectly, by the desire to avoid or promote a particular sequence in one of the other reading frames. For example, particular codon use near the starting point of protein synthesis can have the effect of strengthening or weakening the ribosome binding site of the mRNA. Rare codons usually correspond to rare tRNAs. This in turn will result in a pause in protein synthesis, while the ribosome-mRNA complex waits for such a tRNA to appear. This sort of pausing appears to be built into the sequence of many messages, since pausing at specific places, like domain boundaries, will assist the newly synthesized peptide chain to fold into its proper three-dimensional configuration.

Codon usage can also serve to encourage or avoid certain patterns of DNA sequence. For example, DNA sequences with successive A's spaced one helical repeat apart tend to be bent. This may be desirable in some regions and not in others. Inadvertent promoter-like sequences in the middle of actively transcribed genes are probably best avoided, as are accidental transcription terminators. The flexibility of the genetic code that results from its degeneracy allows organisms to synthesize whatever proteins they need, while avoiding these potential complications. For example, a continuous stretch of T's forms part of one of the transcription termination signals in *E. coli*. The resulting triplet, UUU, which codes for phenylalanine occurs only a third as often as the synonymous codon, UUC. Another example is seen with codons that signal termination of translation. UAA and UAG are two commonly used termination signals. The complements of these signals are UUA and CUA. These are both codons for leucine; however, CUA is the rarest leucine codon, and UUA is also rarely used.

There are many additional constraints on coding sequences beyond the statistics of codon usage. In real protein sequences there are some fairly strong patterns in the occurrence of adjacent amino acids. For example, in a beta sheet structure, adjacent amino acid side chains will point in opposite directions. Where the beta sheet forms part of the structural core of a protein domain, usually one side will be hydrophobic and face in toward the center of the protein, while the other side will be hydrophilic and face out toward the solvent. Thus codons that specify nonpolar residues will tend to alternate to some extent with codons that specify charged or polar residues. Alpha helices will have different patterns of alternation between polar and nonpolar residues because they have 3.4 residues per turn. In many protein structures one face of many of the alpha helices will be polar and one face will be nonpolar. These are not absolute rules; however, alpha helices and beta sheets are the predominant secondary structural motifs in proteins, and as such, they cast a strong statistical shadow over the sorts of codon patterns likely to be represented in a sequence that actually codes for a bona fide mRNA.

Quite a few other characteristics of known proteins seem to be general enough to affect the pattern of bases in coding DNA. Certain dipeptides like trp-trp and pro-pro are usually very rare; an exception occurs in collagenlike sequences, but then the pattern pro-pro-gly will be overwhelmingly dominant. Repeats and simple sequences tend to be rare inside of coding regions. Thus Alu sequences are unlikely to be seen within true coding regions; blocks like AAATTTCCCGGG . . . are also conspicuously rare. VNTRs are also usually absent in coding sequences, although there are some notable exceptions like the androgen receptor which contains three simple sequence repeats. Such exceptions may have important biological consequences, but we do not understand them yet. All of these statistically nonrandom aspects of protein sequence imply that we ought to be able to construct some rather elaborate and sophisticated algorithms for predicting ORFs and splice junctions. Seven of these that were used in the first successful algorithms for finding genes are described below.

### Frame Bias

In a true open reading frame (ORF), the sequence is parsed so that every fourth base must be the beginning of a codon. If we represent a reading frame as $(\ )_n$, in a true reading frame the bases in each position should tend to be those consistent with the preferred codon usage in the particular species observed. The other possible reading frames should tend to be those with poor codon usage. In this way the possibility of accidentally reading a message in the wrong frame will be minimized.

### Fickett Algorithm

This is an amalgam of several different tests. Some of these examine the 3-periodicity of each base versus the known properties of coding DNA. The 3-periodicity is the tendency for the same base to recur in the 1st, 4th, 7th, . . . positions. Other tests look at the overall base composition of the sequence.

### Fractal Dimension

Some dinucleotides are rare, while others are common. The fractal dimension measures the extent to which common codons are clustered with other common ones, and rare codons are clustered with other rare ones. Clustering of similar codon classes is characterized by a low fractal dimension, while alternation will lead to a high fractal dimension. It turns out that exons have low fractal dimensions, while introns have high fractal dimensions. Thus this test combines some features of codon usage, common dipeptide sequences, and simple sequence rejection.

## Coding Six-Tuple Word Preferences

A six-tuple is just a set of six continuous DNA bases. There are $4^6$ possible six-tuples in DNA. Since we have tens of millions of base pairs of DNA to examine for some species, we can make reasonable projections of the likely occurrence of each of these six-tuples in coding sequences versus noncoding sequences. An appropriately weighted sum of these predictions will allow an estimate of the chances that a given segment is coding.

## Coding Six-Tuple In-Frame Preferences

In this algorithm one computes the relative occurrence of preferred six-tuples in each possible reading frame. For true coding sequence, the real reading frame should show an excellent pattern of preferences, while in the other possible reading frames, when actual coding sequences have been examined, the six-tuple preferences appear to be fairly poor. This presumably aids in the selection and maintenance of the correct frame by the ribosome. This particular test turns out to be a very powerful one.

## Word Commonality

This test is also based on the statistics of occurrences of six-tuples. Introns tend to use very common six-tuples; exons tend to use rare six-tuples. Note that here we are talking about the overall frequency of occurrence of six-tuples and not their relative frequency in coding or noncoding regions.

## Repetitive Six-Tuple Word Preferences

This test looks specifically at the six-tuples that are common in the major classes of repeating DNA sequences. These six-tuples will also tend to be rare in true coding sequences.

The large list of tests just outlined raises an obvious dilemma: which one should be picked for best results? However, this is not an efficient way to approach such a problem. Instead, what one aims to do is find the optimal way to integrate the results of all of these tests to maximize the discrimination between coding and noncoding DNA. One must be prepared for the fact that the optimum measure will be species dependent; in addition it may well be dependent on the context of the particular sequence actually being examined. In other words, no simple generally applicable rule for combining the test results into a single score representing coding probability is likely to work. Instead, a much more sophisticated approach is needed. One such approach, which has been very successful, uses the methodology of artificial intelligence algorithms.

## NEURAL NET ANALYSIS OF DNA SEQUENCES

A neural net is one form of artificial intelligence. It is so named because, with neural net algorithms, one attempts to mimic the behavioral characteristics of networks of neurons.

We know that such networks can be trained (i.e., adjusted) to respond to signals or stimuli and to integrate the input from many different sources or sensors. Here the basic properties of neural nets will be illustrated, and then examples of how they have been applied to the analysis of DNA sequences will be shown.

The basic element in a neural net is a node, as shown in Figure 15.3a. This node receives input from one or more sensors, and it delivers output to one or more other nodes or a detector. The behavior of nodes is quantized. The signal input from each sensor is continuously scanned. It is recorded as positive if is above some threshold; otherwise, it is scored as negative (Fig. 15.3b). An input can be stimulatory or inhibitory. A node receiving a stimulatory input will send out the same sign signal. A node receiving an inhibitory signal will send out the opposite sign signal. By analogy, a nerve cell receiving a stimulatory impulse fires, while one receiving an inhibitory impulse does not fire.

Neural nets are collections of nodes wired in particular ways. They are generalizations of simple logical circuits. The variables in a neural net are the signal thresholds and the nature of the response of the nodes. We will illustrate this with three cases of increasing complexity. Consider the simple two-input node shown in Figure 15.3. Suppose that it operates under the following rules: If both sensors are positive, the node sends a positive output. Otherwise, it sends a negative output. This node is operating as the logical and function. It is behaving like a neuron that needs two simultaneous positive inputs in order to fire.

As a second case, consider the same node in Figure 15.3, but now imagine that the node sends a positive output if either input or both inputs are positive. The only way the node sends a negative output is if both sensors are reading negative. This node is acting like the logical and/or function. It stimulates a nerve cell that needs only one positive stimulus to fire.

The third case we will consider is a node that sends a positive signal if either input sensor is positive but not if both sensor inputs are positive. It is difficult to represent this behavior by a single node with simple $+/-$ binary logical properties. Instead, we can represent the behavior by a slightly more complex network with three nodes, as shown in Figure 15.4. Here the two sensors input their signal directly to two of the nodes. Each of these nodes views one input as stimulatory and the other input as inhibitory. Thus each node will fire if and only if it receives one positive and one negative signal. The two nodes feed stimulatory inputs into the third node. This node will be directed to fire if it receives a positive input from either one of the two nodes that precede it. One way to view the structure of the simple neural network shown in Figure 15.4 is that there is hidden

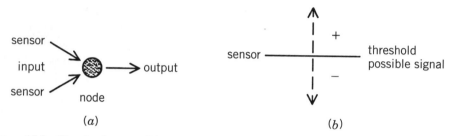

(a)                                                          (b)

**Figure 15.3**    The simplest possible neural net. This net can perform the logical operations "and" and "and/or." (a) Coupling of two inputs to a single output. (b) Effect of sensor threshold on signal value.

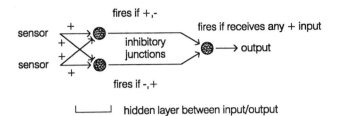

**Figure 15.4**   A more complex neural net which can perform the logical operation either but not both.

layer of nodes between the sensors and the final output node. In this particular case the hidden layer has a very simple structure; yet it is already capable of executing a complicated logical operation.

To use a neural net, one constructs a fairly general set of nodes and connections with one or more hidden layers, as shown in Figure 15.5. This is trained on sequences with known properties. The net is cycled through the training set of data, and weighting factors for each of the connections are adjusted to try to achieve the highest positive output scores for desired input characteristics and the lowest ones for undesired characteristics. A neural net could be used to examine DNA sequence directly, but this would take a very complex net, and the resulting training period would be computationally very intensive. Instead, what works quite satisfactorily is to use as sensor inputs, not individual bases, but instead the seven-sequence analysis algorithms described in the previous section. These sensors are each allowed to scan the DNA sequence over 10-base intervals. The net result of each scan is computed in a 99-base window. This is the length of sequence that is scanned and input into the net. Then the sequence is frameshifted by one base, and the analysis is repeated. The result is scaled, and then each sensor is fed into the neural net. The actual net structure used is shown in Figure 15.6. It consists of the 7 input sensors, 14 hidden nodes in a first layer, 5 hidden nodes in a second layer, and a single output node.

Edward Uberbacher and Robert Mural at Oak Ridge National Laboratory trained the neural net shown in Figure 15.6 on 240 kb of human DNA sequence data, adjusting thresholds, signs, and weighting until the performance of the net appeared to be optimum (1991). The result is a sequence analysis program called GRAIL. The detailed pattern of input into GRAIL from each of seven sensors for a particular DNA sequence is shown in Figure 15.7. Each plot shows the relative probability that the given 99-base window is an exon with coding potential. It is apparent that some sensors like coding six-tuple in frame preferences have much more powerful discrimination than others. However, when the input from all seven sensors is combined by the neural net, the result is a truly striking pattern of prediction of clear exons and introns. This is shown in Figure 15.8. GRAIL works

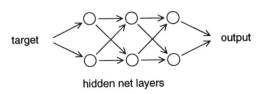

**Figure 15.5**   A still more complex neural net, with several hidden layers.

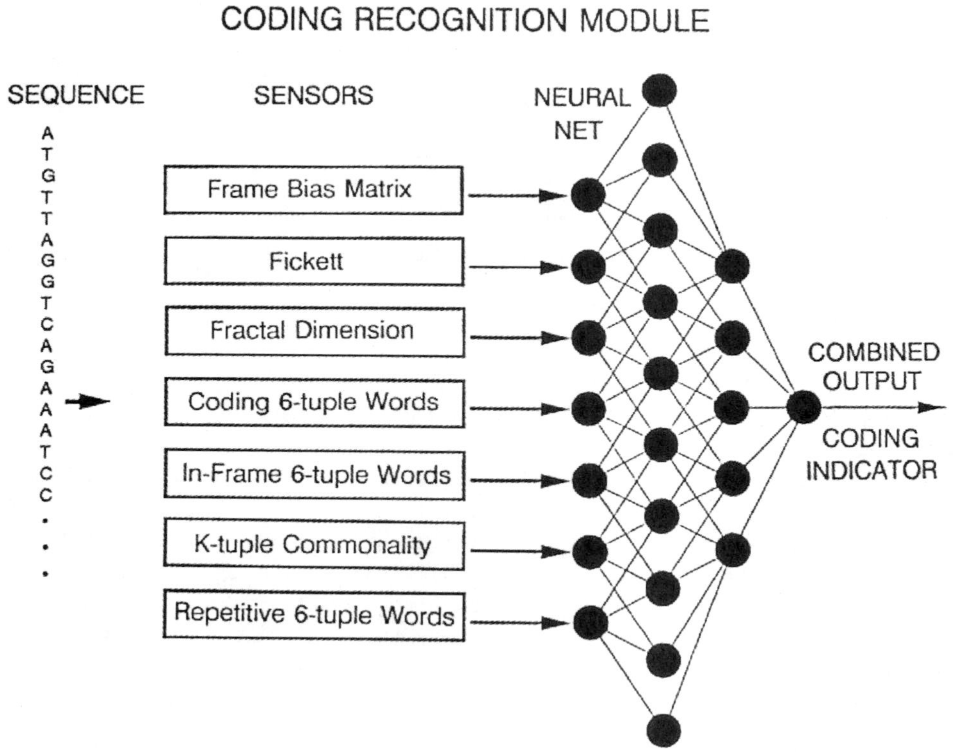

**Figure 15.6**  The actual neural net used in GRAIL analysis of DNA sequences. Adapted from Uberbacher and Mural (1991).

on many different types of human proteins that were not included in the original training set. A number of examples are shown in Figure 15.9. Some caution is needed, however, because not all human genomic sequence is handled well by GRAIL. For example, the human T-cell receptor gene cluster is not readably amenable to GRAIL analysis. The program also has difficulty in finding very small exons, which is not surprising in view of the 99-base window used.

Neural net approaches similar to GRAIL appear to have great promise in other complex problems in biological and chemical analysis. These include prediction of protein secondary and tertiary structure, correction of DNA sequencing errors, and analysis of mass spectrometric chemical fragmentation data. Note, however, that neural nets are only one of a number of different types of algorithmic approaches applicable to such problems, and the vote is still out on which will eventually turn out to be the most effective for particular classes of analysis. However, for the past half-decade, GRAIL has proved to be an extremely useful tool for most applications to human DNA sequence analysis, and it is readily accessible via computer networks, to all interested users.

Since the introduction of GRAIL, improvements have been made on the original algorithms to produce GRAIL 2. Other approaches to gene finding have been proposed, including a linear discriminant method (Solovyev et al., 1994) and, most recently, a quadratic discriminant method (Zhang, 1997). These methods take into account additional factors like the compatibility of the reading frames of adjacent exons and consensus sequences to the intron segment that forms a branched structure as an intermediate step in

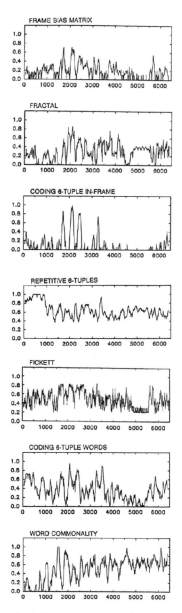

**Figure 15.7** Performance of each of the seven sensors of the net shown in Figure 15.6 on one particular DNA sequence. The vertical axis indicates the probability that each sliding segment of DNA sequence is a coding exon. Taken from Uberbacher and Mural (1991).

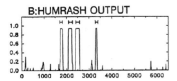

**Figure 15.8** The output of the neural net, based on its optimal evaluation of the sensor results shown in Figure 15.7. Adapted from Uberbacher and Mural (1991).

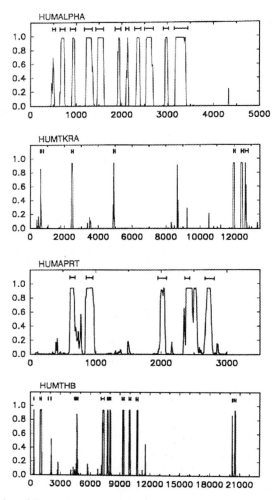

**Figure 15.9**    Examples of the performance of the neural net of Figure 15.6 on a set of different genomic DNA sequences. Adapted from Uberbacher and Mural (1991).

splicing. When tested in a large number of sequences, the three algorithms all perform well, but they are still far from perfect (Table 15.4).

**TABLE 15.4    Success of Exon Prediction: Exons Found by Three Different Schemes**

| Scheme | Sensitivity TP/(TP + FN) | Specificity TP/(TP + FP)S |
|---|---|---|
| GRAIL 2 | 0.53 | 0.60 |
| Linear discriminant analysis | 0.73 | 0.75 |
| Quadratic discriminant analysis | 0.78 | 0.86 |

*Source:* Adapted from Hong (1997)

Note: True positives (TP) are true positives correctly predicted. False positives (FP) are true negatives predicted to be positive. False negatives (FN) are true positives predicted to be negative. Sensitivity is the fraction of true positives found. Specificity is the fraction of positives found that is true.

# SURVEY OF PAST LARGE-SCALE DNA SEQUENCING PROJECTS

Most early large-scale DNA sequencing projects involved a pre-selected gene of particular interest. An example is the enzyme HPRT (57 kb). These projects are milestones in the history of DNA sequencing, but it is difficult to extrapolate the results of such projects to the situation that will apply in most genomic sequencing efforts. In such efforts, which will form the overwhelming bulk of the human genome project, one will be faced with large expanses of relatively uncharted DNA. While the regions selected may contain a few mapped genes, and many cDNA fragments, much of the rationale for looking at the particular region will have to come a posteriori, after the sequence has been completed. To try to get some impression of the difficulties in assembling the sequence, and making a first pass at its interpretation, it is useful to examine the first few efforts at sequencing segments of DNA without a strong functional pre-selection. Here we summarize results from seven projects: the complete sequence of *H. influenzae, M. genitalium,* partial sequences of *E. coli, S. cerevisiae, C. elegans,* and *D. melanogaster,* and several human cosmid DNAs. These sequence data and all other genomic sequence data currently reside in a set of publicly accessible databases. A description of these valuable resources, and how they can be accessed, is provided in the Appendix. A summary of all complete genome sequences publicly available in February 1997 is given in Table 15.5.

The complete DNA sequences of *Haemophilus influenzae* and *Mycoplasma genitalium* both correspond to relatively small bacterial genomes. As expected, they are very rich in genes, and they are especially rich in genes whose function can be surmised by comparison to other sequences in the available genome databases. *M. genetalium* has a 580,070 bp genome with 470 ORFs. These occur on average one per 1235 bp. The average ORF is 1040 bp. Overall the genome is 80% coding. Seventy-three percent of the ORF's correspond to previously known genes.

*H. influenza* has a genome size of 1,830,137 bp. This contains 1743 coding regions, an average of one every 1042 bp. The average gene is 900 bp long. Overall, 85% of the genome is coding. Currently 1007 (58%) of the coding regions can be assigned a functional role. Of the remainder, 385 are new genes that show no significant matches to the databases, while the others match known sequences of unknown function. At an average direct cost of $0.48 per base this project is probably representative of other large-scale efforts using similar technology.

Both the *H. influenzae* and *M. genetalium* sequencing projects were carried out at a single location totally by automated fluorescent DNA sequencing. In contrast, one of the

**TABLE 15.5  Completed Genome Sequences**

| Species | DNA Molecules | kb DNA | Largest DNA (kb) | Open Reading Frames | Genes for RNA |
|---|---|---|---|---|---|
| *M. genitalium* | 1 | 580 | 580 | 470 | 38 |
| *M. pneumonia* | 1 | 816 | 816 | 677 | 39 |
| *M. janneschii* | 3 | 1740 | 1665 | 1738 | ~45 |
| *H. influenza* | 1 | 1830 | 1830 | 1743 | 76 |
| *Synechoncystis sp.* | 1 | 3573 | 3573 | 3168 | ? |
| *E. coli* | 1 | 4639 | 4639 | 4200 | ? |
| *S. cerevisiae* | 16 | 12,068 | 1532 | 5885 | 455 |

efforts to sequence major sections of the *E. coli* genome, directed by Fred Blattner in Madison, Wisconsin, started as basically low-technology, manual DNA sequencing, employing a large number of relatively unskilled workers, and concentrated on relatively simple protocols. The initial result was a 91.4 kb contig. The region contained 82 predicted ORFs or roughly one per kb. The ORFs constituted about 84% of the total sequence. If we scale the properties of this region to the entire 4.7 Mb *E. coli* genome, we can predict that

$$\frac{4.7 \text{ Mb} \times 82 \text{ ORFs}}{0.0914 \text{ Mb}} = 4200 \text{ genes}$$

This is larger than estimates of the number of genes in *E. coli* based on the appearance of protein spots in two-dimensional electrophoretic separations. Past sampling of *E. coli* regions has revealed fairly uniform gene density except for areas around the terminus of replication. Hence the preliminary sequencing results on *E. coli* suggest that a significant number of new and interesting genes remain to be discovered. A more recent report of additional *E. coli* sequences is quite consistent with the earlier observations within a 338,500 base contig, 319 ORFs were found—one per 1060 bases. Of these, 46% are potentially new genes. The complete *E. coli* DNA sequence has just became available, and it contains 4300 genes, in 4.54 Mb, quite consistent with predictions based on partial sequencing results.

The early major accomplishments in *S. cerevisiae* sequencing derive from a very different organizational model than the work on *E. coli*. The approach was still mostly very low technology. It was mostly the result of a dispersed European effort among more than 30 different laboratories, coordinated through a common data collection center in France. The complete DNA sequence of one of the smallest *S. cerevisiae* chromosomes, number III, was the first one determined. At 315 kb it represented the longest continuous stretch of DNA sequence known at the time. The chromosome III sequence was originally reported to contain 182 ORFs. After this was corrected by a more rigorous examination, carried out by Christian Sander in Heidelberg, 176 ORFs remained. These occur at roughly one per 2 kb or half of the density seen in the three bacteria discussed above. The ORFs cover 70% of the DNA sequence; this is not too much lower than the total density of coding sequence in *E. coli*. We can make a rough estimate the number of genes in *S. cerevisiae* by scaling these results to the 12.1 Mb total size of the yeast genome. The result is

$$\frac{12.1 \text{ Mb} \times 176 \text{ ORFs}}{0.315 \text{ Mb}} = 6760 \text{ ORFs}$$

The total number of genes in *S. cerevisiae* will be slightly less than the number of ORFs because occasional genes in yeast consist of more than one exon. In addition, for both bacteria and yeast, we have to add in genes for rRNAs, tRNAs, and other nontranslated species (Table 15.5).

The complete DNA sequences of several other *S. cerevisiae* chromosomes reported were consistent with the results for chromosome III. For example, chromosome VIII has 562,698 bp. It contains 269 ORFs, or 1 per 2 kb. Of these, 124 (46%) corresponded to genes of known function. Chromosome VI has 270 kb. It contains 129 ORFs, again about 1 per 2kb. Of these, 76 (59%) correspond to genes with previously known function. The total sequence of *S. cerevisiae* is now completed. First estimates place the number of ORFs at 5885; doubtless this will change with further analysis.

In the case of *C. elegans* DNA sequencing, we are dealing not with continuous genomic sequence but with the sequence of selected cosmids. The effort, directed by John Sulston of Cambridge, England, and Robert Waterston of St. Louis, Missouri, is also state-of-the art fluorescent DNA sequencing technology with a great deal of automation. The strategy is mostly shotgun, with directed sequencing relegated mostly to closure of gaps between contigs. The first 21.14 Mb of *C. elegans* DNA sequence reported contained a total of 3980 genes of 1 per 4.8 kb on the autosomes and 1 per 6.6 kb on the X chromosome. Only 46% of these matched sequences already in the DNA databases. About 28% of the total DNA is coding; 50% of *C. elegans* is genes, including both exons and introns. This is a sharp drop from the density of coding sequences in simple organisms. The total number of genes in the nematode genome is estimated to be $13,000 \pm 500$. This is a number close to most contemporary expectations for the sizes of the genomes of typical multicellular, highly differentiated organisms like the nematode.

The remaining two DNA sequencing projects that we will discuss illustrate some of the frustrations in detailing with the genomes of higher organisms. The complete DNA sequence of a 338,234 bp region of *D. Melanogaster,* containing the bithorax complex, important in development, has been reported by groups at Caltech and Berkeley. This region is less than 2% coding. It contains only six genes. The final sequencing project we will discuss is a relatively early effort that involved several cosmids from the tip of the short arm of human chromosome 4, a region known to contain the gene responsible for Huntington's disease. The region is band 4p16.3. It is estimated to contain a total of 2.5 Mb of DNA. A 225-kb subset of this region was sequenced. This yielded 13 transcripts in 225 kb or one per 18 kb on average. Another estimate of gene density could be obtained by determining the number of HTF islands in the region. This will be a minimum estimate for the number of genes, since perhaps only half to two-thirds of all genes have HTF islands nearby. In fact, in the 225 kb region, one HTF island was found on average per 28 kb. By comparison, when HTF islands were mapped to a different section of chromosome 4, a 460 kb region near the marker D4S111, the frequency of occurrence of these gene-associated sequences was one per 30 kb. All of these estimates of gene density are remarkably consistent. If we scale these expected gene densities to the entire Huntington's disease region, we obtain an estimate of

$$\frac{2.5 \text{ Mb} \times 13 \text{ genes}}{0.225 \text{ Mb}} = 143 \text{ genes}$$

This makes it clear why finding the gene for Huntington's disease was not an easy task.

The first DNA sequencing effort in band 4p16.3 was carried out in Bethesda, Maryland, under the direction of Craig Ventor. It involved a total of 58 kb of DNA sequence in three cosmids. Three genes were found, each has an HTF island. The average gene density in this relatively small region is one per 19 kb, which is quite consistent with expectations. Less than 10% of the region is coding sequence. The number of Alu repeats in the region is 62, or roughly one per kb. This is comparable to what has been seen in the DNA sequence of two other gene rich, G + C-rich regions. In the human growth hormone region 0.7 Alu's were found per kb; in the HRPT region 0.9 Alu's were found per kb. In stark contrast, in the globin region which is G + C poor, there are only 0.1 Alu's per kb. These results illustrate the mosaic nature of the human genome rather dramatically.

Unlike simple genomes, with relatively uniform DNA compositions, mammalian genomes have mosaic compositions which is reflected in chromosome banding patterns. Scaling of a regional gene density to estimate the total number of genes, must take into account regional characteristics. Long before large-scale DNA sequencing or genome mapping was underway, Georgio Bernardi developed a method of fractionating genomes into regions with various G + C content. This was done by equilibrium ultracentrifugation in density gradients (Chapter 5). The resulting fractions were called isochores. Altogether, Bernardi obtained evidence for five distinct human DNA classes; these could be divided into three easily separated and manipulated fractions. Their properties are summarized below:

| CLASS | GENE DENSITY | GENOME FRACTION | LOCATION |
|---|---|---|---|
| L1,L2 | 1 | 62% | Dark bands |
| H1,H2 | 2 | 31% | Light bands |
| H3 | 16 | 7% | Telomeric light bands |

Several aspects of these results deserve comment. Gene density means the relative number of genes, based on cDNA library comparisons. The genome fraction is estimated from the total amount of material in the density-separated fractions. The telomeric light bands have very special properties, that we have alluded to before. Figure 15.10 illustrates the actual locations seen when DNAs from Bernardi's fraction H3 are mapped by FISH. The preferential location of these sequences on just a small subset of human chromosomal regions is really remarkable.

The Huntington's disease region is known to be a gene-rich light band, so we can pretty much exclude the L1 and L2 classes from consideration. In the Huntington's region, there is one gene on average per 18 kb. If this region is an H3 region, then we can estimate the number of genes in the human genome as

| H3 | 11,700 genes |
|---|---|
| H1,H2 | 6500 genes |
| L1,L2 | 6500 genes |

for a total of 24,700 genes. This estimate is less than twice the number of genes in *C. elegans,* which seems far too low. If we assume that the Huntington's disease region is an H1,H2 region, then the estimate of the number of genes in the human genome becomes

| H3 | 92,000 genes |
|---|---|
| H1,H2 | 51,100 genes |
| L1,L2 | 51,100 genes for a total of 194,200 genes. |

This is a depressingly large number, much larger than previous estimates. This example illustrates how difficult it is to know from very fragmentary data what the real target size of the human genome project is. Perhaps the Huntington's disease region is somewhere between the properties of the H3, and H1 plus H2 fractions, and the gene number somewhere mercifully between the two rather upsetting extremes we have computed. More recent estimates of the number of human genes range from 65,000 to 150,000, which is not too different from the average of our original estimates.

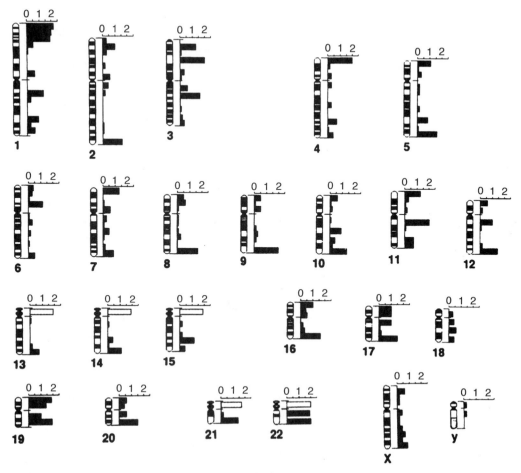

**Figure 15.10**  Distribution of extremely G + C-rich sequences in the human genome. Solid bars show relative hybridization of the H3 dark fraction. Open bars show rRNA-encoding DNA. Taken from Saccone et al. (1992).

## FINDING ERRORS IN DNA SEQUENCES

Quite a few different kinds of errors contaminate data in existing DNA sequence banks. As the amount of data escalates, it will become increasingly important to audit these data continuously. Suspect data need to be flagged before they propagate and affect the results of many sequence comparisons or experimental scientific efforts. For example, an error in one of the earliest complete DNA sequences, the plasmid pBR322, produced a spurious stop codon in one of the proteins coded for by this plasmid. This confounded many researchers who were using this plasmid as a cloning and expression system, since a protein band with an unexplainable size was frequently seen.

Some common errors in DNA sequence data are quite easy to find and correct; others are almost impossible. A major class of error is incorporation of a totally inappropriate sequence. This can come about if, as is not uncommon, DNA samples are mixed up in the laboratory prior to sequencing. It can arise from cloning artifacts. A clone may have picked up bacterial DNA rather than the intended mammalian insert. A number of simple schemes exist that can help to find such errors. Putative genomic or cDNA sequences should be screened against all known common vector sequences. A very frequent error is to include pieces of vector inadvertently as part of the supposed insert. The presence of common repeats like Alu should be searched for in putative cDNAs or exons. Except in the rarest cases, these sequences should not be present there; finding them suggests that a cloning artifact may have occurred.

Rearrangements in cosmid clones and YACs are fairly common. The best way to find these errors at the DNA sequence level is to compare the sequences with other clones in available contigs. A major justification for the additional DNA sequencing required to examine a tiling set of cosmids is that there will be frequent overlaps which can help catch errors caused by rearrangements. Small sequencing errors are still about 1% in automated or manual sequencing. In many past efforts, considerable amounts of data were entered into sequence databases manually. It is vital that this be verified by a process of double entry and comparison. If not, except in the hands of the most compulsively careful individuals, typographical errors will abound.

When a single base is miscalled, either by misreading raw sequence data or by mistranscription in manipulating that data, the error is extremely difficult to detect. However, when a base is inserted or deleted, especially within an ORF, the error is sometimes easily caught. One way to do this is a procedure developed by Janos Posfai and Richard Roberts. In the course of searching a DNA database, to examine possible homology between a new sequence and all preexisting sequences, one can ask whether potential strong sequence homology (usually after the DNA has been translated into protein) is blocked by a frame shift. Where this occurs, a DNA sequencing error is almost always responsible. Several examples of the power of this approach in spotting sequencing errors are shown in Figure 15.11.

An unsolved problem is how to alert the community when errors are found. Given the size of the community and the complexity of the queries it makes against the sequence databases, this is an enormous problem. At some point the databases will have to be intel-

```
D: Transposase TNP A          E: Transposon Tn917

suspected DNA sequence error between        suspected DNA sequence error between nt
positions 3998 and 4032                     positions 4091 and 4103
        473           570                           587           616
D  LRNHIRSGDI.............PRIKLTDLLI         D  LIEVASWTGF.............GTNIGLTKMA
   LRNNIRSGDI........................          LIEVDSWTNF........................
E  .........................PRIKLSDLLI       E  .........................GMNIGLEKMA
        481           565                           582           611

N suspected DNA sequence error between nt   suspected DNA sequence error between nt
positions 4374 and 4395                     positions 4993 and 5056
        681           732                           885           909
D  GTTSSSDGMR...........LDGLLHHETD          D  LNKGEAINAL.............ASALNIIINA
   GKTTASDGMR........................          LNKTEAINAL........................
E  .........................LDGLLYDETD      E  .........................ASALNVLINA
        676           728                           881           904
```

**Figure 15.11** Finding frameshift errors by comparing a new sequence with sequences preexisting in the databases. Adapted from Posfai and Roberts (1992).

ligent enough to be able to evaluate the effect of corrections on past queries and alert the initiators of those queries that might now be subject to altered outcomes. If this cannot be done, inevitably people will begin to repeat queries over and over again to guard against the effects of errors. A second potential unsolved problem is how to deal with fraudulent sequences. Research journals are increasingly reluctant to publish DNA sequence results, and it is almost impossible to publish the raw data supporting DNA sequencing results. Because of this, much sequence data are submitted directly to databases without editorial review of the actual experimental data. This entails the risk that databases might become contaminated willfully or accidentally by the deposit of sequences marred by artifacts or totally artificial. Just how these sequences could be detected and removed remains a serious dilemma. Ultimately it may be necessary to link the databases to archives of raw data so that validation of a suspected artifact is feasible.

## SEARCHING FOR THE BIOLOGICAL FUNCTION OF DNA SEQUENCES

The major thrust of biological research is to understand function. From the viewpoint of the genome, this search for function can occur at two very different levels: individual genes or patterns of gene organization. We first discuss the genome from this latter vantage point. An overview of the arrangement of sequences in the genome may provide patterns of information that offer a clue to global aspects of function. These may be domains of gene activity or gene type that reflect biological processes we have not yet discovered. For example, most similar or related genes are not clustered. Some small clusters are seen, such as the globin genes (Fig. 2.10). The pattern of arrangement of the genes in these clusters presumably reflects an ancient gene duplication, which separated the alpha and beta families, and more recent duplications that evolved the more closely related members of these families. What is striking, and not yet explained, is that the order of the genes in each of these families accurately corresponds to the temporal order in which the genes are expressed during human development.

Another example of intriguing patterns of gene arrangement is the hox gene family in man and the mouse, shown in Figure 15.12. The genes in this family code for factors that determine the segmental pattern of organization of the developing embryo. The family is

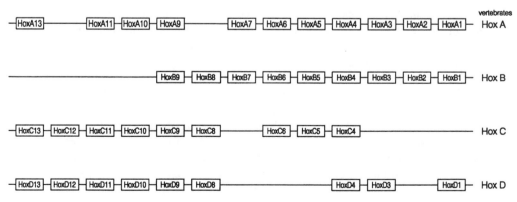

**Figure 15.12** Organization of homeobox (hox) genes in the mammalian genomes. All genes in all four clusters are transcribed from left to right.

complex, and dispersed on a number of different chromosomes. Two aspects of the organization of the family are striking. First, it is so well conserved between the two species. Second, the spatial order of the genes within the family is the same as the order of the segments in the embryo that these genes affect. It is as if, for some totally unexplained reason, the map of the structure of the gene family is an image of the map of the function of that family.

A final example of functionally interesting gene arrangements is seen in a number of the members of the immunoglobulin superfamily including the light and heavy chains of antibodies, and several chains of the T-cell receptor. Here large numbers of related genes are grouped together, mostly in a single continuous segment of a chromosome. The reason for this is probably to assist the rearrangement of these genes, which takes place by DNA splicing to form mature expressed genes for antigen-specific proteins. If other regions of the genome are found with very large clusters of similar genes, one may well suspect that somatic DNA rearrangement or some other unusual biological mechanism will be at play with these genes.

A totally different view of global function afforded by complete physical maps and DNA sequences is the ability to compare these physical structures of DNA with the genetic map. An example is shown in Figure 15.13 for yeast chromosome III. There are clearly some regions where meiotic recombination is much more frequent than average and others where it is greatly suppressed. We do not yet understand the origin of these effects. One possibility is just the presence or absence of local DNA sequences that constitute recombination hot spots. However, there are other more global possibilities. Recombination may correlate with overall transcriptional activity, since highly transcribed chromatin is more open and accessible to all types of enzymes including those responsible for recombination. Thus there may be positional relationships between gene function and recombination, and thus gene evolution, that we still know nothing about today.

## SEARCHING FOR THE BIOLOGICAL FUNCTION OF GENES

Most biologists, when they think of biological function in the context of the genome project, are referring to the function of individual genes. A common criticism of the genome project is that it is relatively useless to know the DNA sequences of genes without strong prior hints about their function. Most traditional molecular genetics begins with a function of interest and attempts to find the genes that determine or affect that function. This traditional view of biology is contrasted with the challenge posed by genome research in Figure 14.1, which can well serve as a paradigm for all of biology. In genome research we will discover DNA sequences with no a priori known function. Our current ability to translate these DNA sequences correctly into protein sequences is excellent, as we showed earlier, by using GRAIL or other powerful algorithms. Our current ability to take these protein sequences and draw immediate inferences about their possible function is well illustrated by the example in Figure 15.14a. Except for those rare readers of this book who are conversant in Dutch, this passage is largely unreadable. However, the frustrating aspect is that the passage is not totally unreadable. Because a number of scientific terms are cognates in Dutch and English, certain features stand out—one knows the passage has something to do with protein structure, but the full impact of the message is completely lost.

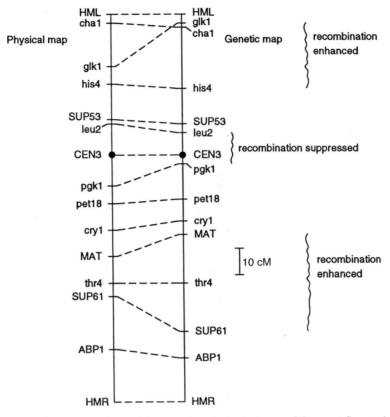

**Figure 15.13** A comparison of the genetic and physical map of the yeast *S. cerevisiae.*

Ingewikkelde en grote biologische
macro-moleculen kunnen spontaan
in hun meest stabiele conformatie vouwen.
Helaas, ontbreekt ons de kennis om dit
proces te voorspellen want de gevouwen
strutuur kan belangrijke aanwijzingen
over die functie van het molecuul bevatten.

(*a*)

We know that large biological molecular can fold
into their most stable state spontaneously but
we really have little ability at present to predict
this folding. Our ignorance is most unfortunate
since the folded structure may contain
important clues on how the molecule functions.

(*b*)

**Figure 15.14** An analogy for the current (*a*) and desired future (*b*) ability to interpret DNA sequence in terms of its likely biological function.

When the passage in Figure 15.14*a* is translated into English, it provides an important clue to one direction that can help find functional clues (Fig. 15.14*b*). Considerable experience to date shows that protein three-dimensional structures are better conserved during evolution than protein sequences. A great deal of current research effort is being devoted to improving our ability to infer possible protein structures from the sequences of sets of related proteins, provided that at least one of them has a known three-dimensional structure. As our ability to do this improves, and as the number of different classes of protein structures has one or more members successfully studied at high-resolution by X-ray crystallographic or nuclear magnetic resonance techniques, the prospects of stepping quickly from a sequence to a realistic, if not exact, model of the structure should improve markedly. However, just knowing a three-dimensional structure does not immediately provide definitive clues to function. It simply makes comparisons between a protein of unknown function and the set of proteins of known function more powerful and more likely to yield useful insights.

Today, when a new segment of DNA sequence is determined, the first thing that is almost always done with it is to compare it to all other known DNA sequences. The purpose is to see if it is related to anything already known. By related, we mean, that there is a statistically significant similarity to one or more preexisting DNA sequences. The definition of what statistically significant means in the context of sequence comparisons is not universally accepted despite decades of work in this area. Obviously, at one extreme, one may find that a new sequence is virtually identical to a preexisting one. Unless the two sequences derive from very similar but not identical organisms, the finding of near identity means true identity with the differences due to sequencing errors, or a new member of a gene family, or an example of proteins very strongly conserved in evolution, like the histones. At the other extreme, a new sequence may match nothing to within whatever local standards of minimal homology are considered operative.

Most often, however, when a new DNA sequence is compared with the current data base of more than 1000 Mb of DNA, some slight or significant sequence homology is found. For coding sequences, it is usually much more powerful to search after translation of DNA to protein. This translation loses very little functional information; it gains considerable statistical power because the noise caused by the degeneracy of the genetic code is blanked out. Thus consider, for example, two arginine codons like AGG and CGA in a corresponding place on two sequences; the only evidence for similarity is the G in position 2, which has roughly one chance in four of occurring randomly. In contrast, posing an arginine opposite an arginine at the same place in a protein sequence has, very crudely, only one chance in 20 of occurring randomly. (In reality the statistical differences are not this great because amino acids with six possible codons, like arginine, also tend to occur much more often than average.)

A statistically significant match between a new sequence and some preexisting sequence implies some or all of the following possibilities: similar function, similar structure, or evolutionary relatedness. It is not easy to sort out these different effects. However, one encouraging feature of such global sequence searches is that their effectiveness appears to be increasing markedly and rapidly as the database grows. Ten years ago Russell Doolittle noted that a new protein sequence had a 25% chance of matching something else in the databases. Currently the odds are considerably better than this. From the first bacterial sequencing projects described earlier in this chapter, between 54% and 78% of the ORFs found showed hints of homology in structure or function with something else in the data base. With the *S. cerevisiae* ORFs on chromosome III, 42% gave hints of homologous structure or function of which 14% were deemed really quite strong. In the case of *C. elegans,* where more extensive data are available, 45% of the ORFs were reported to be relatable to existing databases. It seems likely that in a few years it will be the odd new sequence that does not immediately match something known. While it is too early to be sure how rapidly this goal will be achieved, there is room for considerable optimism at present.

## METHODS FOR COMPARING SEQUENCES

Entire books have been written about the relative merits of different approaches to aligning sequences and testing their relatedness (Waterman, 1995; Gribokow and Deveraux, 1991). The topic is actually quite complex because the nonrandom nature of natural DNA sequences greatly confounds attempts to construct simple statistical tests of relatedness. Here our goal will be to present the basic notions of how sequences are compared and what these comparisons mean. Sequences are strings of symbols. Any two strings can be compared by direct alignment and the use of scoring criteria for similarity. For two strings of length $n$ and $m$ there are $2(n + m - 1)$ possible continuous alignments, by which we mean that no gaps are allowed in either string. Of course many of these alignments are fairly trivial and uninteresting because the strings will barely overlap. The moment gaps are allowed on one or both strings, the number of alignments rises in a combinatorial manner to reach heights that can test the power of the fastest existing supercomputers if the problem is not handled intelligently.

An example of a very simple case in which two very similar DNA sequences are aligned is shown in Figure 15.15. In this case the alignment needed to maximize the apparent similarity between the two sequences is obvious. What is less obvious is the sort of score to give such an alignment. The simplest scoring scheme is black and white: Grade all identities the same and all differences the same. However, this makes little sense from either a biological or a statistical vantage point. As far as biology is concerned, if, for example, we are looking at the functional relatedness of proteins coded for by these sequences, or if we are looking at possible evolutionary relationships between them, transversions (interchange of a purine and a pyrimidine) should be weighted as more consequential differences than transitions (interchange between two pyrimidines or two purines). This is because the rate of transversion mutations is much less than the rate of transitions, and the genetic code appears to have evolved so that effects of transversions on the resulting amino acids are more functionally disruptive than the effect of transitions. For example, many synonymous codons are related by a transition in their third position. But the example goes much deeper; for example, codons for different hydrophobic amino acids are also related mostly by transitions.

```
                    AGCTTACGCAAACC
                    GCTCACGGTTGCCA
        identities  I I I0I I I0I I0I I
        mutations   0 00S000V00S00
```

**Figure 15.15**   A simple example of a comparison between two putatively related nucleic acid sequences and two ways in which their relatedness could be scored, $S$ = transition and $V$ = transversion.

To take statistical factors into account in estimating the significance of a mismatch or a match purely at the DNA level, we have to consider the relative frequency of each residue in the strings being compared. For example, sequences rich in A's will show large numbers of A's matched with A's, just by chance. In order to take this into account, and to add issues like transitions and transversions, one needs to employ a scoring matrix. This is illustrated in Figure 15.16*a*. The 4 × 4 scoring matrix for nucleic acid comparisons allows for any possible weight to be assigned to a particular set of bases at an alignment position. Generally, the same scoring matrix is used for every alignment position, although there is no reason why one should have to do this, nor is there any reason why it is desirable except for simplicity. Think ahead to the alignment of protein sequences where residues on exterior loops can be quite variable without perturbing the overall structure. Therefore, if one had some way of knowing a priori that a residue was in a loop as opposed to a helix or sheet, one could adjust the weighting factors accordingly. This example illustrates the complex interplay between sequence and structure information that really has to occur in very robust comparison algorithms.

The simplest possible DNA scoring matrix, corresponding to the rule used in Figure 15.15 is just a set of identities with no correction for overall base composition (Fig. 15.16*b*). The general case would consist of a set of elements $a_{ij}$ that are all different, except that the matrix should be symmetrical; each $a_{ij} = a_{ji}$ since we have no way, in comparing just two proteins, to favor one sequence over another. The elements $a_{ij}$ must incorporate all of our biological and statistical prejudices. When protein sequences are compared, the scoring matrices can become more complicated. First of all, the matrix must be 20 × 20 instead of 4 × 4. It can be as simple as an identity matrix, just as in the case for nucleic acids, but a much more accurate picture will incorporate statistical information about the relative frequency of amino acids. This immediately raises one serious problem: Does one use the amino acid composition of the two proteins in question to construct the scoring matrix, or does one use the amino acid compositions of all known proteins, or all known proteins from the particular species involved? One can elaborate the problem even further by asking whether the nonrandomness of dipeptide frequencies should be considered in making statistical evaluations for the scoring matrix. There are no simple answers to these questions.

Most commonly, with protein sequence comparisons, one incorporates information about amino acid physical properties into the values of the elements of the scoring matrix. Thus, for example, interchanges among ile, leu, and val, or ser and thr, among proteins known to be related in structure and function are very commonly seen and are presumably mostly innocuous. Examples of two real scoring matrices are shown in Figure 15.17.

$$\text{sequence 1}\quad \begin{array}{c} \\ A \\ G \\ T \\ C \end{array}\overset{\text{sequence 2}}{\overset{\begin{array}{cccc} A & G & T & C \end{array}}{\begin{pmatrix} X_{AA} & X_{AG} & X_{AT} & X_{AC} \\ X_{GA} & X_{GG} & X_{GT} & X_{GC} \\ X_{TA} & X_{TG} & X_{TT} & X_{TC} \\ X_{CA} & X_{CG} & X_{TC} & X_{CC} \end{pmatrix}}}
\qquad
\text{sequence 1}\quad \begin{array}{c} \\ A \\ G \\ T \\ C \end{array}\overset{\text{sequence 2}}{\overset{\begin{array}{cccc} A & G & T & C \end{array}}{\begin{pmatrix} 1 & 0 & 0 & 0 \\ 0 & 1 & 0 & 0 \\ 0 & 0 & 1 & 0 \\ 0 & 0 & 0 & 1 \end{pmatrix}}}$$

**Figure 15.16** Comparison matrices between two nucleic acid sequences. *(a)* A general matrix. *(b)* The simplest possible matrix.

|   | A | C | D | E | F | G | H | I | K | L | M | N | P | Q | R | S | T | V | W | Y |
|---|---|---|---|---|---|---|---|---|---|---|---|---|---|---|---|---|---|---|---|---|
| A | 4 | . | . | . | . | . | . | . | . | . | . | . | . | . | . | . | . | . | . | . |
| C | 0 | 9 | . | . | . | . | . | . | . | . | . | . | . | . | . | . | . | . | . | . |
| D | -2 | -3 | 6 | . | . | . | . | . | . | . | . | . | . | . | . | . | . | . | . | . |
| E | -1 | -4 | 2 | 5 | . | . | . | . | . | . | . | . | . | . | . | . | . | . | . | . |
| F | -2 | -2 | -3 | -3 | 6 | . | . | . | . | . | . | . | . | . | . | . | . | . | . | . |
| G | 0 | -3 | -1 | -2 | -3 | 6 | . | . | . | . | . | . | . | . | . | . | . | . | . | . |
| H | -2 | -3 | -1 | 0 | -1 | -2 | 8 | . | . | . | . | . | . | . | . | . | . | . | . | . |
| I | -1 | -1 | -3 | -3 | 0 | -4 | -3 | 4 | . | . | . | . | . | . | . | . | . | . | . | . |
| K | -1 | -3 | -1 | 1 | -3 | -2 | -1 | -3 | 5 | . | . | . | . | . | . | . | . | . | . | . |
| L | -1 | -1 | -4 | -3 | 0 | -4 | -3 | 2 | -2 | 4 | . | . | . | . | . | . | . | . | . | . |
| M | -1 | -1 | -3 | -2 | 0 | -3 | -2 | 1 | -1 | 2 | 5 | . | . | . | . | . | . | . | . | . |
| N | -2 | -3 | 1 | 0 | -3 | 0 | 1 | -3 | 0 | -3 | -2 | 6 | . | . | . | . | . | . | . | . |
| P | -1 | -3 | -1 | -1 | -4 | -2 | -2 | -3 | -1 | -3 | -2 | -2 | 7 | . | . | . | . | . | . | . |
| Q | -1 | -3 | 0 | 2 | -3 | -2 | 0 | -3 | 1 | -2 | 0 | 0 | -1 | 5 | . | . | . | . | . | . |
| R | -1 | -3 | -2 | 0 | -3 | -2 | 0 | -3 | 2 | -2 | -1 | 0 | -2 | 1 | 5 | . | . | . | . | . |
| S | 1 | -1 | 0 | 0 | -2 | 0 | -1 | -2 | 0 | -2 | -1 | 1 | -1 | 0 | -1 | 4 | . | . | . | . |
| T | 0 | -1 | -1 | -1 | -2 | -2 | -2 | -1 | -1 | -1 | -1 | 0 | -1 | -1 | -1 | 1 | 5 | . | . | . |
| V | 0 | -1 | -3 | -2 | -1 | -3 | -3 | 3 | -2 | 1 | 1 | -3 | -2 | -2 | -3 | -2 | 0 | 4 | . | . |
| W | -3 | -2 | -4 | -3 | 1 | -2 | -2 | -3 | -3 | -2 | -1 | -4 | -4 | -2 | -3 | -3 | -2 | -3 | 11 | . |
| Y | -2 | -2 | -3 | -2 | 3 | -3 | 2 | -1 | -2 | -1 | -1 | -2 | -3 | -1 | -2 | -2 | -2 | -1 | 2 | 7 |

*(a)*

|   | A | C | D | E | F | G | H | I | K | L | M | N | P | Q | R | S | T | V | W | Y |
|---|---|---|---|---|---|---|---|---|---|---|---|---|---|---|---|---|---|---|---|---|
| A | 4 | . | . | . | . | . | . | . | . | . | . | . | . | . | . | . | . | . | . | . |
| C | -2 | 10 | . | . | . | . | . | . | . | . | . | . | . | . | . | . | . | . | . | . |
| D | -1 | -5 | 5 | . | . | . | . | . | . | . | . | . | . | . | . | . | . | . | . | . |
| E | -1 | -5 | 1 | 5 | . | . | . | . | . | . | . | . | . | . | . | . | . | . | . | . |
| F | -2 | -2 | -5 | -4 | 7 | . | . | . | . | . | . | . | . | . | . | . | . | . | . | . |
| G | 0 | -4 | -1 | -2 | -5 | 5 | . | . | . | . | . | . | . | . | . | . | . | . | . | . |
| H | -2 | -5 | -1 | -1 | -1 | -2 | 8 | . | . | . | . | . | . | . | . | . | . | . | . | . |
| I | -2 | -4 | -4 | -3 | 0 | -5 | -4 | 5 | . | . | . | . | . | . | . | . | . | . | . | . |
| K | -1 | -5 | -1 | 1 | -3 | -2 | 0 | -3 | 5 | . | . | . | . | . | . | . | . | . | . | . |
| L | -2 | -4 | -5 | -3 | 1 | -4 | -2 | 2 | -2 | 5 | . | . | . | . | . | . | . | . | . | . |
| M | -1 | -2 | -5 | -2 | 1 | -4 | -2 | 1 | -1 | 2 | 7 | . | . | . | . | . | . | . | . | . |
| N | -1 | -4 | 2 | 0 | -3 | -1 | 1 | -4 | 0 | -3 | -2 | 5 | . | . | . | . | . | . | . | . |
| P | -1 | -5 | -1 | -1 | -4 | -2 | -2 | -3 | -1 | -4 | -4 | -2 | 7 | . | . | . | . | . | . | . |
| Q | -1 | -3 | 0 | 1 | -3 | -2 | 1 | -3 | 1 | -2 | 0 | 0 | -2 | 6 | . | . | . | . | . | . |
| R | -2 | -3 | -2 | 0 | -4 | -3 | 0 | -3 | 2 | -2 | -2 | -1 | -2 | 1 | 7 | . | . | . | . | . |
| S | 0 | -3 | 0 | -1 | -3 | -1 | -2 | -4 | -1 | -3 | -2 | 0 | -1 | 0 | -1 | 4 | . | . | . | . |
| T | -1 | -3 | -1 | -1 | -3 | -3 | -2 | -2 | 0 | -2 | -1 | 0 | -1 | 0 | -1 | 1 | 5 | . | . | . |
| V | 0 | -2 | -4 | -2 | -1 | -4 | -3 | 3 | -2 | 1 | 0 | -3 | -3 | -2 | -3 | -2 | 1 | 5 | . | . |
| W | -3 | -5 | -5 | -4 | 2 | -4 | -3 | -2 | -3 | -1 | 0 | -4 | -5 | -3 | -1 | -4 | -5 | -3 | 10 | . |
| Y | -2 | -4 | -3 | -2 | 3 | -4 | 0 | -1 | -2 | -1 | 0 | -2 | -3 | -2 | -2 | -2 | -2 | -2 | 2 | 7 |
| - | -7 | -9 | -7 | -6 | -8 | -7 | -8 | -9 | -7 | -8 | -8 | -7 | -6 | -7 | -7 | -7 | -8 | -8 | -9 | -8 |

*(b)*

**Figure 15.17** An example of actual scoring matrices for protein sequences that takes into account the similar properties of certain types of amino acids. *(a)* the Blosum G2 matrix used by BLAST (Henikoff and Henikoff, 1993). *(b)* The structural (STTR) matrix of Shpaer et al. (1996).

The values of these elements obviously vary over a wide range. However, despite their different origins, the two matrices are fairly similar.

There is still one additional complication that must be dealt with. This is especially serious when one wishes to estimate the evolutionary relatedness of two proteins or DNAs. Here a yardstick that is often used as a time scale for evolutionary divergence is the probable average number of mutations needed to convert one sequence into the other. Such comparisons among very similar proteins or nucleic acids are relatively simple. Differences seen are presumably real, and similarities are also presumed real. However, when more distant sequences are compared, an apparent similarity has an increasing chance of just being a statistical event, or a reversion. For example, as shown in Figure 15.18 two matching A's could be a true identity (no mutations) or a reversion (a minimum of two mutations). The more distantly related the two sequences, the more the latter possibility has to be weighted. Ways of doing this for simple identity comparison matrices were developed several decades ago by Jukes and Cantor, and later elaborated considerably to take into account statistical effects and similarities in residue properties. The kind of matrix needed in a very simple case is shown in Figure 15.19. It adjusts the relative weights of comparisons as a function of the average extent of differences between the two sequences. The problem of choosing an ideal comparison matrix, which deals with all of these interrelated issues, is still not a simple one.

Once a comparison matrix is chosen, it can be used to evaluate the relative similarity seen in all possible alignments between two strings. When gaps (caused by a putative insertion or deletion, or a pure statistical artifact) are allowed, the problem of actually enumerating and testing all possible comparisons becomes computationally extremely de-

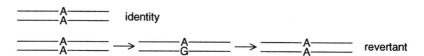

**Figure 15.18**   Difficulties in sequence comparisons when the goal is to estimate the probable number of mutations that have occurred to derive one sequence from another (or both from a common ancestor).

|   | A | C | G | T |
|---|---|---|---|---|
| A | 1-3a | a | a | a |
| C | a | 1-3a | a | a |
| G | a | a | 1-3a | a |
| T | a | a | a | 1-3a |

**Figure 15.19**   A simple scoring matrix that takes into account the average differences between two sequences and allows for the possibility of revertants. Where $a = 1/4(1 - e^{-4d/3})$. The parameter $d$ is a measure of the true evolutionary distance between two sequences being compared. It is the average number of mutations per site that separate one sequence from the other. In the limit $d \rightarrow 0$ the matrix becomes equal to the right-hand panel of Figure 15-16. In the limit $d \rightarrow \infty$ all of the elements of the matrix become equal to 1/4. This means that the sequences have diverged so much that one is essentially comparing two random strings.

manding. Figure 15.20 shows a very simple example. The issue is how to test the likelihood that the postulated gap results in a statistically significant improvement in the alignment score of the two sequences. Obviously there must be a statistical penalty attached to the use of such a gap, since it greatly increases the number of possible comparisons, and thus the chance of finding, at random, a comparison with a score better than some arbitrary value.

From a practical point of view, it is impossible to test all possible gap numbers and locations. One way to deal with this problem is to compare two sequences through smaller windows, sets of successive residues, rather than globally (Fig. 15.21). With two strings of length $n$ and $m$, and a window of length $L$, there are $(n - L + 1)(m - L + 1)$ possible comparisons to be done. This is not a major task for strings the sizes of typical genes. For each choice of window, two substrings of length $L$ are compared, without gaps. The score for this comparison is calculated as the sum over the matrix elements $a_{ij}$ for each of the $L$ residues pairs. To provide a visual overview of the comparison, it is usually convenient to plot all scores above some threshold value as a dot in a rectangular field formed by writing one sequence along the horizontal axis and the other along the vertical axis. Any point in the field corresponds to an alignment of $L$ residues positioned at particular residue positions in the two sequences. This kind of dot matrix plot is shown, schematically in Figure 15.22, and a real example of a sequence comparison at the DNA level for two closely related viruses, SV40 and polyoma is given in Figure 15.23. Any regions with

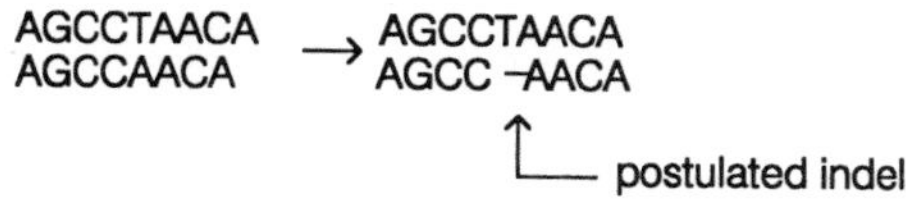

```
AGCCTAACA  →  AGCCTAACA
AGCCAACA      AGCC –AACA
                     ↑
                     └── postulated indel
```

**Figure 15.20**    A simple case of two sequences potentially related by an insertion or a deletion.

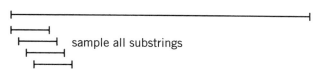

sample all substrings

**Figure 15.21**    Window selection on a single sequence assists in comparisons.

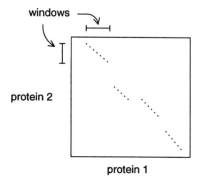

**Figure 15.22**    An example of the comparison of two proteins or DNAs using windows on each, evaluated with a scoring matrix. Shown as dots are all comparisons that score above a selected threshold.

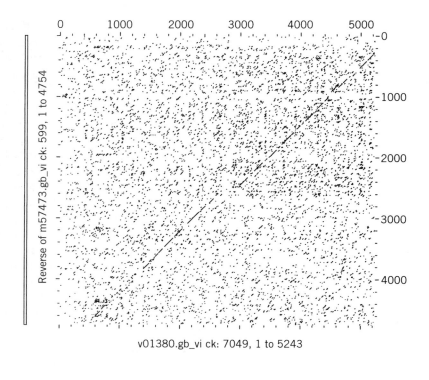

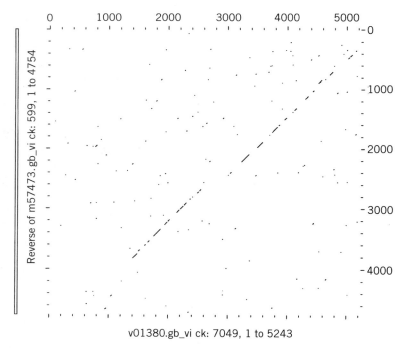

**Figure 15.23**   Results of two actual comparisons between polyoma and SV40 viral DNA sequences, computed as described for the hypothetical example in Figure 15.22. Dots show DNA windows with more than 55% identity within a window (top) or more than 66% (bottom). Provided by Rhonda Harrison.

strong homology show up as a series of dots along a diagonal in these plots. One can piece together the set of diagonals found to construct a model for a global alignment involving all the residues. However, it is not easy to show if a particular global alignment reached in this manner is the best one possible.

## DYNAMIC PROGRAMMING

Dynamic programming is a method that improves on the dot matrix approach just described because it has the power, in principle, to find an optimal sequence alignment. The method is computationally intensive, but not beyond the capabilities of current supercomputers or specialized computer architectures or hardware designed to perform such tasks rapidly and efficiently (Shpaer et al., 1996). The basic notion in dynamic programming is illustrated in Figure 15.24. Here an early stage in the comparison of two sequences is shown. Gaps are allowed, but a statistical penalty (a loss in total similarity score) is paid each time a gap is initiated. (In more complex schemes the penalty can depend on the size of the gap.) We already discussed the fiercely difficult combinatorial problem that results if one tries out all possible gaps. However, with dynamic programming, in the example shown in Figure 15.24, one argues that the best alignment achievable at point $n$ ($n$ is the total number of residues plus gaps that have been inserted into either of the sequences since start of the comparison) must contain the best alignment at point $n - 1$, plus the best scoring of the following three possibilities:

1. Align the two residues at position $n$.
2. Gap the top sequence.
3. Gap the bottom sequence.

The argument behind this approach is not totally rigorous. There is no reason why the optimal alignment of two sequences should be co-linear. DNA rearrangements can change the order of exons, invert short exons, or otherwise scramble the linear order of genetic information. Dynamic programming, as conventionally used, can only find the optimal co-linear alignment. However, one can circumvent this problem, in principle, by attempting the alignments starting at selected internal positions in both possible orientations, and determining if this alters the best match obtained.

Best alignment up to n-1     One of these three is the best alignment up to n

```
                    n-1    n

          ACCG–AACGCCCA    T
          TCCGTAATG–GGA    C

          ACCG–AACGCCCA    –T
          TCCGTAATG–GGA    C

          ACCG–AACGCCCA    T
          TCCGTAATG–GGA    –C
```

**Figure 15.24**  Basic methodology used in comparison of two sequences by dynamic programming.

Sequence alignment by dynamic programming is conveniently viewed by a plot as shown in Figure 15.25. The coordinates of the plot are the two sequences, just as in the dot matrix blot discussed earlier. However, what is plotted at each position is not a score from the comparison of two sequence windows; instead, it is a cumulative score for a path through the two sequences that represents a particular comparison. In the simple example in Figure 15.25, the scoring is just for identities. In a real case, the elements of a scoring matrix would be used. From the example in Figure 15.25a it can be seen that a particular alignment is just a path of arrows between adjacent residues. The three possible steps listed above at a given point in the comparison just correspond to:

1. Advancing along a diagonal
2. Advancing horizontally
3. Advancing vertically

In dynamic programming one must consider all possible paths through the matrix. A particular comparison is a continuous set of arrows. The best comparisons will give the highest scores. Usually there are multiple paths that give equal, optimal alignments. These can be found by starting with the point in the plot with the highest score and tracing back toward the beginning of the sequence comparison to find the various routes that can lead to this final point (Fig. 15.25b).

Rigorous dynamic programming is a very computationally intensive process because there are so many possible paths to be tested. Remember that for each new protein sequence, one wishes to test alignments with all previously known sequences. This is a very

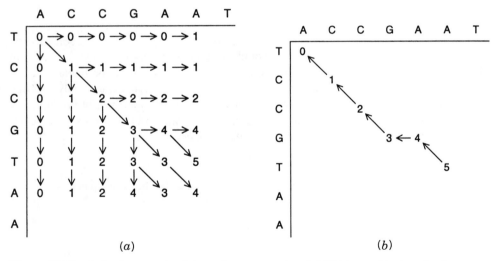

(a)                                                    (b)

**Figure 15.25**  A simple example of dynamic programming. (a) Matrix used to describe the scores of various alignments between two sequences. Note that successive horizontal and vertical moves are forbidden, since this introduces a gap on both strands which is equivalent to a diagonal move—not a gap at all. In each cell only the route with the highest possible score is shown. In practice, each cell can be approached in three different ways—diagonally, horizontally, and vertically. (b) The trace back through the matrix from the highest scoring configuration to find all paths (alignments) consistant with that final configuration. Only the path with the highest score is kept.

large number of dynamic programming comparisons. A number of different procedures can be used to speed this up. Some pay the price of a loss of rigor, so no guarantee exists that the very best alignment will be found.

A very popular program for global database searching has been developed by David Lipman and his coworkers. This is called FASTA for nucleic acid comparisons, FASTP, for protein (or translated ORF) comparisons. The test protein is broken into $k$-tuple words. These are searched through the entire database, in looking for exact matches and scoring them. Some of these best scoring comparisons will form diagonal regions of dot plots with a high density of exact matches. The 10 best of these regions are rescored with a more accurate comparison matrix. They are trimmed to remove residues not contributing to the good score. Then nearby regions are joined, using gap penalties for the inevitable gaps required in this process. The resulting sets of comparisons are examined to find those with the best, provisional fits. Then each of these is examined by dynamic programming within a relatively narrow band of 32 residues around the sites of each of the provisional matches. Even faster algorithms exist, like Lipman's BLAST which looks only at short exact fits but makes a rigorous statistical assessment of their significance. This is used to winnow down the vast array of possible comparisons before more rigorous, and time-consuming, methods are employed.

A key point in favor of the dynamic programming method is that for each comparison cell, as shown in Figure 15.26, one only needs to consider three immediate precursors. This allows the computation to be broken into a set of steps that can be computed in parallel. The comparison can proceed down the diagonal of the plot in waves, with each cell computed independently of what is happening in the other cells along that diagonal (Fig. 15.26). Thus parallel processing computers are well adapted to the dynamic programming method. The result is an enormous increase in our power to do global sequence comparisons. Using a parallel architecture, it is possible, with the existing protein database, to do a complete inversion. That is, all sequences are compared with all sequences by dynamic programming. The resulting set of scores allows the database to be factored into sequence families, and new, unsuspected sets of families have been found in this way. As an alternative to parallel processing, specialized computer chips have been designed and built that perform dynamic programming sequences comparisons, but they do this very quickly because each sequence needs to be passed through the chip only once.

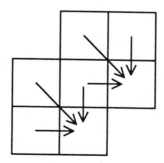

**Figure 15.26**    Parallel processing in dynamic programming. Only three preceding cells are needed to compute the contents of each new cell. Computation can be carried out on all cells of the next diagonal, simultaneously.

## GAINING ADDITIONAL POWER IN SEQUENCE COMPARISONS

With increases in available computation power occurring almost continuously, it is unde-
niably tempting to carry sequence comparisons even further to make them more sensitive.
The goal is to identify, as related, two or more sequences that lie at the edge of statistical
certainty by ordinary methods. There are at least two different ways to attempt to do this.
One can try to introduce information about possible protein three-dimensional structure
for the reasons described earlier in this chapter. Alternatively, one can attempt to align
more than two sequences at a time.

If two proteins are compared at the amino acid sequence level, but one of these proteins is of known three-dimensional structure, the sensitivity of the scoring used can be
improved considerably, as we hinted earlier. One can postulate that if the two proteins are
truly related, they should have a similar three-dimensional structure. In the interior of this
structure, spatial constraints are very important. Insertions and deletions there will be
very disruptive, and thus they should be assigned very costly comparison weights. The effects will be much less serious on the protein surface. Changes in any charged residues on
the interior of the protein are also likely to be very costly. Interchanges among internal
hydrophobic residues are likely to be tolerated, especially if there are compensating increases in size at some points and decreases in size elsewhere. Changes that convert internal hydrophobic residues to charged residues will be devastating. A small number of external hydrophobic residues may not be too serious. Changes in sequence that seriously
destabilize prominent helices and beta sheets are obviously also costly.

A systematic approach to such structure-based sequence comparisons is called threading. Here the sequence to be tested is pulled through the structure of the comparison sequence. At each stage some structure variations are allowed to see if the new sequence is
compatible in this alignment with the known structure. The effectiveness of threading
procedures in improving homolog matching and structure prediction is still somewhat
controversial. However, in principle, this appears to be a very powerful method as more
and more proteins with known three-dimensional structures become available.

As originally implemented by Christian Sander, by attempting to fit proteins of unknown structure to the various known protein structures as just described above, one
gained 5% to 10% in the fraction of the database that showed significant protein similarity matches. Clearly this approach will improve significantly as the body of known protein structures increase in size. Another aspect of the protein structure fitting problem de-
serves mention here. As more and more proteins with known three-dimensional structure
become available, it is of interest to ask whether any pairs of proteins have similar three-
dimensional structure, outside of any considerations about whether they have similar
amino acid sequences. Sander has constructed algorithms that can look for similar structures, essentially by comparing plots of sets of residues that approach each other to within
less than some critical distance. Such comparisons have revealed cases where two very
similar three-dimensional structures had not been recognized before because they were
being viewed from different angles that obscured visual recognition of their similarity. In
the future we must anticipate that the number of structures will be so great that no visual
comparison will suffice, unless there is first a detailed way to catalog or group the structures into potentially similar objects.

The second approach to enhanced protein sequence comparison is multiple alignments. This allows weak homology to be recognized if a family of related proteins is assembled. Suppose, for example, that one saw a sequence pattern like

**TABLE 15.6  One-Letter Code for Amino Acids**

| One-Letter Code | Three-Letter Code | Amino Acid Name | Mnemonic |
|---|---|---|---|
| A | ala | alanine | *a*lanine |
| C | cys | cysteine | *c*ysteine |
| D | asp | aspartate | aspar*d*ate |
| E | glu | glutamate | glutamat*e* |
| F | phe | phenylalanine | *f*enylalanine |
| G | gly | glycine | *g*lycine |
| H | his | histidine | *h*istidine |
| I | ice | isoleucine | *i*soleucine |
| K | lys | lysine | *k* near l |
| L | leu | leucine | *l*eucine |
| M | met | methionine | *m*ethionine |
| N | asn | asparagine | asparagi*n*e |
| P | pro | proline | *p*roline |
| Q | gln | glutamine | "cute" amine |
| S | ser | serine | *s*erine |
| T | thr | threonine | *t*hreonine |
| V | val | valine | *v*aline |
| W | trp | tryptophan | t*w*yptophan |
| Y | tyr | tyrosine | t*y*rosine |

—P—AHQ—L—

in two proteins. Here we use the convenient one letter code for amino acids summarized in Table 15.6. Such a pattern would be far too small to be scored as statistically significant. However, seeing this pattern in many proteins would make the case for all of them much stronger. Currently most multiple alignments are done by making separate pairwise comparisons among the proteins of interest. The extreme version of this is the database inversion described previously. The use of pairwise comparisons throws away a considerable amount of useful information. However, it saves massive amounts of computer time. The alternative, for rigorous algorithms, would be dynamic programming in multiple dimensions. This is not a task to be approached by the faint of heart (or short of budget) with current computer speeds. Future computer capabilities could totally change this.

## DOMAINS AND MOTIFS

From an examination of the proteins of known three-dimensional structure, we know that much of these structures can be broken down into smaller elements. These are domains, independently folded units, motifs, and structural elements or patterns that describe the basic folding of a section of polypeptide chain, to form either a structural backbone or framework for a domain or a binding site for a ligand. Examples are various types of barrels formed by multiple-stranded beta sheets, zinc fingers and helix-turn-helix motifs, both of which are nucleic acid binding elements, Rossman folds and other ways of binding common small molecules like NAD and ATP, serine esterase active sites, transmembrane helices, kringles, and calcium binding sites. A few of these motifs are illustrated schematically in Figure 15.27. The motifs are not exact, but many can be recognized at the level of the amino acid sequence by conserved patterns of particular residues like cysteines or hydrophobic side chains.

Several aspects of the existence of protein motifs should aid our ability to analyze the function of newly discovered sequences once we understand these motifs better. First, there may well be just a finite number of different protein motifs, perhaps just several hundred, and eventually we will learn to recognize them all. Our current catalog of motifs derives from a rather biased set of protein structures. The current protein data base is composed mostly of materials that are stable, easy to crystallize, and obtainable in large quantities. Eventually the list should broaden considerably.

Each protein motif has a conserved three-dimensional structure, and so once we suspect a new protein of possessing a particular motif, we can use that structure for enhanced sequence comparisons as described above. If, as now seems likely, proteins are mostly composed of fixed motifs, any new sequence can be dissected into its motifs, and these can be analyzed one at a time. This will greatly simplify the process of categorizing new structures. However, the key aspect of motifs that makes them an attractive path from sequences to biological understanding is that motifs are not just structural elements, they

## Proteins are assembled from motifs

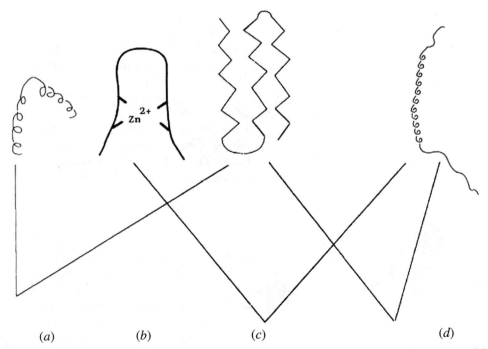

|  (a) | (b) | (c) | (d) |

**Figure 15.27** Some of the striking structural and functional motifs found in protein structures. (*a*) Helix-turn-helix. (*b*) Zinc finger. (*c*) Beta barrel. (*d*) Transmembrane helical domain.

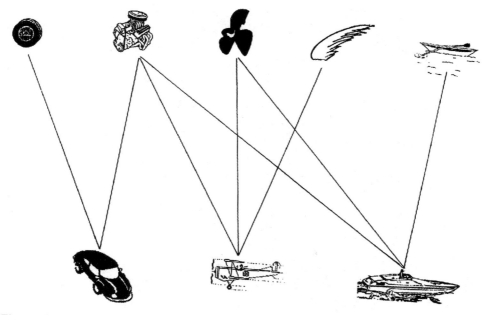

**Figure 15.28**    An example of how motifs (found in transportation vehicles) can be used to infer or relate the structures of composites of motifs.

are functional elements. There must be a relatively finite set of rules that determines what combinations of motifs can live together within a protein domain. Once we know these rules, our ability to make intelligent guesses about function from structural information alone should improve markedly.

One way to view motifs is to consider them as words in the language of protein function. What we do not yet know, in addition to many of the words in our dictionary, is the grammar that combines them. Another way to view protein motifs is the far-fetched, but perhaps still useful, analogy shown in Figure 15.28. This portrays various transportation vehicles and some of the structure-function motifs that are contained in them. If we knew the function of the individual motifs, and had some idea about how they could be combined, we might be able to infer the likely function of a new combination of motifs, even though we had never seen it before; for example, we should certainly be able to make a good de novo guess about the function of a seaplane.

## INTERPRETING NONCODING SEQUENCE

Most of the genome is noncoding, and arguments about whether this is garbage, junk, or potential oil fields have been presented before. The parts of the noncoding region that we are most anxious to be able to interpret are the control sequences which determine when, where, and how much of a given gene product will be made. Our current ability to do this is minimal, and it is confounded by the fact that most of the functional control elements appear to be very short stretches of DNA sequence. These sequences are not translated, so we have only the four bases to use for comparisons or analyses. This leads to a situation with very poor signal to noise. In eukaryotes, control elements appeared to be used in complex combinations like the example shown schematically in Figure 15.29. To analyze of stretch of DNA sequence for all possible combinations of small elements that might be able to come together in such a complex seems intractable by current methods. Perhaps, after we have learned much more about the properties of some of these control regions, we will be able to do a better job. The true magnitude of this problem, currently, is well illustrated by the globin gene family where one key, short control element required for correct tissue-specific gene expression was found (experimentally, and after much effort) to be located more than 20-kb upstream from the start of transcription. Thus finding control elements is definitely not like shooting fish in a barrel.

## DIVERSITY OF DNA SEQUENCES

We cannot be interested in just the DNA sequence of a single individual. All of the intellectual interest in studying both inherited human diseases and normal human variations will come from comparisons among the sequences of different individuals. This is already done when searching for disease genes. Its ultimate manifestations may be the kind of large-scale diagnostic DNA sequencing postulated in Chapter 13.

How different are two humans? From recent DNA sequencing projects (hopefully on representative regions, but there is no way we can be sure of this) differences between any two homologous chromosomes occur at a frequency of about 1 in 500. Scaled to the entire human genome, this is $6 \times 10^6$ differences between any two chromosomes of a single individual, and potentially four times this number when two individuals are compared, since each homolog in one genome must be compared with each pair of homologs in the other. To simply make a catalog of all of the differences for any potential individual against an arbitrary reference haploid genome would require $12 \times 10^6$ entries.

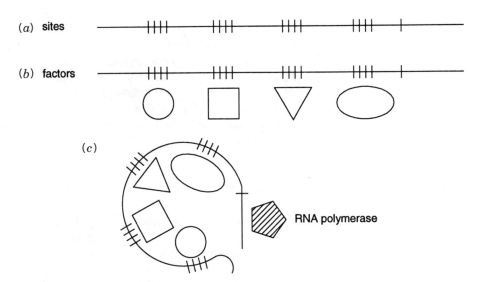

**Figure 15.29** Structure of a typical upstream regulatory region for a eukaryotic gene. *(a)* Transcription factor-binding sequences. *(b)* Linear complex with transcription factors. *(c)* Three-dimensional structure of transcription factor-DNA complex that generates a binding site for RNA polymerase.

The current population of the earth is around $5 \times 10^9$ individuals. Thus, to record all of current human diversity would require a database with $12 \times 10^6 \times 5 \times 10^9 = 6 \times 10^{16}$ entries. By today's mass storage standards, this is a large database. The entire sequence of one human haploid genome could be stored on a single CD rom. On this scale, storage of the catalog of human diversity would require $2 \times 10^7$ CD roms, hardly something one could transport conveniently. However, computer mass storage has been improving at least by $10^4$ per decade, and there is no sign that this rate of increase is slackening. As a result, by the time we have the tools to acquire the complete record of human diversity, some 15 to 20 years from now, we should be able to store it on whatever is the equivalent, then, of a few CD roms. Genome analysis is unlikely to be limited now, or in the future, by computer power.

Why would anyone want to view all of human diversity? There are two basic features that make this information compellingly interesting. First is the complexity of a number of interesting phenotypes that are undoubtedly controlled by many genes. Included are such items as personality, affect, and facial features. Given the limited ability for human experimentation, if we want to dissect some of the genetic contributions to these aspects of our human characteristics, we will have to be able to compare DNA sequences with a complex set of observations. The more sequence we have, and the broader the range of individuals represented, the more likely it seems we should be able to make progress on problems that today seem totally intractable.

The second motivation for studying the vast expanse of human diversity is a burgeoning field called molecular anthropology. A considerable amount of human prehistory has left its traces in our DNA. Within contemporary sequences should be a record of who mated with whom as far back as our species goes. Such a record, if we could untangle it, would provide information about our past that probably cannot be reconstructed so easily and accurately in any other way. Included will be information about mass migrations, like the colonization of the new world, population expansions in prehistory, catastrophes, like plagues, or major climatic fluctuations. A tool that is useful for such studies is called principle component analysis. It looks at DNA or protein sequence variations across geographic regions and attempts to factor out particular variants with localized geographic distribution, or with gradients of geographic distribution.

Many of the results obtained thus far in molecular anthropology are controversial and open to alternative explanations. As the body of sequence data that can be fed into such analyses grows, the robustness of the method and acceptance of its results will probably increase. Junk DNA may be especially useful for these kinds of analyses. It is much more variable between individuals than coding sequences, and thus the amount of information it contains about human prehistory should be all that greater. One final, intriguing source of information is available from occasional DNA samples found still intact in mummies, human remains trapped in ice, or other environments that preserve DNA molecules. Such information will provide benchmarks by which to test the extrapolations that otherwise have to be made from contemporary DNA samples.

## SOURCES AND ADDITIONAL READINGS

Adams, M. D., Kelley, J. M., Gocayne, J. D., Dubnick, M., Polymeropoulos, M. H., et al. 1991. Complementary DNA sequencing: Expressed sequence tags and human genome project. *Science* 252: 1651–1656.

Berks, M. The C. elegans genome sequencing project. *C. elegans* Genome Mapping and Sequencing Consortium. 1995. *Genome Research* 5: 99–104.

Bult, C. J., White, O., Olsen, G. J., Zhou, L., Fleischmann, R. D., Sutton, G. G., Blake, J. A., FitzGerald, L. M., Clayton, R. A., Gocayne, J. D., et al. 1996. Complete genome sequence of the methanogenic archaeon, *Methanococcus jannaschii. Science* 273: 1058–1073

Burland, V., Plunkett III, G., Sofia, H. J., Daniels, D. L., and Blattner, F. R. 1995. Analysis of the *Escherichia coli* genome VI: DNA sequence of the region from 92.8 through 100 minutes. *Nucleic Acids Research* 23: 2105–2119.

Carlock, L., Wisniewski, D., Lorincz, M., Pandrangi, A., and Vo, T. 1992. An estimate of the number of genes in the Huntington disease gene region and the identification of 13 transcripts in the 4p16.3 segment. *Genomics* 13: 1108–1118.

Chothia, C. 1992. One thousand families for the molecular biologist. *Nature* 357: 543–544.

Daniels, D. L., Plunkett III, G., Burland, V., and Blattner, F. R. 1992. Analysis of the *Escherichia coli* genome: DNA sequence of the region from 84.5 to 86.5 minutes. *Science* 257: 771–778.

Fleischmann, R. D., Adams, M. D., White, O., Clayton, R. A., Kirkness, E. F., Kerlavage, A. R., Bult, C. J., Tomb, J. F., Dougherty, B. A., Merrick, J. M., et al. 1995. Whole-genome random sequencing and assembly of *Haemophilus influenzae. Science* 269: 496–512.

Fraser, C. M., Gocayne, J. D., White, O., Adams, M. D., Clayton, R. A., Fleischmann, R. D., Bult, C. J., Kerlavage, A. R., Sutton, G., Kelley, J. M., et al. 1995. The minimal gene complement of *Mycoplasma genitalium. Science* 270: 397–403.

Gardiner, K. 1996. Base composition and gene distribution: Critical patterns in mammalian genome organization. *Trends in Genetics* 12: 519–524.

Goffeau, A. 1995. Life with 482 genes. *Science* 270: 445–446.

Goffeau, A., Barrell, B. G., Bussey, H., Davis, R. W., Dujon, B., Feldman, H., Gailbert, F., Hoheisel, J. D., Jacq, C., Johnston, M., Louis, E. J., Mewes, H. W., Murakami, Y., Philippsen, P., Tettelin, H., and Oliver, S. G. 1996. Life with 6000 genes. *Science* 274: 546–567.

Gribskow, M., and Devereaux, J. 1991. *Sequence Analysis Primer.* Oxford: Oxford University Press.

Harper, R. 1994. Access to DNA and protein databases on the Internet. *Current Opinion in Biotechnology* 5: 4–18.

Himmelreich, R., Hilbert, H., Plagens, H., Pirkl, E., Li, B.-C., and Herrmann, R. 1996. Complete sequence analysis of the genome of the bacterium *Mycoplasma pneumoniae. Nucleic Acids Research* 24:4420–4449.

How, G.-F., Venkatesh, B., and Brenner, S. 1996. Conserved linkage between the puffer fish *(Fugu rubripes)* and human genes for platelet-derived growth factor receptor and macrophage colony-stimulating factor receptor. *Genome Research* 6: 1185–1191.

Johnston, M., Andrews, S., Brinkman, R., Cooper, J., Ding, H., Dover, J., Du, Z., Favello, A., Fulton, L., Gattung, S., et al. 1994. Complete nucleotide sequence of *Saccharomyces cerevisiae* chromosome VIII. *Science* 265: 2077–2082.

Kaneko, T., et al. 1996. Sequence analysis of the genome of the unicellular cyanobacterium *Synechocystis* sp. strain PCC 6803. II. Sequence determination of the entire genome and assignment of potential protein-coding regions. *DNA Research* 3: 109–136.

Karlin, S., and Mrazek, J. 1996. What drives codon choices in human genes? *Journal of Molecular Biology* 262: 459–472.

Lewis, E. B., Knafels, J. D., Mathog, D. R., and Celniker, S. E. 1995. Sequence analysis of the *cis-*regulatory regions of the bithorax complex of *Drosophila. Proceedings of the National Academy of Sciences USA* 92: 8403–8407.

Lopez, R., Larsen, F., and Prydz, H. 1994. Evaluation of the exon predictions of the GRAIL software. *Genomics* 24: 133–136.

Maier, W. G. 1997. Bonsai genomics: Sequencing the smallest eukaryotic genomes. *Trends in Genetics* 13: 46–49.

Martin, C. H., Mayeda, C. A., Davis, C. A., Ericsson, C. L., Knafels, J. D., Mathog, D. R., Celniker, S. E., Lewis, E. B., and Palazzolo, M. J. 1995. Complete sequence of the bithorax complex of *Drosophila*. *Proceedings of the National Academy of Sciences USA* 92: 8398–8402.

McCombie, W. R., Martin-Gallardo, A., Gocayne, J. D., FitzGerald, M., et al. 1992. Expressed genes, Alu repeats and polymorphisms in cosmids sequenced from chromosome 4p16.3. *Nature Genetics* 1: 348–353.

McFadden, G. I., Gilson, P. R., Douglas, S. E., Cavalier-Smith, T., Hofmann, C. J. B., and Murakami, Y., Naitou, M., Hagiwara, H., Shibata, T., Ozawa, M., Sasanuma, S.-I., Sasanuma, M., Tsuchiya, Y., Soeda, E., Yokoyama, K., Yamazaki, M., Tashiro, H., and Eki, T. 1995. Analysis of the nucleotide sequence of chromosome VI from *Saccharomyces cerevisiae*. *Nature Genetics* 10: 261–264.

Mushegian, A. R., and Koonin, E. V. 1996. A minimal gene set for cellular life derived by comparison of complete bacterial genomes. *Proceedings of the National Academy of Sciences USA* 93: 10268–10273.

Oliver, S. G., van der Aart, Q. J. M., Agostoni-Carbone, M. L., Aigle, M., et al. 1992. The complete DNA sequence of yeast chromosome III. *Nature* 357: 38–46.

Oliver, S. 1995. Size is important, but. . . . *Nature Genetics* 10: 253–256.

Oshima, T., et al. 1996. A 718-kb DNA sequence of the *Escherichia coli* K-12 genome corresponding to the 12.7-28.0 min region on the linkage map. *DNA Research* 3: 137–155.

Pearson, W. 1995. Comparison of methods for searching protein-sequence databases. *Protein Science* 4: 1145–1160.

Perlin, M. W., Lancia, G., and Ng, S.-K. 1995. Toward fully automated geneotyping: Genotyping microsatellite markers by deconvolution. *American Journal of Human Genetics.* 57: 1199–1210.

Peters, R., and Sikorski, R. S. 1996. Nucleic acid databases on the Web. *Nature Biotechnology* 14:1728–1729.

Posfai, J., and Robert, R. J. 1992. Finding errors in DNA sequences. *Proceedings of the National Academy of Sciences USA* 89: 4698–4702.

Rost, B., and Sander, C. 1996. Bridging the protein sequence-structure gap by structure predictions. *Annual Review of Biophysics and Biomolecular Structure* 25: 113–136.

Saccone, S., DeSario, A., Delle Valle, G., and Bernardi, G. 1992. The highest gene concentrations in the human genome are in telomeric bands of metaphase chromosomes. *Proceedings of the National Academy of Sciences USA* 89: 4913–4917.

Sharkey, M., Graba, Y., and Scott, M. P. 1997. Hox genes in evolution: Protein surfaces and paralog groups. *Trends in Genetics* 13: 145–151.

Shpaer, E. G., Robinson, M., Yee, D., Candlin, J. D., Mines, R., and Hunkapiller, T. 1996. Sensitivity and selectivity in protein similarity searchers: A comparison of Smith-Watermanin hardware to BLAST and FASTA. *Genomics* 38: 179–191.

Shevchenko, A., Jensen, O. N., Podtelejnikov, A. V., Sagliocco, F., Wilm, M., Vorm, O., Mortensen, P., Shevchenko, A., Boucherie, H., and Mann, M. 1996. Linking genome and proteome by mass spectrometry: Large-scale identification of yeast proteins from two dimensional gels. *Proceedings of the National Academy of Sciences USA* 93: 14440–14445.

Shoemaker, D. D., Lashkari, D. A., Morris, D., Mittman, M., and Davis, R. W. 1996. Quantitative phenotypic analysis of yeast deletion mutants using a highly parallel molecular bar-coding strategy. *Nature Genetics* 14: 450–456.

Smith, R. 1996. Perspective: Sequence data base searching in the era of large-scale genomic sequencing. *Genome Research* 6: 653–660.

Smith, V., Chou, K. N., Lashkari, D., Botstein, D., and Brown, P. O. 1996. Functional analysis of the genes of yeast chromosome V by genetic footprinting. *Science* 274: 2069–2074.

Sulston, J., Du, Z., Thomas, K., Wilson, R., et al. 1992. The *C. elegans* genome sequencing project: A beginning. *Nature* 356: 37–41.

Strauss, E. J., and Falkow, S. M. 1997. Microbial pathogenesis: Genomics and beyond. *Science* 276: 707–712.

Uberbacher, E. C., and Mural, R. J. 1991. Locating protein-coding regions in human DNA sequences by a multiple sensor-neural network approach. *Proceedings of the National Academy of Sciences USA* 88: 4698–4702.

Waterman, M. S. 1995. *Introduction to Computational Biology: Sequences, Maps and Genomes.* London: Chapman and Hall.

Wilson, R. et al. 1994. 2.2 Mb of contiguous nucleotide sequence from chromosome III of *C. elegans. Nature* 363: 32–36.

Wu, R. 1993. Development of enzyme-based methods for DNA sequence analysis and their applications in the genome proejcts. In A. Meister, ed., *Advances in Enzymology.* New York: Wiley.

Zhang, M. Q. 1997. Identification of protein coding regions in the human genome by quadratic discriminant analysis. *Proceedings of the National Academy of Sciences USA* 94: 565–568.

# APPENDIX
# Databases

GenBank: A public database built and distributed by the National Center for Biotechnology Information.
http://www.ncbi.nlm.nih.gov/
ftp:ncbi.nlm.nih.gov

EMBL Nucleotide Sequence Database: A central activity of the European Bioinformatics Institute; an EMBL outstation.
http://www.ebi.ac.uk
ftp:ftp.ebi.ac.uk

DNA Data Bank of Japan (DDBJ): Activities began in earnest in 1986 in collaboration with EMBL and GenBank.
http://www.ddbj.nig.ac.jp/
ftp:ftp2.ddbj.nig.ac.jp

Genome Sequence DataBase (GSDB): Designed to meet the community-wide challenges of managing, interpreting, and using DNA sequence data at an ever increasing rate.
http://www.ncgr.org/gsdb

Protein Information Resource (PIR) and the PIR-International Protein Sequence Database:
http://www.nbrf.georgetown.edu
ftp:nbrf.georgetown.edu

MIPS: A database for protein sequences, homology data, and yeast genome information:
www.mips.embnet.org

SWISS-PROT protein sequence data bank and its supplement TrEMBL:
www.nlm.nih.gov (or EBI server)

Metabolic pathway collection:
www.cme.msu.edu/WIT/

NRSub database:
ftp://biom3.univlyon1.fr/pub/nrsub
ftp://ftp.nig.ac.jp/pub/db/nrsub

GDB Human Genome Database:
http://gdbwww.gdb.org/
ftp://ftp.gdb.org/

Radiation Hybrid Database:
http://www.ebi.ac.uk/RHdb
ftp://ftp.ebi.ac.uk/pub/databases/RHdb

Mouse Genome Database (MGD): A comprehensive public resource of genetic, phenotypic, and genomic data.
ftp://www.informatics.jax.org/
ftp://ftp.informatics.jax.org/pub/

Molecular Probe Data Base (MPDB):
http://www.biotech.ist.unige.it/interlab/mpdb.html

Signal Recognition Particle Database (SPRDB):
http://pegasus.uthct.edu/SRPDB/SRPDB.html
ftp://diana.uthct.edu

Viroid and viroid-like sequence database: Addition of a hepatitis delta virus RNA section:
http://www.callisto.si.usherb.ca/~jpperra

Mutation spectra database for bacterial and mammalian genes:
http://info.med.yale.edu/mutbase/

PRINTS protein fingerprint database:
http://www.biochem.ucl.ac.uk/bsm/dbbrowser/
ftp://s-ind2.dl.ac.uk/pub/database/prints

PROSITE datases:
http://expasy.hcuge.ch/
ftp.ebi.ac.uk

Blocks Database servers:
http://blocks.fhcrc.org/
ftp://ncbi.nlm.nih.gov/repository/blocks

HSSP database of protein structure-sequence alignments:
http://www.sander.embl-heidelberg.de/
ftp://ftp.embl-ebi.ac.uk/pub/databases/

Dali/FSSP classification of the three-dimensional protein folds:
http://www.embl-heidelberg.de/dali/fssp/

DEF database of protein fold class predictions:
http://zeus.cs.uoi.gr/neural/biocomputing/def.html

SCOP: A structural classification of proteins database:
http://scop.mrc-lmb.cam.ac.uk/scop/

SBASE protein domain library: A collection of annotated protein sequence segments:
http://www.icgeb.triester.it
http://base.icgeb.triester.it/sbase/(using BLAST)

Codon usage tabulated from the international DNA sequence databases:
ftp://ftp.nig.ac.jp/pub/db/codon/current/

TransTerm: translational signal database:
http://biochem.otago.ac.nz:800/Transterm/homepage.html

REBASE: restriction enzymes and methylases:
http://www.neb.com/rebase

Ribonuclease P database:
http://www.mbio.ncsu.edu/RNaseP/home.html

TRANSFAC, TRRD, and COMPEL: Toward a federated database system on transcriptional regulation:
http://transfac.gbf-braunschweig.de
http://www.bionet.nsc.ru/TRRD

MHCPER: A database of MHC-binding peptides:
http://wehih.wehi.edu.au/mhcpep/
ftp.wehi.edu.au/pub/biology/mhcpep/

Histone and histone fold sequences and structures:
http://www.ncbi.nlm.nih.gov/Basevani/HISTONES/

Directory of P450-containing Systems:
http://www.icgeb.trieste.it/p450
http://p450.abc.hu

O-GLYCBASE: A revised database of O-glycosylated proteins:
http://www.cbs.dtu.dk/databases/OGLYCBASE/
ftp.cbs.dtu.dk/pub/Oglyc/Oglyc.base

*Mitochondria*

MITOMAP: Human mitochondrial genome database:
http://www.gen.emory.edu/mitomap.html

MmtDB: A Metazoa mitochondrial DNA variants database:
http://www.ba.cnr.it/~areamt08/MmtDBWWW.htm

IMGT: International ImMunoGeneTics database:
http://imgt.cnusc.fr:8104

SWISS-2DPAGE database of two-dimensional polyacrylamide gel electrophoresis:
http://expasy.hcuge.ch/

GRBase: A database linking information on proteins involved in gene regulation:
http://www.access.digex.net/~regulate
ftp://ftp.trevigen.com/pub/Tfactors

ENZYME data bank:
http://expasy.hcuge.ch/

*E. coli*

Compilation of DNA sequences of *E. coli* K12: Description of the interactive databases ECD and ECDC:
http://susi.bio.uni-giessen.de/usr/local/www/html/ecdc.html

EcoCyc: Encyclopedia of *E. coli* Genes and Metabolism:
http://www.ai.sri.com/ecocyc/ecocyc.html

GenProtEC: Genes and proteins of *E. coli* K12:
http://www.mbl.edu/html/ecoli.html

*Yeast*

Yeast Protein Database(YPD): A database for the complete proteome of *Saccharomyces cerevisiae:*
http://www.proteome.com/YPDhome.html

LISTA, LISTA-HOP, and LISTA-HON: A comprehensive compilation of protein encoding sequences and its associated homology databases from the yeast

*Saccharomyces:*
http://www.ch.embnet.org/
ftp://bioftp.unibas.ch

*Drosophila*

FlyBase: A *Drosophila* database:
http://flybase.bio.indiana.edu
ftp:flybase.bio.indiana.edu(in /flybase)

GIF-DB: A WWW database on gene interactions involved in *Drosophila melanogaster* development:
http://www-biol.univ-mrs.fr/~lgpd/GIFTS_home_page.html

*RNA*

Compilation of 5S rRNA and 5S rRNA gene sequences:
http://rose.man.poznan.pl/5SData/5SRNA.html
http://www.chemie.fu-berlin.de/fb_chemie/ibc/agerdmann/5S_rRNA.html
ftp.fu-berlin.de/science/biochem/db/5SrRNA

Small RNA database:
http://mbcr.bcm.tmc.edu/smallRNA/smallrna.html

uRNA database:
http://pegasus.uthct.edu/uRNADB/uRNADB.html

guide RNA database:
http://www.biochem.mpg.de/~goeringe/

Ribosomal Database Project (RDP):
http://rdpwww.life.uiuc.edu/
ftp://rdp.life.uiuc.edu

Database on the structure of small ribosomal subunit RNA:
http://rrna.uia.ac.be/ssu/

Database on the structure of large ribosomal subunit RNA:
http://rrna.uia.ac.be/lsu/

RNA modification database:
http://www-medlib.med.utah.edu/RNAmods/RNAmods.html
ftp://medlib.med.utah.edu/library/RNAmods

Expansion of the 16S and 23S ribosomal RNA mutation databases (16SMDB and 23SMDB):
http://www.fandm.edu/Departments/Biology/Databases/RNA.html

Compilation of tRNA sequences and sequences of tRNA genes:
ftp.ebi.ac.uk/pub/databases/tRNA

*Specific gene Haemophilia:*

Factor VIII Mutation Database on the WWW: The Haemophilia A Mutation, Search, Test, and Resource Site:
http://europium.mrc.rpms.ac.uk
ftp.ebi.ac.uk/pub/databases/hamsters

Haemophilia B: Database of point mutations and short additions and deletions:
http://www.umds.ac.uk/molgen/haemBdatabase

Database and software for the analysis of mutations in the human *p53* gene, the human *hprt* gene, and both the *lacI,* and *lacZ* gene in transgenic rodents: http://sunsite.unc.edu/dnam/mainpage.html

*p53* and APC gene mutations: Software and databases

Database of *p53* gene somatic mutations in human tumors and cell lines: http://www.ebi.ac.uk/srs/

*Human*

PAH mutation analysis consortium databases: http://www.cf.ac.uk/uwcm/mg/hgmd0.html

alpha/beta fold family of proteins database and the cholinesterase gene server ESTHER: http://www.ensam.inra.fr/cholinesterase

Marfan Database: Software and database for the analysis of mutations in the human FBN1 gene

BTKbase, mutation database for X-linked agammaglobulinemia (XLA): http://www.helsinki.fi/science/signal/btkbase.html

Human type I collagen mutation database: http://www.le.ac.uk/depts/ge/collagen/collagen.html

*Receptors*

Androgen receptor gene mutations database: http://www.mcgill.ca/andogendb/ ftp.ebi.ac.uk/pub/databases/androgen

Nuclear Receptor Resource Project: http://nrr.georgetown.edu/nrr.html

Glucocorticoid receptor resource: http://biochem1.basic-sci.georgetown.edu/grr/grr.html

# INDEX

Abundance
  of human repeat classes, 508
  of mRNAs, 382
Accuracy
  of raw sequence data, 342
  of DNA fragment lengths, 235
Achilles' heel methods, 476–480
  with PNA, 491
Acrocentric chromosomes, 59
Activation energy of DNA melting, 78
Adapters
  in PCR, 108–110
  use in finding small restriction fragments, 264
  use in genomic sequencing, 391
Adenomous polyposis coli gene, 452–453
Affinity capture
  use in differential cloning, 494
  use in electrophoresis, 486–488
  use in purification of aptamers, 517
Affymetrix, Inc., 407, 421
Agarose gels
  DNA trapping by, 48
  preparation of DNA in, 160
  use in electrophoresis, 137
Aging, role of mitochondria in, 56–57
AIDS, DNA diagnosis in, 463
ALF DNA sequencing instrument, 334
Alignment of sequence, 551–560
Algorithms
  for ORF prediction, 534
  for sequence comparisons, 559
Allele, defined, 170
Allele sizes in fragile X repeats, 444
Allele-specific amplification
  by LCA, 126
  by PCR, 115
Alpha helices, 534
Alpha satellite, 33
Alu repeat(s), 507–515
  absence in coding sequence, 534
  as evidence for human origin of clone, 268
  density in complete human sequences, 543
  for specific detection of human DNA, 250
  influence on renaturation kinetics, 84

  use in in situ hybridization, 221
Alumorphs, 513
Alu-primed PCR
  use in hncDNA preparation, 388
  use in library construction, 303
Alzheimers disease genetics, 448
Ambiguities in DNA diagnosis, 450
Amino acids, one letter code for, 561
2-Aminoadenosine
  use in short primers, 367
  in SBH, 418
Amplification
  as a DNA preparation procedure, 98
  linear, 99
  of detection signals, 92
  of YACs, 289
  with nested primers, 105
  *See also* PCR.
Amplitaq FS, for DNA sequencing, 336
Analysis, of DNA sequences, 530–535
Anatomy by cDNA abundance, 384
Aneuploidy
  in human disease, 51
  partial, 54
Angelman syndrome, 206
Angle between fields in PFG, 146
Annealing temperature in PCR, 105
Ansorge, Wilhelm, 334, 349
Anthropology at the molecular level, 565
Anticipation in fragile X disease, 442
Antigen detection by immuno-PCR, 119–122
Antigene drugs, 521
Antiparallel strands in triplexes, 473
Antisense, 520–521
APC gene in colon cancer, 452–453
Applied Biosystems, Inc., 334
Aptamers, 517–518
  definition of, 15
*Arabidopsis thaliana*
  as model organism, 25
  Sequencing projects, 378
Arbitrary primers in sequencing, 365
Arnheim, Norman, 117
Array hybridization, ideal, 303
Array screening, 296
Arrayed libraries in mapping, 237